Y0-BSD-784

ASYMMETRIC SYNTHESIS

ASYMMETRIC SYNTHESIS

CONSTRUCTION OF CHIRAL MOLECULES USING AMINO ACIDS

Gary M. Coppola
Herbert F. Schuster
Sandoz Research Institute
East Hanover, New Jersey

A WILEY-INTERSCIENCE PUBLICATION
JOHN WILEY & SONS
New York • Chichester • Brisbane • Toronto • Singapore

03422136
CHEMISTRY
repl. B0011322058

Copyright © 1987 by John Wiley & Sons, Inc.

All rights reserved. Published simultaneously in Canada.

Reproduction or translation of any part of this work beyond that permitted by Section 107 or 108 of the 1976 United States Copyright Act without the permission of the copyright owner is unlawful. Requests for permission or further information should be addressed to the Permissions Department, John Wiley & Sons, Inc.

Library of Congress Cataloging in Publication Data:

Coppola, Gary M. (Gary Mark), 1948–
Asymmetric synthesis.

"A Wiley-Interscience publication."
Includes bibliographies and index.
1. Chemistry, Organic—Synthesis. 2. Chirality.
3. Amino acids. I. Schuster, Herbert F. (Herbert Franz), 1949– II. Title.

QD262.C57 1987 547'.2 86-22472
ISBN 0-471-82874-2

Printed in the United States of America

10 9 8 7 6 5 4 3 2

QD 269
C571
1987
Chem

To
my wife, Joanne
and my children,
Matthew and Laura

G.M.C.

To
my wife, Maro
my daughter, Kristiana
and my parents,
Frank and Heidi Schuster

H.F.S.

PREFACE

Within the past 20 years asymmetric synthesis has forged to the forefront of organic chemistry. Ingenious synthetic methodologies that employ members of the "chiral carbon pool" comprise a significant portion of this technology. To the organic chemist, this chiral pool, which is composed mainly of naturally occurring amino acids, terpenes, sugars, and carbohydrates, is an invaluable source of stereochemically pure molecules. The majority of these compounds are commercially available, and many are inexpensive. Volume 4 of *Asymmetric Synthesis* by J. D. Morrison and J. W. Scott (Academic Press, 1984) contains a list of 375 chiral building blocks along with their commercial availability, approximate price, and literature references.

Obviously, a compilation of various aspects of asymmetric synthesis using these 375 chiral carbon fragments would easily fill several books. We therefore chose to limit this book to a specific class of compounds, the α-amino acids and second-generation intermediates that can be derived from them. We further chose to limit the scope of the book to include only those chemical transformations that would be of general interest to the organic chemist, in either the industrial or the academic environment. Special attention has been paid to the asymmetric synthesis of key pharmaceutical agents, agrochemicals, and a host of natural products including alkaloids, terpenoids, carbohydrates, and insect pheromones.

Two avenues of asymmetric synthesis will be explored. The first incorporates the chiral carbon (or carbons) of the amino acid directly into a target molecule. The second uses the inherent chirality of an amino acid (as either an internal feature of a molecule or an external chiral auxiliary) to induce asymmetry in a reaction that would not normally be asymmetrically biased.

Our aim is to demonstrate to the chemist the synthetic utility of amino acids for the generation of chiral reagents, intermediates, and final products, and to present the material in a logical fashion to enable the reader to retrieve information in an expeditious manner.

This book is divided into nine chapters, each of which is devoted to a particular amino acid or a family of similar amino acids. The chapters start with alanine, the simplest member of the α-amino acids, and progress to amino acids of gradually increasing complexity. Although the main focus of the book is on the naturally occurring L-amino acids, significant attention is also paid to manipulations of their corresponding unnatural D-isomers. Of the 19 common amino acids (excluding glycine), chemistry associated with 17 individual members of the group is presented.

Extensive schemes and reactions containing over 1900 structures are provided to illustrate the varied assortment of chiral intermediates that can be generated from amino acids and their derivatives. Some sequences that are too long and cumbersome for the allocated space are condensed by means of bold arrows, which allow a smooth transition from one structural type to another.

References include papers published up to December 1985. Articles that have appeared in the literature subsequent to the preparation of the original manuscript are summarized in an addenda section at the end of the appropriate chapter.

The authors wish to thank Sonya Beasley, Jean McCarthy, Carol Patterson, and Rosalie Piegario for efficient typing of the manuscript and Gertie Williams, Henry Mah, Mirinisa Myers, and Eva Contos of the Sandoz library staff for their assistance in gathering literature references.

GARY M. COPPOLA
HERBERT F. SCHUSTER

East Hanover, New Jersey
July 1986

CONTENTS

ABBREVIATIONS

Ac	Acetyl
Ar	Aryl
Boc	*t*-Butoxycarbonyl
Bop	Benzotriazolyloxytris[dimethylamino]phosphonium hexafluorophosphonate
BPPM	(2*S*,2*R*)-*N*-Boc-4-diphenylphosphino-2-diphenylphosphinomethylpyrrolidine
CAN	Ceric ammonium nitrate
Cbz	Benzyloxycarbonyl
CDI	*N*,*N*′-Carbonyldiimidazole
COD	1,5-Cyclooctadiene
Cp	Cyclopentadiene
DBN	1,5-Diazabicyclo[4.3.0]non-5-ene
DBU	1,8-Diazabicyclo[5.4.0]undec-7-ene
DCA	Dicyclohexylamine
DCC	1,3-Dicyclohexylcarbodiimide
de	Diastereomeric excess
DEAD	Diethyl azodicarboxylate
DHP	Dihydropyran
DIAD	Diisopropyl azodicarboxylate
DIBAH	Diisobutylaluminum hydride
DMAP	4-Dimethylaminopyridine
DMF	*N*,*N*-Dimethylformamide

DMP	2,2-Dimethoxypropane
DMPU	3,4,5,6-Tetrahydro-1,3-dimethyl-2(1*H*) pyrimidinone
DMSO	Dimethyl sulfoxide
DPPA	Diphenylphosphoryl azide
ee	Enantiomeric excess
EEDQ	1-Ethoxycarbonyl-2-ethoxy-1,2-dihydroquinoline
Et	Ethyl
Fmoc	9-Fluorenylmethyloxycarbonyl
HBT	1-Hydroxybenzotriazole
HMPA	Hexamethylphosphoramide
KHMDS	Potassium hexamethyldisilazide
LDA	Lithium diisopropylamide
LiHMDS	Lithium hexamethyldisilazide
LTMP	Lithium tetramethylpiperidide
MCPBA	*m*-Chloroperoxybenzoic acid
Me	Methyl
MEM	2-Methoxyethoxymethyl
MMC	Magnesium methyl carbonate
MMT	(4-Methoxyphenyl)diphenylmethyl
MOM	Methoxymethyl
Ms	Mesyl (methanesulfonyl)
MVK	Methyl vinyl ketone
NaHMDS	Sodium hexamethyldisilazide
Nb	*N*-Hydroxy-5-norbornene-*endo*-2,3-dicarboxamide
PCC	Pyridinium chlorochromate
Ph	Phenyl
Pht	Phthaloyl
Pic	Picoline
PMS	*p*-Methylbenzylsulfonyl
PNB	*p*-Nitrobenzyl
PNP	*p*-Nitrophenyl
PPE	Polyphosphoric ester
PPTS	Pyridinium *p*-toluene sulfonate
PTSA	*p*-Toluenesulfonic acid
PTSCl	*p*-Toluenesulfonyl chloride
Py	Pyridine
RAMP	(*R*)-1-Amino-2-methoxymethylpyrrolidine
SAMP	(*S*)-1-Amino-2-methoxymethylpyrrolidine
SDMP	(*S*)-(−)-Dimethoxymethyl-2-methoxymethylpyrrolidine

TBPS	*t*-Butyldiphenylsilyl
TBS	*t*-Butyldimethylsilyl
TDI	*N*,*N′*-Thionyldiimidazole
TEA	Triethylamine
Tf	Triflate
TFA	Trifluoroacetic acid
THF	Tetrahydrofuran
THP	2-Tetrahydropyran
TMEDA	*N*,*N*,*N′*,*N′*-Tetramethylenediamine
TMS	Trimethylsilyl
Tr or Trt	Trityl
Troc	Trichloroethoxy
Ts	Tosyl (*p*-Toluenesulfonyl)
TsCl	*p*-Toluenesulfonyl chloride
WSC	Water-soluble carbodiimide(1-ethyl-3-[3-(dimethylamino)propyl]carbodiimide hydrochloride)
Δ	Heat

ASYMMETRIC SYNTHESIS

INTRODUCTION

Compounds that occur in nature are optically active because living organisms tend to produce only a single enantiomer of a given molecule. The asymmetry of these molecules arises from the inherent chirality of the enzymes that are responsible for their production.

Receptor sites in biological systems, which are also optically active, have the ability to differentiate between two enantiomers of a specified molecule. Although the apparent physical differences between two enantiomers may seem small, the spatial orientation of a single functional group drastically affects the properties of the compound. This has strong implications for the human body.

For example, our senses of taste and smell are highly sensitive to subtle stereochemical differences in molecules that stimulate them. A classic illustration is our olfactory response to the enantiomeric forms of the terpene carvone. (*R*)-Carvone has the odor of spearmint, whereas (*S*)-carvone smells like caraway.[1-3]

(R)-Carvone (S)-Carvone

α-Amino acids themselves exhibit striking dissimilarities in their taste properties. The L-enantiomorphs of leucine, phenylalanine, tyrosine, and tryptophan taste bitter, whereas their corresponding D-enantiomorphs are sweet.[4]

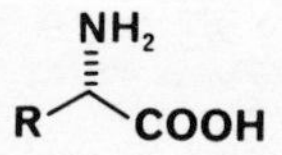

L-amino acid 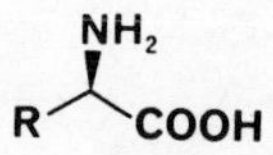 D-amino acid

As one can imagine, the development of an inexpensive sweet-tasting organic compound as a food additive has tremendous potential in the marketplace. The dipeptide ester aspartame is rapidly gaining an increasing market share as a low-calorie sweetener and is used extensively in soft drinks. Its backbone is composed of two amino acids; L-aspartic acid, which has no taste; and L-phenylalanine, which is bitter. Together they form a molecule with intensely sweet taste characteristics (approximately 160 times sweeter than sucrose). Substitution of the L-phenylalanine portion of the molecule with its antipode D-phenylalanine, which in itself is sweet tasting, causes the resulting dipeptide to taste bitter.[5]

Aspartame

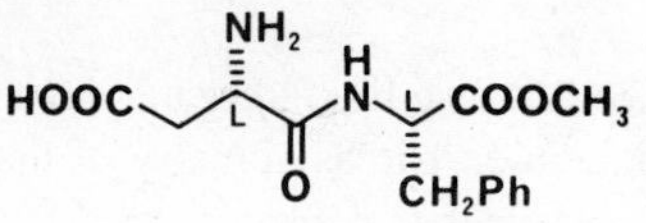

sweet 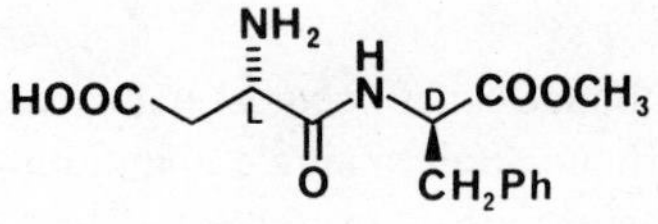 bitter

Whenever a compound is introduced into the body, as either a food additive or a drug, the question of toxicity always arises. With molecules possessing one or more asymmetric centers, adverse toxicologic properties can sometimes be attributed to one enantiomer and not the other.

Take, for example, the amino acid penicillamine. Its D-antipode is a chelating agent used to remove heavy metals from the body. Toxicity from this isomer is seldom severe. It is particularly useful in the treatment of Wilson's disease and biliary cirrhosis, where serum and liver copper concentrations, respectively, are excessively high. It is also an efficacious antidote for lead, gold, or mercury poisoning.[6] In contrast, the L-antipode of penicillamine causes optic atrophy, which can lead to blindness.[7]

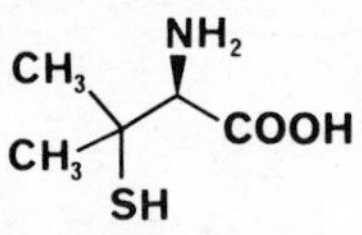

D-penicillamine 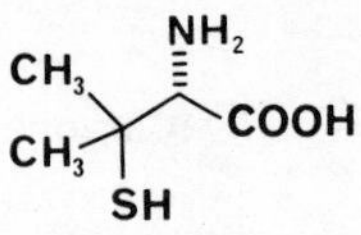 L-penicillamine

Who can forget the tragic consequences brought about by the drug thalidomide? In the early 1960s it was used therapeutically as a sedative and a hypnotic. Even though the thalidomide molecule contains an asymmetric center, the drug was administered in its racemic form. Although the drug appeared relatively inocuous, its use by pregnant women resulted in a high incidence of fetal deaths, neonatal deaths, and congenital malformations.[8] The teratogenicity has subsequently been found to be a property of only the (*S*)-(−)-enantiomer.[9]

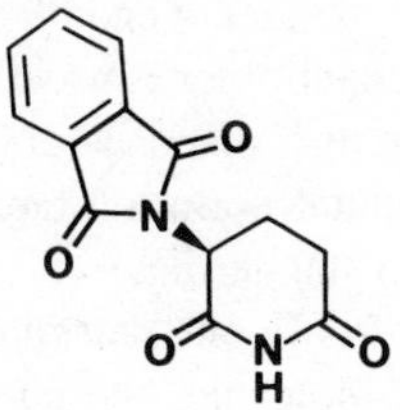

(S)-Thalidomide 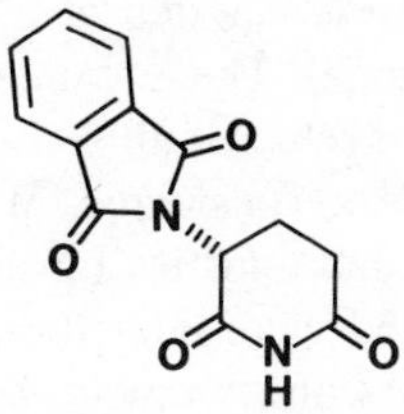 (R)-Thalidomide

These are only a few of literally hundreds of examples where biological systems, whether in the plant, animal, or insect kingdom, react differently to one enantiomeric form of a certain molecule. It is thus highly desirable, if not mandatory, to prepare molecules in enantiomerically pure form for meaningful studies on their physical or biological properties.

For preparation of enantiomerically homogeneous molecules, the chemist basically has two options. The molecules can either be synthesized in racemic form and resolved into their optical antipodes, or the synthesis can be performed in an enantioselective (or preferably enantiospecific) fashion so as to produce chirally enriched products.

The technique of resolving enantiomeric pairs of molecules has been used for well over a century. In 1848 Louis Pasteur mechanically separated crystals of each optical isomer of sodium ammonium tartrate. Obviously this method is not suitable as a general resolution technique; however, the idea of forming quaternary ammonium salts of a racemate and selectively crystallizing one desired enantiomer is still used today as a method of resolution. Consequently, this dictates that the compound to be separated contain either an acid or amine functionality. One must then find a suitable amine (usually an alkaloid) or acid-containing member of the "chiral pool" to act as a resolving agent that will impart the desired degree of crystallinity in the resulting salt in order to effect efficient separation.

Other methods of resolution involve the formation of a covalent bond between the racemic substrate and a chirally pure molecule. The resulting pair of diastereomers, in many instances, can be separated by chromatographic techniques and the desired enantiomer regenerated from the appropriate diastereomer by chemical manipulations. This, again, requires that the substrate possess some sort of "handle" that is capable of reacting with the resolving agent.

Advances have recently been made in resolution techniques by chromatographing racemates on chiral adsorbants. The stationary phase contains ionically bound optically active amino acids[10] (usually D-phenylglycine or L-leucine) or proteins.

Enantiomeric separation is achieved through a combination of π-acidity, hydrogen bonding, and steric interactions. The method is applicable to a wide range of compound types[11] and is a useful tool for rapid analytical separations. In selected cases preparative separations have been achieved on gram quantities of racemate.[12]

Each of the methods of resolution described above suffers from the major disadvantage that half of the mixture is the wrong enantiomer. A quantitative separation in the resolution step at best allows for the recovery of only half of the synthesized material. The remaining undesired enantiomer is usually discarded. On industrial-scale syntheses this could mean that the disposal of tons of organic substances must be considered. With environmental pollution rapidly becoming a factor, the cost and logistics for disposal can be staggering.

Thus far we have considered the separation of only one pair of enantiomers. As additional centers of asymmetry are added to a molecule, the number of enantiomers increases exponentially by a factor of 2^n. This means that two asymmetric centers generate four enantiomers, three asymmetric centers generate eight enantiomers, and so forth. Consequently, the separation of one particular enantiomer from the mixture by resolution techniques is impractical, if not impossible, to achieve.

The solution to the aforementioned problems is asymmetric synthesis. Strategically, this can be accomplished by two basic approaches. First, the synthesis of a target molecule can be designed so as to incorporate a chiral fragment whose absolute stereochemistry is already established. In this respect, nature's "chiral pool" offers a tantalizing array of these fragments that can judiciously be used to one's advantage. Second, the asymmetry in the target molecule can be induced by means of an external chiral auxiliary reagent that, under ideal conditions, is recoverable and recyclable. Again, nature furnishes the source of chirality inherent in these auxiliaries.

Amino acids can be used to fulfill the criteria for both approaches to asymmetric synthesis. Of all the members of the "chiral pool," the amino acids as a family are the most versatile. Nearly all of the 19 common α-amino acids (glycine excluded) are available (many commercially) in two enantiomeric forms. Therefore, synthetic routes can be planned to produce either enantiomer of a specified molecule.

L-Amino acids, which are commonly found in nature, are the more abundant and hence less expensive of the two forms. If an unnatural D-amino acid is required in a synthesis and cost is a factor, in many instances it can be prepared efficiently from a less expensive L-amino acid. Examples of this are presented in various chapters of this book.

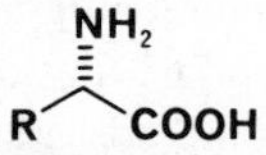

NH_2 (hashed bond) — R, COOH

L-amino acid

NH_2 (wedged bond) — R, COOH

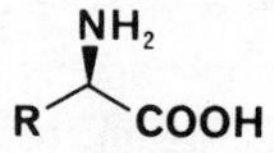

D-amino acid

We have chosen to use the classical L and D descriptors only in referrence to the configuration of amino acids because this is the convention most widely used by commercial sources. The descriptors for stereochemical centers of all other molecules will conform to the Cahn–Ingold–Prelog (CIP) rules[13,14] and be designated either (*R*) or (*S*). As a point of information, the naturally occurring L-amino acids all have the (*S*) configuration with the exception of L-cysteine (R = CH_2SH), where it is (*R*). Although the absolute stereochemistry of L-cysteine is the same as that of the other L-amino acids, the configuration changes from (*S*) to (*R*) as a result of the prioritization of sulfur over oxygen in the CIP sequence rules.

To perform useful chemical transformations using amino acids, it is often necessary to protect either or both reactive functionalities of the molecule (the amine and carboxylic acid groups). The carboxylic acid group can easily be converted to an ester by standard esterification techniques. The amine must be protected with a group that can withstand a wide range of chemical manipulations and yet should also be easily removable under mild conditions. Two such groups that have emerged as versatile protective functionalities are the *t*-butoxycarbonyl (Boc) and benzyloxycarbonyl (Cbz) groups.

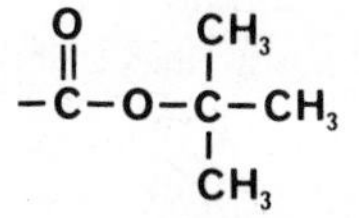

$$-\overset{\overset{\displaystyle O}{\|}}{C}-O-\overset{\overset{\displaystyle CH_3}{|}}{\underset{\underset{\displaystyle CH_3}{|}}{C}}-CH_3 \qquad\qquad -\overset{\overset{\displaystyle O}{\|}}{C}-O-CH_2-C_6H_5$$

Boc group 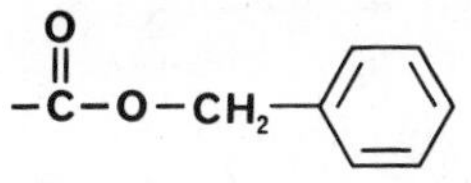 **Cbz group**

The Boc group is able to withstand catalytic hydrogenation, sodium in liquid ammonia, alkali, and hydrazine[15] but is rapidly cleaved under mild acid conditions, usually trifluoroacetic acid. On the other hand, the Cbz group is relatively stable to acidic conditions but is easily removed by catalytic hydrogenation. Consequently, this allows selective removal of one protecting group in the presence of the other.

Throughout this book the reader will observe extensive use of the Cbz and Boc groups in a wide variety of syntheses. Although nearly all of the common amino acids are commercially available in their Boc- or Cbz-protected forms, it may be economically advantageous for the chemist to perform the protection either directly on the less expensive amino acid itself or at a later stage in the synthetic sequence. Rather than repetitiously describe the techniques of these protections in each chapter of this book, we would like to briefly summarize at this point several general methods, which have withstood the test of time, for the protection of amino acids with either the Boc or Cbz functionality.

The Cbz group is probably the simpler of the two groups to attach. This is performed by simultaneously adding 1 equivalent of benzyl chloroformate and 4 *N* aqueous sodium hydroxide to a solution of 1 equivalent of the amino acid in 2 *N* sodium hydroxide. After approximately 30 min the mixture is washed with an organic solvent and the aqueous phase acidified to Congo red to precipitate the product.[16] Other bases such as sodium bicarbonate[17] or potassium bicarbonate[18] have also been used with equal success to furnish the Cbz–amino acid in greater than 85% yield.

$$\underset{\text{RCHCOOH}}{\overset{\text{NH}_2}{|}} + \text{ClCOOCH}_2\text{Ph} \xrightarrow{\text{base}} \underset{\text{RCHCOOH}}{\overset{\text{NHCOOCH}_2\text{Ph}}{|}}$$

The Boc group cannot be introduced in the same manner as the Cbz because *t*-butyl chloroformate is thermally unstable. Instead, a general method has been developed using di-*t*-butyl dicarbonate (Boc anhydride) as the *t*-butoxy carbonylating reagent. The protection proceeds as follows; to a stirred mixture of 0.01 mol of an amino acid derivative and 0.01 mol of an organic base such as triethylamine, tetramethylguanidine, or Triton B in 25 mL of DMF is added 0.011 mol of di-*t*-butyl dicarbonate. After stirring at room temperature for 10 min the mixture is washed with dilute acetic acid and the solvent evaporated to give the Boc–amino acid derivative in high yield.[19]

Alternatively, the reaction can be performed on an amino acid ester in a two-phase system using chloroform (20 mL) as the organic phase and an aqueous solution of sodium bicarbonate (0.01 mol) and sodium chloride (2.0 g) in 15 mL of water. After refluxing for 90 min the organic layer is separated and the chlorform evaporated to furnish the Boc–amino acid ester also in high yield.[20]

$$\underset{\text{RCHCOOR}'}{\overset{\text{NH}_2}{|}} + t\text{-BuO}\overset{\text{O}}{\overset{\|}{\text{C}}}\text{-O-}\overset{\text{O}}{\overset{\|}{\text{C}}}\text{O}t\text{-Bu} \xrightarrow{\text{base}} \underset{\text{RCHCOOR}'}{\overset{\text{NHCOO}t\text{-Bu}}{|}}$$

Two additional reagents that have been utilized as *t*-butoxycarbonylating reagents are *t*-butyl azidoformate (Boc–azide)[21–23] and *t*-(butoxycarbonyloxyimino)-2-phenylacetonitrile (Boc–ON).[24,25] Due caution should be exercised when working with Boc-azide, as it has been known to explode on distillation.[26]

$$(\text{CH}_3)_3\text{C-O-}\overset{\text{O}}{\overset{\|}{\text{C}}}\text{-N}_3$$

Boc-azide

$$(\text{CH}_3)_3\text{C-O-}\overset{\text{O}}{\overset{\|}{\text{C}}}\text{-O-N=C(CN)(C}_6\text{H}_5)$$

Boc-ON

REFERENCES

1. M. Windholz (Ed.), *The Merck Index*, 9th ed., Merck, Rahway, NJ, 1976, p. 239.
2. G. F. Russel and J. I. Hills, *Science*, **172,** 1043 (1971).
3. L. Friedman and J. G. Miller, *Science*, **172,** 1044 (1971).
4. J. Solms, L. Vuataz, and R. H. Egli, *Experientia*, **21,** 692 (1965).

5. R. H. Mazur, J. M. Schlatter, and A. H. Goldkamp, *J. Am. Chem. Soc.*, **91,** 2684 (1969).
6. A. Osol (Ed.), *Remington's Pharmaceutical Sciences, 16th ed.*, Mack Publishing, Easton, PA, 1980, p. 1170.
7. T. Z. Csaky, *Cutting's Handbook of Pharmacology, 6th ed.*, Appleton-Century-Crofts, New York, 1979, p. 161.
8. G. W. Mellin and M. Katzenstein, *New Engl. J. Med.*, **267,** 1184 (1962).
9. G. von Blaschke, H. P. Kraft, K. Finkentscher, and F. Köhler, *Arzneim.-Forsch./Drug Res.*, **29,** 1640 (1979).
10. W. H. Pirkle, J. M. Finn, J. L. Schreiner, and B. C. Hamper, *J. Am. Chem. Soc.*, **103,** 3964 (1981).
11. I. W. Wainer and T. D. Doyle, *LC Magazine*, February 1984.
12. W. H. Pirkle and J. M. Finn, *J. Org. Chem.*, **47,** 4037 (1982).
13. R. S. Cahn, C. Ingold, and V. Prelog, *Angew. Chem., Int. Ed.*, **5,** 385 (1966).
14. V. Prelog and G. Helmchen, *Angew. Chem., Int. Ed.*, **21,** 567 (1982).
15. M. Bodanszky, Y. S. Klausner, and M. A. Ondetti, *Peptide Synthesis*, Wiley, New York, 1976, p. 26.
16. H. E. Carter, R. L. Frank, and H. W. Johnston in E. C. Horning, Ed., *Organic Synthesis, Vol. 3*, Wiley, New York, 1955, p. 167.
17. D. R. Hwang, P. Helquist, and M. S. Shekhani, *J. Org. Chem.*, **50,** 1264 (1985).
18. C. H. Levenson and R. B. Meyer, Jr., *J. Med. Chem.*, **27,** 228 (1984).
19. L. Moroder, A. Hallett, E. Wunsch, O. Keller, and G. Wersin, *Hoppe-Seyler's Z. Physiol. Chem.*, **357,** 1651 (1976).
20. D. S. Tarbell, Y. Yamamoto, and B. M. Pope, *Proc. Natl. Acad. Sci.* (USA), **69,** 730 (1972).
21. E. Schnabel, *Liebigs Ann. Chem.*, **702,** 188 (1967).
22. K. P. Polzhofer, *Tetrahedron Lett.*, 2305 (1969).
23. A. Ali, F. Fahrenholz, and B. Weinstein, *Angew. Chem., Int. Ed.*, **11,** 289 (1972).
24. M. Itoh, D. Hagiwara, and T. Kamiya, *Tetrahedron Lett.*, 4393 (1975).
25. M. Itoh, D. Hagiwara, and T. Kamiya, *Bull. Chem. Soc. Jpn.*, **50,** 718 (1977).
26. H. Yajima, H. Kawatani, and Y. Kiso, *Chem. Pharm. Bull*, **18,** 850 (1970).

Amino Acid Reviews

27. K. Drauz, A. Kleeman, and J. Martens, *Angew. Chem., Int. Ed.*, **21,** 584 (1982).
28. J. Martens, *Topics Curr. Chem.*, **125,** 165 (1984).
29. A. Kleemann, *Chem. Zt.*, **106,** 151 (1982).

CHAPTER ONE

ALANINE

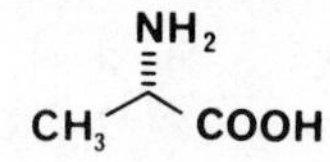

(*S*)-2-AMINOPROPANOIC ACID

Alanine, the simplest member of the amino acid family, is quite abundant in nature. It is one of the main components of silk fibroin and, together with glycine, occurs in relatively large amounts in human blood plasma.[1] It can also be found in soils, stones, shellfish, and volcanic ash. Its natural occurrence, however, is not limited to earth alone. It, as well as other amino acids, has been detected in lunar soil and in a variety of meteorites. Commercially, it can be produced in reasonable quantities by either enzymatic or fermentative processes.

When designing syntheses of complex molecules as either pharmaceutical agents or natural products, it is often necessary to introduce a methyl group in a particular configuration with respect to the remainder of the molecule. Since alanine bears a chiral methyl substituent with an *S* configuration for the naturally occurring material and an *R* configuration for the unnatural isomer, it is ideally suited for elaboration to the target compound. The commercial availability of both enantiomers makes alanine an attractive intermediate for asymmetric synthesis.

1.1 L-ALANINE

One of the most common transformations of amino acids is the formation of *N*-carboxyanhydrides. This serves to internally protect its two functional groups as well as activate the carbonyl of the acid toward nucleophilic attack.

The *N*-carboxyanhydride (**2**) of L-alanine (**1**) is easily prepared in high yield by the reaction of **1** with either an excess of phosgene at room temperature[2,3] or trichloromethyl chloroformate at 40°C.[4]

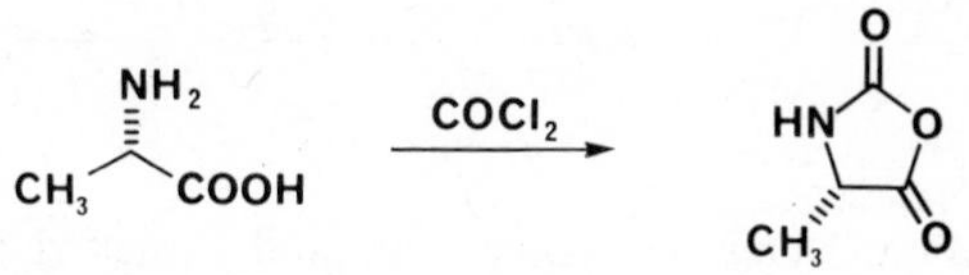

As a demonstration of its reactivity, the oxazolinedione **2** is readily opened with ethyl glycinate in chloroform at −65°C, and L-alanylglycine (**3**) is isolated in 68% yield after 3 hr.[2]

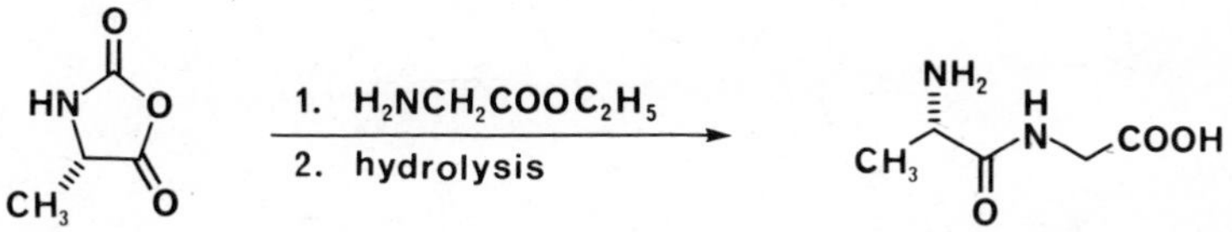

Chiral oxazolidines can also be constructed by the condensation of amino acids with aldehydes under acid catalysis. Thus when the commercially available Fmoc-protected L-alanine and an aldehyde are refluxed in toluene in the presence of *p*-toluenesulfonic acid, the (4*S*)-4-methyl-5-oxazolidinone **5** is formed in good yield. Recently it has been found that exposure of **5** to triethylsilane and trifluoroacetic acid causes reductive cleavage of the heterocyclic ring to give *N*-alkylated Fmoc-L-alanines **6** in high yield.[5] The Fmoc group can be easily removed with either piperidine or diisopropylamine–DMF. The procedure is also applicable for other amino acids such as phenylalanine, valine, methionine, lysine, serine, and histidine and thus provides an easy entrance to *N*-alkylated amino acids, some of which are biologically important.

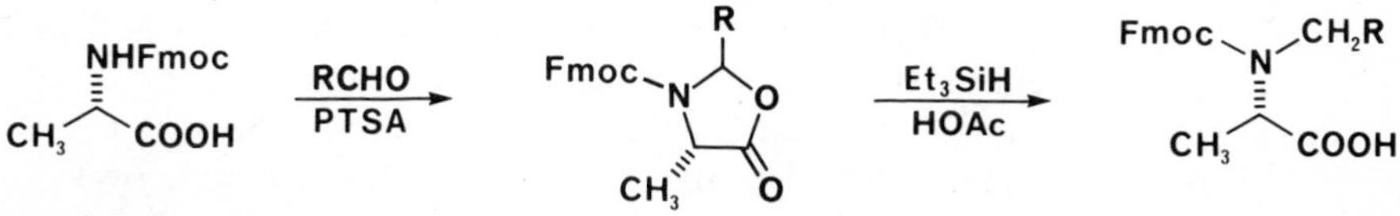

R	Yield **5** (%)	Yield **6** (%)
H	96	98
CH_3	79	74
Cyclohexyl	77	91
$CH_2C_6H_5$	30	95

Other heterocyclic systems that possess a chiral *C*-methyl substituent can be constructed by using the difunctionality of the alanine molecule. The amino group of **1** is sufficiently nucleophilic to undergo a conjugate addition to α-naphthoquinone (**7**), which furnishes benzindole derivative **8** on cyclization.[6]

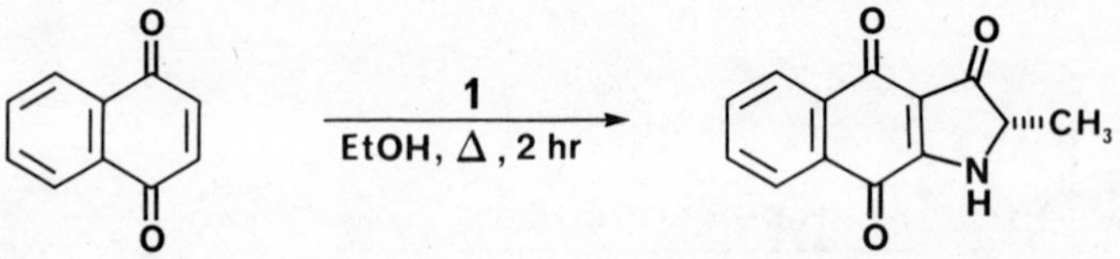

1
EtOH, Δ, 2 hr

CH_3

7 8

The condensation of 5-chlorosalicaldehyde (**9**) with **1** in DMF (room temperature, 24 hr) initially forms the Schiff base of the amino acid. Subsequent addition of DCC to the reaction mixture causes cyclodehydration to the chiral benzoxazepinone **10** within 2 hr.[7]

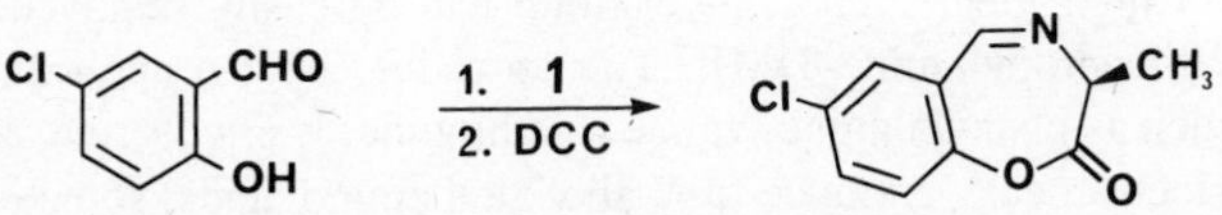

Cl CHO OH
1. 1
2. DCC
Cl N CH_3 O O

9 10

1,4-Benzodiazepin-2-ones (**14**) are readily accessible according to the short sequence shown in Scheme 1. The first step in the synthesis requires the formation of the optically active amide **13** from 2-amino-5-chlorobenzophenone (**11**) and a protected version of L-alanine **12** (both commercially available). After removal of the protecting group, the free amine cyclizes with the ketone when the pH is adjusted to 7–8.[8] Other 3-substituted benzodiazepinones are prepared analogously by using phenylalanine, tyrosine, valine, threonine, and tryptophan in place of alanine.

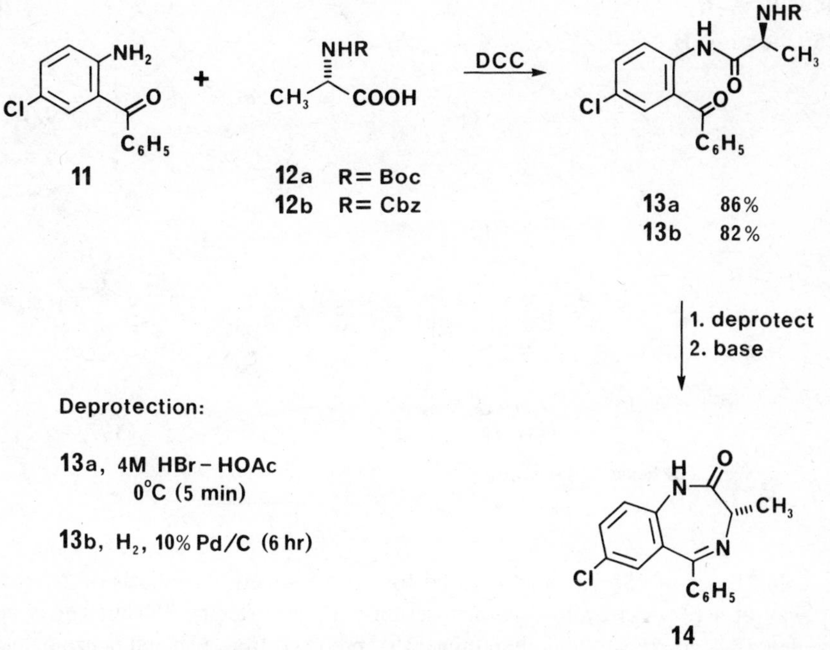

11 + 12a R= Boc; 12b R= Cbz —DCC→ 13a 86%; 13b 82%

1. deprotect
2. base
→ 14

Deprotection:

13a, 4M HBr – HOAc
0°C (5 min)

13b, H_2, 10% Pd/C (6 hr)

Scheme 1

An interesting asymmetric synthesis of secondary amines involves a transamination of the amino group of an optically active amino acid ester to a ketone[9] (Scheme 2). The Schiff base **16** is prepared by refluxing phenylacetone and L-alanine ester **15** in benzene by means of a Dean–Stark trap. The imine bond is catalytically hydrogenated to give **17** in 59% overall yield. The purified ester **17** is then sequentially treated with *t*-butyl hypochlorite, sodium ethoxide, and sulfuric acid to give (*S*)-2-amino-1-phenylpropane (*d*-amphetamine) (**18**) in 37% yield based on phenylacetone.

The size of the ester group has a dramatic influence on the optical yield of the product. The use of **15a** (ethyl ester) affords an optical yield of 66%, whereas **15b** (*t*-butyl ester) increases the enantiomeric purity to 85%. The size of the α substituent on the amino acid has little effect on the asymmetric transformation. For example, L-valine *t*-butyl ester only increases the optical yield to 87% vs. 85% for **15b**).

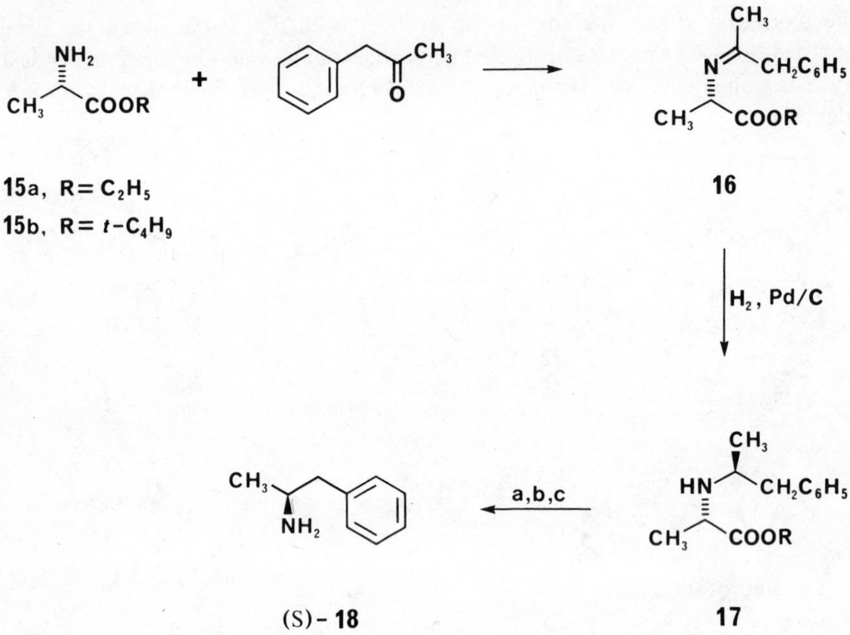

NH2
CH3 COOR
+
CH3
O
CH3
N
CH2C6H5
CH3 COOR
15a, R= C2H5
15b, R= t-C4H9
16
H2, Pd/C
CH3
HN
CH2C6H5
CH3 COOR
17
a,b,c
CH3
NH2
(S)- 18

Scheme 2 (a) t-C_4H_9OCl, ether; (b) $NaOC_2H_5$; (c) 5% H_2SO_4.

Schiff bases of **15b** can also be used for the asymmetric synthesis of β-lactams by way of a [2 + 2] cycloaddition reaction with azidoketene.[10] Thus (S)-N-benzylidene-1-t-butoxycarbonylethylamine (**19**), prepared from **15b** and benzaldehyde, when treated with azidoacetyl chloride and triethylamine, affords a diastereomeric mixture of β-lactams **20a** and **20b**.

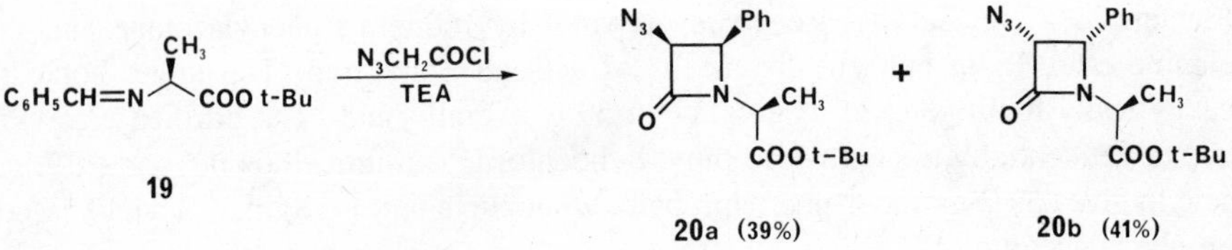

CH3
C6H5CH=N COO t-Bu
19
N3CH2COCl
TEA
N3
Ph
O
N
CH3
COO t-Bu
20a (39%)
+
N3
Ph
O
N
CH3
COO t-Bu
20b (41%)

These diastereomers are readily separable by column chromatography. The azide of **20a** is reduced to an amino group by catalytic hydrogenation, and the resulting 3-amino-β-lactam **21** is condensed with benzaldehyde to give a Schiff base. Reaction of this imine with azidoketene then gives chiral bis-β-lactam **22** as the only isomer. Reduction of the azide function followed by acetylation furnishes **23** (Scheme 3). The same series of reactions was also performed on **20b**.

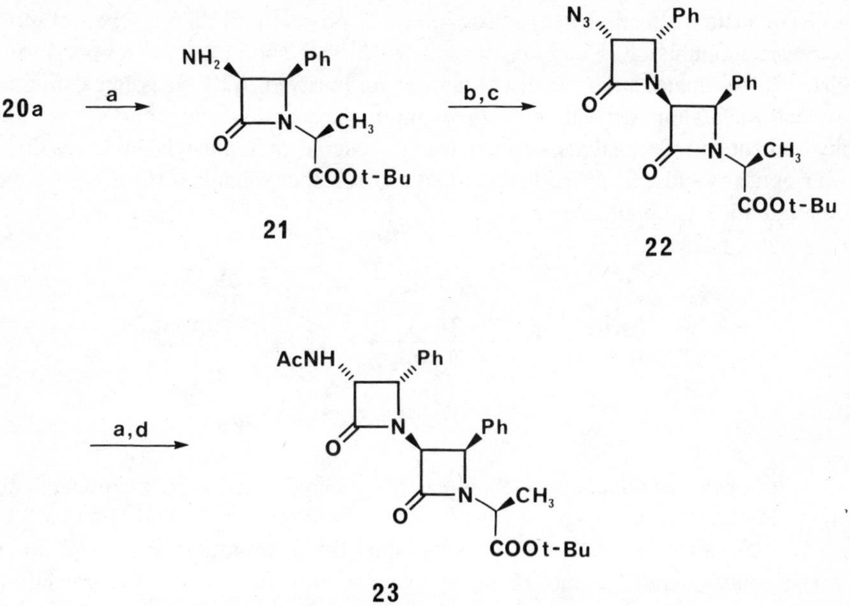

Scheme 3 (a) H_2 (1 atm), 5% Pd/C; (b) C_6H_5CHO, $MgSO_4$ (96% from **20a**); (c) N_3CH_2COCl, Et_3N (74%); (d) Ac_2O, *N*-methylmorpholine (85%).

In an effort to minic NADH, Endo and co-workers[11] have constructed a chiral model for the reduction of ethyl benzoylformate. This example of a redox reaction makes use of amino acid containing 1,4-dihydropyridines in the presence of magnesium perchlorate. The preparation of the 1-benzyl-1,4-dihydronicotinamide derivative of L-alanine is outlined in Scheme 4.

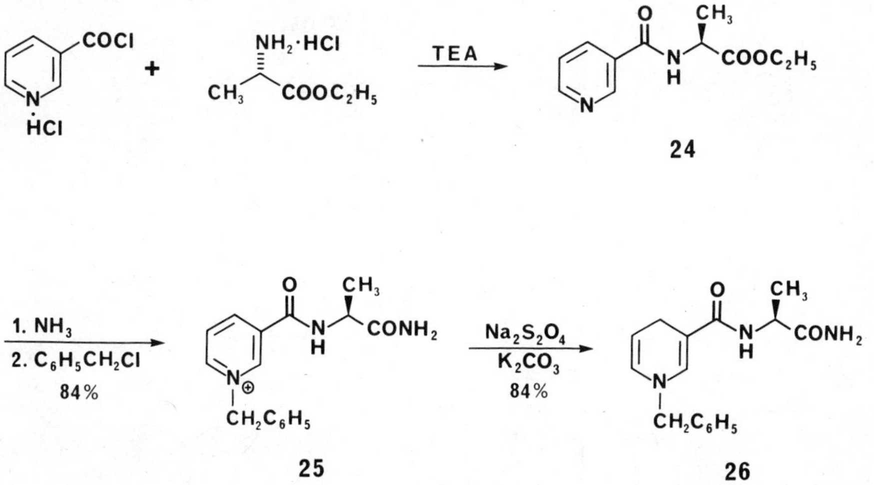

Scheme 4

The reduction of ethyl benzoylformate (**27**) proceeds in the presence of stoichiometric amounts of **26** and magnesium perchlorate and produces (*R*)-ethylmandelate (**28**) in quantitative yield with an optical purity of 47%. It is interesting to note that the alanine derivative **26** gives the highest enantiomeric excess. Similar dihydronicotinamide analogs derived from L-leucine or L-phenylalanine result in lower optical yields, 26% for leucine and 5% for phenylalanine. (In this case the *S* configuration is obtained.)

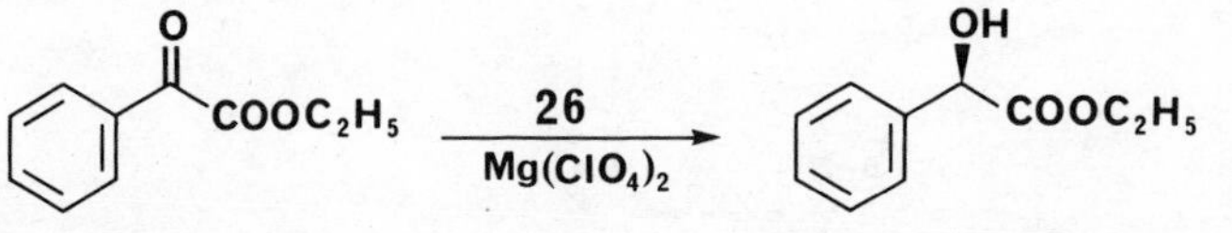

$$\text{PhCOCOOC}_2\text{H}_5\ (\mathbf{27}) \xrightarrow[\text{Mg(ClO}_4)_2]{\mathbf{26}} \text{PhCH(OH)COOC}_2\text{H}_5\ (\mathbf{28})$$

L-*N*-Ethoxycarbonylalanine (**29**) is easily prepared from **1** by treatment with ethyl chloroformate in the presence of sodium hydroxide.[12] The addition of 300 mol % of allyllithium to **29** cleanly forms either the β, γ-unsaturated ketone **30** or the *E*-α,β-unsaturated ketone **31** by simple variation in the isolation procedure (Scheme 5). The former is produced when the reaction mixture is quenched with 40% H_3PO_4, whereas the latter is obtained by saturated NH_4Cl workup.[13]

These transformations occur with complete retention of asymmetry. The first 100 mol % of the lithium reagent generates the lithium carboxylate of **29**, whereas the second 100 mol % of lithium reagent abstracts the hydrogen from the protected nitrogen. This abstraction is probably responsible for the preservation of asymmetry since formation of the carbamate anion insulates the α-hydrogen and prevents its removal, thus averting racemization.[13] The third 100 mol % of lithium reagent is then responsible for reaction with the carboxyl carbonyl to give the observed products.

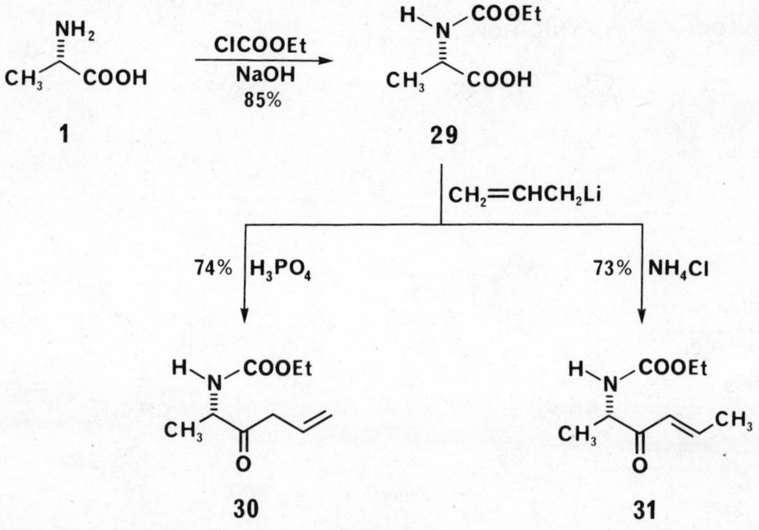

$$CH_3CH(NH_2)COOH\ (\mathbf{1}) \xrightarrow[\text{NaOH, 85\%}]{\text{ClCOOEt}} CH_3CH(NHCOOEt)COOH\ (\mathbf{29})$$

$$\mathbf{29} \xrightarrow{CH_2{=}CHCH_2Li} \begin{cases} \xrightarrow{H_3PO_4,\ 74\%} CH_3CH(NHCOOEt)COCH_2CH{=}CH_2\ (\mathbf{30}) \\ \xrightarrow{NH_4Cl,\ 73\%} CH_3CH(NHCOOEt)COCH{=}CHCH_3\ (\mathbf{31}) \end{cases}$$

Scheme 5

The synthesis of optically pure α-aminoalkyl aryl ketones is of great importance in that it provides a route to a variety of biologically significant compounds containing the phenylethylamine grouping in their active, asymmetric configurations.[12] An interesting method for the preparation of these systems involves a Friedel–Crafts acylation procedure. Carbamate **29** is readily converted to its acid chloride **32** by treatment with oxalyl chloride in the presence of a catalytic amount of DMF. Reaction of **32** with benzene and $AlCl_3$ (200 mol %) results in a high yield of the ketocarbamate[12] **33** in 96–97% optical purity.[14]

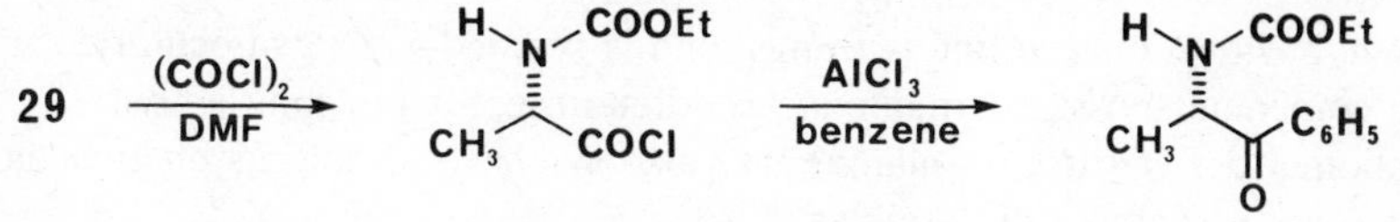

29 → [(COCl)₂, DMF] → 32 → [AlCl₃, benzene] → 33

By a variety of reductive procedures, **33** can produce a host of optically pure ephedrines and amphetamines (Scheme 6).

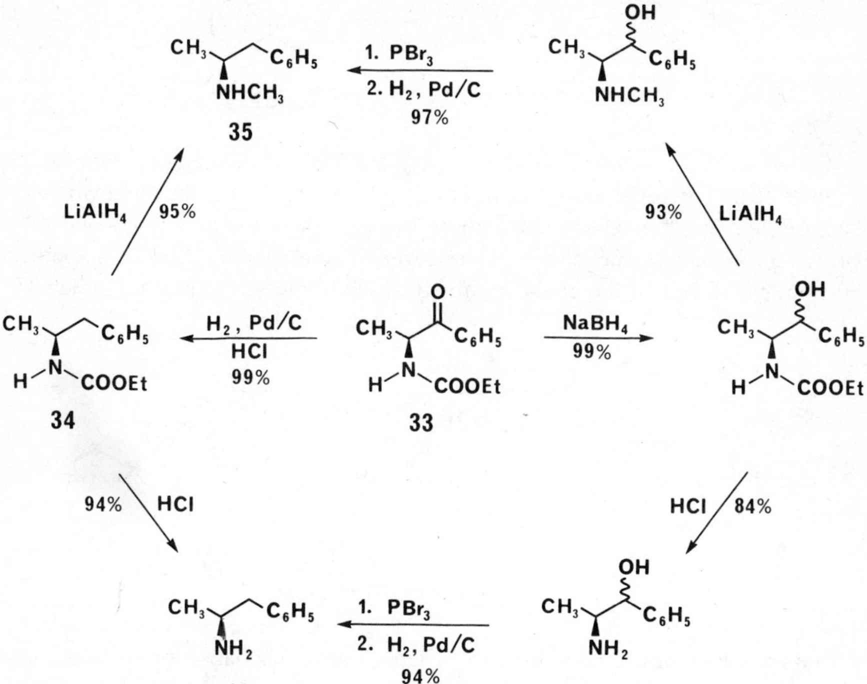

33 → [NaBH₄, 99%] → alcohol carbamate → [LiAlH₄, 93%] → N-methyl amino alcohol → [1. PBr₃, 2. H₂, Pd/C, 97%] → 35
33 → [H₂, Pd/C, HCl, 99%] → 34 → [LiAlH₄, 95%] → 35
34 → [HCl, 94%] → 18
alcohol carbamate → [HCl, 84%] → amino alcohol → [1. PBr₃, 2. H₂, Pd/C, 94%] → 18

Scheme 6 (Reprinted with permission from T. P. Buckley, III and H. Rapoport, *J. Am. Chem. Soc.*, **103**, 6157 (1981). Copyright 1981 American Chemical Society.)

Chiral 1,2-propylenediamines are readily accessible from L-alanine ethyl ester (**15a**). Exposure of **15a** to methanol saturated with ammonia (0°C, 4 days) furnishes

L-alanine amide (**36**). Reduction of the amide with diborane gives (*S*)-(−)-1,2-diaminopropane (**37**) isolated as its dihydrochloride salt.[15]

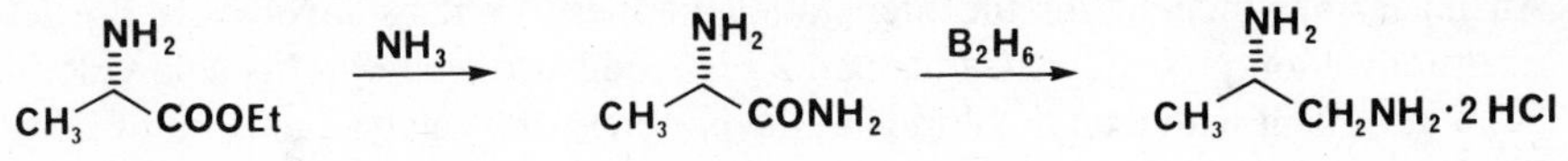

When **37** is allowed to react with imidate ester **38**, (*S*)-(−)-4-methyl-2-(1-naphthylmethyl) imidazoline is formed and is isolated as its hydrochloride salt **39**. This compound possesses moderate α-adrenoreceptor blocking activity.[15,16] The enantiomer derived from D-alanine was also synthesized, and its pharmacological profile was essentially the same as that for **39**.

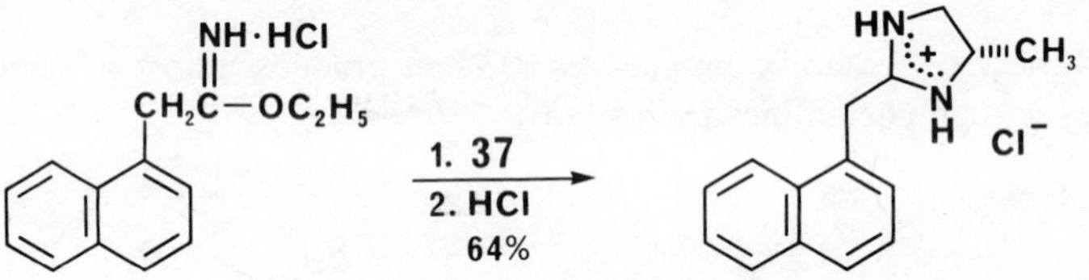

Occasionally, it may be desirable to have a chiral α-disubstituted amino acid as an intermediate in organic synthesis. Direct alkylation of an amino acid would not be expected to produce any asymmetry because once the planar enolate **40** is generated, any enantiofacial bias is destroyed. The alkylating agent can approach **40** from either face of the enolate with equal facility to afford racemic products.

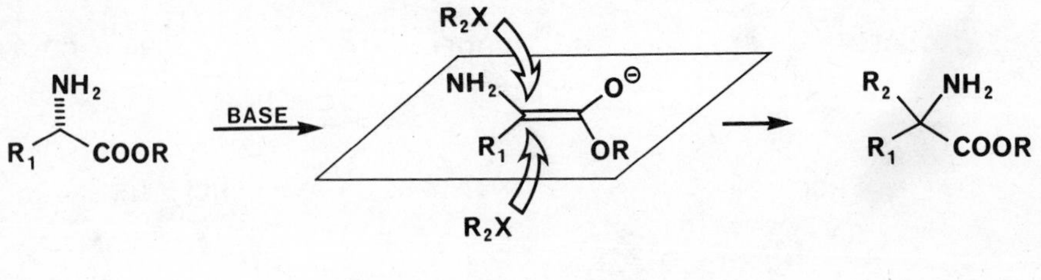

We are aware of only one instance where direct α-alkylation of an amino acid derivative gives any optically active products. This occurs with di-*t*-butyl L-(+)-*N*-formyl aspartate, where an enantiomeric excess of approximately 60% is realized.[17] The reason for the asymmetric induction is still not clear.

One method particularly effective for inducing asymmetry in amino acid alkylation is to attach a bulky chiral auxiliary to the nitrogen atom, which thus will direct the approach of the alkylating agent from one side of the enolate. Scheme 7 illustrates such an approach. The Schiff base (**43**) of 1,2;3,4-di-*O*-cyclohexylidene-α-D-*galacto*-hexo-1,5-dialdopyranose (**41**) and L-alanine methyl ester (**42**) is formed in high yield by refluxing the two in benzene in the presence of 4-Å molecular sieves and a trace of *p*-toluenesulfonic acid.

Metallation of **43** occurs smoothly with LDA at −70°C, and, after the addition of HMPA and the alkyl bromide, the mixture is allowed to warm to room temperature. Hydrolysis of the alkylated product **44** with hydrochloric acid in methanol furnishes (*S*)-α-methyl amino acid methyl esters **45** with an enantiomeric excess ranging from 44 to 85% (see Table 1.1). The use of HMPA in the reaction is critical for induction of the desired (α*S*) configuration in intermediate **44**. In the absence of HMPA, the opposite configuration (α*R*) is produced (de 44%). Other bases such as potassium *t*-butoxide or sodium hydride can be used in the metallation of **43**; however, these lead to lower asymmetric induction.[18]

The complementary (*R*)-α-methylamino acid methyl esters (**49**) are independently accessible by an ingenious method devised by Schöllkopf et al.[19] (Scheme 8). Heating of L-alanine methyl ester (**42**) at 100–110°C (18 hr) affords its cyclic dimer, cyclo-(L-Ala–L-Ala) (**46**) in good yield. Treatment of **46** with trimethyloxonium tetrafluoroborate gives (3*S*, 6*S*)-2,5-dimethoxy-3,6-dimethyl-3,6-dihydropyrazine **47** (93–95% optically pure).[20] Lithiation of **47** occurs smoothly at −70°C with either *n*-butyllithium or LDA, and alkylation produces **48** (78–93% yield),

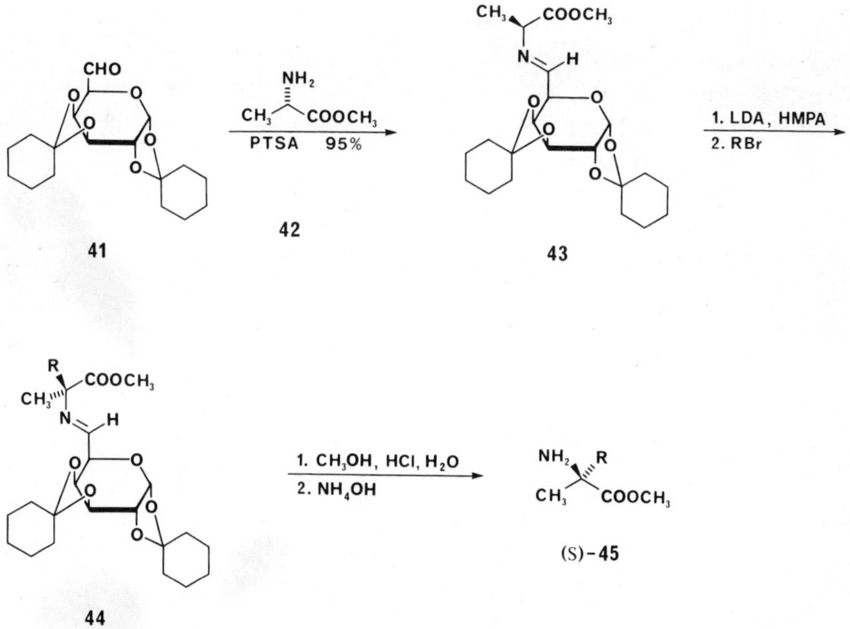

Scheme 7

TABLE 1.1. α-Alkylated Alanine Methyl Esters Obtained from Procedures Outlined in Schemes 7 and 8

R	Yield (S)-**45** (%) [%ee] (NH_2, R, CH_3, $COOCH_3$)	Yield **48** (%)	Yield (R)-**49** (%) [%ee] (R, NH_2, CH_3, $COOCH_3$)
$C_6H_5CH_2$	72 [83]	88	91 [93]
2-$C_{10}H_7CH_2$	70 [78]	88	87 [90]
3,4-$(CH_3O)_2C_6H_3CH_2$	84 [85]	93	87 [90]
$C_6H_5CH{=}CHCH_2$	65 [60]	85	82 [84]
$CH_2{=}CHCH_2$	73 [70]	88	78 [90]
$HC{\equiv}C{-}CH_2$	69 [44]	–	–
2-quinolyl-CH_2	–	78	90 [90]
3-pyridyl-CH_2	–	91	76 [90]

where the *R* configuration is induced at C-6. The asymmetric induction of this process ranges between 92 and 95%. Hydrolysis with 0.5 *N* HCl gives a mixture of the (*R*)-ester **49** and the regenerated ester **42.** These can be separated by either distillation or chromatography. A list of various (*R*)-esters **49** can be found in Table 1.1.

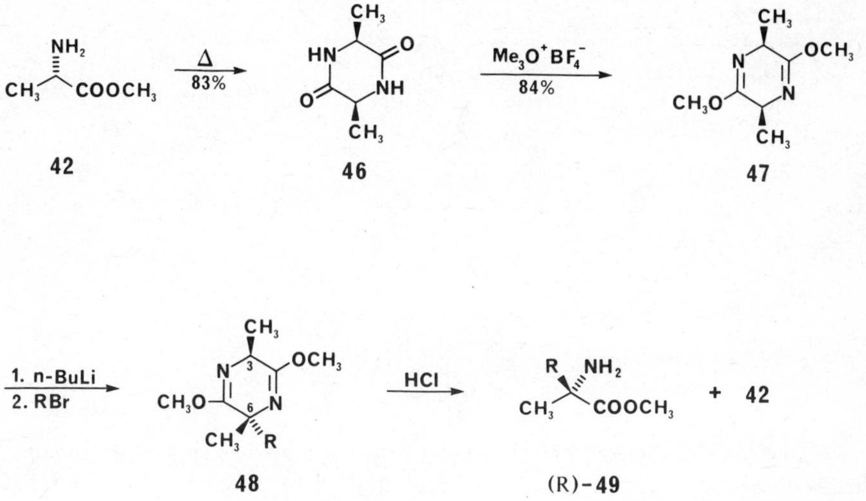

Scheme 8

The high asymmetric induction in the **47** → **48** reaction can be rationalized as shown in structure **50.** The angular methyl at C-3 encumbers the bottom face sufficiently to direct approach of the alkylating agent from the top. This also places the more bulky R group in a favorable *trans* position to the methyl at C-3 (**51**).

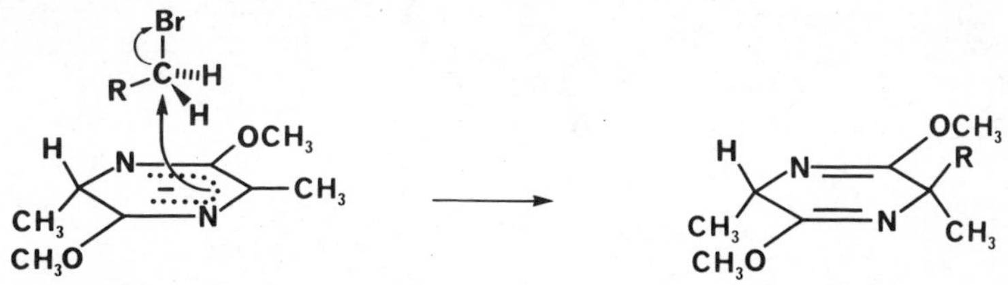

Not only is lactim ether **47** effective for inducing asymmetry in alkylation reactions, but it also provides respectable induction of chirality in reactions with carbonyl compounds.[21] Again, the *R* configuration is induced at C-6 (**52**) by the same rationale evoked previously in the transition structure **50**. Reactions proceed best in dimethoxyethane to give inductions ranging from 82% to greater than 95%. When prochiral carbonyl compounds are used, asymmetric induction at C-7 is also observed (52–74%). Subsequent hydrolysis of adduct **52** leads to (*R*)-α-methylserine esters **53**.

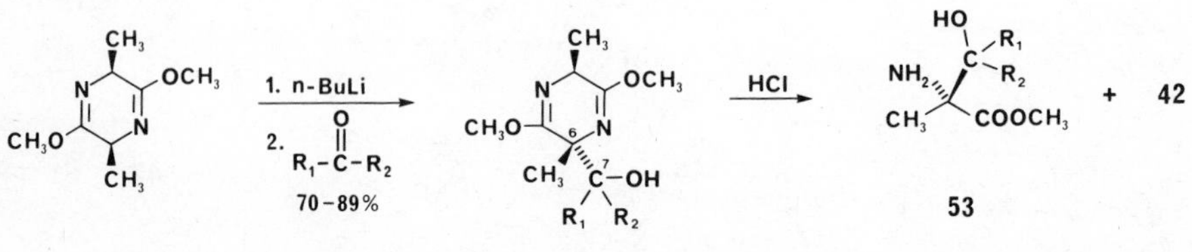

R_1 = H, C_6H_5, 4-$OCH_3C_6H_4$

R_2 = H, CH_3, C_6H_5

Using this technique, a French group of investigators has applied Schöllkopf's methodology to a short synthesis of a 2-amino-2-deoxypentose.[22] Initial condensation of lithiated **47** with L-2,3-O-isopropylidene glyceraldehyde (**54**) produces a mixture consisting mainly (80%) of **55** plus its respective (6*S*)- and (7*S*)-diastereomers. Exposure of this mixture to aqueous trifluoroacetic acid gives a quantitative yield of aminolactones out of which pure L-2-methyl-2-aminoxylono butyrolactone (**56**) can be crystallized.

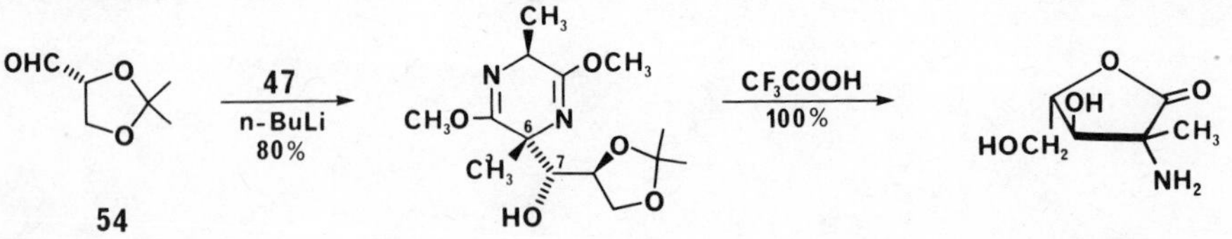

The (3*S*, 6*S*)-2,5-diketopiperazine heterocycle, which is structurally similar to **46**, can be found in nature. It is an integral feature of the alkaloid echinulin (**59**). Its total synthesis has been accomplished in a convergent manner.[23] The key step in the synthesis is a condensation of gramine derivative **58** with diketopiperazine **57**. The optically active portion of the molecule is derived from L-alanine.

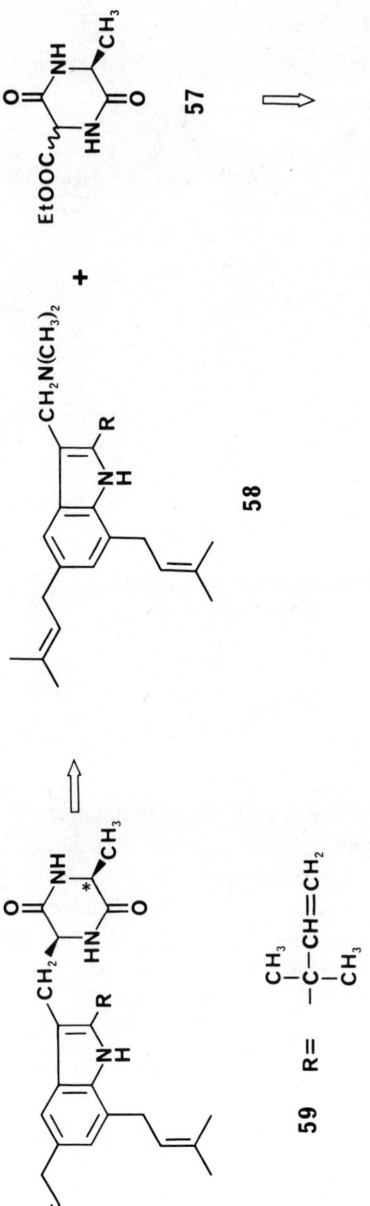

CH_2 NH HN CH_3 O O * R N H

$\Rightarrow$

$CH_2N(CH_3)_2$ R N H **58**

+

EtOOC NH HN CH_3 O O **57**

$\Downarrow$

NH_2 CH_3 * COOH

59 R= $-C(CH_3)_2-CH{=}CH_2$

The chiral heterocycle **57** is easily prepared from **12b** in three steps (Scheme 9), using methodology developed for the analogous racemic heterocycle.[24]

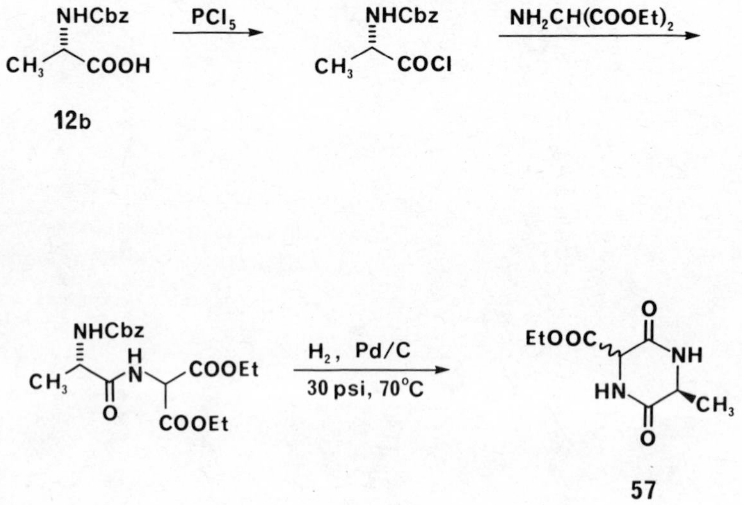

NHCbz; CH_3; COOH — **12b** — PCl_5 → NHCbz; CH_3; COCl — $NH_2CH(COOEt)_2$ →

NHCbz; CH_3; H; N; O; COOEt; COOEt — H_2, Pd/C; 30 psi, 70°C → EtOOC; O; NH; HN; CH_3; O — **57**

Scheme 9

The condensation of **57** with **58** is carried out in the presence of powdered KOH and requires 86 hr to complete. The product **60** is isolated in 50% yield and is composed of a 2:1 mixture of *cis* and *trans* isomers. Hydrolysis of the ester is accomplished in 0.1 *N* NaOH, and decarboxylation is effected in refluxing dioxane (1 hr). The resulting isomers are then separated by column chromatography, with echinulin (**59**) obtained in 42% yield and its unnatural isomer **61** in 33% yield.

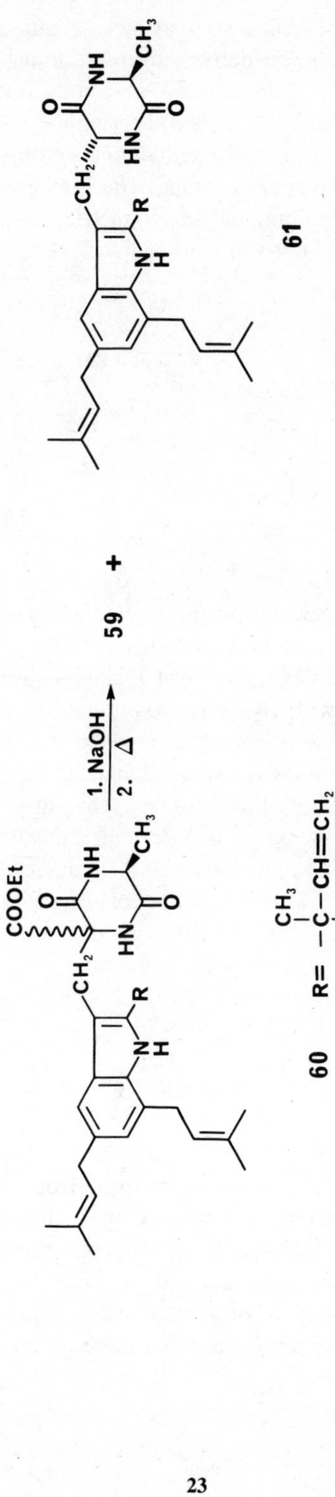

COOEt
CH_2
NH
HN
O
O
CH_3
R
N
H

$$\mathbf{60} \xrightarrow[\text{2. } \Delta]{\text{1. NaOH}} \mathbf{59} + \mathbf{61}$$

60 R = $-\overset{CH_3}{\underset{CH_3}{C}}-CH{=}CH_2$

CH_2
O
NH
HN
O
CH_3
R
N
H

61

In some syntheses, at first glance, it may not be immediately obvious that a chiral center in a molecule can be derived from an amino acid. One fascinating example is the translation of L-alanine to α-L-mycaminoside (**70**) (Scheme 10). In the product, the asymmetric carbon (labeled) bearing the methyl group is inherited from **1**; however, the nitrogen of the dimethylamino group is not associated with the amino acid. It is introduced at a later stage. The synthesis of **70** is quite amazing in that the chirality of its five contiguous asymmetric centers is controlled by the one asymmetric carbon that is derived from alanine.

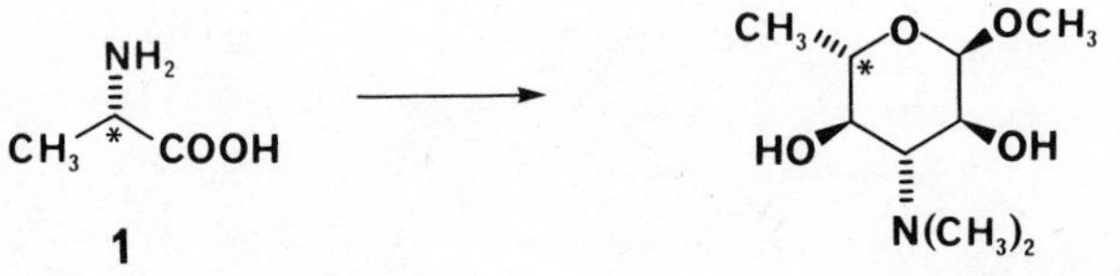

NH_2
CH_3 * COOH
1

CH_3 O OCH_3
HO OH
$N(CH_3)_2$
70

Mycaminoside, as well as several other interesting sugars illustrated in Scheme 10, are synthesized from a common intermediate **65.** The first step in the preparation of **65** is a nitrous acid deamination of L-alanine, where (*S*)-2-acetoxypropanoic acid (**62**) is produced with 96% retention of configuration.[25] The compound (*S*)-**62** is then converted to the anomeric mixture of 2-hexenopyranosid-4-uloses **65** and **66** (ratio 2:1) in five steps (60% overall yield from **63**). The two diastereomers are readily separable by column chromatography to afford the optically pure α-anomer **65** in 30% yield from **63**. Catalytic hydrogenation of **65,** followed by reduction of the ketone with lithium aluminum hydride, furnishes methyl α-L-amicetoside (**67**) in 57% yield. Direct reduction of **65** with lithium aluminum hydride gives the unsaturated alcohol **68** (74% yield), which can than be converted to either α-L-mycaminoside (**70**) or α-L-oleandroside (**73**) as shown in Scheme 10.[26]

1.2 D-ALANINE

Occasionally, in designing a synthetic route using amino acids to a particular target molecule, the point of chirality cannot be derived from the naturally occurring enantiomer. The opposite configuration must be used and thus requires a D-amino acid. Many D-amino acids are commercially available; however, they are substantially more expensive than the more common L-amino acids. For example, one supplier that lists both enantiomers of alanine sells L-alanine for 25¢/g, whereas D-alanine costs $1.72/g. There are even larger price differentials between the two enantiomers of other amino acids.

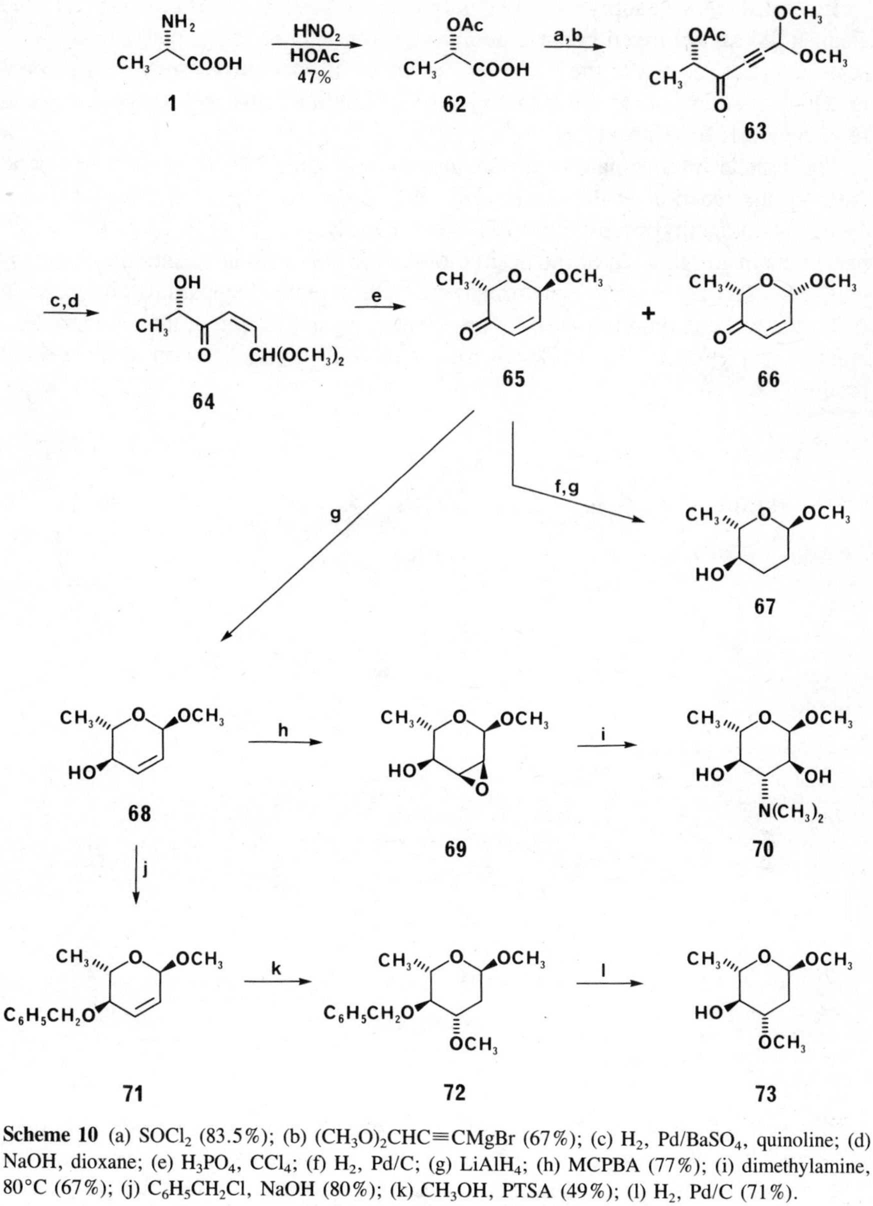

Scheme 10 (a) $SOCl_2$ (83.5%); (b) $(CH_3O)_2CHC{\equiv}CMgBr$ (67%); (c) H_2, Pd/$BaSO_4$, quinoline; (d) NaOH, dioxane; (e) H_3PO_4, CCl_4; (f) H_2, Pd/C; (g) $LiAlH_4$; (h) MCPBA (77%); (i) dimethylamine, 80°C (67%); (j) $C_6H_5CH_2Cl$, NaOH (80%); (k) CH_3OH, PTSA (49%); (l) H_2, Pd/C (71%).

If a substantial quantity of the D-amino acid is required in a synthesis and the chemist has an unlimited budget, purchasing from a chemical supplier would present the easiest access to these starting materials. If not, one can rely on a process in which L-amino acids are transformed to D-amino acids in high optical yield (e.g., alanine, Scheme 11).

The critical intermediate in the sequence, imidazoline **78,** is prepared in good yield by the reaction of the (*S*)-imidate **76**[27] (prepared from L-alanine) with (*S*)-(+)-2-(aminomethyl)pyrrolidine (**77**). Acid hydrolysis of **78** (6 *M* HCl, 110°C, 20 hr) results in a high yield of the desired D-alanine (**79**) with an enantiomeric excess of 93.8%. The asymmetric transformation can be explained by partial epimerization at the exocyclic carbon followed by preferential crystallization of that epimer. This epimeric depletion in solution keeps the equilibrium in the direction of the desired product **79.**[28,29]

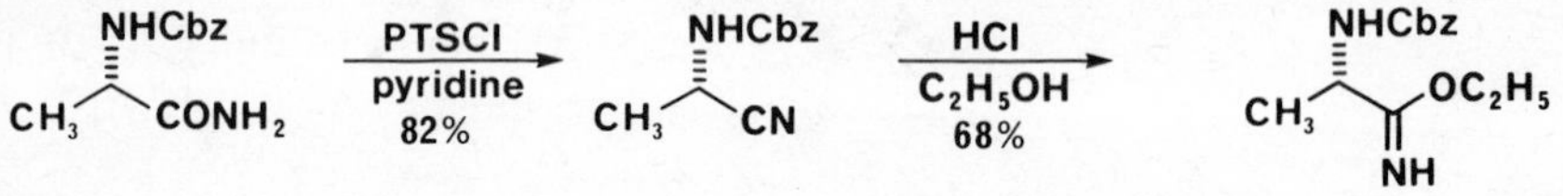

NHCbz CH$_3$ CONH$_2$ **74** —PTSCl, pyridine, 82%→ NHCbz CH$_3$ CN **75** —HCl, C_2H_5OH, 68%→ NHCbz CH$_3$ OC_2H_5 NH **76**

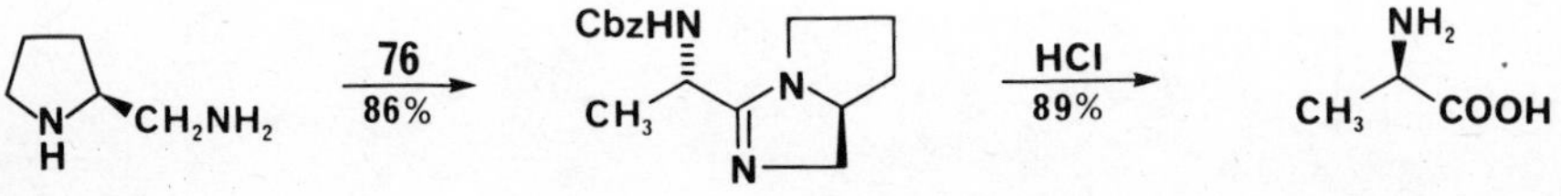

N H CH_2NH_2 **77** —**76**, 86%→ CbzHN CH$_3$ N N **78** —HCl, 89%→ NH$_2$ CH$_3$ COOH **79**

Scheme 11

Bleomycin, whose structure is shown in Figure 1, is an important antitumor antibiotic clinically used in the treatment of squamous cell carcinoma and malignant lymphoma.[30] This highly complex molecule is interesting in that the chirality in several portions of its structure can be derived from various amino acids, discussed in Chapters 5 and 7 of this book.

The total synthesis of bleomycin basically is accomplished by a series of coupling reactions of seven building blocks: pyrimidoblamic acid (**80**), *erthro*-β-hydroxy-L-histidine (**81**), (2*S*, 3*S*, 4*R*)-4-amino-3-hydroxy-2-methylpentanoic acid

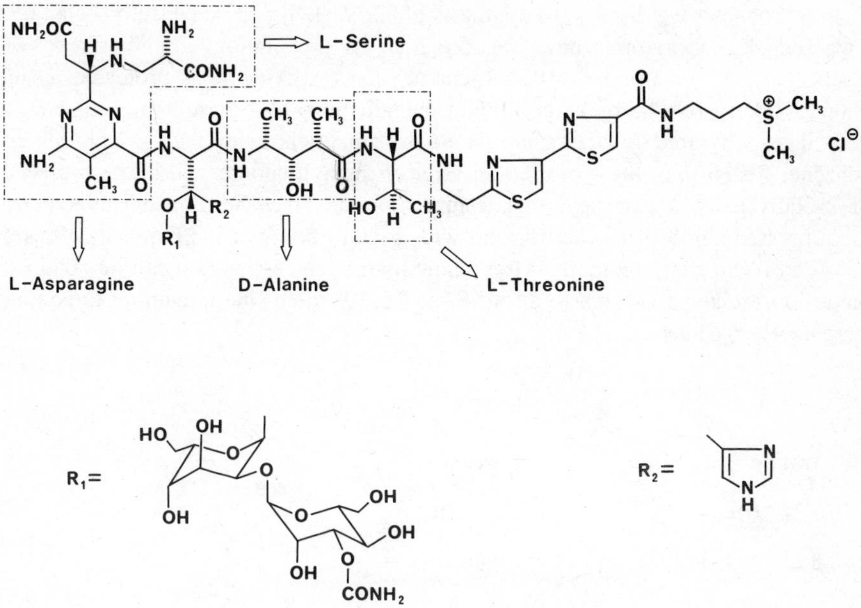

L-Serine
L-Asparagine
D-Alanine
L-Threonine

Figure 1. Amino acids used to construct bleomycin.

(**82**), L-threonine (**83**), 2′-(2-aminoethyl)-2,4′-bithiazole-4-carboxylic acid (**84**), and the 2-O-(3-O-carbamoyl-α-D-mannopyranosyl)-α-L-gulopyranosyl residue (see Figure 1).[31–34]

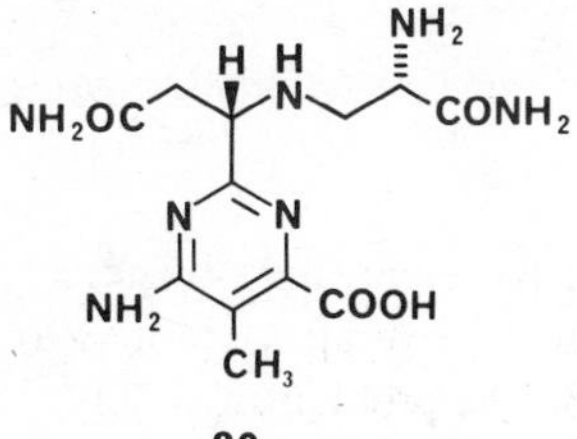

80 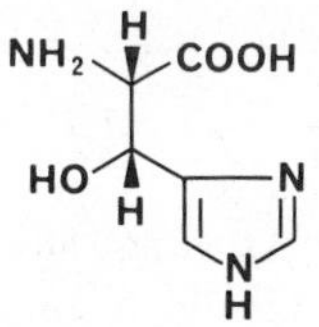 **81** 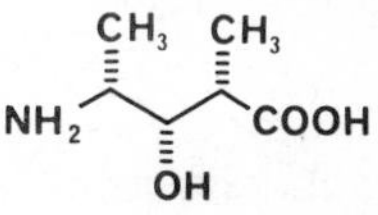 **82**

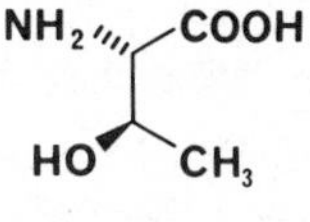

83 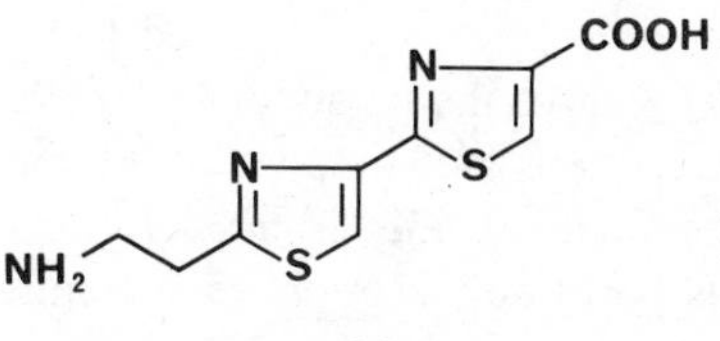 **84**

In this chapter we discuss the synthesis of **82,** since it is derived from D-alanine. Two separate laboratories have succeeded in its preparation. The first reported synthesis of **82** occurred in 1974 (Scheme 12).[35] D-Alanine is protected as its phthaloyl derivative **85** by reaction with phthalic anhydride in refluxing toluene.[36] It is then converted to acid chloride **86** by treatment with thionyl chloride in benzene. Reaction of **86** with the half ester of methylmalonic acid in the presence of 2 equivalents is isopropyl magnesium bromide gives **88** as a diastereomeric mixture. Reduction of the keto group with sodium borohydride produces **89** as a mixture of four diastereomers. After acidic hydrolysis, separation can be obtained on an ion-exchange column to afford **82** in 12.2% yield (the remaining three diastereomers = 61%).

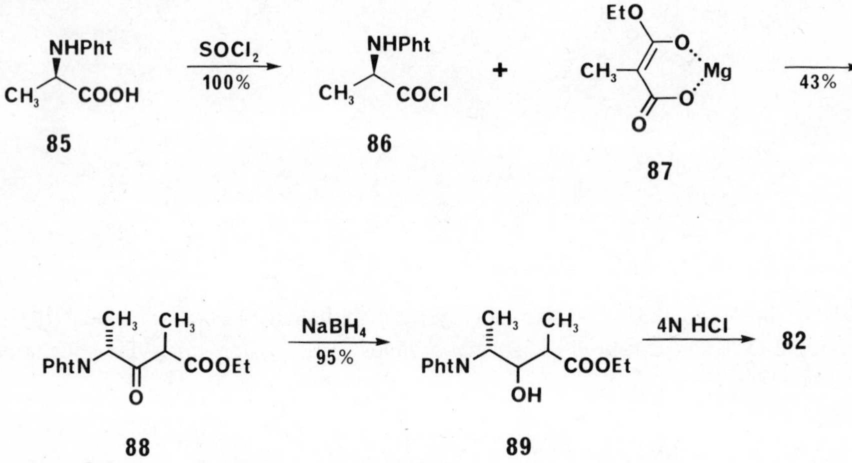

Scheme 12

It is obvious that this process is not an efficient one; therefore, a more stereoselective synthesis of **82** is desired in order to increase its overall yield and to alleviate the tedious separation of diastereomers. In 1982 the problem was solved by Ohno and co-workers[37] (Scheme 13). They converted D-alanine derivatives **90** to the chiral *R*-aminoaldehydes **93** by first preparing 3,5-dimethylpyrazoles **92**[38] and then reducing them under mild conditions with lithium aluminum hydride. These aldehydes are then subjected to an aldol condensation with (*E*)-vinyloxyboranes **94** to give the phenylthioesters **95** in good yield.

These reactions are complete in 30 min at 0°C and give **95a** as a 20:1 mixture of diastereomers or **95b** as a 35:1 mixture, clearly an acceptable ratio. If the reaction is performed at room temperature, the chemical yield increases to 77%, however, the isomer ratio of **95b** drops to 20:1. This procedure enjoys the additional advantage that the products **95** are activated esters and can be readily employed in the coupling reaction leading to bleomycin.[34] The thioester can also be hydrolyzed to the free acid **82** by treatment with mercury trifluoroacetate.

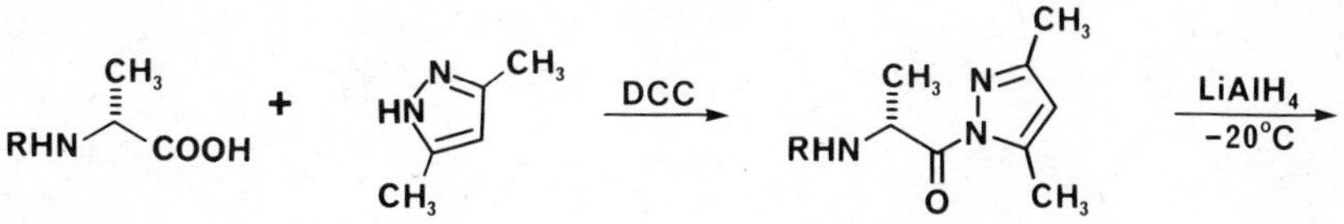

90a R= Boc
90b R= Cbz

91

92a 95%
92b 79%

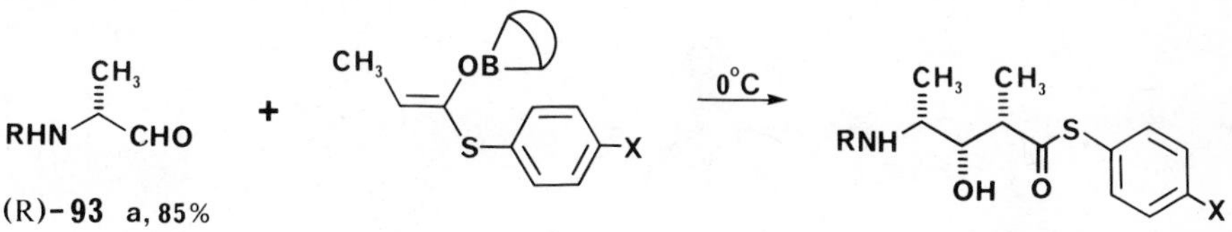

(R)-**93** a, 85%
b, 95%

94 (X= H, NO2)

95 a, 62%
b, 60%
(X= H)

Scheme 13

Chiral piperidine heterocycles comprise the backbone of a variety of alkaloids. The synthesis of (3*S*, 6*R*)- and (3*S*, 6*S*)-deoxycassine methoxymethyl ethers (**102** and **103**) has been accomplished in nine steps starting from D-alanine (Scheme 14).[39] As in the previously described scheme, an *R*-aminoaldehyde (**96**) plays an important role in the introduction of the methyl substituent in the proper absolute configuration. Aldehyde **96** is obtained from **85** by a two-step sequence involving the formation of acid chloride **86** followed by a Rosenmund reduction.

Aldehyde **96** is reacted with diallylzinc, and the resulting alcohol is protected with a MOM group to give **97** as a 5:1 mixture of the 3β- and 3α-diastereomers. The 3β-isomer is hydroborated and then treated with acid to form alcohol **98.** Oxidation of the primary alcohol to an aldehyde followed by a Wittig reaction affords the (Z)-alkene **100.** Intermediate **100** can also be prepared in a shorter route by a Grignard reaction of **96** with (3Z)-pentadecenylmagnesium bromide. The resulting tertiary alcohol **99** is isolated as a 3:1 mixture of 3β- and 3α-diastereomers. Protection with a MOM group then gives **100**. The synthesis is completed by deprotection to the free amine **101** followed by an intramolecular aminomercuration to produce the isomeric mixture of **102** and **103.**

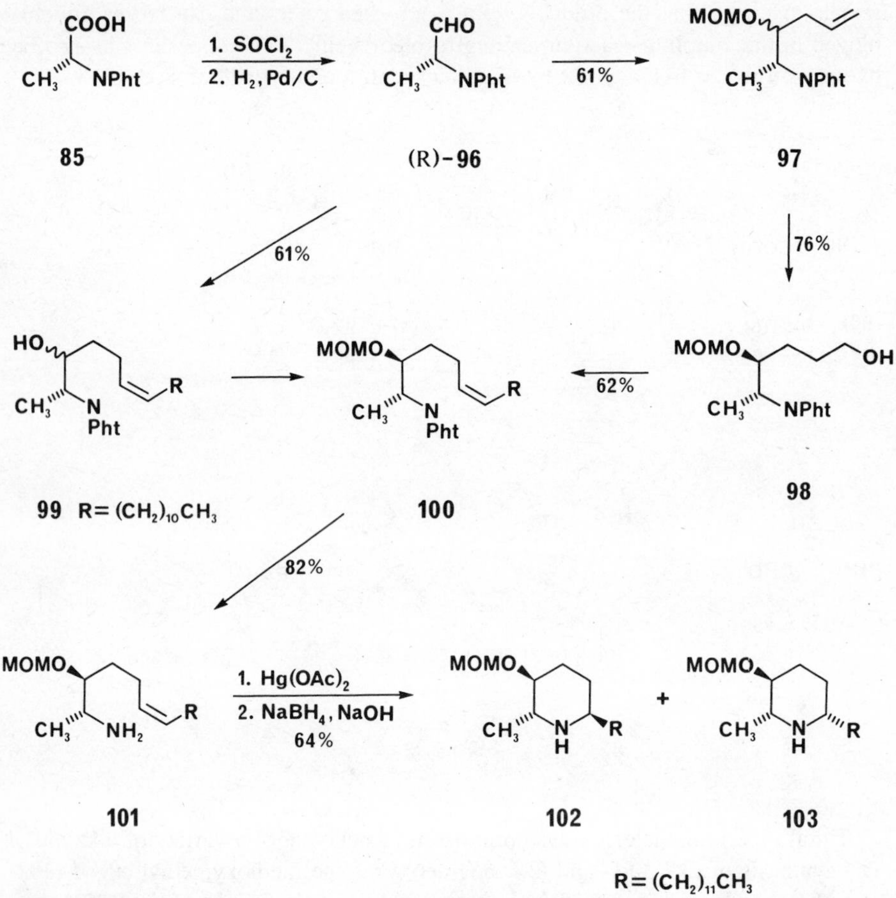

COOH, CH_3, NPht — 1. $SOCl_2$, 2. H_2, Pd/C → CHO, CH_3, NPht — 61% → MOMO, CH_3, NPht

85 (R)-96 97

61% 76%

HO, CH_3, N, Pht, R → MOMO, CH_3, N, Pht, R ← 62% — MOMO, OH, CH_3, NPht

99 R= $(CH_2)_{10}CH_3$ 100 98

82%

MOMO, CH_3, NH_2, R — 1. $Hg(OAc)_2$, 2. $NaBH_4$, NaOH, 64% → MOMO, CH_3, N, H, R + MOMO, CH_3, N, H, R

101 102 103

R= $(CH_2)_{11}CH_3$

Scheme 14

Now that we have examined the piperidine ring, let us turn our attention to the morpholine heterocycle. To date, it appears that simple chiral morpholines have been prepared only in order to study their chiroptical properties. One interesting derivative that has been synthesized for that purpose is (3*R*)-3-methylmorpholine (**108**). If **108** is retrosynthetically analyzed, one can envision that the (*R*) configuration of the 3-methyl substituent can be derived from D-alanine (**79**).

The requisite (*R*)-aminoalcohol **105** is obtained from **79** by sequentially benzoylating its amino group and then reducing both functional groups of **104** with lithium aluminum hydride.[40] Compound **105** is then reacted with ethylene oxide in the presence of 1 equivalent of phenol to produce diol **106.** Cyclization to **107** is effected with 70% sulfuric acid at 140°C for 15 hr and then hydrogenolysis of the *N*-benzyl group over 10% Pd/C at 60–80 atm and 80°C furnishes the desired (3*R*)-morpholine[41] (Scheme 15). No yields were reported for any of the latter steps.

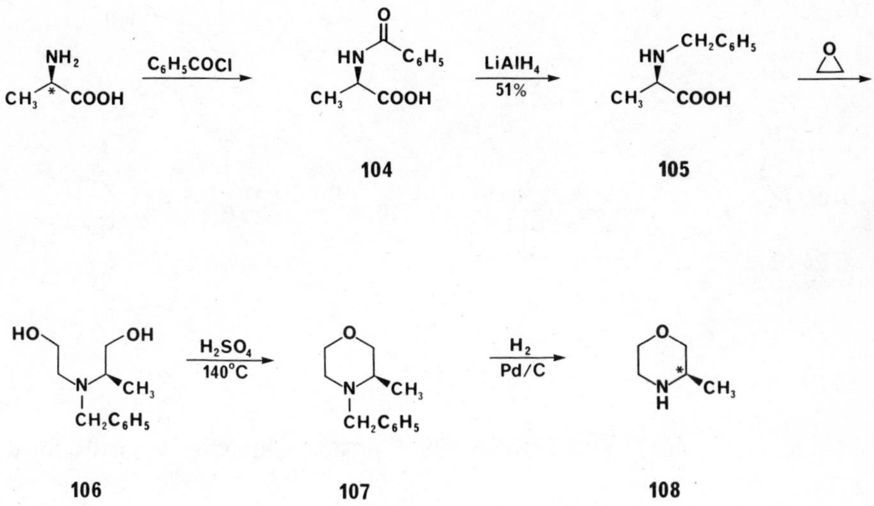

Scheme 15

1.3 ADDENDA

Since the original preparation of the manuscript for this book, several interesting articles that further utilize alanine derivatives for asymmetric synthesis have appeared in the literature.

Earlier in this chapter we described how L-alanine can effectively be α-alkylated to produce either (*S*)- or (*R*)-α-methyl amino acid esters **45** and **49** by relying on chiral auxiliaries. Seebach et al.[42,43] recently disclosed a procedure that requires no chiral auxiliary and affords α-branched amino acids by self-reproduction of chirality.

This is accomplished by converting L-alanine (via its ester **15a**) to imidazolidinones **111** as shown in Scheme 16. A noteworthy advantage of this process is that either (2*S*)-**111a** or (2*R*)-**111b** can be produced simply by varying the conditions for the cyclization of **110**. Cold methanolic HCl produces an *N*-unsubstituted (2*S*, 5*S*)-imidazolidinone (90% diastereoselective) that is subsequently benzoylated to **111a,** whereas heating **110** with benzoic anhydride at 130°C (4 hr) results in the direct formation of (2*R*, 5*S*)-**111b** (71% diastereoselective).[42] Phenylalanine, phenylglycine, valine, and methionine behave similarly in these transformations.

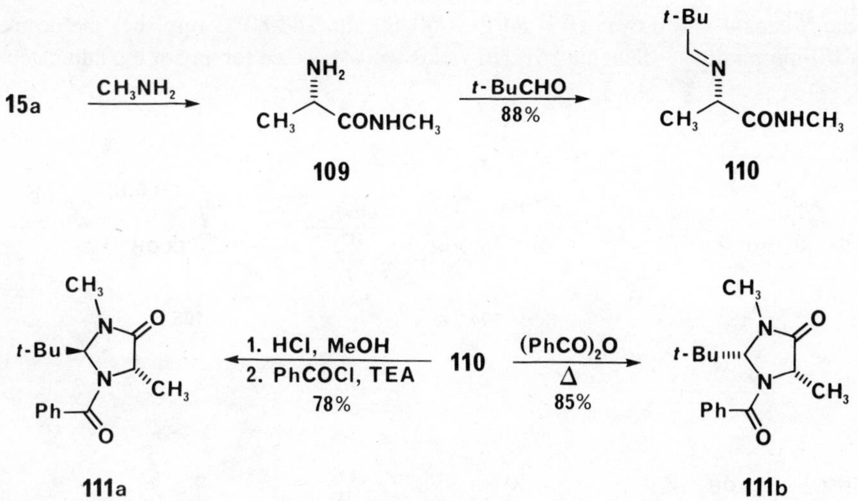

Scheme 16

Deprotonation of **111** with LDA at −78°C produces bright orange–red-colored solutions of the corresponding enantiomeric enolate (see Scheme 17). These anions are cleanly alkylated from the face opposite the *t*-butyl group to give the 5,5-disubstituted imidazolidinones **113a** or **113b** in 70–90% yields. Only one diastereomer is formed. The amino acid **114** is then liberated by heating **113** in 6 *N* HCl at 175–185°C in a sealed tube (4 hr). In this manner both (*R*)- and (*S*)-α-methyldopa are easily prepared from L-alanine.[43]

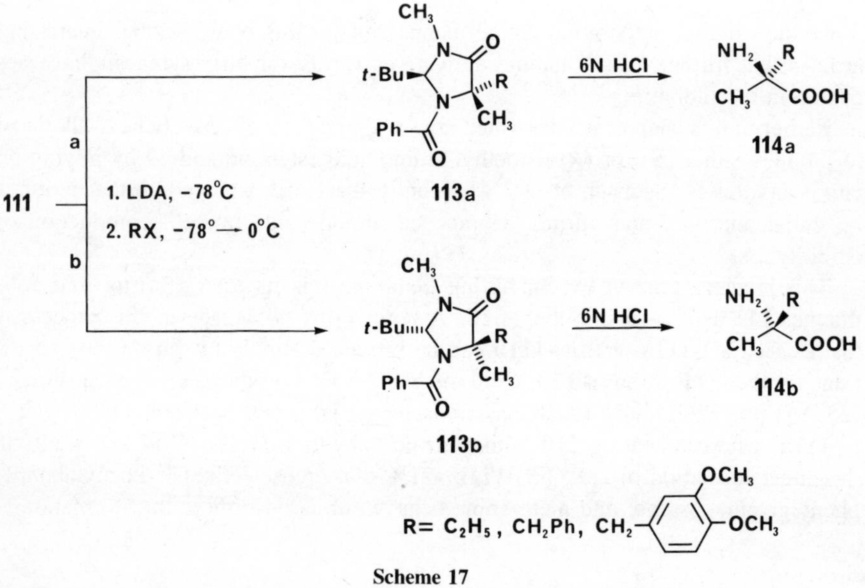

Scheme 17

The interesting cyclopeptide ulicyclamide **115** contains features associated with five of the common amino acids: alanine, phenylalanine, isoleucine, threonine, and proline.

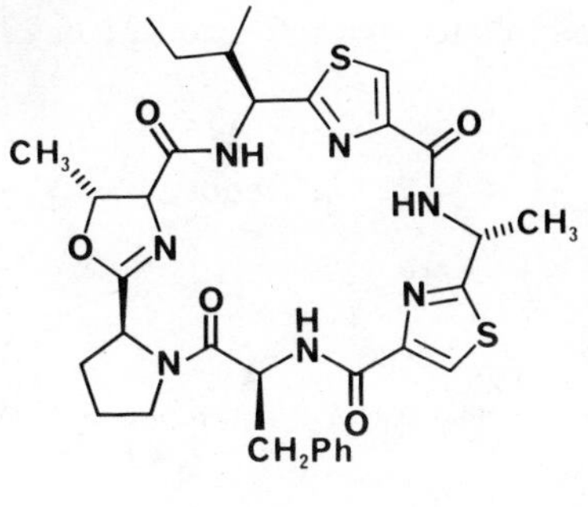

115

The synthesis of **115** is of interest because it does not involve only the coupling of these five amino acids. The key fragment (**124**) of the macrocycle requires the preparation of two distinct chiral thiazole derivatives, **119** and **122**, whose genesis can be traced to L-alanine and L-isoleucine, respectively.

The first thiazole **119** is prepared from **1** as shown in Scheme 18. After formation of the thiazole ring[44] (**118**), the amino group is introduced into the side chain by displacement of the acetoxy by azide under Mitsunobu conditions. The concomitant inversion of configuration (*S* → *R*) places the methyl group in the correct naturally occurring geometry of **115.** Reduction of the azide to an amine is effected by catalytic hydrogenation.

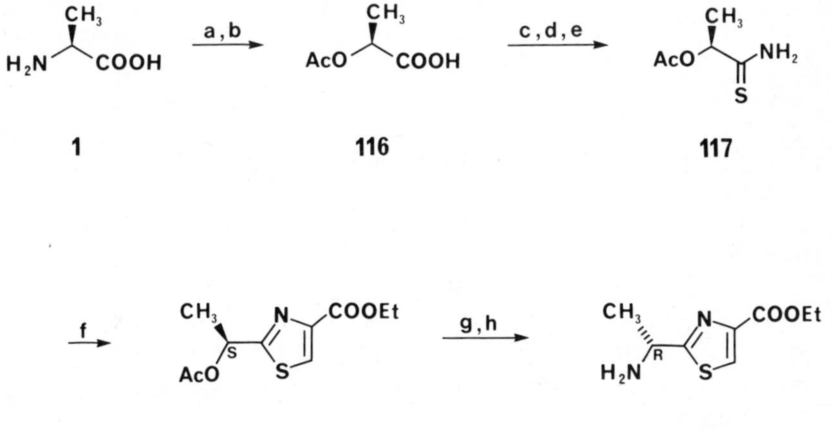

1 → (a,b) → **116** → (c,d,e) → **117** → (f) → **118** → (g,h) → **119**

Scheme 18 (a) Diazotization; (b) acetylation; (c) $SOCl_2$; (d) NH_3; (e) Lawesson's reagent; (f) ethyl bromopyruvate; (g) DEAD, PPh_3, HN_3 (inversion); (h) H_2/Pd.

In an analogous fashion, L-isoleucine (**120**) is converted to thiazole **121.** Again, the configuration of the α-carbon must be inverted in order to prepare for subsequent manipulations at this site. This is accomplished cleanly by displacement of the acetoxy with benzoic acid under Mitsunobu conditions. Base hydrolysis of the intermediate benzoate affords the hydroxy acid **122** in 29% overall yield.

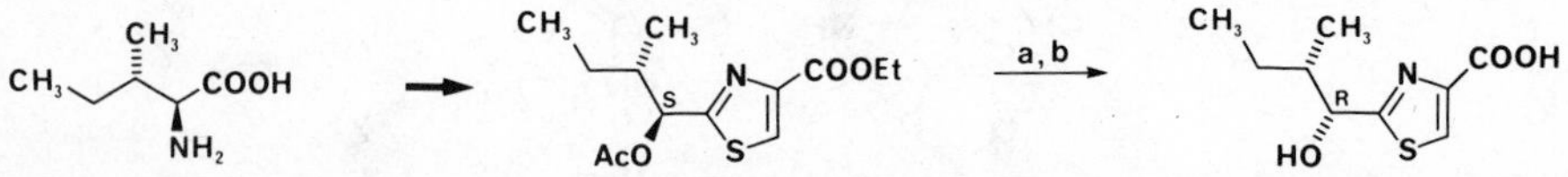

(a) DEAD, PPh_3, PhCOOH (inversion); (b) NaOH, H_2O.

The two thiazoles **119** and **122** are coupled by a redox condensation (dipyridyl disulfide, PPh_3) to give the amide **123.** Conversion of the hydroxyl group into an amine functionality (**124**) is accomplished as previously described for the preparation of **119.** The resulting inversion of configuration from (*R*) to (*S*) places the isobutyl group in the naturally occurring stereochemistry.[45]

The remainder of the ulicyclamide molecule is assembled in a stepwise fashion by attaching the remaining amino acids in a linear manner to the amino group of **124.** The last step requires a macrocyclization under high-dilution conditions.

(a) DEAD, PPh_3, HN_3 (inversion); (b) H_2/Pd.

REFERENCES

1. T. Scott and M. Brewer, *Concise Encyclopedia of Biochemistry*, Walter de Gruyter, New York, 1983, p. 14.
2. J. L. Bailey, *J. Chem. Soc.*, 3461 (1950).
3. W. D. Fuller, M. S. Verlander, and M. Goodman, *Biopolymers*, **15,** 1869 (1976).
4. M. Oya, R. Katakai, and H. Nakai, *Chem. Lett.*, 1143 (1973).
5. R. M. Freidinger, J. S. Hinkle, D. S. Perlow, and B. H. Arison, *J. Org. Chem.*, **48,** 77 (1983).
6. A. M. Osman, M. A. El-Maghraby, Z. H. Khalil, and Kh. M. Hassan, *Egypt. J. Chem.*, **18,** 993 (1975).

7. M. Bodanszky, U. S. Patent 3,880,838 (1975); *Chem. Abstr.*, **83,** 43746d (1975).
8. V. Sunjic, F. Kajfez, I. Stromar, N. Blazevic, and D. Kolbah, *J. Heterocycl. Chem.*, **10,** 591 (1973).
9. S. Yamada, N. Ikota, and K. Achiwa, *Tetrahedron Lett.*, 1001 (1976).
10. N. Hatanaka and I. Ojima, *J. Chem. Soc., Chem. Commun.*, 344 (1981).
11. T. Endo, Y. Hayashi, and M. Okawara, *Chem. Lett.*, 391 (1977).
12. T. F. Buckley, III and H. Rapoport, *J. Am. Chem Soc.*, **103,** 6157 (1981).
13. C. G. Knudsen and H. Rapoport, *J. Org. Chem.*, **48,** 2260 (1983).
14. D. E. McClure, B. H. Arison, J. H. Jones, and J. J. Baldwin, *J. Org. Chem.*, **46,** 2431 (1981).
15. D. D. Miller, F. Hsu, R. R. Ruffolo, Jr., and P. N. Patil, *J. Med. Chem.*, **19,** 1382 (1976).
16. F. Hsu, A. Hamada, M. E. Booker, H. Fuder, P. N. Patil, and D. D. Miller, *J. Med. Chem.*, **23,** 1232 (1980).
17. D. Seebach and D. Wasmuth, *Angew. Chem., Int. Ed.*, **20,** 971 (1981).
18. I. Hoppe, U. Schöllkopf, and R. Tolle, *Synthesis*, 789 (1983).
19. U. Schöllkopf, W. Hartwig, and U. Groth, *Angew Chem., Int. Ed.*, **18,** 863 (1979).
20. U. Schöllkopf, W. Hartwig, U. Groth, and K. Westphalen, *Liebigs Ann. Chem.*, 696 (1981).
21. U. Schöllkopf, W. Hartwig, and U. Groth, *Angew. Chem., Int. Ed.*, **19,** 212 (1980).
22. J. Depezay, A. Dureault, and T. Prange, *Tetrahedron Lett.*, 1459 (1984).
23. S. Inoue, N. Takamatsu, and Y. Kishi, *Yakugaku Zasshi*, **97,** 558 (1977); *Chem. Abstr.*, **87,** 102505g (1977).
24. H. E. Zaugg, M. Freifelder, H. J. Glenn, B. W. Horrom, G. R. Stone, and M. R. Versten, *J. Am. Chem. Soc.*, **78,** 2626 (1956).
25. P. Brewster, F. Hiron, E. D. Hughes, C. K. Ingold, and P. A. D. S. Rao, *Nature*, **166,** 179 (1950).
26. K. Koga, S. Yamada, M. Yoh, and T. Mizoguchi, *Carbohydr. Res.*, **36,** C9 (1974).
27. Y. Hirotsu, T. Shiba, and T. Kaneko, *Bull. Chem. Soc. Jpn.*, **40,** 2945 (1967).
28. S. Shibata, H. Matsushita, M. Noguchi, M. Saburi, and S. Yoshikawa, *Chem. Lett.*, 1305 (1978).
29. S. Shibata, H. Matsushita, K. Kato, M. Noguchi, M. Suburi, and S. Yoshikawa, *Bull. Chem. Soc. Jpn.*, **52,** 2938 (1979).
30. H. Umezawa, *Progr. Biochem. Pharmacol.*, **11,** 18 (1976).
31. T. Takita, Y. Umezawa, S. Saito, Y. Muraoka, M. Suzuki, M. Otsuka, S. Kobayashi, and M. Ohno, *Tetrahedron Lett.*, 671 (1981).
32. T. Takita, Y. Umezawa, S. Saito, H. Morishima, T. Tsuchiya, T. Miyake, S. Kageyama, S. Umezawa, Y. Muraoka, M. Suzuki, M. Otsuka, M. Narita, S. Kobayashi, and M. Ohno, *Tetrahedron Lett.*, 521 (1982).
33. Y. Aoyagi, K. Katano, H. Suguna, J. Primeau, L. Chang, and S. M. Hecht, *J. Am. Chem. Soc.*, **104,** 5537 (1982).
34. S. Saito, Y. Umezawa, H. Morishima, T. Takita, H. Umezawa, M. Narita, M. Otsuka, S. Kobayashi, and M. Ohno, *Tetrahedron Lett.*, 529 (1982).

35. T. Yoshioka, T. Hara, T. Takita, and H. Umezawa, *J. Antibiot.*, **27,** 356 (1974).
36. R. C. Job and T. C. Bruice, *J. Am. Chem. Soc.*, **96,** 809 (1974).
37. M. Narita, M. Otsuka, S. Kobayashi, M. Ohno, Y. Umezawa, H. Morishima, and S. Saito, *Tetrahedron Lett.*, 525 (1982).
38. R. Nishizawa, T. Saino, T. Takita, H. Suda, T. Aoyagi, and H. Umezawa, *J. Med. Chem.*, **20,** 510 (1977).
39. Y. Moriyama, Y. Saitoh, Y. Igarashi, T. Sugimoto, T. Takahashi, and H. Khuong, *Koen Yoshishu–Tennen Yuki Kagobutsu Toronkai 22nd*, 532 (1979); *Chem. Abstr.*, **93,** 168451s (1980).
40. J. H. Hunt and D. McHale, *J. Chem. Soc.*, 2073 (1957).
41. G. Bettoni, C. Franchini, R. Perrone, and V. Tortorella, *Tetrahedron*, **36,** 409 (1980).
42. R. Naef and D. Seebach, *Helv. Chim. Acta*, **68,** 135 (1985).
43. D. Seebach, J. D. Aebi, R. Naef, and T. Weber, *Helv. Chim. Acta*, **68,** 114 (1985).
44. U. Schmidt and R. Utz, *Angew. Chem., Int. Ed.*, **23,** 725 (1984).
45. U. Schmidt and P. Gleich, *Angew. Chem., Int. Ed.*, **24,** 569 (1985).

CHAPTER TWO

THE PHENYLALANINE FAMILY

Phenylalanine is an aromatic proteogenic amino acid related to alanine simply by the presence of an additional phenyl group. It is essential in the animal diet.[1] Two other amino acids in this family are tyrosine (4-hydroxyphenylalanine) and DOPA (3,4-dihydroxyphenylalanine). Both of these are nonessential in humans since they can be derived from phenylalanine by a series of metabolic hydroxylation reactions. Because of their similarities, these amino acids are discussed in this chapter.

2.1 PHENYLALANINE

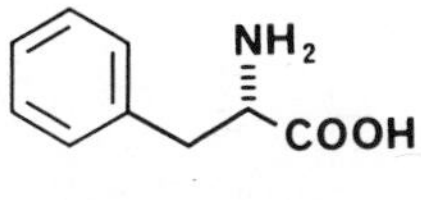

NH_2

COOH

(*S*)-α-Aminobenzenepropanoic acid (**1**)

In Chapter 1 we examined reactions in which an amino acid is incorporated into the skeletal framework of a molecule. Before continuing along those lines, we begin this section with quite a remarkable reaction, the aldol condensation, in which the mere presence of an amino acid influences the stereochemical outcome of the transformation.

The aldol condensation has shown great versatility in the synthesis of natural products as well as other complex molecules. In 1971, a significant breakthrough occurred in the steroid field in which these molecules could be prepared in optically active form without resorting to a resolution. This required the development of methodology for the synthesis of optically active 7a-alkyl-tetrahydro-1,5-indandiones (**3**), which are important CD ring fragments in the construction of the steroidal framework.

Hajos and Parrish[2,3] and Eder et al.[4] independently discovered that prochiral triketones **2**, when exposed to L-amino acids in the presence of perchloric acid, undergo an intramolecular aldol cyclization to produce **3** in good chemical and optical yields. It is fortuitous that the naturally occurring amino acids induce chirality that translates to the desired 13β-alkyl configuration in the steroid nucleus. On further study of the **2** → **3** conversion, Danishefsky and Cain[5] have found that the nature of the R group dictates which amino acid is required for the optimum induction of asymmetry. When R ≠ H, L-phenylalanine is the amino acid of choice, whereas when R = H, L-proline is preferred. (For a detailed discussion on proline-mediated aldol cyclizations, please see Chapter 8.)

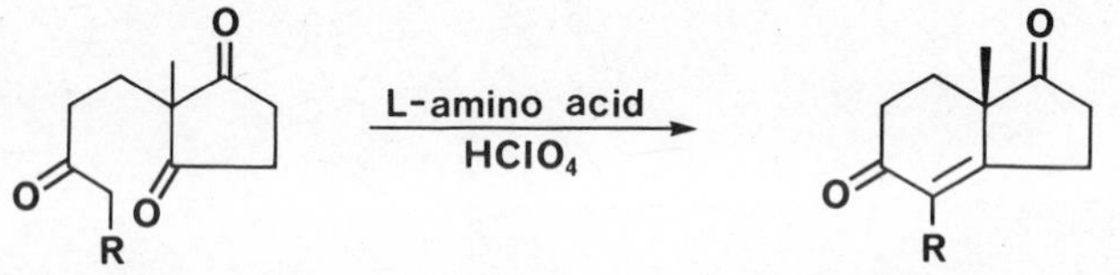

L-amino acid
$HClO_4$

2 3

Mechanistically, it was originally speculated that induction of asymmetry occurs as a result of initial enamine formation (**4**), in which subsequent cyclization followed by hydrolysis of the resulting immonium intermediate produces **3** while simultaneously regenerating the amino acid.[5,6] However, a later study concludes that species **4** is not formed. Rather, a molecular complex **5** stabilized by both hydrogen bonding and nonbonding interactions is more likely.[7]

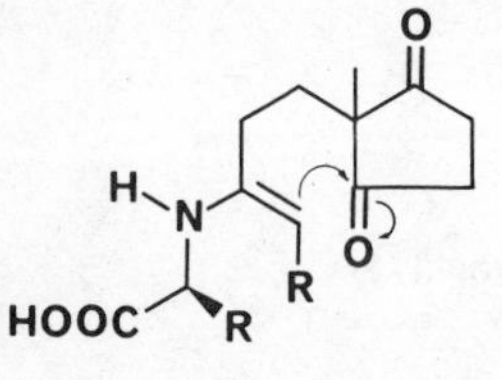

4 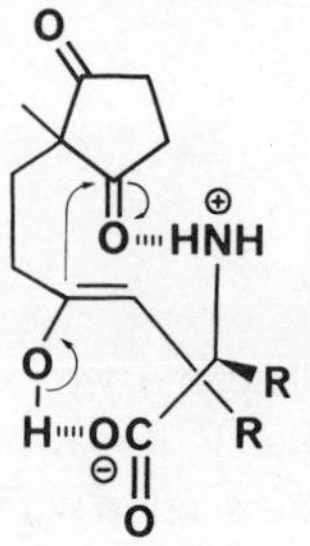 5

Several interesting sytheses exemplifying the use of phenylalanine to induce chirality in the key aldol cyclization step are shown below. Estrone (**12**), one of the female sex hormones, is secreted by the ovaries. Only its (+)-enantiomer exhibits estrogenic hormonal activity; the opposite antipode is inactive. The chiral aldol approach seems ideally suited for the generation of the required 13β-methyl configuration of the steroid. Danishefsky and Cain,[5,8] in their unique synthesis of (+)-estrone (Scheme 1), use a phenylalanine-controled aldolization of triketone **6** to introduce the asymmetric center in enedione **7.** The only problem step in the entire sequence is the conversion of **8** to **9,** where a mixture of diastereomers (37% βH:44% αH) is formed. The desired αH isomer, however, is separable by column chromatography and is then carried on to (+)-estrone.

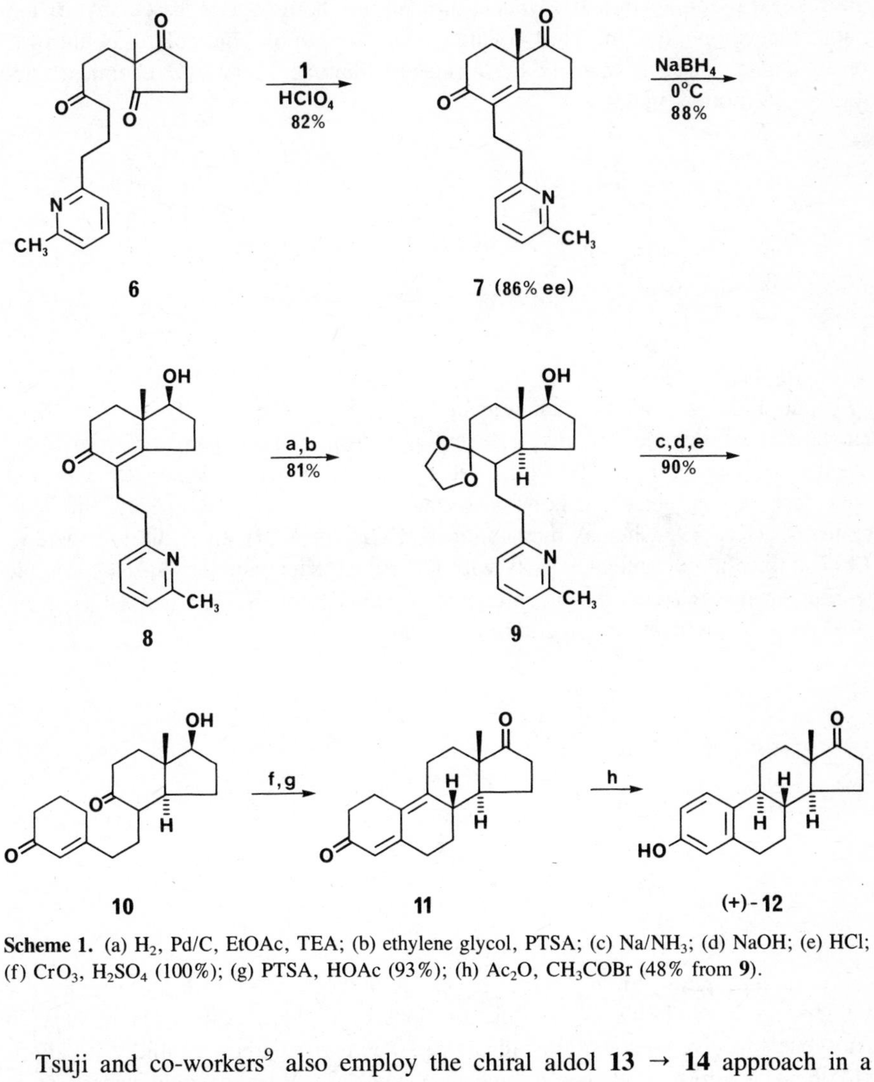

Scheme 1. (a) H_2, Pd/C, EtOAc, TEA; (b) ethylene glycol, PTSA; (c) Na/NH_3; (d) NaOH; (e) HCl; (f) CrO_3, H_2SO_4 (100%); (g) PTSA, HOAc (93%); (h) Ac_2O, CH_3COBr (48% from **9**).

Tsuji and co-workers[9] also employ the chiral aldol **13** → **14** approach in a synthesis of optically active (+)-19-nortestosterone (**15**).

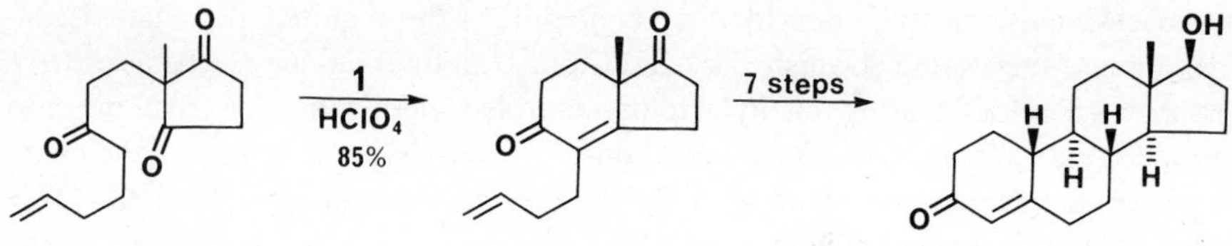

1, $HClO_4$, 85%; 7 steps

13 14 (76% ee) 15

The chiral aldol is not restricted to the formation of 1,5-indandiones, but also affords excellent asymmetric inductions in the homologous 8a-octahydro-1,6-naphthalenedione system. Thus, treatment of **16** with an equimolar amount of L-phenylalanine in acetic acid (2 hr) furnishes diketone **17** in 80% chemical yield and 87.3% optical yield.[10]

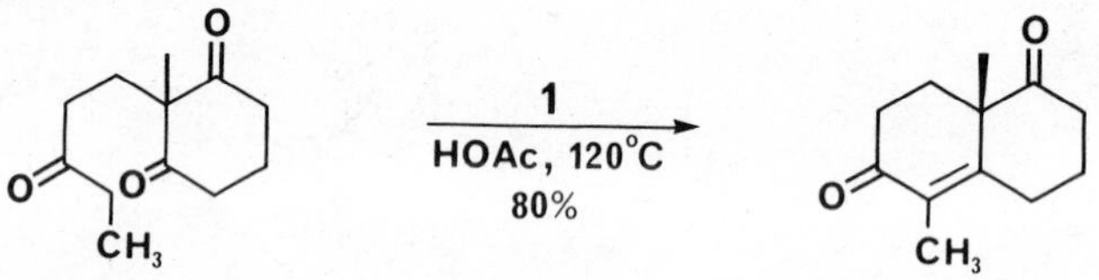

1, HOAc, 120°C, 80%

16 17

Dione **17** plays an important role in the mechanistic study of oxy-Cope rearrangements of bicyclic systems. The required carbinol **18** is prepared from **17** by reaction with lithium acetylide followed by partial reduction. When **18** is treated with base, a ring-enlarged dione is formed. The use of KH in THF furnishes optically active **19,** whereas methanolic $NaOCH_3$ or KOH gives the racemate of **19.** The retention of optical activity with KH suggests a concerted [3,3]sigmatropic rearrangement, whereas the racemization observed with $NaOCH_3$ or KOH is indicative of a fragmentation–recombination mechanism.

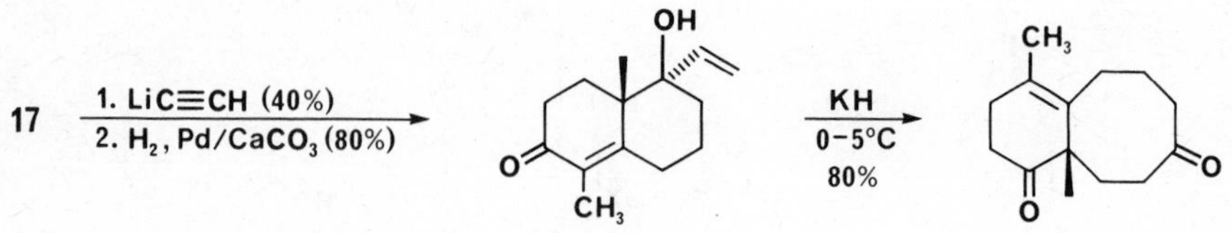

17 — 1. LiC≡CH (40%); 2. H_2, Pd/$CaCO_3$ (80%) — KH, 0–5°C, 80%

18 19

Let us now return to chemistry that integrates the amino acid into a molecular framework. Two simple cyclization reactions of L-phenylalanine (**1**) occur with phosgene[11] to give its *N*-carboxyanhydride **20** and with carbon disulfide[12] to afford (4*S*)-benzyl-2-thioxo-5-thiazolidinone (**21**). As with *N*-carboxyanhydrides, **21** is useful in peptide synthesis because of the activation of the acid carbonyl toward nucleophilic attack. The reaction of **21** with sodium glycinate in methanol results in a 61% yield of L-phenylalanylglycine.

$COCl_2$ ← **1** → CS_2, 62%

20 **1** **21**

The unusual tricyclic system **23** is also an internally activated, protected form of L-phenylalanine. It is prepared by reacting 3-formyl-*N*-hydroxyphthalimide (**22**) with **1** (which gives a Schiff base) and then lactonizing the intermediate in the presence of ethoxyacetylene.[13] The reaction of **23** with amino acid esters readily promotes peptide bond formation.

22 —**1**→ **23**

A third method of activating phenylalanine toward aminolysis is the formation of (4*S*)-2-benzyloxy-4-benzyloxazol-5(4*H*)-one (**26**). The cyclodehydration of Cbz–L-phenylalanine (**24**) with phosgene was originally reported by Miyoshi[14,15] to give an aziridione. Unfortunately, the structure was erroneously assigned as **25**, and it was not until 6 years later that Jones and Witty[16,17] unambiguously elucidated the structure as **26**. The method for its preparation was also improved and made more repeatable with the use of phosphorus pentachloride in place of phosgene. Optically active peptides are easily obtained from **26** in 29–84% yield simply by reaction with an amino acid ester in chloroform at −20°C.[18]

25 ←$COCl_2$, 48%— **24** —PCl_5, 43%→ **26**

For alkylation of amino acids to occur with retention of asymmetry, a cyclic derivative is required in order to establish a diastereofacial bias for the approach of the alkylating agent to the preformed enolate. The alkylation of phenylalanine can be effected by using just this approach (Scheme 2). Cbz–L-phenylalanine (**24**), when condensed with benzaldehyde or 2,4-dichlorobenzaldehyde, affords a 9:1 mixture of diastereomeric oxazolidinones **27** and **28** (separable by chromatography). Alkylation of the potassium enolates of **27** or **28** occurs stereospecifically and yields a single diastereomer in 80% yield. In either case, alkylation proceeds from the side opposite the aryl group and results in either retention of configuration for **27** → **29** or inversion for **28** → **30.** (*S*)-α-Methylphenylalanine (**31**) or its (*R*)-antipode **32** is then obtained from the oxazolidinones by a two-step hydrolytic–reductive process.

Since (*R*)-**32** is obtained from the minor diastereomer **28,** it is accessible in higher overall yield starting with Cbz–alanine (**33**). The condensation of **33** with 2,4-dichlorobenzaldehyde furnishes a mixture of oxazolidinones in a 4 : 1 ratio. The desired *cis*-**34** (major isomer) is alkylated as described above with retention of configuration. Hydrolysis and reduction then gives (*R*)-**32.**[19]

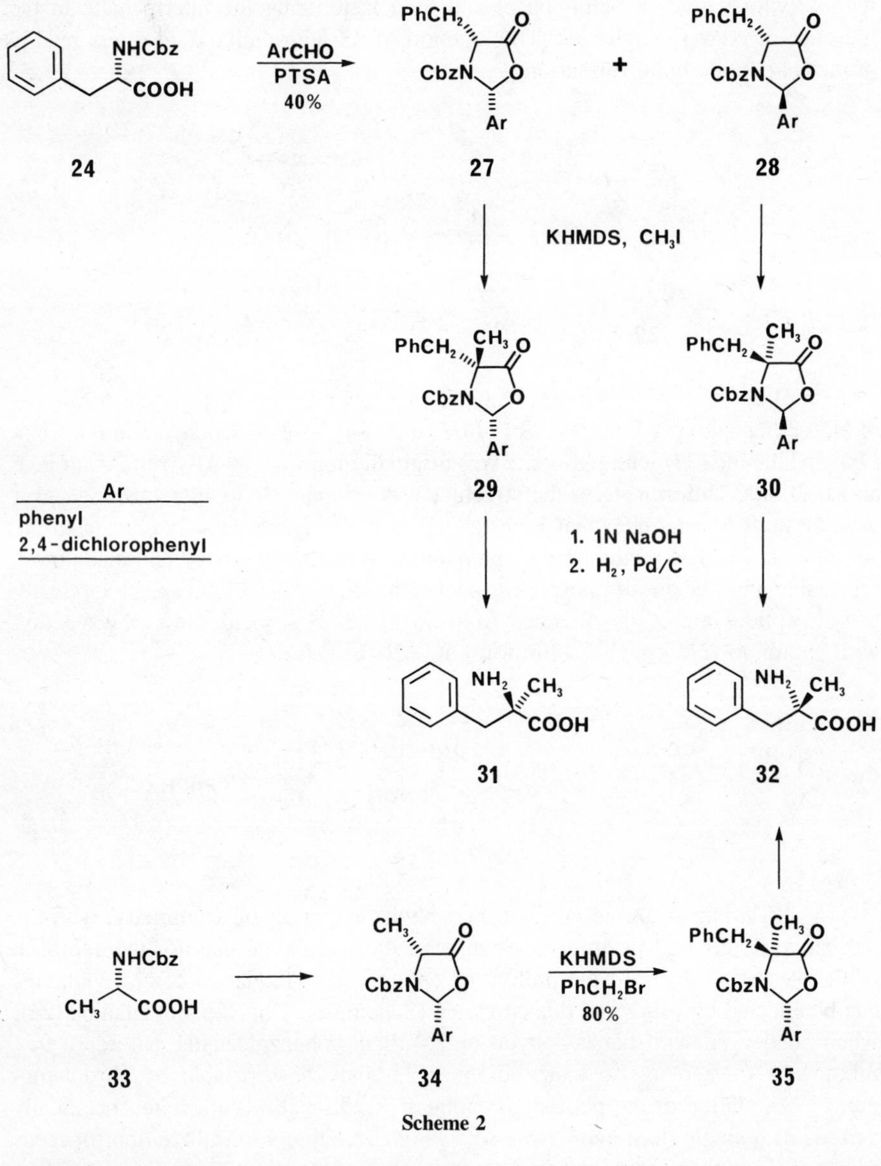

Ar
phenyl
2,4-dichlorophenyl

Scheme 2

Chiral isoquinoline derivatives can be generated from phenylalanine by cyclization between its amine and phenyl groups. The condensation of **1** with formaldehyde in concentrated HCl results in the formation of (3*S*)-carboxy-1,2,3,4-tetrahydroisoquinoline (**36**) in 50–80% yield. By a three-step sequence shown in Scheme 3, **36** is converted to (3*S*)-carboxydihydroisocarbostyril (**39**) in approximately 10% overall yield.[20]

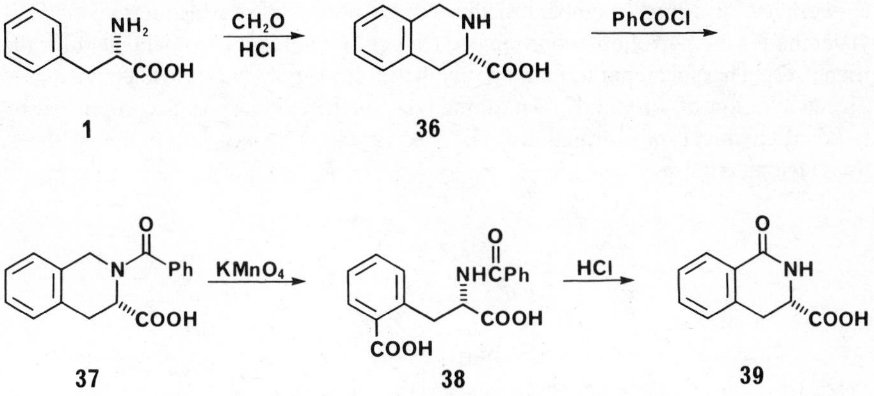

Scheme 3

Chiral indanones can be created from phenylalanine by another mode of cyclization, this time between the phenyl and acid groups. This is accomplished by converting the *N*-carbomethoxy derivative **40** to its acid chloride and then performing an intramolecular Friedel–Crafts acylation (Scheme 4). The product, **41,** is produced in 55–75% yield and is greater than 98% enantiomerically pure. The Cbz derivative **24** cannot be used in this reaction and results only in tar formation, presumably as a result of generation of benzyl carbonium ions.

Reduction of the prochiral ketone group of **41** with either diborane or sodium borohydride gives **42** as an 86:14 mixture of enantiomers (due to partial racemization). Further reduction with diborane or lithium aluminum hydride furnishes the *N*-methyl derivative **43** also in the same enantiomeric ratio. Chirally pure **43,** however, can be obtained directly (80–90% yield) from **41** by way of lithium aluminum hydride reduction. The primary amino derivative **44** is prepared from **42** by reaction with trichlorosilane–triethylamine followed by acid hydrolysis.[21]

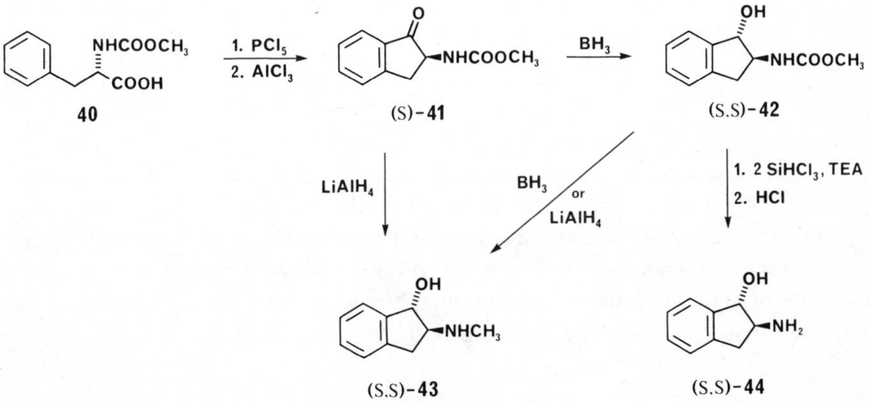

Scheme 4

Cytochalasin B (phomin) (**45**), a metabolic product of the microorganism *Phoma* (strain S298), is a member of a class of naturally occurring cytostatic compounds. Its remarkable biological profile[22] makes it a tantalizing synthetic target for the organic chemist.

In an elegant converget synthesis of this highly complex molecule, Stork and co-workers[23] use three members of the "chiral pool" in its construction; L-phenylalanine for the pyrrolidone portion **46,** (+)-citronellol and (+)-malic acid for the triene **47.** The focal point in their synthetic strategy is a regioselective [4 + 2]cycloaddition of **46** and **47.** The remainder of the synthesis is accomplished by standard chemical transformations. The macrocycle is formed late in the synthesis by a lactonization reaction.

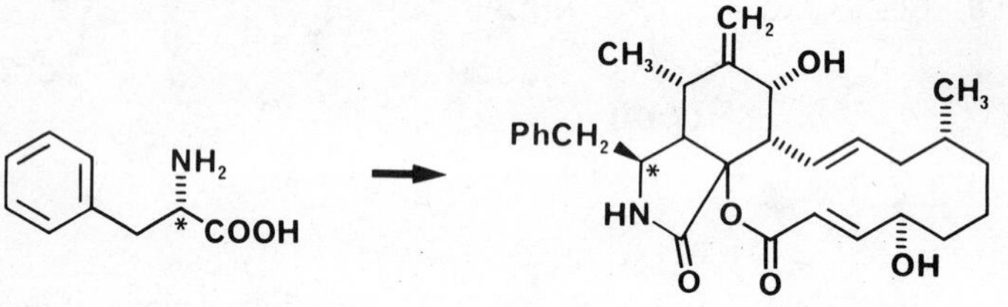

45

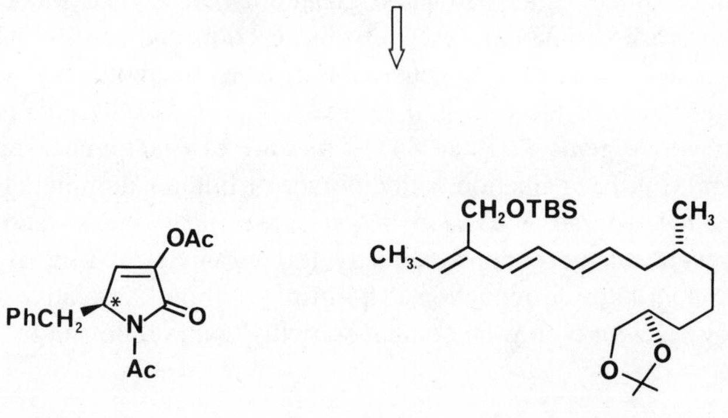

46 **47**

The initial step for the synthesis of **46** requires the homologation of L-phenylalanine to (*S*)-3-amino-4-phenylbutyric acid (**49**) (Scheme 5). An Arndt–Eistert reaction on Cbz–phenylalanine (**24**) followed by reductive deprotection nicely accomplishes the transformation.[24] Condensation of **49** with methyl oxalate in the presence of sodium methoxide forms the heterocycle **50** and *O*,*N*-diacetylation, and hydrolytic decarboxylation completes the preparation of **46.**

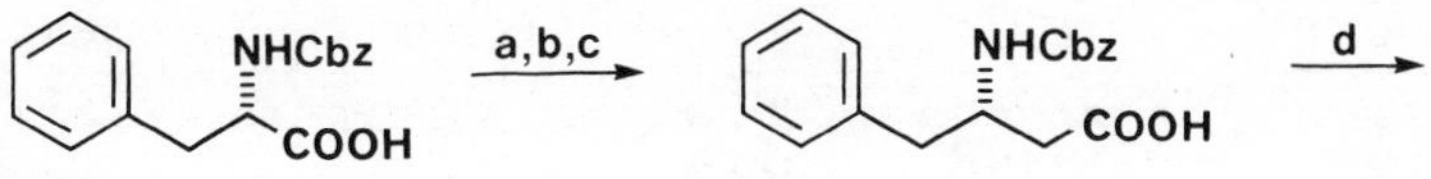

NHCbz COOH —a,b,c→ NHCbz COOH —d→

24 48

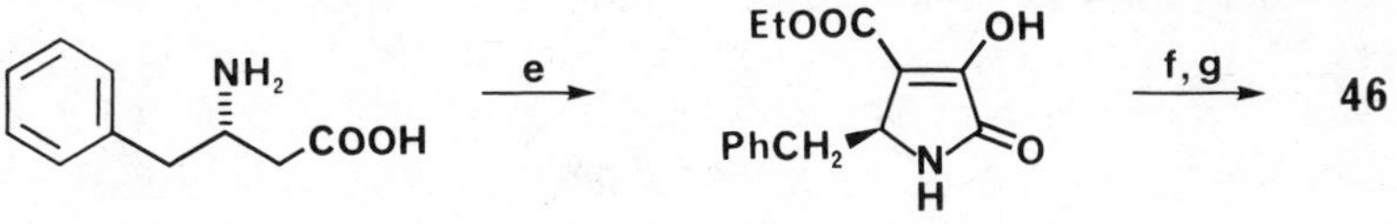

NH_2 COOH —e→ EtOOC OH $PhCH_2$ N H O —f, g→ 46

49 50

Scheme 5. (a) CH_2N_2; (b) Ag_2O; (c) NHO_3 (57% overall yield); (d) H_2, Pd/C; (e) $(COOCH_3)_2$; (f) Ac_2O, DMAP, pyridine; (g) DMSO, NaCl, H_2O.

In designing syntheses using members from the chiral pool, it is sometimes found that the required chiral building block is either not commercially available or is extremely expensive. In certain instances it may be advantageous to convert a more common, less expensive chiral pool intermediate into one of considerable rarity. A practical example illustrating such an approach is the conversion of L-phenylalanine to (*S*)-(−)-3-phenyllactic acid (**52a**). If purchased, the lactic acid derivative is 50 times more expensive than phenylalanine.

Reaction of α-amino acids with nitrous acid are known to give α-hydroxy acids with retention of configuration due to the participation of the neighboring carboxyl group.[25] L-Phenylalanine, when deaminated with nitrous acid (3 hr at 0°C), behaves similarly and affords (*S*)-**52** in 40% yield.[26] With **1,** however, a second mode of reaction can be put into operation if a strongly ionizing acid is employed in the deamination step. Trifluoroacetic acid causes the phenyl group to participate in preference to the carboxylate and give the phenyl migration product **53c** as the major constituent of the reaction mixture. The optical purity, however, is severely degraded.[27] Hydrolysis of the trifluoroacetate group of **53c** affords tropic acid (**53a**), the acid component of the ester alkaloids hyoscyamine, and hyoscine isolated from *Datura stramonium* plants.[28]

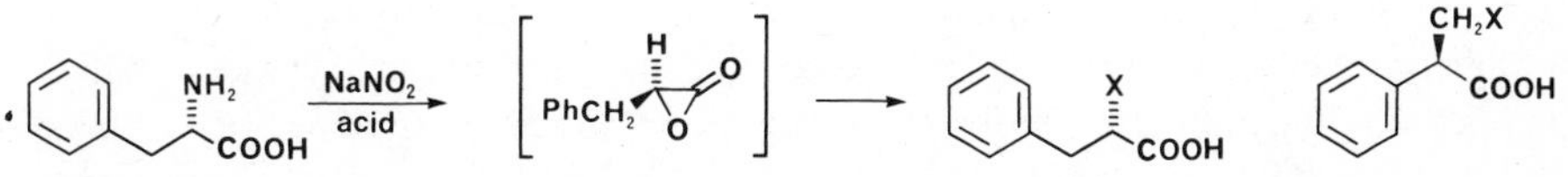

NH_2 COOH —$NaNO_2$, acid→ [H $PhCH_2$ O O] → X COOH CH_2X COOH

1 51 (S)-52 (S)-53

No.	Acid	X	Percent **52**	Percent **53**
a	H_2SO_4	OH	78 (91)	2
b	CH_3COOH	$OCOCH_3$	81 (84)	<0.1
c	CF_3COOH	$OCOCF_3$	25 (65)	63 (27)

Yields in parentheses denote optical yield.

The yield and optical purity of tropic acid can be vastly improved by trifluoroacetic acid solvolysis of *o*-nitrobenzenesulfonate (*S*)-**54,** which is derived from (*S*)-**52a** (Scheme 6). In this manner, optically pure (*R*)-**58** can be produced in 87% yield. Similarly, acetolysis of **54** affords optically pure (*R*)-(+)-3-phenyllactic acid ethyl ester (**56**) in 82% yield.[29] As you have probably discovered by now, both of these transformations proceed with complete inversion of configuration and produce the unnatural isomers of tropic acid and 3-phenyllactic acid (as their esters). Direct deamination of L-phenylalanine ethyl ester gives inferior results and produces (*R*)-tropic acid ethyl ester (34% optical yield) contaminated with six other products.[30]

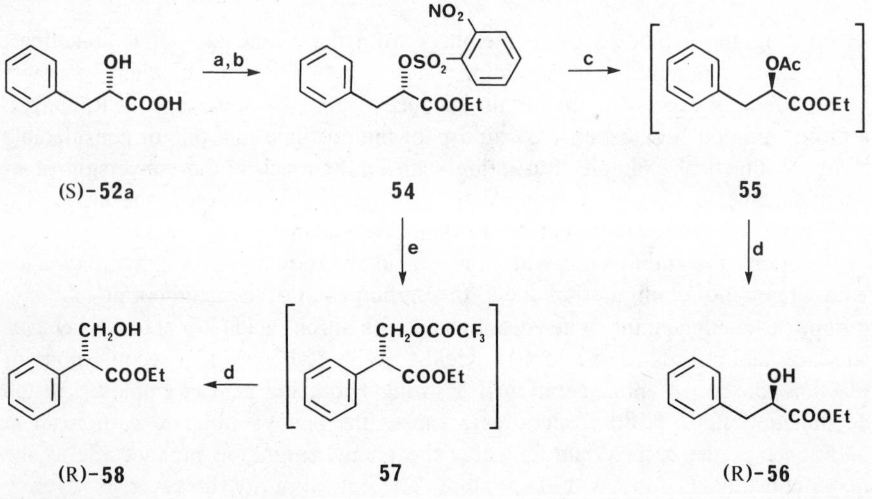

(S)-**52a** → (a,b) → **54** → (c) → [**55**] → (d) → (R)-**56**

54 → (e) → [**57**] → (d) → (R)-**58**

Scheme 6. (a) Esterify; (b) *o*-nitrobenzenesulfonyl chloride; (c) NaOAc, HOAc; (d) HCl, EtOH (82%); (e) CF_3COONa, CF_3COOH (87%).

To generate natural tropic acid using the protocol outlined in Scheme 6, one must start with D-phenylalanine (**59**). Deamination of **59** in the presence of nitrous acid yields (*R*)-(+)-3-phenyllactic acid (**60**). Esterification, conversion to the *o*-nitrobenzenesulfonate ester, and trifluoroacetolysis completes the synthesis. Alternately, (*R*)-**56** can be used in the sequence L-**1** → (*R*)-**56** → (*S*)-**53a.** (*R*)-3-Phenyllactic acid (**60**) is incorporated in the simulated biosyntheses of the tropane alkaloids (*R*)-(−)-littorine (**61**) and (*S*)-(−)-hyoscyamine (**62**). Esterification of **60** with α-tropine affords natural littorine (**61**), and then tosylation followed by trifluoroacetolysis and hydrolysis results in the formation of **62** as a consequence of phenyl migration[29] (Scheme 7).

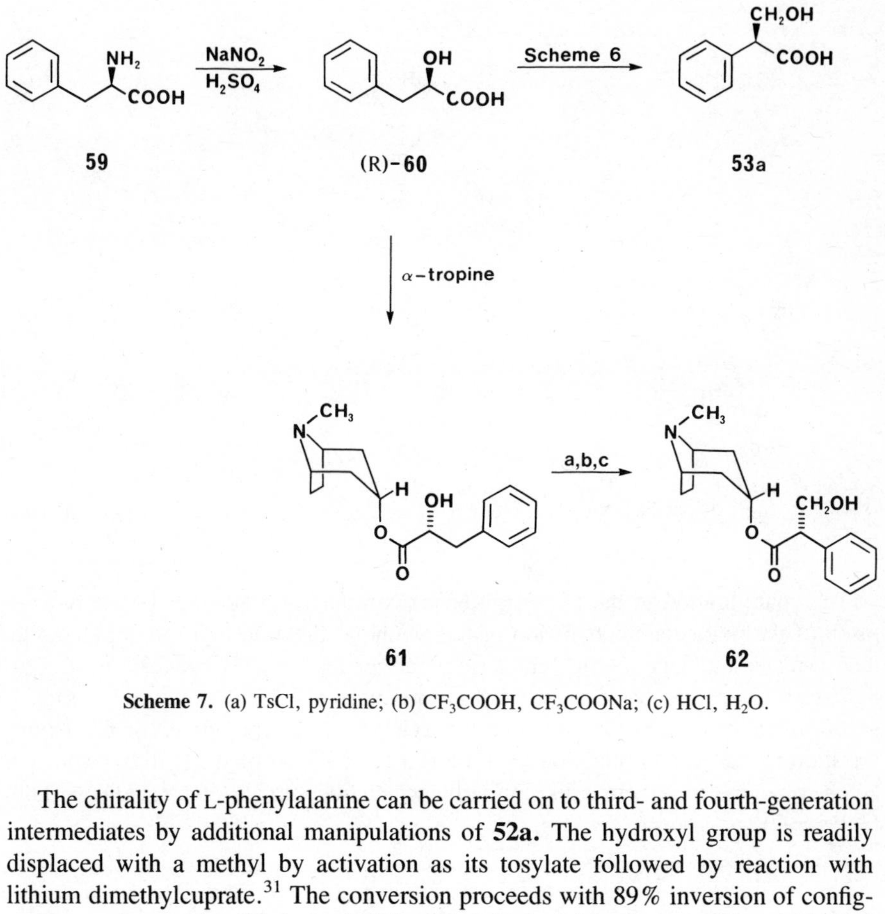

Scheme 7. (a) TsCl, pyridine; (b) CF_3COOH, CF_3COONa; (c) HCl, H_2O.

The chirality of L-phenylalanine can be carried on to third- and fourth-generation intermediates by additional manipulations of **52a.** The hydroxyl group is readily displaced with a methyl by activation as its tosylate followed by reaction with lithium dimethylcuprate.[31] The conversion proceeds with 89% inversion of configuration and provides (*R*)-(−)-ethyl-2-methyl-3-phenylpropionate (**64**) in 55–63% yield.

Organometallics such as methyllithium preferentially react with the carbonyl group of either **52a** or **63** to produce *vicinal* diols **66** (Scheme 8). Reaction with tosylate **63** proceeds with full inversion at the asymmetric center and produces (*R*)-(+)-3-methyl-1-phenyl-2,3-butanediol [(*R*)-**66**] in 81% yield. Conversely, esterification of **52a** followed by reaction with methyllithium produces (*S*)-**66** (98% yield) as a result of retention of configuration. Both reactions also give similar stereochemical results with *n*-butyllithium.[32,33]

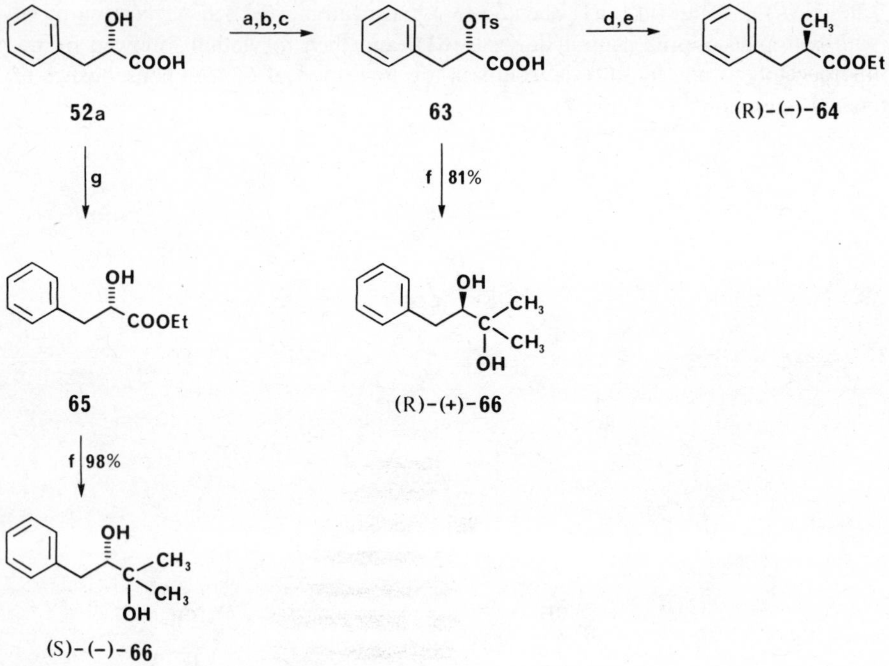

52a → (a,b,c) → 63 → (d,e) → (R)-(−)-64

52a → (g) → 65 → (f, 98%) → (S)-(−)-66

63 → (f, 81%) → (R)-(+)-66

Scheme 8. (a) $C_6H_5CH_2OH$, PTSA (100%); (b) TsCl, pyridine (89%); (c) H_2, Pd/C (95%); (d) $(CH_3)_2CuLi$; (e) C_2H_5OH, $SOCl_2$ (55–63% for the two steps); (f) CH_3Li, ether, −10°C; (g) C_2H_5OH, PTSA.

The configuration in the **65** → (*S*)-**66** transformation is retained probably as a result of the immediate conversion of the alcoholic function to a lithium alkoxide that insulates the asymmetric center from further reaction. The inversion of configuration [**63** → (*R*)-**66**] can arise by an intramolecular displacement of the tosyl group by either lithium alkoxide or carboxylate to form epoxide **67** or **68**. Isomerization to the α-keto alkoxide **69** followed by addition of methyllithium to the ketone produces the inverted dilithium salt, which liberates (*R*)-**66** on acidic workup (Scheme 9).

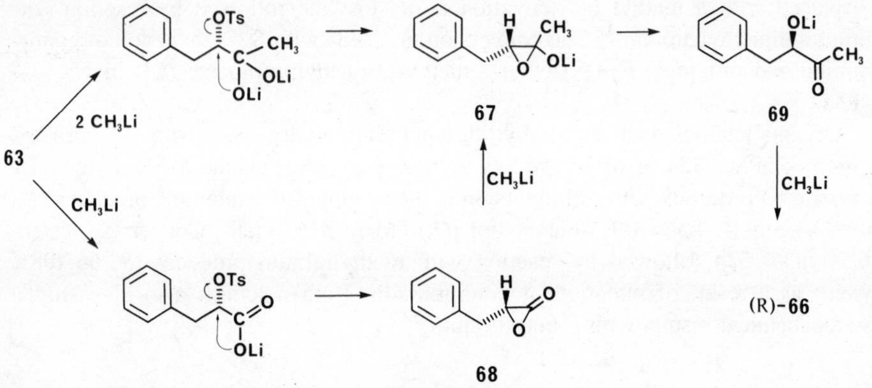

63 → (2 CH_3Li) → → 67 → 69 → (CH_3Li) → (R)-66

63 → (CH_3Li) → → 68 → (CH_3Li) → 67

Scheme 9

(*S*)-(−)-4(2,2,4,4-Tetramethyl-1,3-dioxolan-5-yl)-butan-2-one (**72**) is a versatile intermediate for the synthesis of optically active terpenes such as (*R*)-(+)-epoxygeraniol (**73**), (*R*)-(+)-marmin (**74**), (*R*)-(+)-epoxyaurapten[34] (**75**), (*R*)-(+)-epoxyfarnesol (**76**), and (*R*)-(+)- and (*S*)-(−)-squalene-2,3-oxide (**77**).[35,36] It is readily prepared from (*S*)-(−)-**66** by protection of the *vicinal*-diol function as a cyclic carbonate (**70**) followed by conversion of the phenyl group to a methyl ester (**71**). Standard chemical manipulations (e)–(j), which homologates the ester to an acetonyl moiety, produces the desired **72**.[32,33] Scheme 10 depicts the transference of chirality from L-phenylalanine [through **72** (labeled carbon)] to the terpenes **73**–**77**.

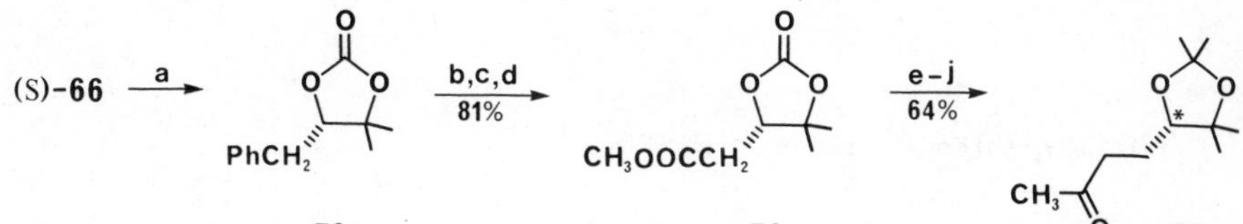

(a) $(EtO)_2CO$, NaOEt (100%); (b) O_3; (c) H_2O_2; (d) CH_2N_2; (e) $LiAlH_4$; (f) acetone, PTSA (94% overall); (g) TsCl, pyridine (89%); (h) KCN, DMF, 60°C (99%); (i) 20% NaOH (91%); (j) CH_3Li (85%).

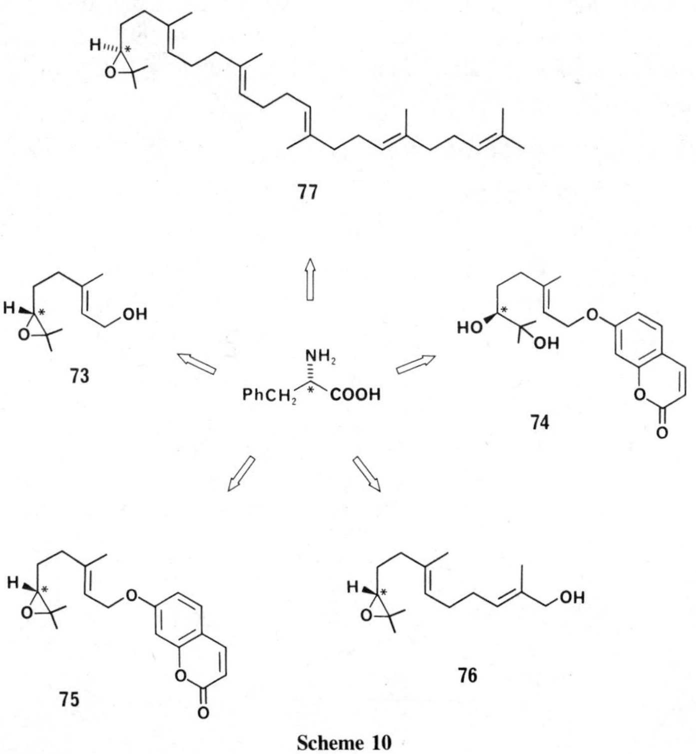

Scheme 10

Homogenous asymmetric hydrogenation of prochiral α-acylaminoacrylic acids with soluble chiral rhodium complexes can provide the drug industry with substantial quantities of optically active amino acids. Recently, a ditertiary bisphosphine ligand (**79**), derived from L-phenylalanine, has been developed according to the reaction sequence L-**1** → **52a** → **78** → **79**.[37] When complexed with bis[cyclooctadienrhodium]chloride, **79** forms an excellent chiral rhodium catalyst **80** for asymmetric hydrogenations.

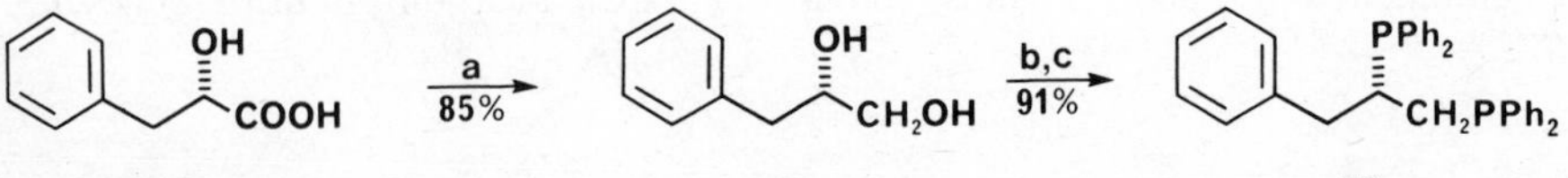

OH COOH —a, 85%→ OH CH_2OH —b,c, 91%→ PPh_2 CH_2PPh_2

52a **78** **79**

(a) $LiAlH_4$; (b) TsCl, pyridine; (c) $NaPPh_2$

$$[Rh(COD)Cl]_2 + \mathbf{79} + NaBF_4 \longrightarrow [Rh(COD)\mathbf{79}]^+ BF_4^-$$

80

(*Z*)-α-Acetylaminoacrylic acids (**81**) are enantioselectively hydrogenated to (*S*)-α-acetylamino acids (**82**) in the presence of **80**, which is prepared *in situ* from **79**.[37] Optical yields are excellent and range from 84.5 to 99%. Other phosphine ligands based on alanine,[38] valine,[37] and 4-hydroxyproline[39–43] form rhodium com-

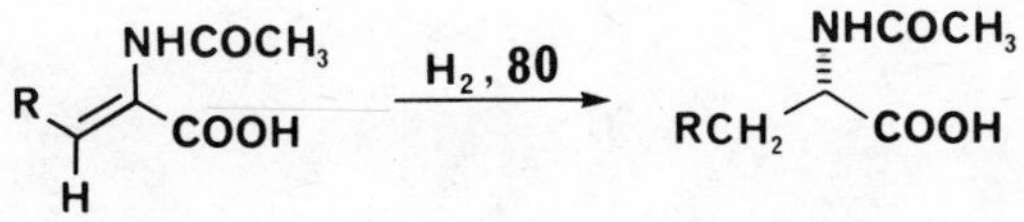

$NHCOCH_3$ R COOH H —H_2, **80**→ $NHCOCH_3$ RCH_2 COOH

81 **82**

R	yield (%)	optical yield (%)
H	99	84.5
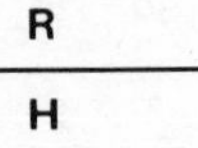	99	99.0
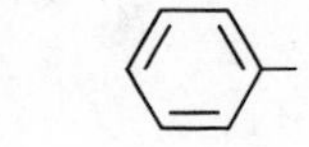 AcO–	99	96.0
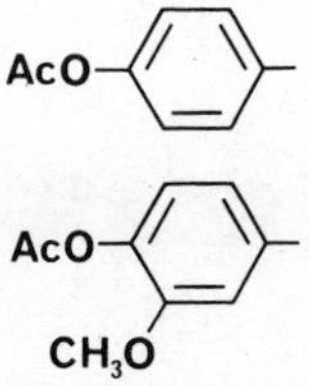 AcO–, CH_3O	99	90.4

plexes that also catalyze asymmetric hydrogenations of phochiral α-aminoacrylic acid derivatives and give the corresponding α-amino acid in 16–99% ee.

2.1.1 Phenylalaninal Derivatives

Suitably protected phenylalanine derivatives can be partially reduced to the corresponding phenylalaninal. Thus, when **83** is allowed to react with diisobutylaluminum hydride in toluene at −50°C, **84** is formed in 55% yield.[44] The product obtained is optically pure; however, extreme care must be taken in the purification stage. On exposure to silica gel, a considerable amount of racemization occurs through a keto–enol tautomerism.

NHCbz COOEt **83** —DIBAH 55%→ NHCbz CHO **84**

Stork and Nakamura[45] use this α-amino aldehyde for an improved, simplified synthesis of cytochalasin B (**45**). In his original synthesis Stork employs an intermolecular [4 + 2]cycloaddition to join both chiral pieces and resorts to a lactonization late in the synthesis to form the macrocyclic ring. The nature of the intermolecular cycloaddition does not control the regiochemistry and produces a 4:1 mixture of both desired and undesired regioisomers.[23]

In his improved version, the lactam ring is formed in one step from the reaction of **84** with phosphonate **85.** The product, **86,** is obtained with an enantiomeric excess of 80%.

OTHP $(CH_3O)_2P(O)$ $COOCH_3$ **85** —**84** 70%→ OTHP $PhCH_2$ N O Cbz **86**

After protecting group manipulation, the lactam is coupled with a chiral acid derived from (+)-pulegone to give tetraene **87.** Now, an intramolecular Diels–Alder reaction only produces one regioisomer **88.** The stereochemistry in the cyclohexane ring is not totally controlled and gives a 4:1 mixture of the desired isomer (shown as **88**) and its opposite diastereomer (methyl group and olefin above the plane of the paper).[45] Intermediate **88** is then converted to cytochalasin B (**45**) by standard synthetic manipulations.

PhCH$_2$ N Ac O O O OTBS

87

180–190°C
35%

PhCH$_2$ AcN O O OTBS

88

→ 45

In a somewhat similar study, Fuchs and co-workers[46] apply this concept for a potential approach to cytochalasin C. One crucial difference in Fuch's approach is to use a (*Z*)-diene in the intramolecular Diels–Alder cyclization. The (*Z*)-diene can attain only a single transition state in the reaction, and for chiral dienes this translates into asymmetric induction. A simple model system for such a hypothesis is easily constructed from L-phenylalanine.

The *N*-tosyl derivative **89** is reduced with borane–dimethyl sulfide, and the resulting alcohol is *N*-benzylated to give the protected chiral alcohol **90,** and oxidation with PCC smoothly affords the phenylalaninal derivative **91.**

NHTs COOH

89

→

R N Ts CH_2OH

90

PCC →

R N Ts CHO

91

R= CH_2Ph

Aldehyde **91** is then converted to the triene **92** by a Wittig reaction followed by detosylation and amide formation with 3,3-dimethyl-6-oxocyclohexene-1-carboxylic acid. Heating **92** in toluene for 40 hr produces the optically active tricycle **93** as the only product in nearly quantitative yield.

R N O O CH_3 CH_3

92

110°C
95%

CH_3 CH_3 H H R R N O O

93

R= CH_2Ph

Bestatin (**100**) (Scheme 11), isolated from culture filtrates of *Streptomyces olivoreticuli,* inhibits aminopeptidase B and leucine aminopeptidase but not aminopeptidase A, carboxypeptidases, or endopeptidases.[47] Its synthesis incorporates still another method for the preparation of phenylalaninal derivatives. The (3*R*) configuration of the product dictates the use of D-phenylalanine as the starting material.

Cbz–D-phenylalanine (**94**) can be activated as a 3,5-dimethylpyrazolide (**95**). Partial reduction with lithium aluminum hydride at −15 to −20°C furnishes Cbz–D-phenylalaninal (**96**). Immediate treatment with cold aqueous sodium hydrogen sulfite produces a solid adduct that, when treated with potassium cyanide, affords the cyanohydrin **97.** Hydrolysis with 6 *N* HCl gives a diastereomeric mixture of (2*R*,3*R*)- and (2*S*,3*R*)-acids (55% overall yield from D-phenylalanine). The desired (2*S*,3*R*)-isomer **98** is separable from the (2*R*,3*R*)-isomer by fractional crystallization from ethyl acetate. Coupling of **98** with L-leucine benzyl ester in the presence of DCC and 1-hydroxybenzotriazole furnishes **99,** which, on removal of the benzyl group by catalytic hydrogenation, gives bestatin (**100**).[48]

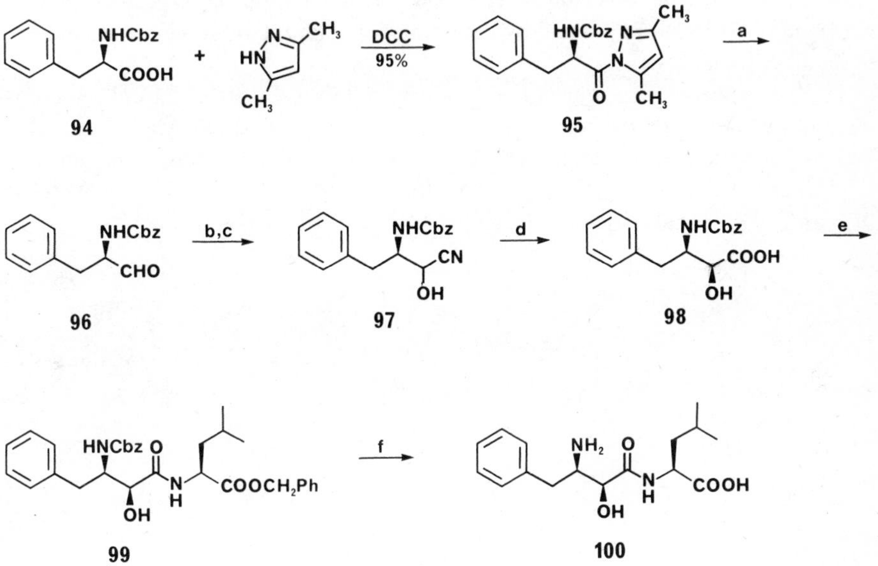

Scheme 11. (a) $LiAlH_4$, −20°C; (b) $NaHSO_3$; (c) KCN; (d) 6 *N* HCl; (e) L-leucine benzyl ester, DCC/HBT (82%); (f) H_2, Pd/C (88%).

2.1.2 Phenylalaninol

The carboxylate group of phenylalanine (**1**) or its ethyl ester is easily reduced to the primary alcohol **101** (Scheme 12) with a variety of standard reducing agents. Probably the most convenient method consists of treatment of **1** with boron trifluoride etherate followed by borane–dimethyl sulfide.[49] The presence of BF_3 increases the rate of reduction of the carboxylic acid and eliminates the need for a large excess of borane complex because of competing boron–nitrogen bond-forming reactions.

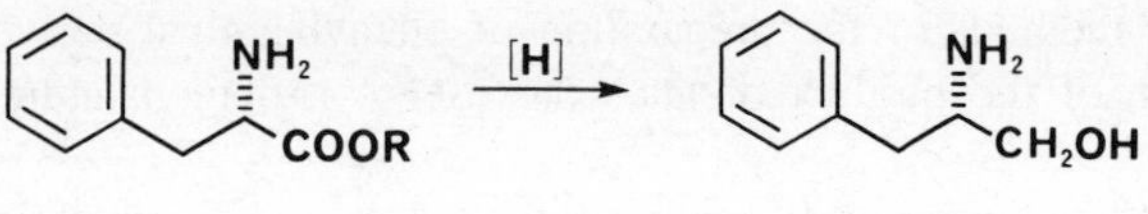

R	Reducing agent	Percent **101**	Ref.
C_2H_5	$LiAlH_4$	75	50
C_2H_5	$NaBH_4$	84	51
H	$(CH_3)_2S \cdot BH_3$	77	49

Asymmetric synthesis by way of chiral oxazoline intermediates has gained increasing importance within the last 10 years. (4*S*)-(−)-2,4-Dibenzyl-2-oxazoline (**103**) is readily prepared by the condensation of (*S*)-phenylalaninol (**101**) with ethyl 2-phenethylimidate (**102**).[52] The benzyl methylene in the 2 position can be alkylated (**104**) in 59–71% yield by deprotonation with *n*-butyllithium and addition of an alkyl iodide.

Further deprotonation with *n*-butyllithium, followed by stereoselective protonation of the anion and oxazoline ring hydrolysis, furnishes 2-phenylalkanoic acids **105** in 29–53% optical yields. The optical purities of the products are directly related to the size of the R group. As the steric bulk of the alkyl substituent increases, the optical yield also increases.

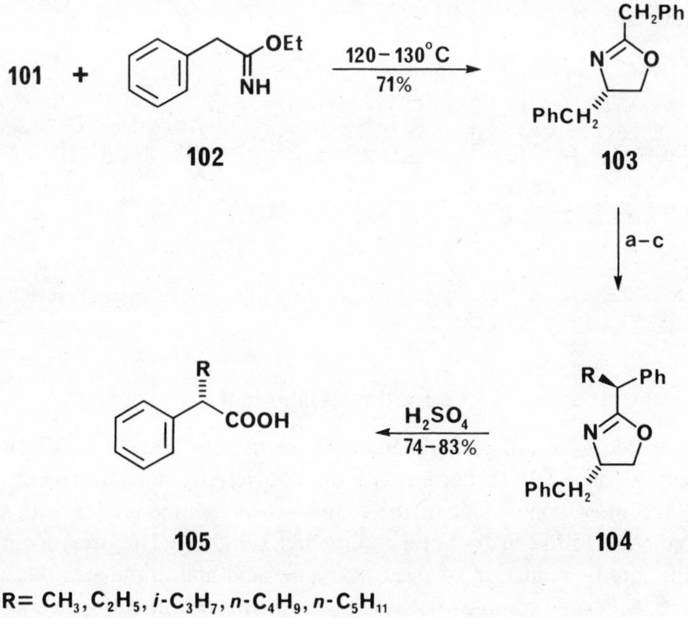

R= CH_3, C_2H_5, i-C_3H_7, n-C_4H_9, n-C_5H_{11}

Scheme 12. (a) n-C_4H_9Li, −78°C; (b) RI; (c) n-C_4H_9Li.

Few reactions can rival the predictability and stereochemical control exhibited by the Diels–Alder cycloaddition. For many years the induction of asymmetry in this reaction has attracted keen interest from a number of research groups. Recently, Evans et al.[53] have investigated the use of (*S*)-phenylalaninol-derived oxazolidinones as chiral auxiliaries for controlling the diastereoselection of the [4 + 2]cycloaddition.

The treatment of **101** with phosgene affords (*S*)-oxazolidinone **106.** Deprotonation with *n*-butyllithium or ethylmagnesium bromide followed by acylation with an acryloyl chloride gives the α,β-unsaturated carboximides **107** in good yields.

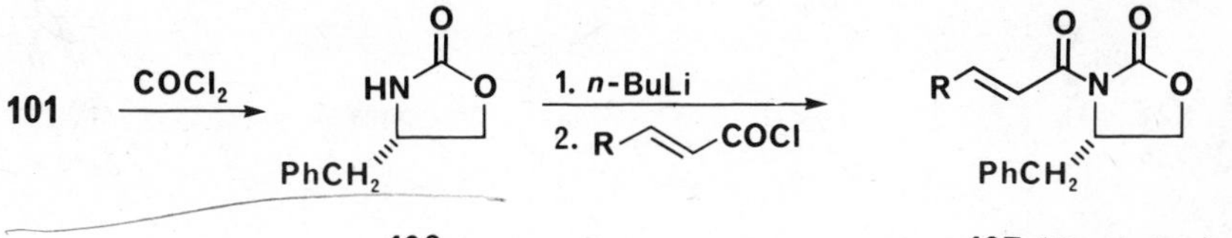

101 $\xrightarrow{COCl_2}$ 106 $\xrightarrow{\text{1. } n\text{-BuLi; 2. R-CH=CH-COCl}}$ 107 (R = H, CH_3)

Exposure of **107** to Lewis acids such as a dialkylaluminum chloride produces an extremely reactive *s-cis* bidentate chelated dienophile **108,** which, when R = H, undergoes cycloaddition even at −100°C. The diethylaluminum chloride-promoted Diels–Alder reaction of **107a** with either isoprene or piperylene produces cycloadducts **109a** (85% yield) and **110a** (84% yield) with diastereoselectivities of 95:5 for **109a** and >100:1 for **110a** (Scheme 13). The less reactive crotonate dienophile **107b** also undergoes cycloaddition with these same dienes at −30°C (6 hr) to afford cycloadducts **109b** (83%) and **110b** (77%) in greater than 94% diastereomeric purity. Cycloadduct **109a** is readily converted to (*R*)-(+)-α-terpineol (**111**) by transesterification with 1.5 equivalents of lithium benzyloxide (93% yield) followed by treatment with methylmagnesium bromide.[53] In this transesterification step, the chiral auxiliary **106** is regenerated.

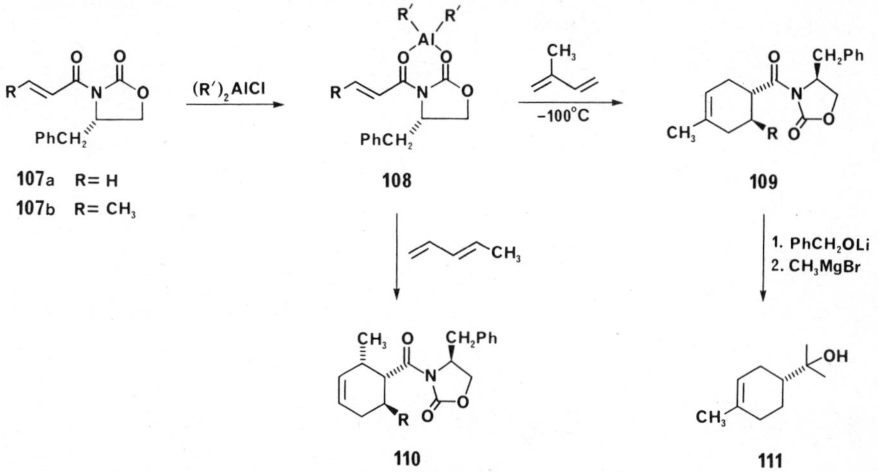

107a R = H; 107b R = CH_3 $\xrightarrow{(R')_2AlCl}$ 108 $\xrightarrow{\text{isoprene, } -100°C}$ 109 $\xrightarrow{\text{1. } PhCH_2OLi;\ \text{2. } CH_3MgBr}$ 111

108 + piperylene → 110

Scheme 13

The methodology is adaptable to the intramolecular Diels–Alder reaction. Reaction of triene carboximides **112** with dimethylaluminum chloride in methylene chloride (−30°C) results in a nearly quantitative yield of diestereomers **113** and **114** in the ratios indicated (Scheme 14). Conversion of the major diastereomer **113** to the respective benzyl ester **115** is effected with lithium benzyloxide.[54]

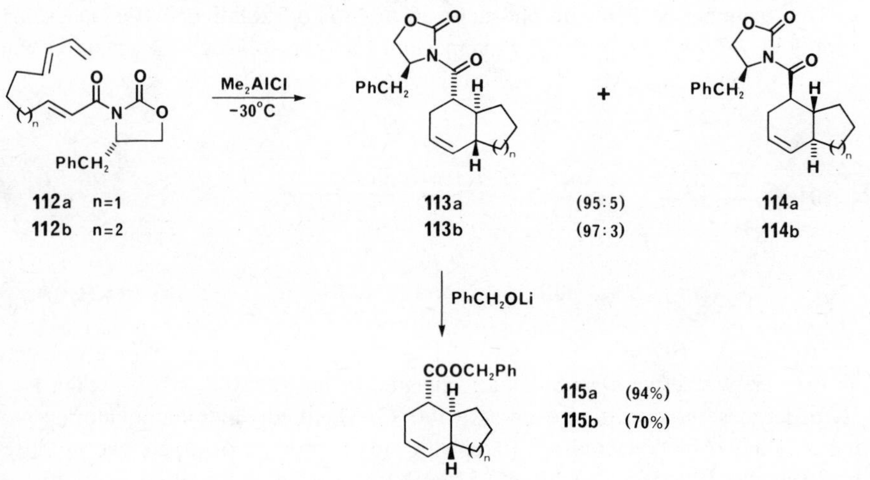

112a n=1
112b n=2

Me_2AlCl
-30°C

113a (95:5) 114a
113b (97:3) 114b

$PhCH_2OLi$

115a (94%)
115b (70%)

Scheme 14

In addition to oxazolines, optically active pyrrolidines are useful in asymmetric synthesis. (*R*)-(−)-2-Benzylpyrrolidine (**120**) can be prepared according to the series of reactions outlined in Scheme 15. (*S*)-Phenylalaninol (**101**) is ditosylated with tosyl chloride and then reacted with diethyl potassiomalonate to give a mixture of **116** and **117** in 54 and 17% yields, respectively. The crude mixture is then treated with 47% HBr to give (*R*)-(−)-4-amino-5-phenylvaleric acid (**118**). Cyclization and reduction completes the synthesis.[55]

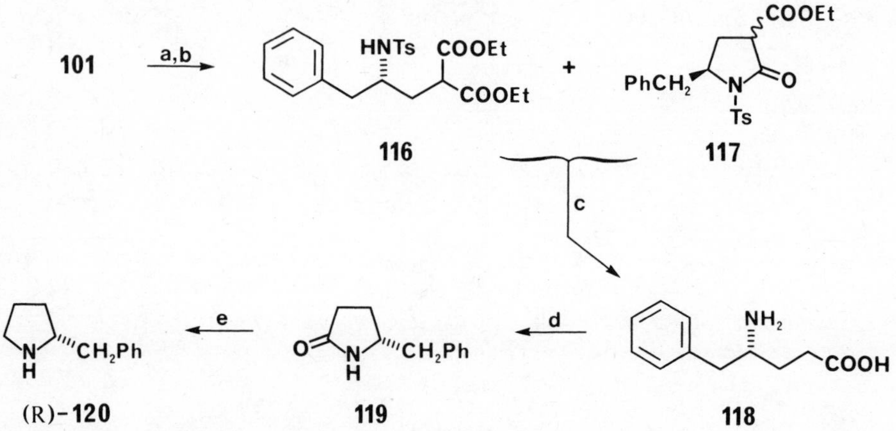

101 → (a,b) 116 + 117 → (c) 118 → (d) 119 → (e) (R)-120

Scheme 15 (a) TsCl (94%); (b) $KCH(COOEt)_2$; (c) 47% HBr (46% from **101**); (d) dioxane (reflux) (100%); (e) $LiAlH_4$ (46%).

Condensation of (*R*)-**120** with 2-phenylpropanal gives enamine **121** (100% yield). Michael addition of **121** to methyl vinyl ketone followed by acid-catalyzed cyclization gives (*S*)-**122** in 43% chemical yield and 31% optical yield. An analogous reaction using a pyrrolidine derived from (*S*)-valinol gives comparable optical purities but only 14% chemical yield.

(R)-**120** + [2-phenylpropanal: $C_6H_5CH(CH_3)CHO$] → **121** (N-pyrrolidinyl enamine, CH_2Ph; $CH=C(C_6H_5)CH_3$) —MVK→ **122** (4-CH_3-4-C_6H_5-cyclohexenone)

It is not necessary to use cyclic chiral auxiliaries to obtain efficient induction of asymmetry in organic reactions. Meyers has developed several acyclic auxiliaries based on amino acids, one of which was derived from L-phenylalanine. It is simply prepared by reduction of **1** to **101** followed by *O*-alkylation with methyl iodide.[56]

101 (CH_2Ph, NH_2, OH) —1. KH; 2. CH_3I; 75%→ **123** (CH_2Ph, NH_2, OCH_3)

The product, **123,** is prone to absorption of carbon dioxide from the atmosphere; therefore, it must be either stored under an inert gas or converted to its hydrochloride salt.

Methoxy amine **123** reacts with aldehydes or ketones to give chiral Schiff bases **124,** which, when metallated, form metalloenamines (**125**). Alkylation of these enamines proceeds with a high degree of diastereoselectivity because of the rigidity imparted as a result of internal chelation of the metal with the methoxy group.

123 + $R_1CH_2COR_2$ → **124** (R_1CH_2, R_2, N, CH_2Ph, OCH_3) —LDA→ **125** (R_2, R_1, N, CH_2Ph, Li←OCH_3)

Enantiomeric (*R*)- or (*S*)-2-methyloctanal (**128**) can be prepared by this technique (Scheme 16). The Schiff bases **126** and **127** are obtained in 95–100% yield by reaction of **123** with either propanal or octanal. Metallation with LDA at −23°C, alkylation with either *n*-hexyl iodide or methyl iodide at −78°C, and acid hydrolysis furnish (*R*)-**128** in 36% yield (42% ee) or (*S*)-**128** in 46% yield (47% ee), respectively.[56] The chiral auxiliary **123** is also recoverable if desired. In this series of reactions, other analogs similar to **123** but derived from valinol, leucinol, or phenylglycinol give lower optical purities of **128.**

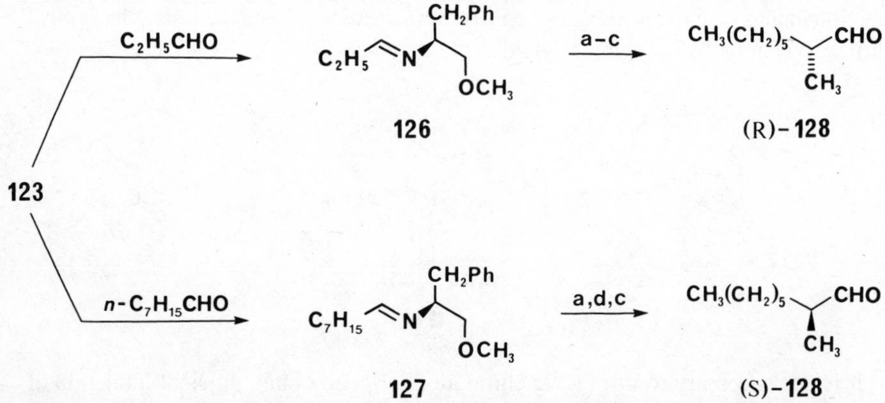

123 → (C_2H_5CHO) → **126** ($C_2H_5CH{=}N$–CH(CH_2Ph)CH_2OCH_3) → (a–c) → $CH_3(CH_2)_5CH(CH_3)CHO$ (R)-**128**

123 → (n-$C_7H_{15}CHO$) → **127** ($C_7H_{15}CH{=}N$–CH(CH_2Ph)CH_2OCH_3) → (a,d,c) → $CH_3(CH_2)_5CH(CH_3)CHO$ (S)-**128**

Scheme 16 (a) LDA, −23°C; (b) n-$C_6H_{13}I$, −78°C; (c) HOAc, NaOAc; (d) CH_3I, −78°C.

Cyclic ketones are better behaved than aldehydes and can be alkylated with a higher degree of diastereoselectivity. Their Schiff bases **129** are metalated with LDA and alkylated with methyl iodide to furnish (*S*)-2-methylcycloalkanones (**130**) in high chemical yields (after hydrolysis). The enantiomeric excess of the products are very high with cyclohexanone and cycloheptanone but drop significantly as the ring size increases. With cyclohexanone, the use of a more bulky alkylating agent (e.g., ethyl or propyl iodide) instead of methyl iodide dramatically increases the optical yield observed in the derived 2-alkylcyclohexanone (2-ethyl, 94% ee; 2-propyl, 99% ee).[57,58]

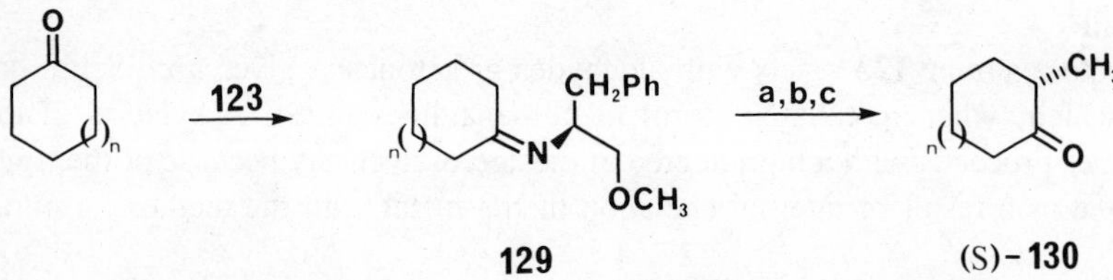

Cycloalkanone → (123) → **129** (N–CH(CH_2Ph)CH_2OCH_3 imine) → (a,b,c) → (S)-**130** (2-methylcycloalkanone)

(a) LDA, −30°C
(b) CH_3I, −78°C
(c) HOAc, NaOAc

n	Yield **130** (%)	Percent ee
1	65	87
2	90	85
3	65	20
5	80	30
7	88	59
10	75	37

In the cases of C_{12} and C_{15} cycloalkanones, the enantioselectivity of methylation is adjustable. At low temperatures, metallation of **124** (R_1 and R_2 are connected) results in the kinetic formation of (*E*)-lithioenamine **125** (R_1 and R_2 *cis*). Alkylation in this configuration produces the previously described C_{12} and C_{15} (*S*)-2-methylcycloalkanones **130.** However, when the lithioenamine is heated, the thermodynamically controlled (*Z*)-lithioenamine **125** (R_1 and R_2 *trans*) is formed. Methylation at −78°C then affords the opposite (*R*)-enantiomer [e.g., (*R*)-(+)-2-methylcyclododecanone, 80% ee]. In a similar fashion, acyclic ketones are alkylated enantioselectively under both kinetic and thermodynamic modes.

2.2 TYROSINE

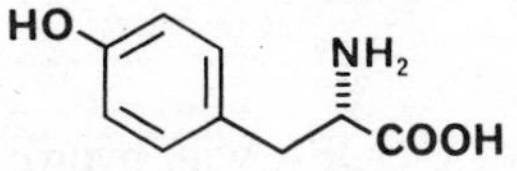

(*S*)-α-Amino-4-hydroxybenzenepropanoic acid

Tyrosine (**131**) is structurally identical to phenylalanine with the exception that it possesses a *para* hydroxy substituent in the aromatic ring. The presence of this hydroxyl function allows certain reactions to occur on the aromatic nucleus that otherwise would not have been possible with phenylalanine.

Thyroxine (**132**) is one of two major hormones secreted by the thyroid gland. This highly unusual iodine-containing amino acid is responsible for the transportation of iodine throughout the body, consequently influencing the rate of general metabolic activity. Its synthesis has been accomplished from L-tyrosine (**131**) by a series of aromatic substitution reactions (Scheme 17) and affords the hormone **132** in 27% overall yield.[59]

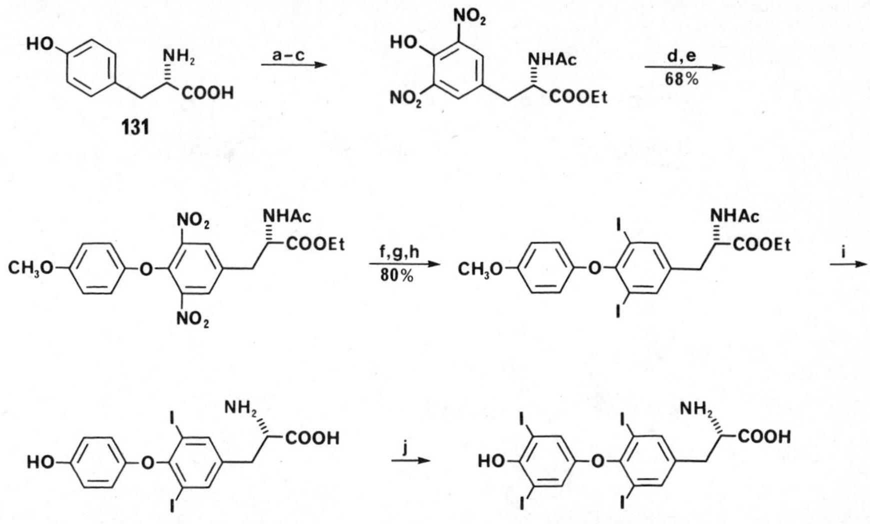

Scheme 17 (a) HNO_3 (87%); (b) Ac_2O (82%); (c) EtOH, PTSA (86%); (d) TsCl, pyridine; (e) 4-methoxyphenol; (f) H_2, Pd/C; (g) HNO_2; (h) I_2, NaI; (i) HI, HOAc (90%); (j) I_2, Et_2NH (89%).

The aromatic ring of tyrosine is also susceptible to Friedel–Crafts acylation. Treatment of **131** with acetyl chloride and aluminum chloride in nitrobenzene affords 3-acetyl-L-tyrosine (**133**), which, on exposure to alkaline hydrogen peroxide, is converted to 3,4-dihydroxy-L-phenylalanine (L-DOPA) (**134**).[60]

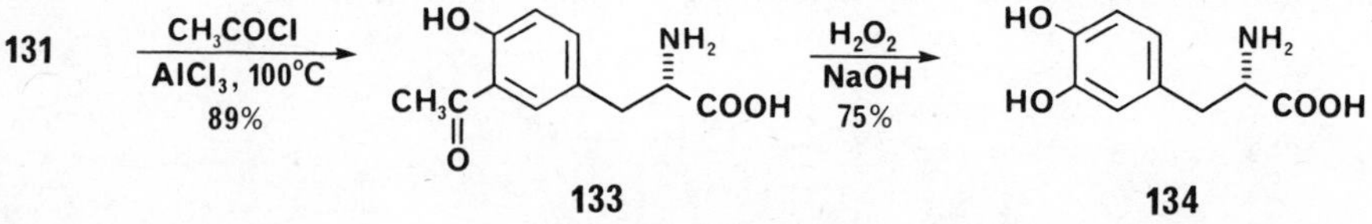

A monoprotected version of L-DOPA with hydroxyl group differentiation (**139**) is also available from L-tyrosine (methyl ester) as shown in Scheme 18. The key introduction of the 3-hydroxy group is accomplished by a sequential Fries rearrangement of **137** followed by a Baeyer–Villiger oxidation.[61] Compounds **134** and **139** are useful intermediates for the synthesis of a variety of interesting alkaloids (see Section 2.3).

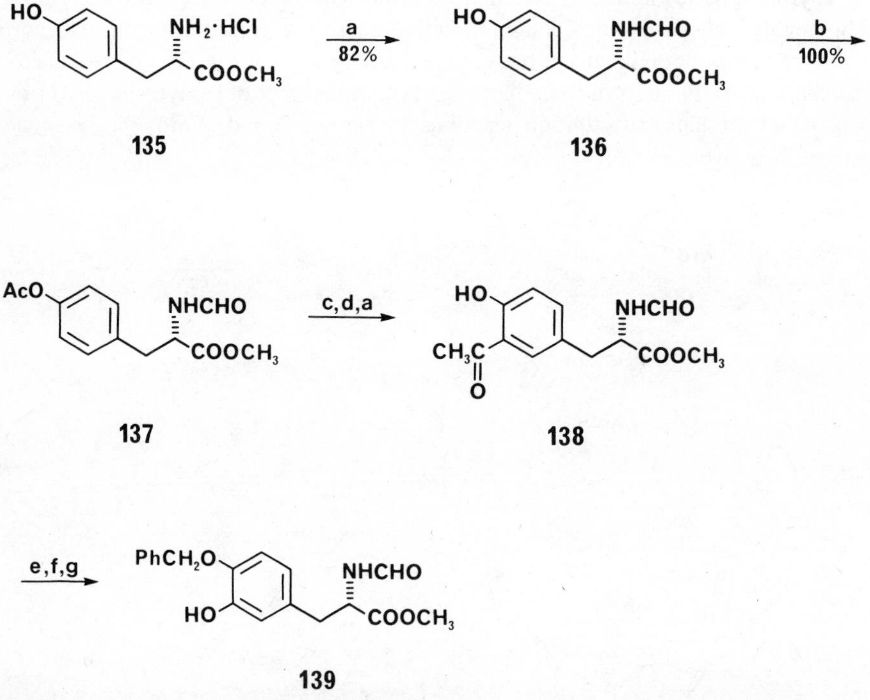

Scheme 18 (A) HCOONa, HCOOH, Ac_2O; (b) Ac_2O, TEA; (c) $AlCl_3$, tetrachloroethane, 130°C (77%); (d) CH_3OH, HCl (60%); (e) $PhCH_2Cl$, K_2CO_3, NaI (83%); (f) MCPBA; (g) K_2CO_3 (67%).

The oxygen substituent on the aromatic ring of tyrosine allows the ring to be reduced under Birch conditions to give a cyclohexenone system. This is beautifully illustrated in the synthesis of (+)-anticapsin (**144**) (Scheme 19). Direct reduction of L-tyrosine-*O*-methyl ether (**140a**) is not possible because of the propensity of amino enones to undergo 1,4-addition forming 6-oxo-octahydroindoles.[62] However, if the amino group is protected as its *N*,*N*-diallyl derivative (**140b**), the intramolecular Michael addition is blocked. Allylation of L-tyrosine-*O*-methyl ether (**140a**) is easily accomplished with an excess of allyl bromide in the presence of diisopropylethylamine (Hünig's base) to give the desired product in 84% yield.

Birch reduction of **140b** followed by esterification of the crude salt furnishes enol ether **141** in 83% yield. Hydrolysis and isomerization of the double bond produces a mixture of diastereomeric enones from which **142** is chromatographically isolable (30–35% yield). Epoxidation gives a 5:2 mixture of *cis* and *trans* epoxides that are chromatographically separated to afford the desired isomer **143** in 23% yield. Deallylation with Wilkinson's catalyst followed by saponification furnishes the natural product **144**.[63]

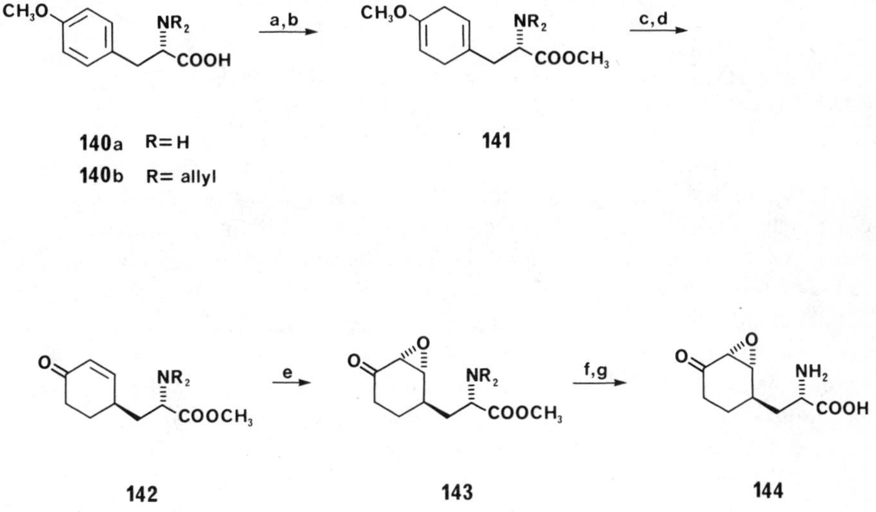

140a R=H
140b R= allyl

141

142 **143** **144**

Scheme 19. (a) Li, NH_3, EtOH; (b) CH_3I, HMPA, H_2O (83% overall); (c) 1% HCl (100%); (d) 0.3 *N* HCl–DMSO; (e) 30% H_2O_2, CH_3OH, 0°C, 1 hr; (f) $[(C_6H_5)_3P]_3RhCl$; (g) NaOH.

The amaryllidaceae alkaloids exhibit a variety of interesting ring systems. Two members of this class, (+)-maritidine and (−)-galanthamine, are biogenetically synthesized from L-tyrosine methyl ester (**135**) by an intramolecular phenolic oxidative coupling of an appropriate norbelladine derivative (Schemes 20 and 21).

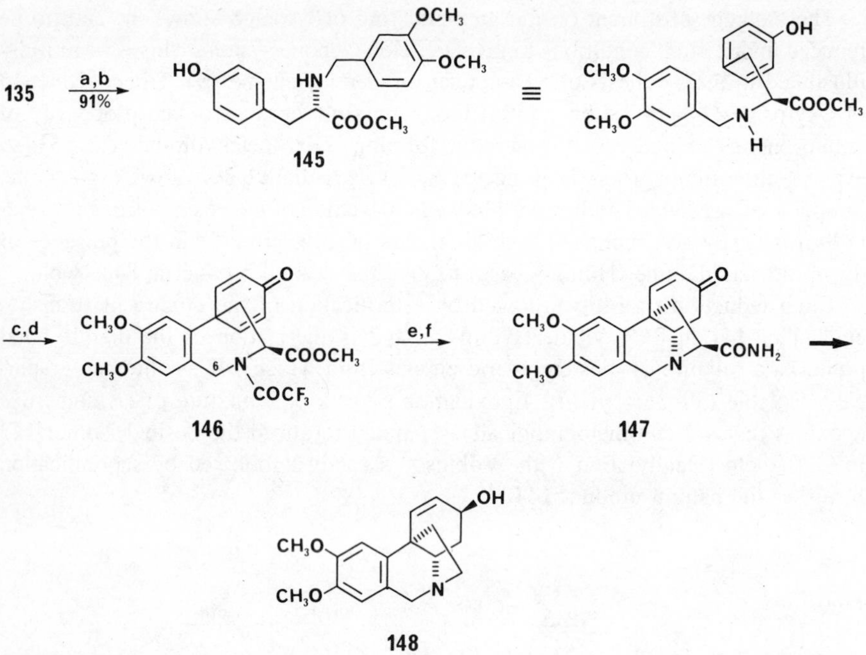

Scheme 20. (a) Veratraldehyde; (b) $NaBH_4$; (c) $(CF_3CO)_2O$, pyridine, $-30°C$ (81%); (d) thallium(III) trifluoroacetate (66.5%); (e) NH_3, CH_3OH (78.5%); (f) NaOH (41%).

In the synthesis of (+)-maritidine (**148**), amine **145** is converted to its *N*-trifluoroacetyl derivative, which, when treated with thallium(III) trifluoroacetate, undergoes the oxidative coupling to give spiro dienone **146** in 66.5% yield. Amidation of **146** with ammonia, followed by removal of the trifluoroacetyl group with sodium hydroxide, results in a spontaneous intramolecular Michael addition of the amine to the enone. This highly specific asymmetric cyclization produces **147** as the only diastereomer. This stereospecificity occurs because the cyclization places the carboxamide group in a sterically favorable position with respect to the C-6 methylene. Therefore, in retrospect, the asymmetric center of L-tyrosine controls the chirality generated in the remainder of the molecule. Removal of the carboxamide group by functional group manipulation, reduction of the enone, and epimerization of the hydroxy group completes the synthesis of **148**.[64,65]

The synthesis of galanthamine (**154**) follows approximately the same route as that described for maritidine. Amine **149** is converted to the *N*-trifluoroacetyl norbelladine derivative **150** with trifluoroacetic anydride in pyridine. Oxidation with 5 equivalents of manganic tris(acetylacetonate) affords the cyclized narwedine-type enone **151** in 49% yield. Functional group manipulation, which includes reductive elimination of the phenolic group (as its phosphate ester), results in the formation of (+)-galanthamine [(+)-**154**]. Since this is the unnatural isomer, a parallel synthesis starting from D-tyrosine should furnish the natural (−)-enantiomer. Intermediate **152,** however, can be epimerized to **153** (in 78% optical purity) according to the series of reactions (c), (h), and (i) outlined in Scheme 21. This intermediate is then carried on to the natural (−)-galanthamine [(−)-**154**].[66,67]

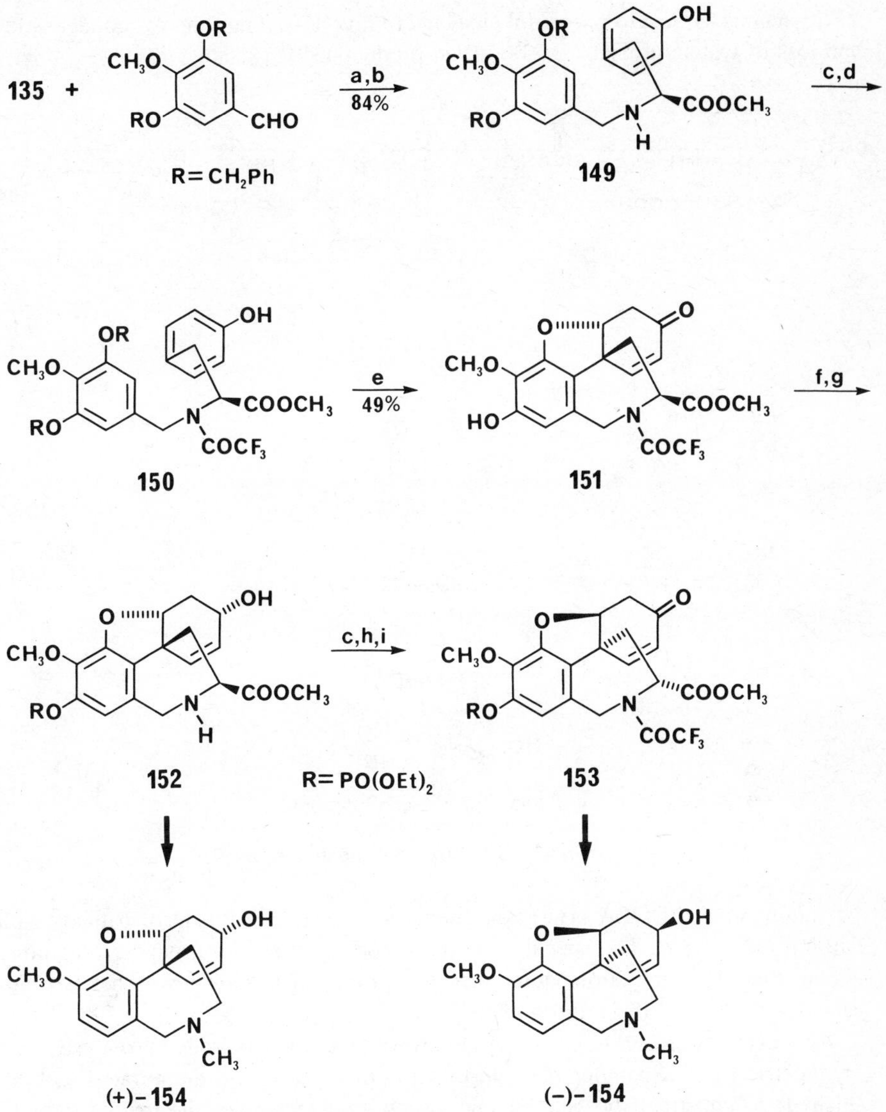

Scheme 21 (a) 4-Å molecular sieves; (b) $NaBH_4$; (c) $(CF_3CO)_2O$, pyridine (90%); (d) H_2, Pd/C (100%); (e) $Mn(acac)_3$, CH_3CN; (f) $(EtO)_2POCl$, TEA (81%); (g) $NaBH_4$ (21%); (h) LDA, TMEDA, HMPA (21%); (i) PCC (72%).

An extremely unusual and interesting reaction of the tyrosine nucleus involves the photocyclization of *N*-chloroacetyl-L-tyrosine methyl ether (**156**). This substrate is easily prepared by the reaction of **155**[68] with chloroacetyl chloride.[69] Irradiation of an aqueous solution of **156** for 2 hr results in a deep-seated photolytic rearrangement to give the pyrroloazepinone **159** in 17% yield.[70] The transformation is probably initiated by homolysis of the C–Cl bond with subsequent attack of the resulting carbene on the aromatic ring. Hydrolysis of the intermediate enol ether produces **157,** which is subject to retroaldol cleavage to give **158.** Transannular condensation and loss of water furnishes the observed product **159**[71] (Scheme 22).

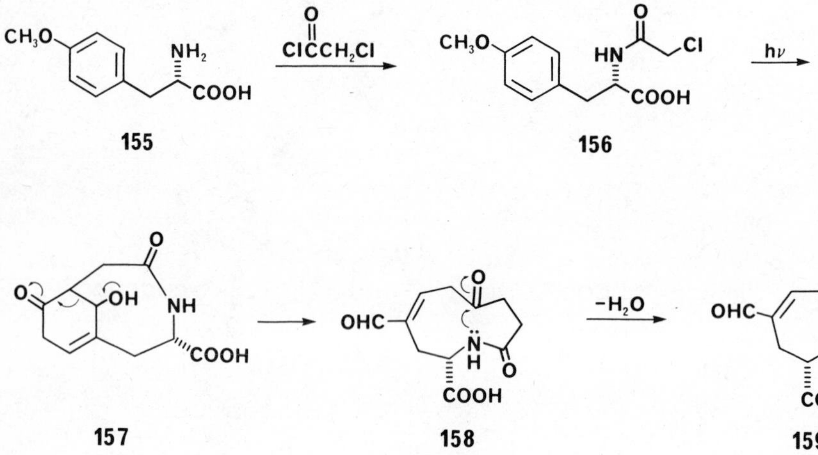

155 → ($ClCCH_2Cl$, with C=O) → **156** → ($h\nu$) → **157** → **158** → ($-H_2O$) → **159**

Scheme 22

2.3 DOPA

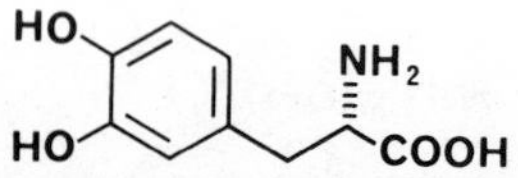

(*S*)-α-Amino-3,4-dihydroxybenzenepropanoic acid

The amino acid L-DOPA (**160**) is a crucial drug in the treatment of patients with Parkinson's disease. The speculation that its decarboxylation product, dopamine, is converted to tetrahydroisoquinolines has prompted several researchers to biogenetically simulate these types of reactions.

An elegant approach to chiral tetrahydroisoquinoline alkaloids makes use of an asymmetric Pictet–Spengler reaction of an optically active α-amino acid with an aldehyde to construct the heterocyclic system. For example, the treatment of L-DOPA (**160**) with formaldehyde in the presence of sulfuric acid affords the natural product (*S*)-**161a** in 87% yield. Alternately, catalytic hydrogenation of a mixture of **160** and formaldehyde directly produces the *N*-methyl derivative (*S*)-**161b** in 78% yield. An analogous reaction of **160** with acetaldehyde in the presence of sulfuric acid gives a mixture of *cis*- and *trans*-isoquinolines **162** and **163** (ratio 95:5)[72] (Scheme 23).

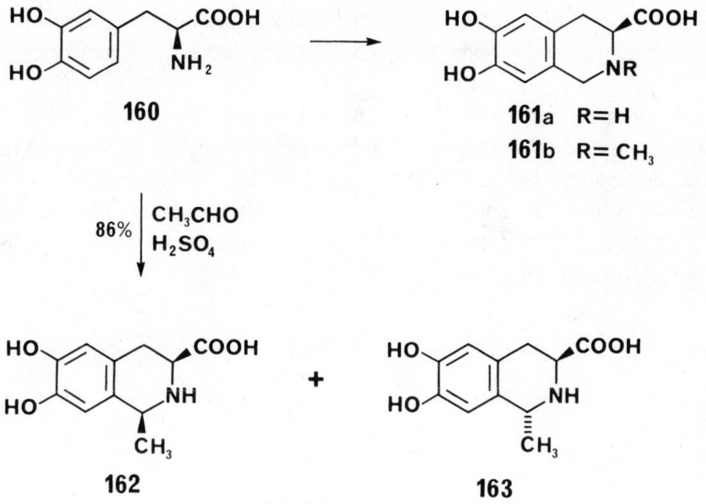

Scheme 23

Compound **162** can be further transformed to (+)-*O*-methylcorytenchirine (**166**), an 8-methylberbine, according to the series of reactions shown in Scheme 24. To perform the **162** → **164** transformation without racemization, it is necessary to protect the nitrogen with a formyl group which is then easily removed in the presence of the ethyl ester. Alkylation of **164** with the phenethyl bromide derivative produces the required intermediate **165** needed for the iminium salt cyclization leading to **166**.[73] The **162** → **166** transformation occurs with 94% retention of asymmetry.

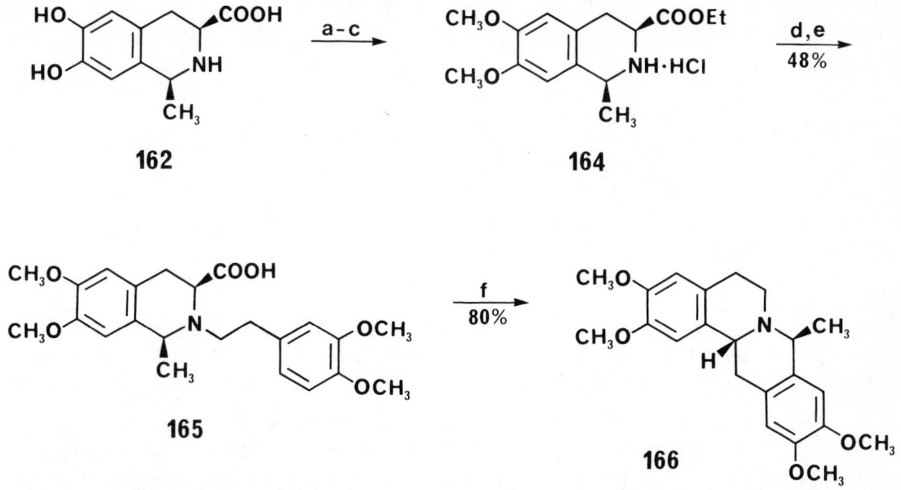

Scheme 24. (a) EtOH, HCl (94%); (b) HCOOK, HCOOH, Ac_2O (88%); (c) $KHCO_3$, $(CH_3)_2SO_4$ (80%); (d) 1-(2-bromoethyl)-3,4-dimethoxybenzene, K_2CO_3; (e) HCl (pH 6); (f) $POCl_3$.

The principle of building isoquinolines from amino acids and aldehydes provides a general entrance to the benzylisoquinoline alkaloid family, and this is illustrated by a relatively compact synthesis of (*S*)-(+)-laudanosine (**171**) (Scheme 25). A Pictet–Spengler reaction of L-DOPA methyl ester hydrochloride (**167**) and sodium 3-(3,4-dimethoxyphenyl)glycidate (**168**) at pH 4 affords a diastereometric mixture of cyclized products **169** and **170** in a ratio of 3.2 : 1. These isomers are chromatographically separable, and the major isomer **169** is converted to the natural product **171** by alkylation and decarbonylation.[74,75] The unnatural (*R*)-(−)-laudanosine is available from the minor isomer **170.**[76]

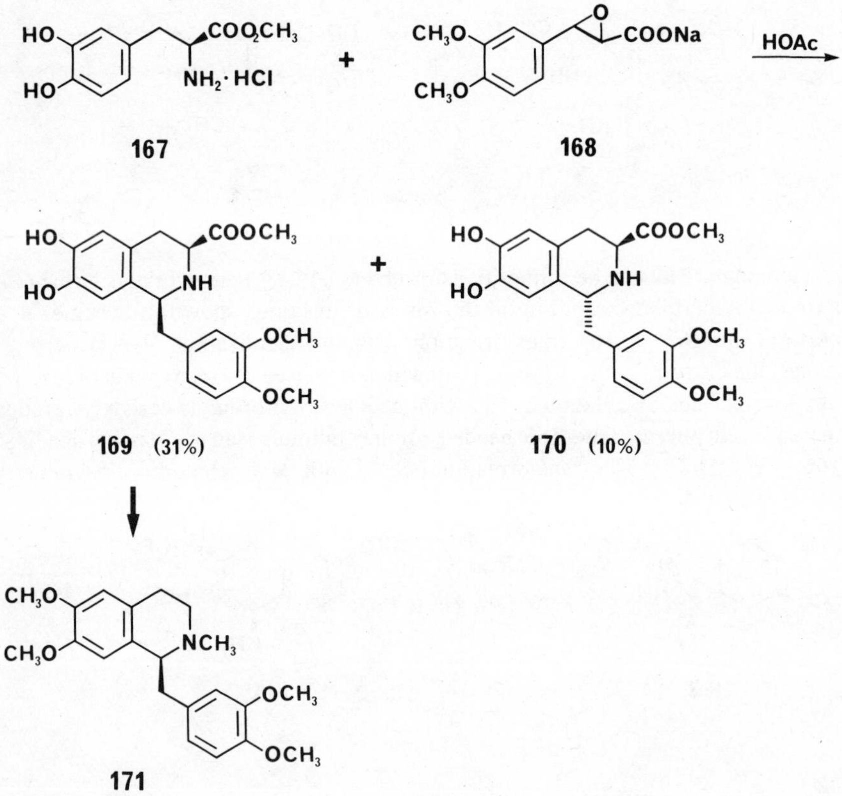

Scheme 25

(*S*)-Reticuline (**178**) is structurally similar to (*S*)-laudanosine with the exception that each aromatic ring possesses one hydroxy and one methoxy group instead of two methoxy groups. If it is to be prepared analogously to the route shown in Scheme 25, the hydroxyl groups of L-DOPA must be selectively differentiated. The required monoprotected derivative **173** is accessible from L-tyrosine by way of **139** (see Scheme 18) or by a shorter route directly from L-DOPA methyl ester hydrochloride (**167**) as shown.

167 →(a) 172 →(b) 139 →(c) 173

(a) HCOONa, HCOOH, Ac_2O (100%); (b) $PhCH_2Cl$, K_2CO_3, NaI (33%); (c) 4.5% HCl, CH_3OH (90%).

A Pictet–Spengler reaction of **173** and **174** at pH 4 produces a mixture of **175** and **176** (ratio 2.1:1). The more stable *cis*-isomer **175** is *O*-methylated with diazomethane to give **177,** and this is converted to (*S*)-reticuline (**178**) in a relatively straightforward manner (Scheme 26).[61] The synthesis of **178** also allows one access to alkaloids belonging to the sinomenine, aporphine, and protoberbine series.

173 + 174 (R = CH_2Ph) →(HOAc) 175 + 176

175 →(CH_2N_2, 83%) 177 → 178

Scheme 26

(*S*)-*O,O*-Dimethyl-α-methyldopa (**179**) is an industrial precursor for the antihypertensive drug α-methyldopa. It is readily transformed to the chiral diketopiperazine **181** by initial formation of its *N*-carboxyanhydride (**180**), followed by peptide formation with methyl glycinate and cyclization in refluxing toluene.[77]

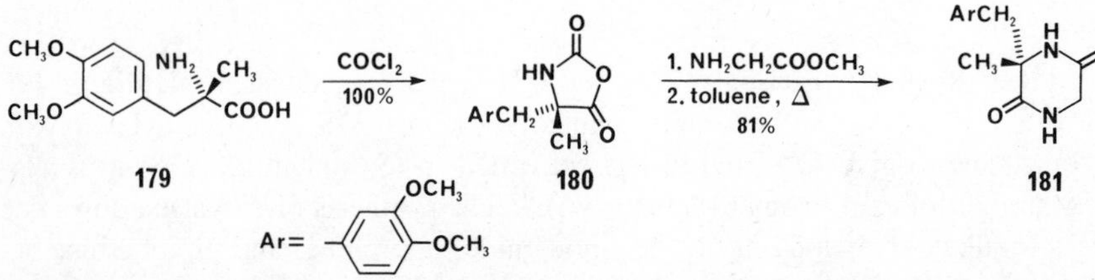

The heterocycle **181** reacts with trimethyloxonium tetrafluoroborate to give the bis-lactim ether **182** in 82% yield, which generates the lithio derivative **183** on deprotonation with *n*-butyllithium at −70°C. Reaction of **183** with a variety of electrophiles (Scheme 27) occurs with extremely high optical induction and produces adducts with diastereometric excess of 93–95%. This high diastereoselection in the C–C bond forming step can be attributed to the anion **183** adopting a "folded" conformation such as **184.** This spacial geometry efficiently shields one face of the anion causing the electrophile to approach C-6 *trans* to the 3,4-dimethoxybenzyl group.

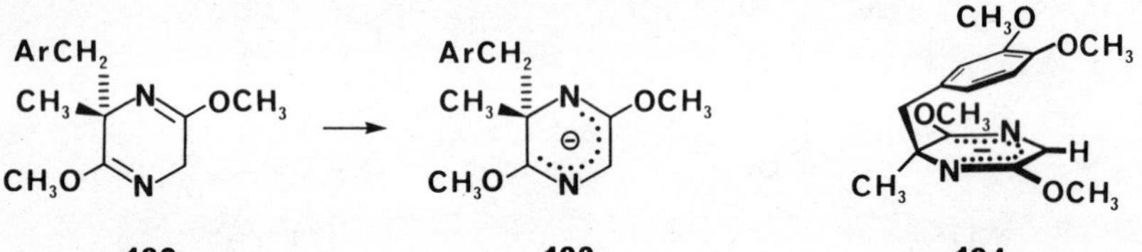

Anion **183** is readily alkylated with alkyl, 2-alkenyl, or 2-alkynyl halides to afford the alkylated products **185** (Scheme 27) in 21–95% yields (80–95% de). These, in turn, can be hydrolyzed to an (*R*)-amino acid (**189**) in 48–84% yield.[77] The chiral auxiliary **179** (as its methyl ester) is regenerated in the reaction and is readily separable from the desired product. When alkylated with chloromethyl benzyl ether, **188** is formed in 74% yield (95% de). Careful acidic hydrolysis produces (*R*)-*O*-benzylserine methyl ester (**190**), which is sequentially converted to (*R*)-*O*-benzylserine (**191**) and (*R*)-serine (**192**).[78]

Condensation of **183** with acetone furnishes either adduct **186** (79% yield, 93% de) or **187** (88% yield, 93% de) depending on whether the intermediate alkoxide is quenched with acetic acid or methyl iodide. Two hydrolytic processes then give (*R*)-(−)-β-hydroxyvaline (**193**).[79]

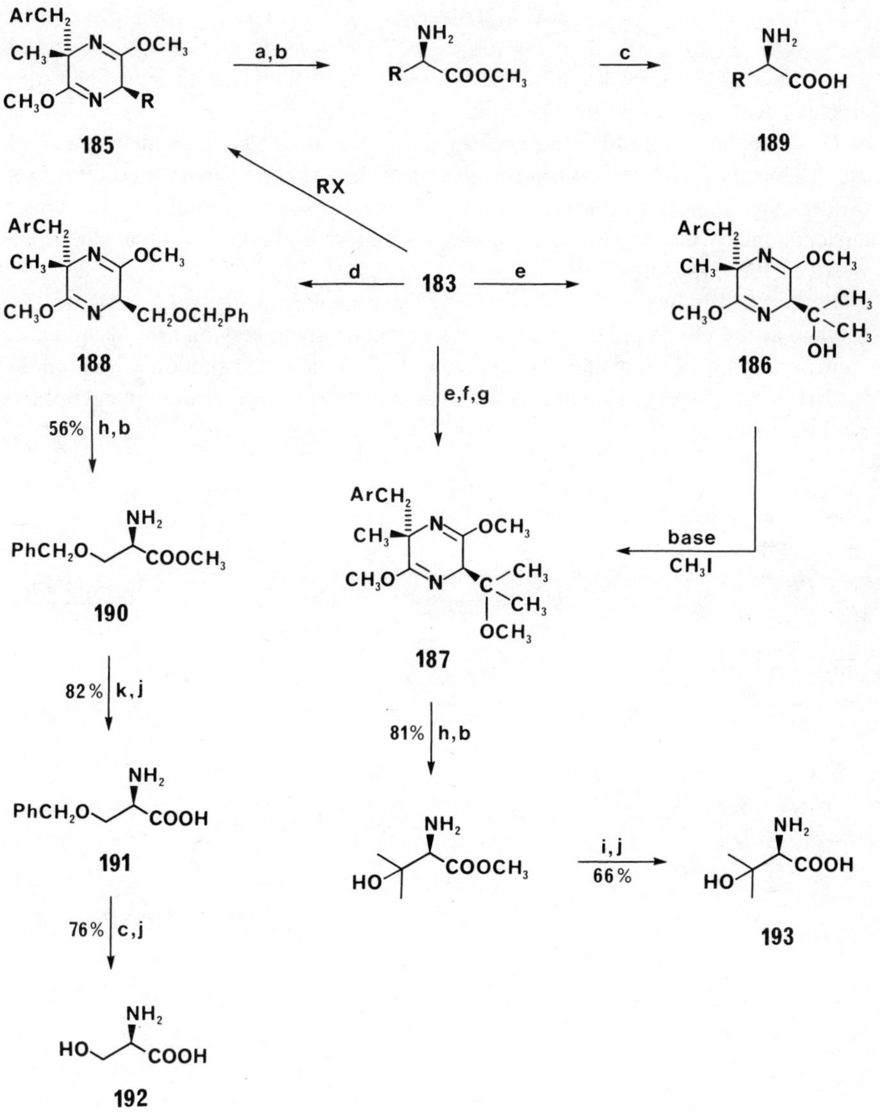

Scheme 27. (a) 0.25 *N* HCl; (b) NH_3; (c) 6 *N* HCl; (d) $ClCH_2OCH_2Ph$; (e) acetone; (f) *t*-C_4H_9OK, HMPA; (g) CH_3I; (h) 0.1 *N* HCl; (i) 48% HBr; (j) propylene oxide; (k) 2 *N* HCl.

2.4 ADDENDA

Since the original preparation of the manuscript for this book, several interesting articles have appeared in the literature.

(*S*)-Phenylalaninol (**101**) forms chiral amino alcohols **194** on condensation with cyclic ketones such as cyclopentanone and cyclohexanone. Reaction of **194** with bis(dimethylamino)-di-*n*-butylstannane generates organotin enamines **195,** which nucleophilically add to electrophilic alkenes to give 2-alkylated cycloalkanone derivatives **196**[80] (Scheme 28).

Best optical inductions are obtained by using pentane as the solvent and methyl acrylate as the electrophile. In analogous reactions using acrylonitrile, asymmetric induction drops off dramatically (e.g., $n = 1$, 36% ee). Substitution of (*S*)-valinol for **101** in the sequence shown in Scheme 28 results in slightly lower optical purities for **196** (93–95% ee).

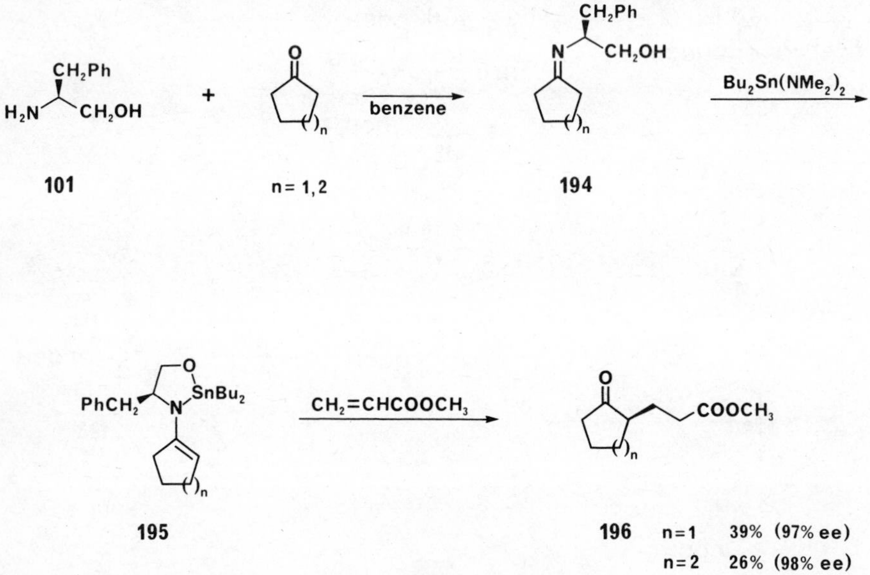

CH_2Ph / H_2N CH_2OH (**101**) + cycloalkanone ($n = 1, 2$) —benzene→ **194** —$Bu_2Sn(NMe_2)_2$→ **195** —$CH_2{=}CHCOOCH_3$→ **196**

196 n = 1 39% (97% ee)
n = 2 26% (98% ee)

Scheme 28

REFERENCES

1. T. Scott and M. Brewer, *Concise Encyclopedia of Biochemistry,* Walter de Gruyter, New York, 1983, p. 341.
2. Z. G. Hajos and D. R. Parrish, *Ger. Patent* 2,102,623 (1971); *Chem. Abstr.*, **76,** 59072x (1972).
3. Z. G. Hajos and D. R. Parrish, *J. Org. Chem.*, **39,** 1615 (1974).

4. U. Eder, G. Sauer, and R. Weichert, *Angew. Chem., Int. Ed.*, **10,** 496 (1971).
5. S. Danishefsky and P. Cain, *J. Am. Chem. Soc.*, **98,** 4975 (1976).
6. G. Sauer, U. Eder, and G. Hoyer, *Chem. Ber.*, **105,** 2358 (1972).
7. A. Sarkar, H. R. Y. Jois, T. R. Kasturi, and D. Dasgupta, *Proc. Indian Acad. Sci. (Chem. Sci.)*, **91,** 475 (1982).
8. S. Danishefsky and P. Cain, *J. Am. Chem. Soc.*, **97,** 5282 (1975).
9. I. Shimizu, Y. Naito, and J. Tsuji, *Tetrahedron Lett.*, 487 (1980).
10. R. Uma, S. Swaminathan, and K. Rajagopalan, *Tetrahedron Lett.*, 5825 (1984).
11. W. D. Fuller, M. S. Verlander, and M. Goodman, *Biopolymers*, **15,** 1869 (1976).
12. J. R. A. Pollock, *Br. Patent Appl.* 2,035,998 (1980); *Chem. Abstr.*, **94,** 84108t (1981).
13. M. Bodanszky, U.S. Patent 3,880,838 (1975); *Chem. Abstr.*, **83,** 43746d (1975).
14. M. Miyoshi, *Bull. Chem. Soc. Jpn.*, **43,** 3321 (1970).
15. M. Miyoshi, *Bull. Chem. Soc. Jpn.*, **46,** 212 (1973).
16. J. H. Jones and M. J. Witty, *J. Chem. Soc., Chem. Commun.*, 281 (1977).
17. J. H. Jones and M. J. Witty, *J. Chem. Soc., Perkin Trans. I.*, 3203 (1979).
18. M. Miyoshi, *Bull. Chem. Soc. Jpn.*, **46,** 1489 (1973).
19. S. Karaday, J. S. Amato, and L. M. Weinstock, *Tetrahedron Lett.*, 4337 (1984).
20. G. E. Hein and C. Niemann, *J. Am. Chem. Soc.*, **84,** 4487 (1962).
21. D. E. McClure, B. H. Arison, J. H. Jones, and J. J. Baldwin, *J. Org. Chem.*, **46,** 2431 (1981).
22. M. Binder and C. Tamm, *Angew. Chem., Int. Ed.*, **12,** 370 (1973).
23. G. Stork, Y. Nakahara, Y. Nakahara, and W. J. Greenlee, *J. Am. Chem. Soc.*, **100,** 7775 (1978).
24. N. C. Chaturvedi, W. K. Park, R. R. Smeby, and F. M. Bumpus, *J. Med. Chem.*, **13,** 177 (1970).
25. P. Brewster, F. Hiron, C. K. Ingold, and P. A. D. S. Rao, *Nature*, **166,** 179 (1950).
26. S. G. Cohen and S. Y. Weinstein, *J. Am. Chem. Soc.*, **86,** 5326 (1964).
27. K. Koga, C. C. Wu, and S. Yamada, *Tetrahedron Lett.*, 2287 (1971).
28. M. L. Louden and E. Leete, *J. Am. Chem. Soc.*, **84,** 4507 (1962).
29. S. Yamada, K. Koga, T. M. Juang, and K. Achiwa, *Chem. Lett.*, 927 (1976).
30. S. Yamada, T. Kitagawa, and K. Achiwa, *Tetrahedron Lett.*, 3007 (1967).
31. S. Terashima, C. C. Tseng, and K. Koga, *Chem. Pharm. Bull.*, **27,** 747 (1979).
32. S. Terashima, M. Hayashi, C. C. Tseng, and K. Koga, *Tetrahedron Lett.*, 1763 (1978).
33. S. Terashima, C. C. Tseng, M. Hayashi, and K. Koga, *Chem. Pharm. Bull.*, **27,** 758 (1979).
34. S. Yamada, N. Oh-hashi, and K. Achiwa, *Tetrahedron Lett.*, 2557 (1976).
35. S. Yamada, N. Oh-hashi, and K. Achiwa, *Tetrahedron Lett.*, 2561 (1976).
36. M. A. Abdallah and J. N. Shah, *J. Chem. Soc., Perkin Trans. I*, 888 (1975).
37. W. Bergstein, A. Kleemann, and J. Martens, *Synthesis*, 76 (1981).
38. K. Kellner, A. Tzschach, Z. Nagy-Magos, and L. Marko, *J. Organomet. Chem.*, **193,** 307 (1980).
39. K. Achiwa, *J. Am. Chem. Soc.*, **98,** 8265 (1976).

40. G. L. Baker, S. J. Fritschel, J. R. Stille, and J. K. Stille, *J. Org. Chem.*, **46,** 2954 (1981).
41. K. Achiwa, *Chem. Lett.*, 777 (1977).
42. U. Schmidt, J. Häusler, E. Oehler, and H. Poisel, *Fortschr. Chem. Org. Naturst.*, **37,** 251 (1979).
43. I. Ojima and N. Yoda, *Tetrahedron Lett.*, 1051 (1980).
44. A. Ito, R. Takahashi, and Y. Baba, *Chem. Pharm. Bull.*, **23,** 3081 (1975).
45. G. Stork and E. Nakamura, *J. Am. Chem. Soc.*, **105,** 5510 (1983).
46. S. G. Pyne, M. J. Hensel, S. R. Byrn, A. J. McKenzie, and P. L. Fuchs, *J. Am. Chem. Soc.*, **102,** 5960 (1980).
47. H. Umezawa, T. Aoyagi, H. Suda, M. Hamada, and T. Takeuchi, *J. Antibiot.*, **29,** 97 (1976).
48. T. Takita, H. Suda, T. Aoyagi, and H. Umezawa, *J. Med. Chem.*, **20,** 510 (1977).
49. C. F. Lane, U.S. Patent 3,935,280 (1976); *Chem. Abstr.*, **84,** 135101p (1976).
50. P. Karrer, P. Portmann, and M. Suter, *Helv. Chim. Acta,* **31,** 1617 (1948).
51. H. Seki, K. Koga, H. Matsuo, S. Ohki, I. Matsuo, and S. Yamada, *Chem. Pharm. Bull.*, **13,** 995 (1965).
52. S. Shibata, H. Matsushita, H. Kaneko, M. Noguchi, M. Saburi, and S. Yoshikawa, *Chem. Lett.*, 217 (1981).
53. D. A. Evans, K. T. Chapman, and J. Bisaha, *J. Am. Chem. Soc.*, **106,** 4261 (1984).
54. D. A. Evans, K. T. Chapman, and J. Bisaha, *Tetrahedron Lett.*, 4071 (1984).
55. C. C. Tseng, S. Terashima, and S. Yamada, *Chem. Pharm. Bull.*, **25,** 29 (1977).
56. A. I. Meyers, G. S. Poindexter, and Z. Birch, *J. Org. Chem.*, **43,** 892 (1978).
57. A. I. Meyers, D. R. Williams, G. W. Erickson, S. White, and M. Druelinger, *J. Am. Chem. Soc.*, **103,** 3081 (1981).
58. A. I. Meyers, D. R. Williams, S. White, and G. Erickson, *J. Am. Chem. Soc.*, **103,** 3088 (1981).
59. J. R. Chalmers, G. T. Dickson, J. Elks, and B. A. Hems, *J. Chem. Soc.*, 3424 (1949).
60. H. Bretschneider, K. Hohenlohe-Oehringer, A. Kaiser, and U. Wölcke, *Helv. Chim. Acta,* **56,** 2857 (1973).
61. M. Konda, T. Shioiri, and S. Yamada, *Chem. Pharm. Bull.*, **23,** 1063 (1975).
62. R. W. Rickards, J. L. Rodwell, and K. J. Schmalzl, *J. Chem. Soc., Chem. Commun.*, 849 (1977).
63. B. C. Laguzza and B. Ganem, *Tetrahedron Lett.*, 1483 (1981).
64. S. Yamada, K. Tomioka, and K. Koga, *Tetrahedron Lett.*, 57 (1976).
65. K. Tomioka, K. Koga, and S. Yamada, *Chem. Pharm. Bull.*, **25,** 2681 (1977).
66. K. Shimizu, K. Tomioka, S. Yamada, and K. Koga, *Chem. Pharm. Bull.*, **26,** 3765 (1978).
67. K. Shimizu, K. Tomioka, S. Yamada, and K. Koga, *Heterocycles,* **8,** 277 (1977).
68. B. R. Baker, J. P. Joseph, and J. H. Williams, *J. Am. Chem. Soc.*, **77,** 1 (1955).
69. E. Ronwin, *J. Org. Chem.*, **18,** 1546 (1953).
70. O. Yonemitsu, B. Witkop, and I. L. Karle, *J. Am. Chem. Soc.*, **89,** 1039 (1967).

71. O. Yonemitsu, T. Tokuyama, M. Chaykovsky, and B. Witkop, *J. Am. Chem. Soc.*, **90,** 776 (1968).
72. A. Brossi, A. Focella, and S. Teitel, *Helv. Chim. Acta,* **55,** 15 (1972).
73. R. T. Dean and H. Rapoport, *J. Org. Chem.*, **43,** 4183 (1978).
74. S. Yamada, M. Konda, and T. Shioiri, *Tetrahedron Lett.*, 2215 (1972).
75. M. Konda, T. Shioiri, and S. Yamada, *Chem. Pharm. Bull.*, **23,** 1025 (1975).
76. M. Konda, T. Oh-ishi, and S. Yamada, *Chem. Pharm. Bull.*, **25,** 69 (1977).
77. U. Schöllkopf, W. Hartwig, K. Pospischil, and H. Kehne, *Synthesis,* 966 (1981).
78. J. Nozulak and U. Schöllkopf, *Synthesis,* 866 (1982).
79. U. Schöllkopf, J. Nozulak, and U. Groth, *Synthesis,* 868 (1982).
80. C. Stetin, B. DeJeso, and J. Pommier, *J. Org. Chem.*, **50,** 3863 (1985).

CHAPTER THREE

VALINE

NH_2, CH_3, COOH, CH_3

(*S*)-2-Amino-3-methylbutanoic acid

Valine, the next-higher homologue of the naturally α-alkylated amino acids, is an essential glucogenic and proteogenic amino acid.[1] The rather bulky isopropyl group adjacent to the asymmetric center makes it an excellent substrate (itself or in derivatives thereof) for the induction of asymmetry in a variety of chemical transformations. These are explored as we proceed through this chapter.

3.1 VALINE AND ITS ESTERS

In previous chapters we have shown that amino acids can be converted to diketopiperazines and that their derivatives undergo a variety of reactions resulting in high asymmetric induction. Valine is no exception. The mixed diketopiperazines cyclo-(L-Val–Gly) (**3a**) and cyclo-(L-Val–D,L-Ala) (**3b**) are easily prepared from L-valine (**1**) in two steps. L-Valine *N*-carboxyanhydride (**2**) is produced in quantitative yield by the treatment of **1** with phosgene. Reaction of **2** with either glycine ethyl ester[2] or D,L-alanine methyl ester[3] forms an intermediate dipeptide that cyclizes to the heterocycle **3** in high yield when refluxed in toluene for 12 hr.

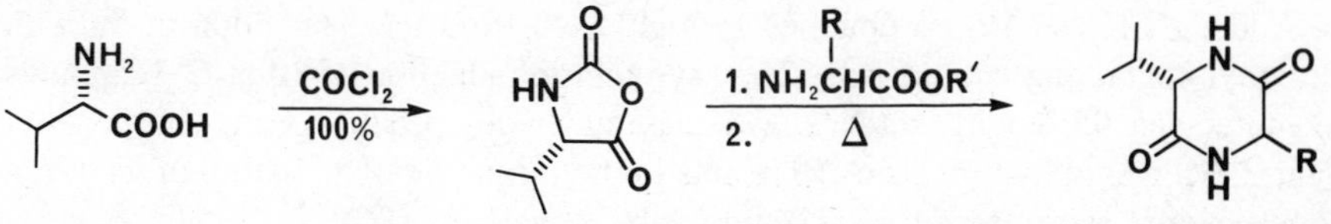

No.	R	R′	Yield (%)
3a	H	C_2H_5	91
3b	CH_3	CH_3	85

The diketopiperazine **3** is then transformed to its bis-lactim ether **4** with trimethyloxonium tetrafluoroborate in methylene chloride. Lithiation of **4** at −70°C occurs regiospecifically at the C-3 portion of the molecule to form the charged species **5**, which reacts with various electrophiles as illustrated in Schemes 1–5.

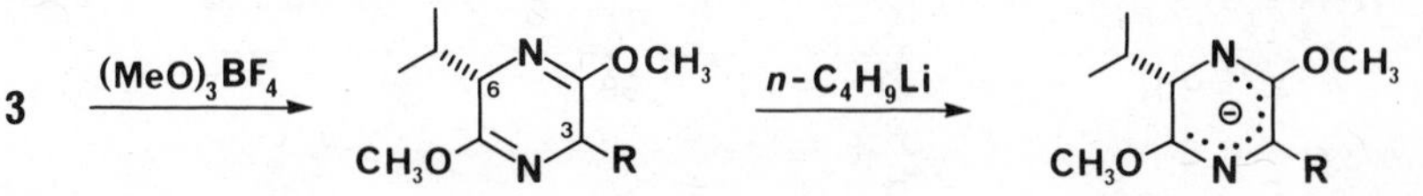

a, R=H

b, $R=CH_3$

4a (82%)

4b (89%)

Alkylation of **5a** with either alkyl bromides[2] or chloromethyl benzyl ether[4] gives the (3*R*)-addition products **6** or **7**, indicating that the alkylating agent approaches C-3 *trans* to the isopropyl group at C-6. Respective diastereoselectivities of up to 95 and 93% are observed in these products. Careful acidic hydrolysis of the adducts produces (*R*)-amino acid esters **8** or (*R*)-*O*-benzylserine methyl ester (**9**) along with the regenerated chiral auxiliary methyl L-valinate (Scheme 1). These are separated by distillation.

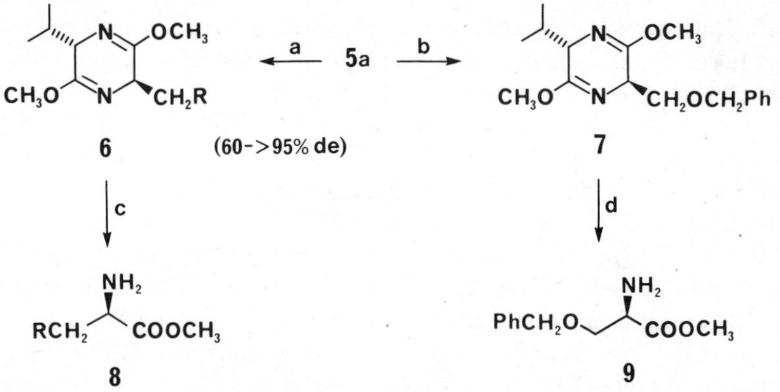

(60->95% de)

R= alkyl, alkenyl, alkynyl, aryl

Scheme 1. (a) RCH_2Br (62–92%); (b) $C_6H_5CH_2OCH_2Cl$ (81%); (c) 0.25 *N* HCl (52–89%); (d) 0.1 *N* HCl (62%).

Adducts of type **10** are obtained in high yield from the interaction of **5a** with carbonyl-containing compounds. The asymmetric induction of **10** at C-3 amounts to more than 95% with ketones, whereas aldehydes produce somewhat smaller values. Mild acid hydrolysis of **10** results in the formation of a mixture of the (2*R*)-serine methyl ester derivatives **11** and methyl L-valinate.[5,6]

5a → (R_1–C(=O)–R_2) → **10** → (H^+) → **11**

R_1	R_2	Percent **10**	Percent **11**
CH_3	CH_3	98	59
i-C_3H_7	H	96	48
t-C_4H_9	H	96	56

Optionally, β-fluoro amino acids, which are potential inhibitors of pyridoxal phosphate-dependent enzymes, are accessible from **10.** When these adducts are treated with diethylaminosulfur trifluoride, the hydroxy group is replaced by fluorine. Subsequent acidic hydrolysis then furnishes the fluoro amino acid ester. In this manner methyl (*R*)-β-fluorovalinate and methyl (2*R*)-β-fluorophenylalaninate are prepared in 46 and 40% yield, respectively (>95 ee).[7]

The scope of the reaction can be extended to include the preparation of β,γ-unsaturated amino acids (Scheme 2). Reaction of **5a** with 2-dimethyl *t*-butylsilyl alkanals (which are masked alkenyl groups) or dialkyl thioketones produces adducts **12** or **13.** Diastereomeric excesses greater than 95% are not uncommon for these condensations. Elimination of the functional groups as shown generates the desired unsaturated amino acid derivatives **16** or **18** in good yields.[8]

The C-3 methyl lithiated species **5b** is capable of entering into many of the same reactions as **5a** to produce a series of α-methyl amino acid derivatives. For example, alkylation of **5b** with primary alkyl, alkenyl, alkynyl, and arylalkyl bromides occurs almost diastereospecifically (>95%) to give products **19** that are then hydrolyzed to the (*R*)-α-methyl-α-amino acid esters **20.**[3] Analogously, (*R*)-*O*-benzyl-α-methylserine methyl ester (**21**) is synthesized by alkylation of **5b** with chloromethyl benzyl ether and acid hydrolysis (Scheme 3). Compound **21** can be further converted to (*R*)-(−)-α-methylserine in 84% yield by refluxing in 6 *N* HCl followed by treatment with propylene oxide.[9]

If **5b** is alkylated with benzyl 2-iodoethyl sulfide and the intermediate subsea 70–80% 5a b,c

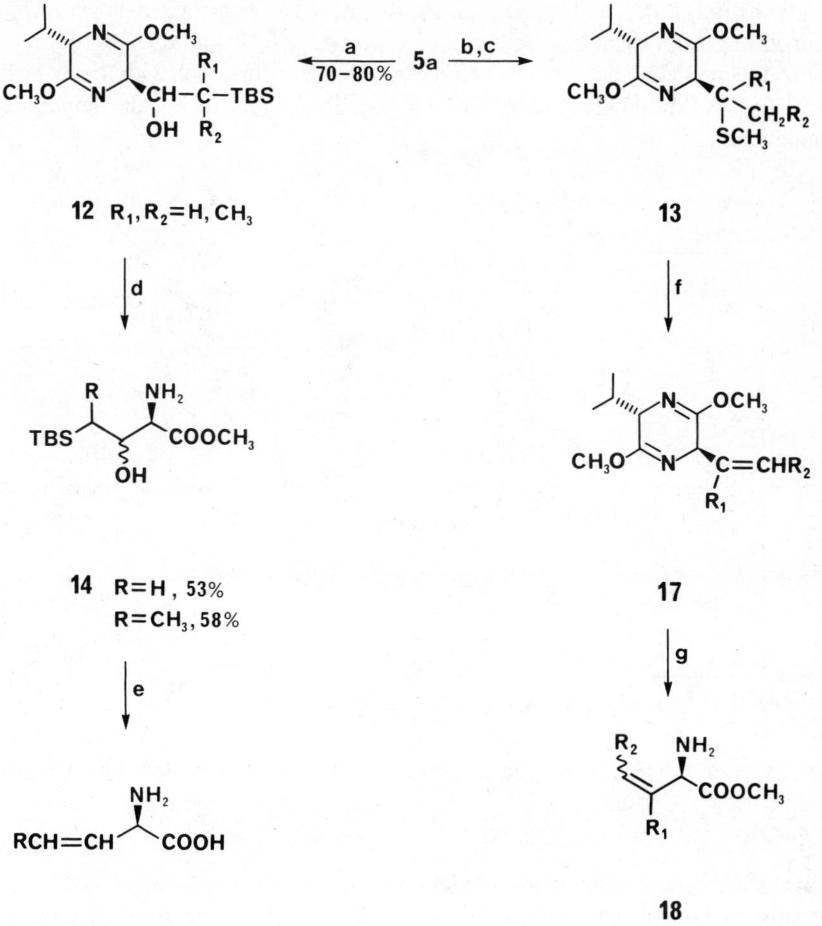

12 $R_1, R_2 = H, CH_3$

13

d

f

14 R=H, 53%
R=CH_3, 58%

17

e

g

18

16 R=H, 62%
R=CH_3, 70%
(E:Z, 1:1)

R_1	R_2	E:Z (**17**)	% **18**
C_2H_5	CH_3	2.5:1	73
$-(CH_2)_4-$		–	76
CH_3	CH_3	1.5:1	64

Scheme 2. (a) TBS–C(R_1)(R_2)–CHO, –70°C; (b) $R_1C(=S)CH_2R_2$, –70°C; (c) CH_3I, 23°C; (d) 0.25 *N* HCl; (e) 5 *N* HCl; (f) Raney Ni (100%); (g) 0.1 *N* HCl.

quently hydrolyzed, methyl (*R*)-2-amino-4-benzylthio-2-methylbutanoate (**22**) is obtained in greater than 95% ee. Further hydrolysis of **22** with 6 *N* HCl furnishes *S*-benzyl-α-methyl homocysteine (71% yield), and desulfurization with Raney nickel gives (*R*)-isovaline (71% yield),[10] which is a component of certain amphiphilic peptide antibiotics.

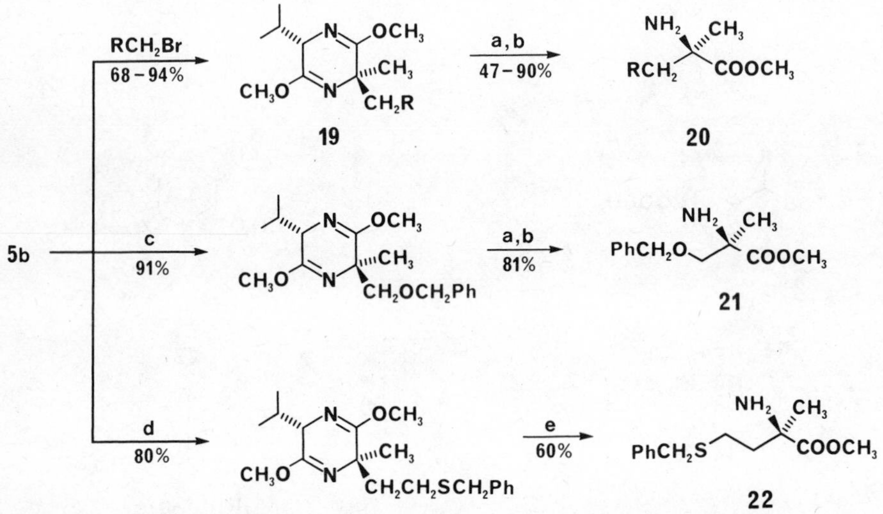

Scheme 3. (a) 0.25 *N* HCl; (b) NH_4OH; (c) $ClCH_2OCH_2Ph$; (d) $ICH_2CH_2SCH_2Ph$; (e) 0.1 *N* HCl.

By a slight modification in the alkylation step, (*R*)-α-methylcysteine methyl ester derivatives **25** are also available (>95% ee) as shown in Scheme 4.[11] Finally, (*R*)-α-alkenylalanine methyl esters (**28**) are prepared by dehydration of the aldol-type adducts **26** followed by mild acid hydrolysis of the olefinic intermediates **27**[12] (Scheme 5).

The use of cyclo-(L-Val–D,L-Ala) (**3b**) in the preparation of α-methyl–α-amino acids has a distinct advantage over cyclo-(L-Ala–L-Ala), which was described in Chapter 1. In that series, only half of the chiral auxiliary (L-alanine methyl ester) was recovered because half of the cyclo-(L-Ala–L-Ala) molecule was alkylated and converted to the α-methyl amino acid ester. With **3b,** the methyl group is not derived from a chiral source; therefore, all the chiral auxiliary L-valine methyl ester is recoverable.

Other six-membered ring heterocycles that possess a 1,4 relationship of the heteroatoms are available from amino acids. Specifically, the interesting chiral 1,4-oxazine-2-ones **32** were prepared in order to study the effect of asymmetric induction at C-5 on catalytic hydrogenation of the azomethine double bond.[13]

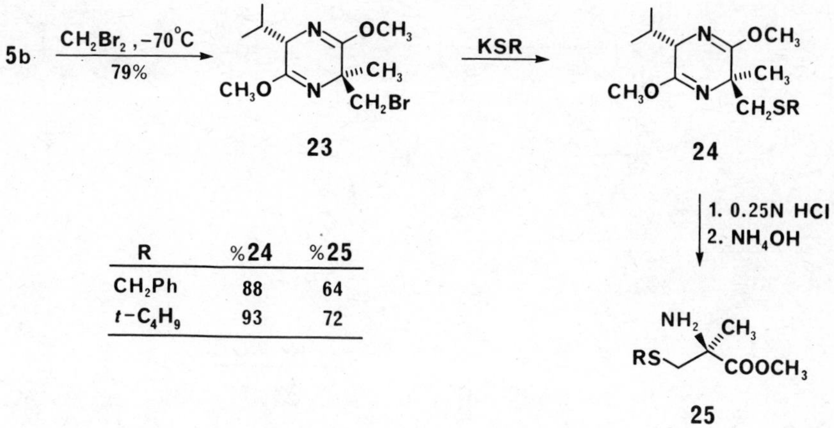

5b $\xrightarrow[79\%]{CH_2Br_2,\ -70^\circ C}$ 23 $\xrightarrow{KSR}$ 24 $\xrightarrow[2.\ NH_4OH]{1.\ 0.25N\ HCl}$ 25

R	%24	%25
CH_2Ph	88	64
t-C_4H_9	93	72

Scheme 4

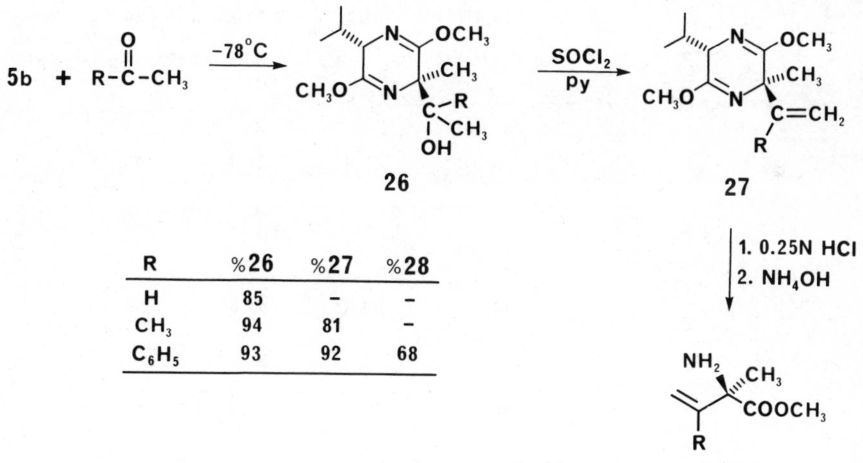

5b + $R-C(=O)-CH_3$ $\xrightarrow{-78^\circ C}$ 26 $\xrightarrow[py]{SOCl_2}$ 27 $\xrightarrow[2.\ NH_4OH]{1.\ 0.25N\ HCl}$ 28

R	%26	%27	%28
H	85	–	–
CH_3	94	81	–
C_6H_5	93	92	68

Scheme 5

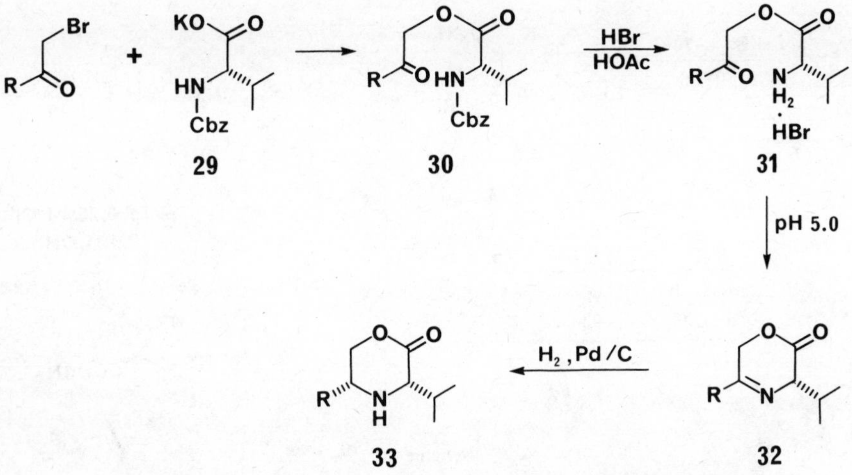

Br, R, O + KO, O, HN, Cbz → R, O, HN, Cbz, O, O → (HBr / HOAc) R, O, N, H_2 · HBr, O, O

29 30 31

↓ pH 5.0

R, N, H, O, O ← (H_2 , Pd / C) R, N, O, O

33 32

R	% 30	% 31	% 32	% 33
	79.5	95.9	67.8	61.4
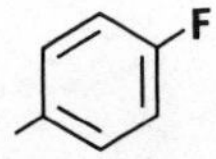 F	74.5	81.3	71.1	34.7
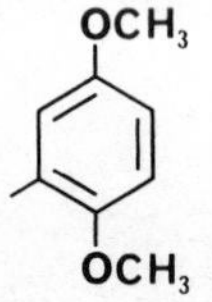 OCH_3, OCH_3	100	82.0	81.4	45.0
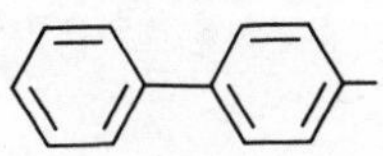	76.8	97.3	83.0	68.9
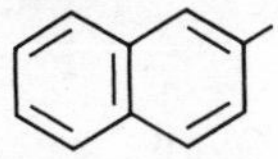	78.9	98.9	88.2	50.6

Scheme 6

The synthetic sequence leading to **32** (Scheme 6) begins by condensation of the potassium salt of Cbz–L-valine (**29**) with an α-halomethyl ketone. The Cbz-protecting group of **30** is then removed with 33% HBr in acetic acid. The standard method of hydrogenolytic cleavage of the Cbz group cannot be applied in this case because of preferential fission of the amino acid ester group. Cyclization of hydrobromide salt **31** is easily accomplished at room temperature in acetate buffer (pH 5.0) to give the desired oxazinones **32** in good yields.

When the azomethine double bond of **32** is hydrogenated at 1 atm in the presence of 10% Pd/C, *cis*-3,5-disubstituted morpholines **33** are obtained in moderate yield. Asymmetric induction in this transformation is extremely high (98–99%) and produces essentially one diastereomer.

Within the last decade considerable interest has been focused on monocylic β-lactams because of their antibiotic properties. Since these molecules contain an asymmetric center in the lactam ring, an efficient method of introducing this center is highly desirable. Classically, a reliable method of producing monocyclic β-lactams involves the cycloaddition of ketenes with imines.[14]

Recently it has been found that the incorporation of a chiral site in the imine component causes tremendous asymmetric induction in the cycloaddition. The requisite chiral imines are prepared by condensing an aldehyde with an L-amino acid methyl ester. Exposure of the imine and dimethylketene methyl trimethylsilyl acetal (**36**) to titanium tetrachloride (−78°C, 24 hr) results in the diastereoselective formation of β-lactam **37.** Of five amino acids studied (alanine, phenylalanine, valine, leucine, and glutamic acid), valine derivatives **35** are vastly superior for introducing the desired stereochemistry at C-4 in the β-lactam.[15]

NH_2–CH(i-Pr)–$COOCH_3$ (**34**) —RCHO→ RCH=N–CH(i-Pr)–$COOCH_3$ (**35**) —**36**, $TiCl_4$→ **37** (R, CH_3, CH_3, O, N, $COOCH_3$)

$(CH_3)_2C{=}C(OCH_3)(OTMS)$

36

R	yield **37** (%)	% ee
C_2H_5	73	94
n-C_3H_7	74	90
n-C_4H_9	77	90
i-C_4H_9	81	98

The high asymmetric induction arises when a "titanium template" is invoked in the transition state of the reaction. The initial kinetic complex **38** that is generated at −78°C forms a titanium enolate on reaction with **36.** This tridentate intermediate (**39**) produces a rigid bicyclic transition state that places the R group of the imine and the *C*-methyl group of the ketene in a sterically favorable relationship and thus produces the excellent diastereoselectivity in the cyclization.

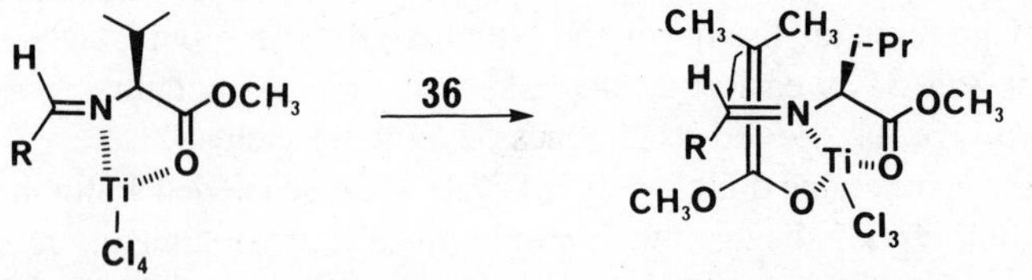

Another asymmetric transformation where a chelated metal forms a rigid transition state leading to high optical induction is the alkylation of α-alkyl β-ketoesters (Scheme 7). In this type of reaction, the β-ketoester is initially converted to a chiral enamine (e.g., **41**) by using L-valine *t*-butyl ester as a chiral auxiliary reagent. The enamine **41** is first lithiated with LDA at −78°C and is then treated with a ligating agent to form the tetracoordinated lithium cation **42.**

Depending on the nature of the ligand, alkylation can be "tuned" to give either enantiomer of **43** (after hydrolysis). In the presence of one equivalent of HMPA, **43a** is isolated in 57% yield (>99% ee), whereas the addition of 2 equivalents of THF leads to a 63% yield of **43b** (92% ee).[16,17]

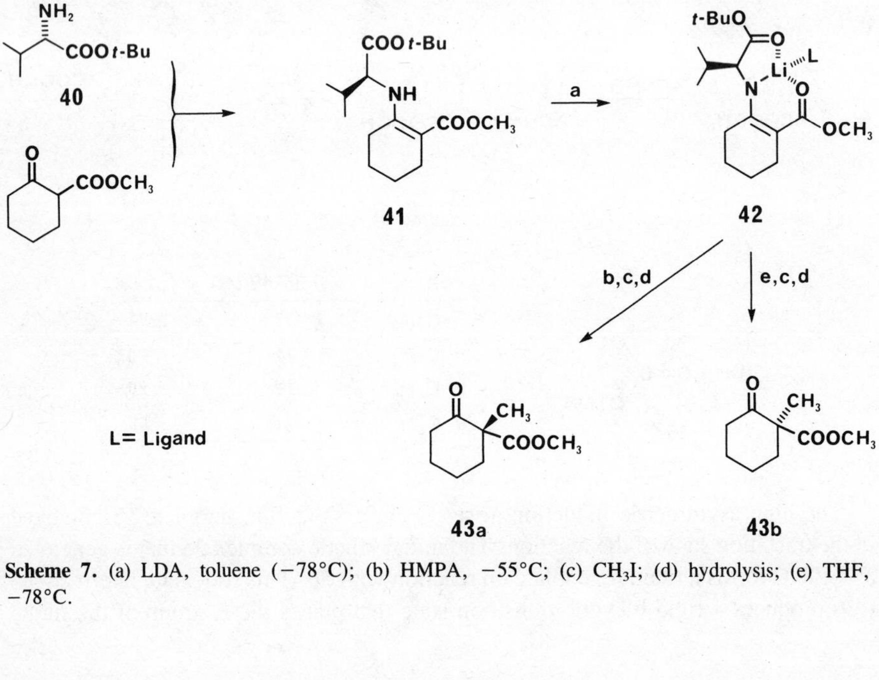

L= Ligand

Scheme 7. (a) LDA, toluene (−78°C); (b) HMPA, −55°C; (c) CH_3I; (d) hydrolysis; (e) THF, −78°C.

A plausible explanation for the adjustable asymmetric induction is the adoption of a *trans*-fused chelated structure of the lithioenamine where the unshared electron pair on nitrogen remains conjugated with the enamine double bond. The bulky and strongly ligating HMPA (L in **42** and in Figure 2) coordinates with the lithium cation to satisfy the tetravalency and results in increaseing the negative charge. This both suppresses bottom side attack and enhances top side attack to give **43a.** On the other hand, the weakly ligating THF does not activate the enamine system enough to simply react with methyl iodide. The interaction of the lithium cation with the alkylating agent results in an increase of concentration of methyl iodide at the bottom side, thus leading to preferential bottom side attack, giving **43b.**[17]

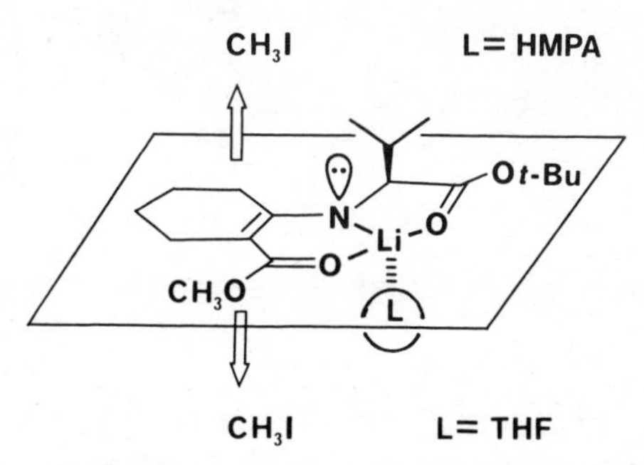

Figure 2. Ligand effect on alkylation trajectory of lithiated enamine **42.** (Reprinted with permission from *Tetrahedron Lett.*, 5677 (1984), K. Tomioka et al. Copyright 1984, Pergamon Press, Ltd.)

In Chapter 1 (Section 1.1) it has been shown that in the presence of Mg^{2+} ions a 1,4-dihydropyridine functionalized in the 3 position with a residue derived from L-alanine is capable of mimicking NADH in the reduction of ethyl benzoylformate to (*R*)-ethyl mandelate (47% ee). Kellogg[18] has extended this concept by embedding the 1,4-dihydropyridine moiety in the periphery of a macrocycle. The rationale for such an interesting approach is that if the magnesium ion is held in a defined position in the macrocycle, the ketone carbonyl oxygen of the ethyl benzoylformate will coordinate with the metal and orient itself on the macrocyclic surface in a sterically unencumbered position. This, in turn, will place the carbon atom to be reduced in a stereoelectronically favorable position relative to the 4 position of the 1,4-dihydropyridine.

The chiral recognition in the reduction in supplied from the steric effects that arise from the bulkiness of asymmetric groups on the skeleton of the macrocycle. These groups are derived from amino acids. L-Valine appears to be the amino acid of choice because of its good solubility and polarity properties.

The construction of the 1,4-dihydropyridine-containing chiral macrocycles is outlined in Scheme 8. After initial reaction of L-valine *t*-butyl ester (**40**) with pyridine-3,5-dicarboxylic acid chloride to give **44,** the free diacid is obtained by deblocking with trifluoroacetic acid. The diacid is converted to its dicesium salt (known not to involve racemization), and reaction with an α,ω-dibromide furnishes the macrocyle **45.** *N*-Methylation with methyl fluorosulfonate followed by reduction with sodium dithionite culminates the synthesis of **46.**[19,20]

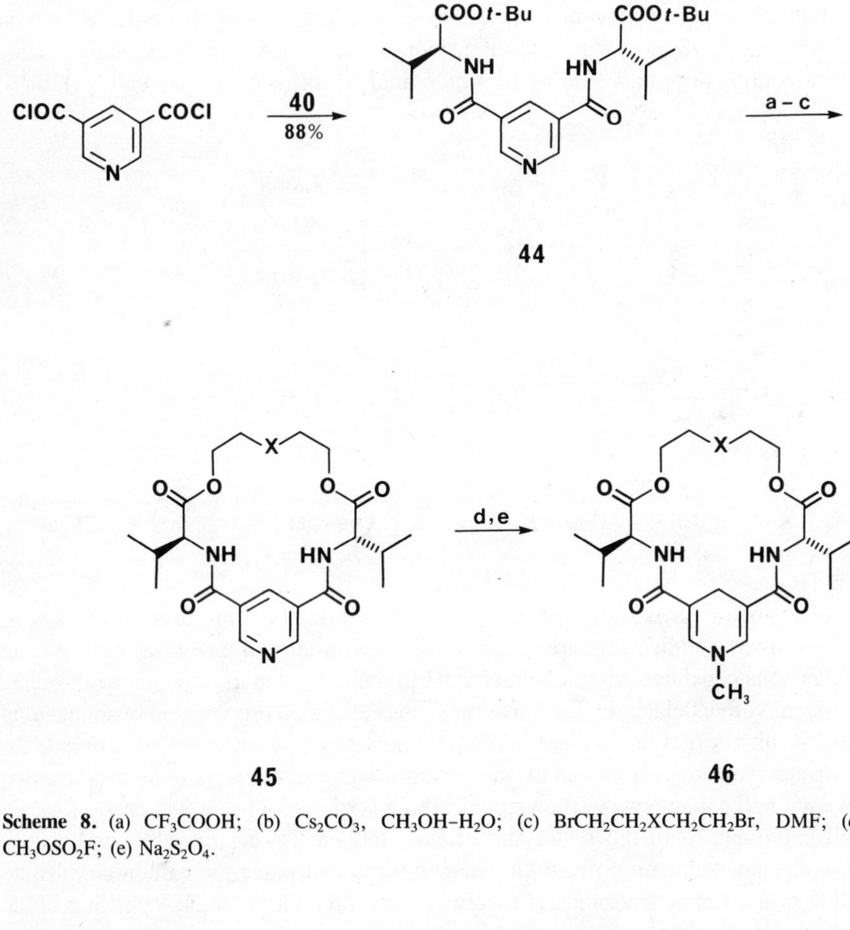

Scheme 8. (a) CF_3COOH; (b) Cs_2CO_3, $CH_3OH–H_2O$; (c) $BrCH_2CH_2XCH_2CH_2Br$, DMF; (d) CH_3OSO_2F; (e) $Na_2S_2O_4$.

The macrocycles **46** are not only aesthetically pleasing to the organic chemist; they also perform the function for which they were designed. When **46,** ethyl benzoylformate (**47**), and magnesium perchlorate are allowed to react in acetonitrile at room temperature (1–2 days), complete reduction occurs, giving (*S*)-ethyl mandelate (**48**) in good chemical yields. The enantiomeric excesses are excellent with macrocyclic ring sizes of 18, 19, and 21 atoms. As the size continues to increase, the optical yields decline to a mediocre level. These results represent some of the highest asymmetric inductions ever achieved with optically active 1,4-dihydropyridines.[20]

COOEt

47

46 | $Mg(ClO_4)_2$

OH

COOEt

(S)-**48**

X in **46**	Yield **48** (%)	Percent ee
CH_2	70	90
$(CH_2)_2$	75	88
$(CH_2)_4$	80	83
$(CH_2)_6$	75	53
$(CH_2)_8$	60	42
O	80	86
$O(CH_2)_2$	—	43
$(OCH_2CH_2)_2$	60	54

3.2 VALINOL

Chiral amino alcohols are gaining increasing importance in the area of asymmetric synthesis. (*S*)-2-Amino-3-methyl-1-butanol (L-valinol) (**49**) is ideally suited for the preparation of chiral auxiliary reagents. Its isopropyl group, which is strategically located adjacent to the versatile amino function, is large enough to enantioselectively direct certain reactions to occur from an area sterically uncongested by this group.

Valinol can either be purchased commercially or prepared from the natural L-valine. If large quantities are needed, it may be economically advantageous to opt for the synthetic route since L-valinol is approximately 36 times as expensive as L-valine. Basically two simple methods are available for the preparation of L-valinol. The first is a classical lithium aluminum hydride reduction of L-valine methyl ester (**34**).[21] The second method affords **49** directly from L-valine (**1**). The procedure consists of treating **1** with a stoichiometric amount of boron trifluoride etherate in THF followed by a slight excess of borane–methyl sulfide complex. In this manner L-valinol is obtained in 44–62% yield.[22,23] In the second route, the saving of one step (the preparation of **34**) is an important factor to consider when planning a synthesis starting with **49.**

$$\text{(CH}_3)_2\text{CH–CH(NH}_2\text{)–COOR} \longrightarrow \text{(CH}_3)_2\text{CH–CH(NH}_2\text{)–CH}_2\text{OH}$$

1 R=H

34 $R=CH_3$

49

The conversion of **49** into useful chiral auxiliaries has been accomplished by several research groups. The first type of auxiliary to be considered is acyclic, that is, where none of the valinol molecule is incorporated into a ring.

Meyers et al.[24] have developed a unique process where isoquinolines can be enantioselectively alkylated in the 1 position with high asymmetric induction. One series of auxiliaries evaluated relies on L-valinol as the source of chirality. This requires the conversion of **49** to dimethylformamidine derivatives **51.** They are generally prepared in high yield by treating **49** with DMF dimethylacetal followed by alkylation of the intermediate alcohol **50.** In the case of **51d, 49** is reacted first with isobutene (to form its *t*-butyl ester) and then with DMF acetal.[25]

49 $\xrightarrow[40°C]{Me_2NCH(OMe)_2}$ **50** ($N{=}CHNMe_2$, CH_2OH) $\xrightarrow[\text{base}]{RX}$ **51** ($N{=}CHNMe_2$, CH_2OR)

51	R	yield (%)
a	TMS	–
b	TBS	97
c	$SiEt_3$	98
d	$t\text{-}C_4H_9$	80
e	CH_3	–

Formamidines **51,** without isolation, are heated with isoquinoline (**52**) to generate the chiral formamidine of the isoquinoline **53** (Scheme 9). Metallation of **53** to its 1-lithio derivative is quantitatively accomplished with LDA at −78°C. Further cooling to −100°C is required for the alkylation step to achieve the best selectivity. Even at this temperature the alkylation (with methyl iodide) proceeds rapidly and is complete within 1 hr. Formamidine **54,** without purification, is readily cleaved with hydrazine in ethanol–acetic acid (pH 8) to give a mixture of

(*S*)-1-methyl-1,2,3,4-tetrahydroisoquinoline (**55**) and the recovered *O*-alkylated valinol derivative.

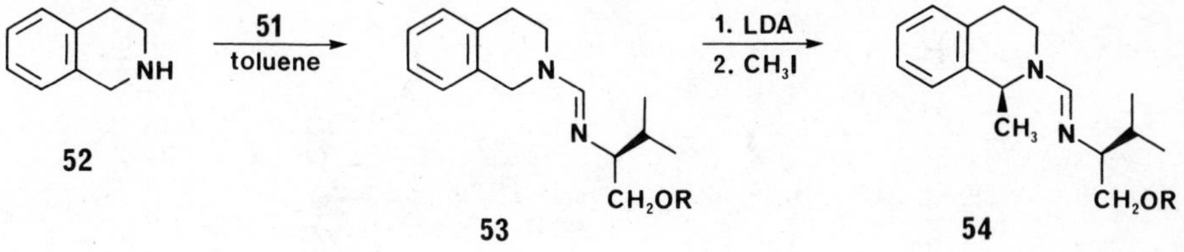

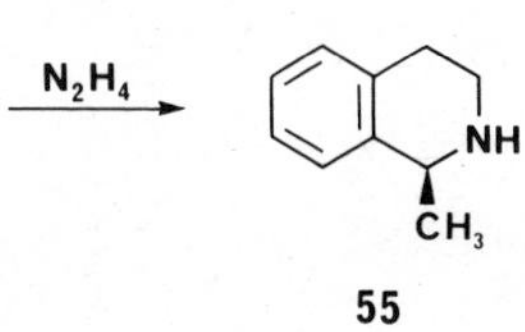

52 → (51, toluene) → 53 → (1. LDA; 2. CH_3I) → 54 → (N_2H_4) → 55

R	%53	%55	%ee
TMS	96	52	88
TBS	93	74	75
$SiEt_3$	96	70	74
t-C_4H_9	95	90	86
CH_3	99	46	84

Scheme 9

Alkylation of **53d** with a difunctional alkylating agent such as 1-bromo-4-chlorobutane affords **56,** which, on hydrazinolysis, cyclizes to the benzoquinolizidine **57,** a degradation product of securine (50–60% yield, 93–95% ee).[25]

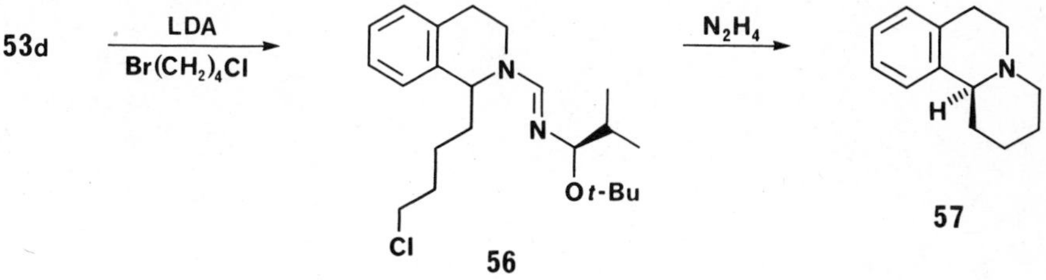

53d → (LDA, $Br(CH_2)_4Cl$) → 56 → (N_2H_4) → 57

Several naturally occurring isoquinoline alkaloids are readily accessible with this methodology (Scheme 10). (−)-Salsolidine (**60**), (*S*)-(+)-homolaudanosine (**61**), and (−)-norcoralydine (**62**) are obtained in a few steps by using the enantioselective alkylation of **59** as the key reaction in the synthetic sequence. The enantiomeric excess for each product usually exceeds 95%.[25]

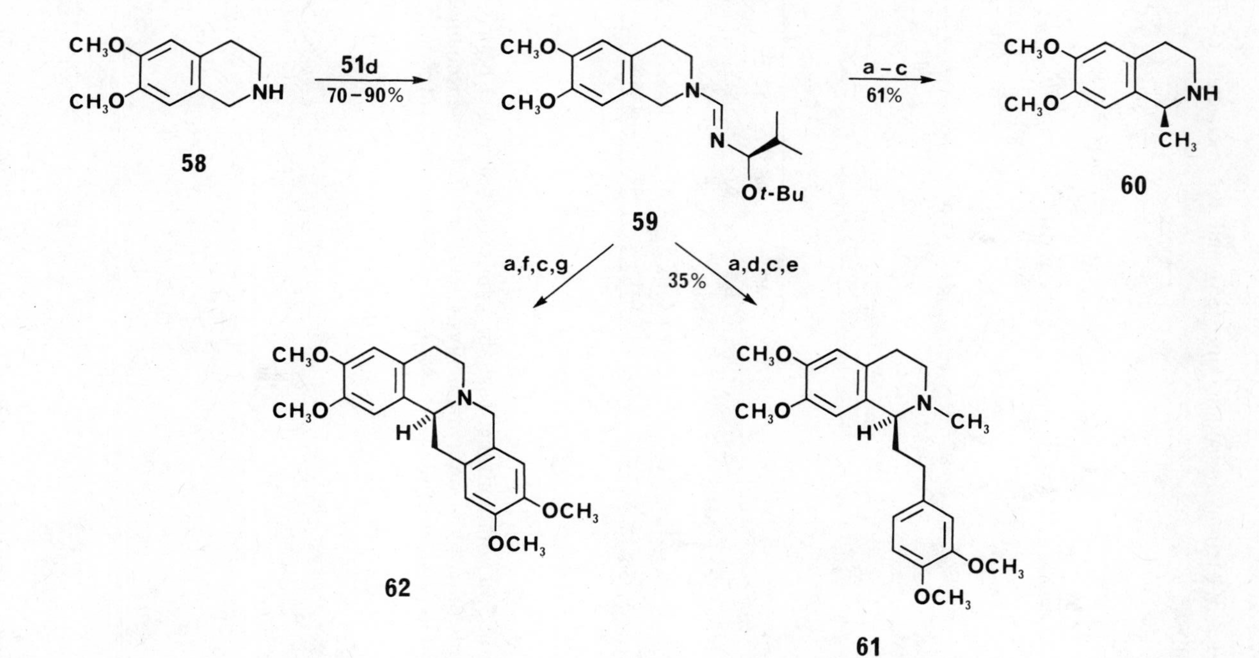

Scheme 10. (a) LDA, −78°C; (b) CH_3I, −100°C; (c) N_2H_4; (d) 3,4-dimethoxyphenethyl iodide, −100°C; (e) HCOOEt, $LiAlH_4$ (46%); (f) 3,4-dimethoxybenzyl bromide, −100°C (75%); (g) CH_2O, 2 *N* HCl.

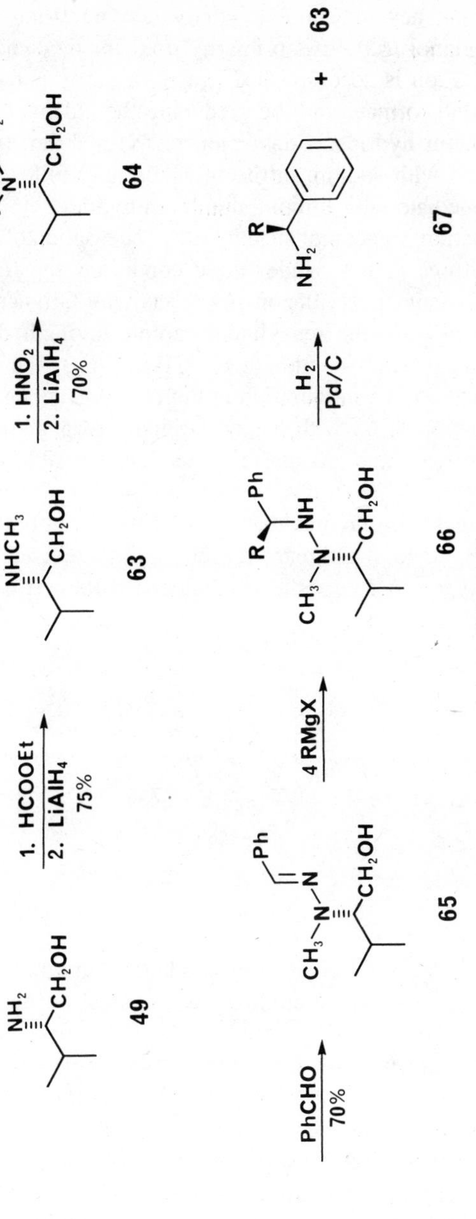

49 → (1. HCOOEt; 2. $LiAlH_4$; 75%) → 63 → (1. HNO_2; 2. $LiAlH_4$; 70%) → 64

→ (PhCHO; 70%) → 65 → (4 RMgX) → 66 → (H_2; Pd/C) → 67 + 63

Scheme 11

Another interesting acyclic chiral auxiliary derived from valinol is (*S*)-*N*,3-dimethyl-2-aminobutanol (**63**). It is primarily used for the synthesis of chiral hydrazones. Its preparation is accomplished in two steps by initially formylating **49** on nitrogen with ethyl formate and then reducing the aldehyde to a methyl group with lithium aluminum hydride. Conversion to (*S*)-*N*,3-dimethyl-2-hydrazinobutanol (**64**) is effected with sodium nitrite in acetic acid followed by reduction of the *N*-nitroso intermediate with lithium aluminum hydride[26] (Scheme 11).

Hydrazone formation is accomplished by the condensation of benzaldehyde with **64** in refluxing benzene (1 hr). Under these conditions the (*E*)-hydrazone **65** is produced as the only isomer. Hydrazone **65** reacts with Grignard reagents in ether or THF (40–45°C, 24 hr) at the benzylidene carbon, giving hydrazines **66** in good yields. With alkylmagnesium halides (e.g., CH_3, C_2H_5), the reaction diastereospecifically induces the (*S*) configuration at the newly generated chiral center. On the other hand, reaction of **65** with benzyl Grignard reagents produces an almost statistical mixture of (*S*)- and (*R*)-diastereomers. A plausible mechanism for obtaining such high optical inductions invokes a chelated six-membered transition state **68** that is energetically favorable because of the equitorially placed isopropyl group. A second molecule of Grignard reagent is then directed by the lone pair of electrons of the oxygen to attack the benzylidene carbon from the *re–si* face to give the observed products **66.**

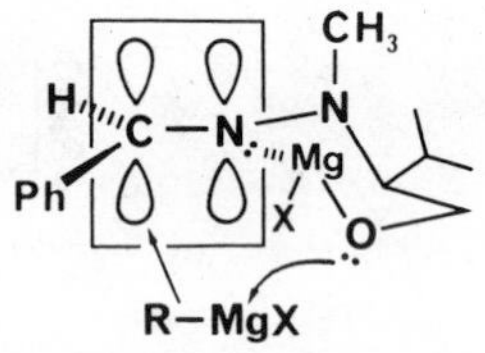

68

Hydrogenolysis of the N–N bond of **66** produces a mixture of (*S*)-α-alkyl benzylamines **67** and the regenerated auxiliary **63.** Optical purities of **67** are greater than 99%.

Few heterocycles can rival the versatility in modern-day asymmetric syntheses such as that of the oxazoline ring. A variety of oxazoline derivatives have been synthesized from members of the chiral pool. Those that are not derived from amino acid sources are beyond the scope of this book and are not discussed here.

Oxazolines are easily prepared from L-valinol derivatives with reagents that are capable of inserting a carbon atom in the ring by cyclization with both its OH and NH_2 groups. Aldehydes, ketones, carboxylic acids, and difunctional carbonyl compounds such as phosgene or diethylcarbonate all succeed in producing the oxazoline ring system.

The reaction of **49** or **63** with aliphatic or aromatic aldehydes takes place readily either in refluxing benzene[27] or at 0–5°C in the presence of $MgSO_4$[28] to produce oxazolidine derivatives **69** in 81–100% yields. In all cases, only one diastereomer is formed.

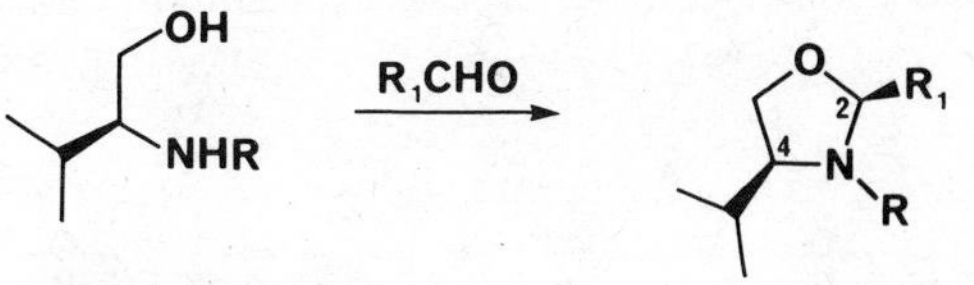

R_1CHO

49 R=H
63 $R=CH_3$

69

$R_1 = C_2H_5$, aryl, cyclohexyl, 3-cyclohexenyl

Reaction of **69** with Grignard reagents occurs regiospecifically at the 2 position to give acyclic amino alcohols **70** in good yields. With ethyl magnesium bromide and **69** ($R = CH_3$, R_1 = aryl), a mixture of diastereomeric products **70a** and **70b** is formed with **70a** representing 79–90% of the two.[27] Even more interesting, when benzylmagnesium chloride derivatives are allowed to react with **69** ($R = H, CH_3$; R_1 = cyclohexyl, 3-cyclohexenyl), only one diastereomer **70a** is formed in 71–83% yields.[28] These chiral phenethylamines exhibit analgesic activity.

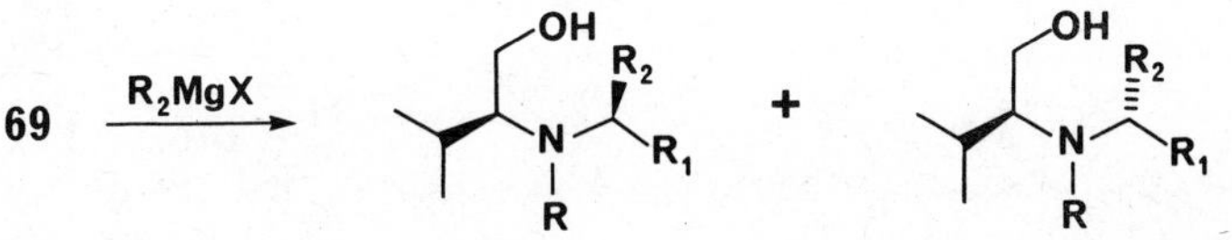

69 R_2MgX → **70a** + **70b**

The high diastereoselectivity of the reaction can be explained by the attachment of the magnesium atom of the Grignard reagent to the lone pair of electrons on oxygen (from the less sterically hindered side) followed by attack of the R_2 anion on C-2 from the same side (see **71**).

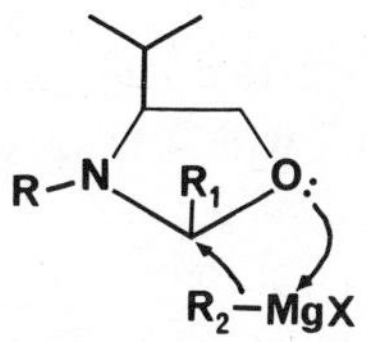

71

Partial asymmetric induction can also be observed in reactions that take place in a remote region from the chiral site in the oxazolidine ring. For example, the reaction of chiral Mannich bases **72** with phenyllithium reagents produces alcohols **73** and **74** in 32–36% diastereomeric excess as a result of 1,5-asymmetric induction[29] (Scheme 12).

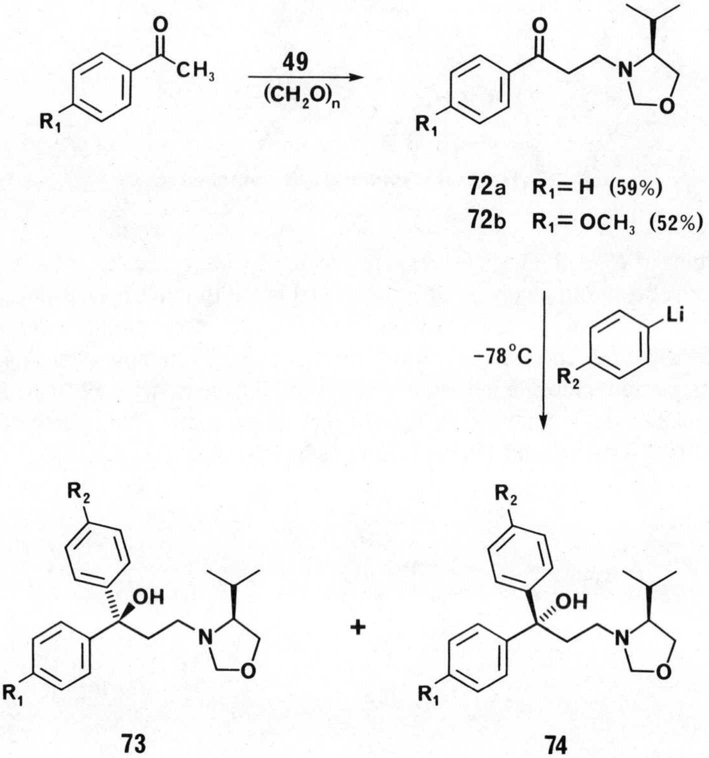

49
$(CH_2O)_n$

72a $R_1 = H$ (59%)
72b $R_1 = OCH_3$ (52%)

−78°C

73 + **74**

R_1	R_2	Yield (%)	Ratio **73**:**74**
H	OCH_3	82	66:34
OCH_3	H	89	61:39

Scheme 12

Bicyclic lactam **76** (Scheme 13) is synthesized as a single diastereomer in one step from the reaction of L-valinol with 3-benzoylpropionic acid (**75a**) or levulinic acid (**75b**) by azeotropic removal of water.[30,31] These versatile intermediates are highly useful for the generation of quaternary chiral centers. The new asymmetric center is formed by metallation of the lactam ring with LDA at −78°C followed by addition of an alkyl halide. The alkylation occurs *endo*-selectively to give alkylated products **77** with *endo*:*exo* ratios ranging from 9 to 30:1. A second alkylation under the same conditions furnishes dialkylated lactams **78** in 49–90% yield. Cleavage of the bicyclic system with 10% H_2SO_4–butanol produces chiral esters **79** (71–92% yield) with enantiomeric purities in excess of 95%. The hydrolysis cannot be performed on the monoalkylated lactams **77** because of the formation of racemic products.

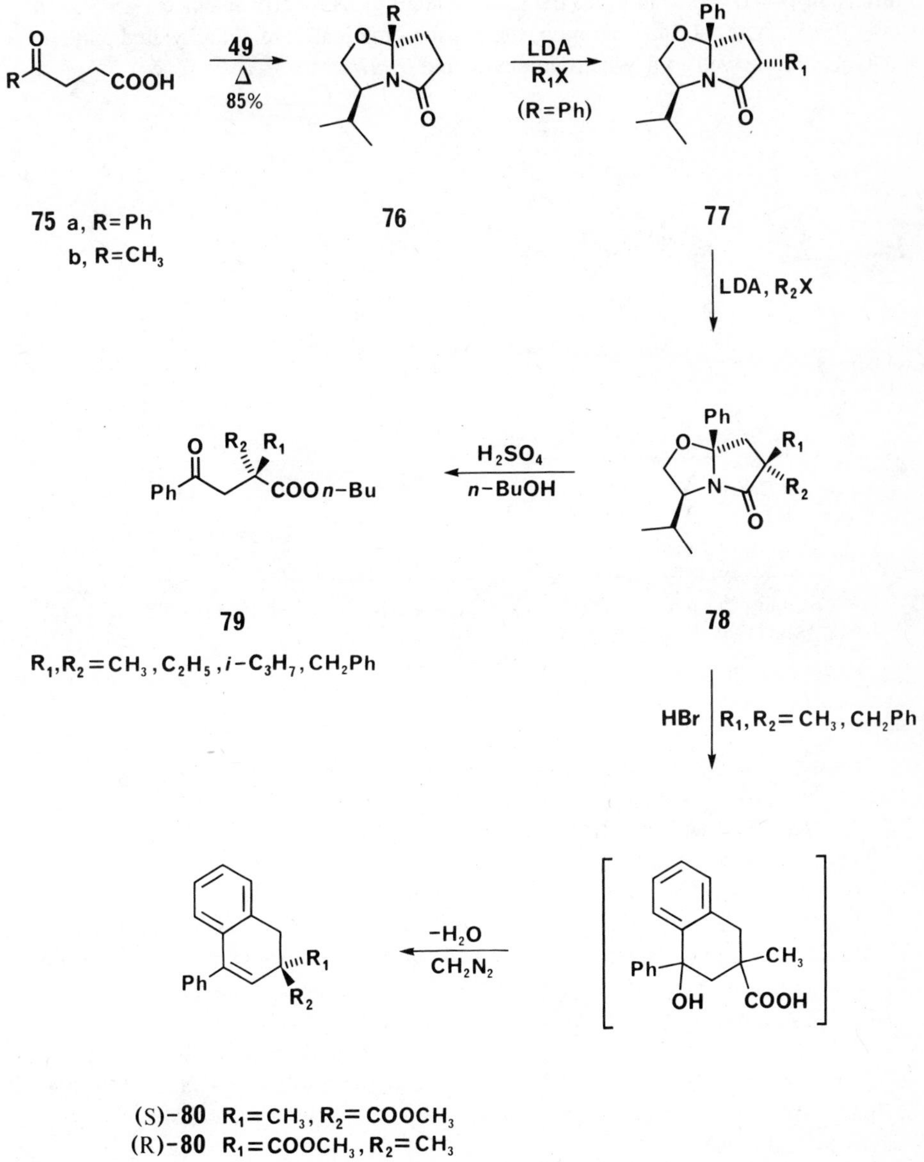

75 a, R=Ph
b, R=CH_3

49, Δ, 85%

76

LDA, R_1X (R=Ph)

77

LDA, R_2X

78

H_2SO_4, n-BuOH

79

$R_1,R_2 = CH_3, C_2H_5, i\text{-}C_3H_7, CH_2Ph$

HBr $R_1,R_2 = CH_3, CH_2Ph$

$-H_2O$, CH_2N_2

(S)-**80** $R_1 = CH_3$, $R_2 = COOCH_3$
(R)-**80** $R_1 = COOCH_3$, $R_2 = CH_3$

Scheme 13

Optionally, lactam **78** (if R_1 or R_2 = benzyl) can be converted to chiral dihydronaphthalenes **80** in 80% yield by sequential treatment with 48% HBr and esterification with diazomethane.

If R in **76** is a methyl group (**76b**), the methodology can be applied to the synthesis of chiral cyclopentenones **81.** Monoalkylation of **76b** occurs with rather poor selectivity (*endo*:*exo* ratio 1–1.5:1); however, when this diasteromeric mixture is alkylated a second time, the stereoselectivity is greatly enhanced (*endo*:*exo* ratio 95:5). After chromatography, the diastereomerically pure dialkylated bicyclic lactams are cleaved to a γ-keto aldehyde and cyclized to **81** (>99% ee).[31]

76b —(a,b; a,c)→ [bicyclic lactam with CH_3, R_1, R_2] —(d,e)→ [keto aldehyde with CH_3, OHC, R_1, R_2] —(f)→ **81**

R_1	R_2	Yield (%) 76b → 81	Configuration
$CH_2{=}CHCH_2$	$PhCH_2$	50	(*S*)
$PhCH_2$	$CH_2{=}CHCH_2$	44	(*R*)
C_2H_5	$PhCH_2$	41	(*S*)

(a) *s*-BuLi; (b) R_1X; (c) R_2X, −100°C; (d) Red-Al, −25°C; (e) $Bu_4NH_2PO_4$; (f) 0.25 equivalents KOH/EtOH.

Carboxylic acids (or imidates) react with L-valinol to form 2-substituted oxazoline derivatives **82** in good yields. Kurth et al.[32] use these oxazolines to prepare *N*-allylketene *N,O*-acetals **84,** which are useful chiral aza-Claisen substrates.

The substrates are straightforwardly produced from oxazolinium salts **83** by deprotonation with 130 mol % of *n*-butyllithium in THF. The excess base is required to neutralize adventitious acid. Subsequent thermolysis of **84** in decalin (3 hr) results in a highly diastereoselective aza-Claisen rearrangement to oxazoline **85** (94% de). It is imperative to perform the deprotonation and rearrangement as a one-pot operation in order to achieve the high diastereoselectivity.

These Claisen products are masked pent-4-enoic acids that can be freed to **86** by acid hydrolysis. In this manner (*R*)-(−)-2-methylpent-4-enoic acid and (*S*)-(+)-2-(phenylmethyl)pent-4-enoic acid can be prepared in high optical and chemical yields (Scheme 14).

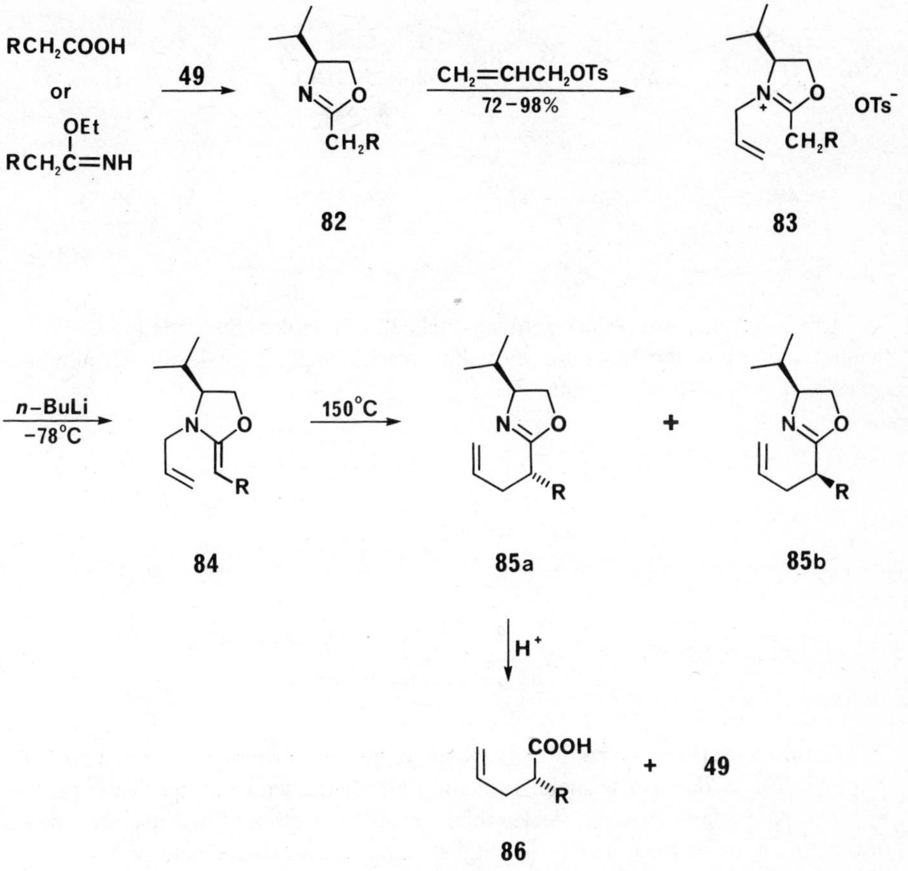

RCH_2COOH or $RCH_2C(OEt)=NH$ → (49) → 82 → ($CH_2=CHCH_2OTs$, 72–98%) → 83 (OTs^-) → (n-BuLi, −78°C) → 84 → (150°C) → 85a + 85b

85a → (H^+) → 86 + 49

R	Yield 82 (%)	Yield 85 (%) (from 82)	Ratio 85a:85b	Yield 86 (%)
CH_3	87	80	97:3	87
CH_2Ph	82	81	97:3	80

Scheme 14

Perhaps one of the most revolutionary advances in asymmetric synthesis is the development of *N*-acyl oxazolidinones as chiral enolate synthons in carbon–carbon bond forming reactions. Pioneered by Evans, these enolates show remarkable diastereoselection in alkylation,[33] acylation,[34] and aldol[35] processes.

One chiral auxiliary, **87,** whose asymmetric center is derived from an amino acid source, is easily prepared in high yield from L-valinol by treatment with either phosgene or diethylcarbonate. Imide derivatives **88** are then obtained simply by lithiating **87** with *n*-butyllithium at −78°C and acylating with an acid chloride.[35]

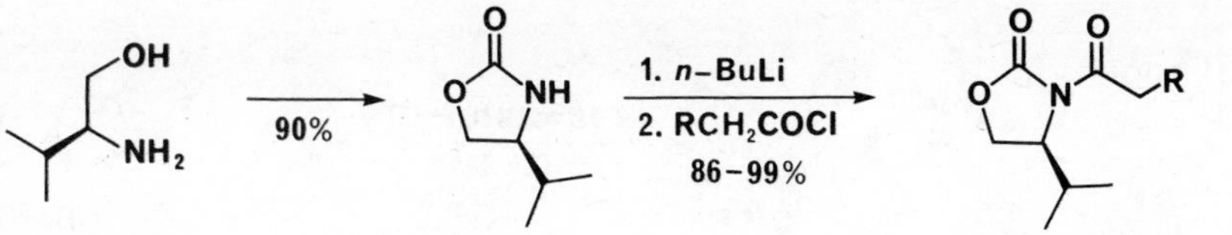

90%

1. *n*-BuLi
2. RCH_2COCl
86–99%

49 **87** **88** a, R=H
b, $R=CH_3$
c, $R=SCH_3$

Metallation of these imides forms a chelated (*Z*)-enolate **89.** This immobilized enolate system is the basis for providing astonishingly high levels of chirality transfer in a variety of reactions.

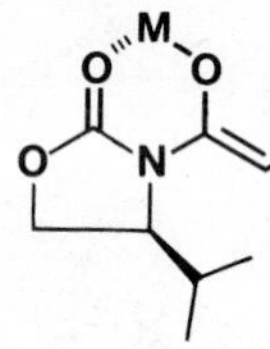

89 M=Li, Na, B(*n*-Bu)$_2$

Alkylation Reactions. The lithium enolate of **88b** is conveniently formed in THF with LDA at −78°C. The major limitation associated with this species is that it is not a particularly reactive nucleophile toward alkylation. Consequently, these reactions are performed at 0°C. At higher temperatures the enolate decomposes. To enhance reactivity, the sodium enolate of **88b** can be generated (NaHMDS, −78°C, THF), and alkylation then occurs readily at −78°C. With either enolate isobutyl bromide fails to react.

Generally, small alkylating agents are less stereoselective than those with larger steric bulk. (Methyl iodide is the least, with approximately 90% diastereoselectivity.) This is of little consequence since the alkylated diastereomers are easily separated by either column chromatography or recrystallization.[33]

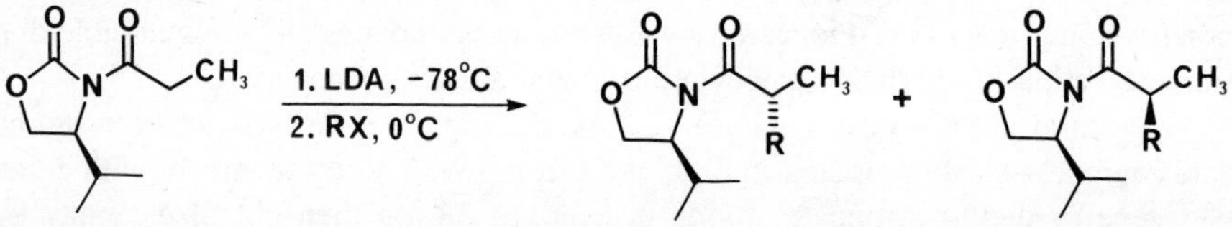

1. LDA, −78°C
2. RX, 0°C

88b **90** + **91**

R	Yield (%)	Ratio **90**:**91**
$PhCH_2$	92	120:1
$CH_2{=}CHCH_2$	71	98:2
$CH_2{=}C(CH_3)CH_2$	75	98:2
$PhCH_2OCH_2$	77	98:2
C_2H_5	36	94:6

The alkylated imides **90** are extremely susceptible to nucleophilic attack at the exocyclic carbonyl, thus rendering the nondestructive removal of the oxazolidone auxiliary a highly versatile process (Scheme 15). Hydrolysis, transesterification, and reduction can easily be performed under mild conditions; however, as the steric demands of the R group increase, competitive nucleophilic attack at the oxazolidone carbonyl becomes significant. This can be circumvented by transesterification with lithium benzyloxide to give the corresponding chiral benzyl esters in greater than 90% yield. To date no method has been devised to cleave imides **90** directly to an aldehyde. Instead, the imide is reduced to the primary alcohol with lithium aluminum hydride and is then oxidized under modified Moffatt conditions to the aldehyde. In all the transformations illustrated in Scheme 15, auxiliary **87** is cleanly regenerated and can be isolated for recycling purposes.

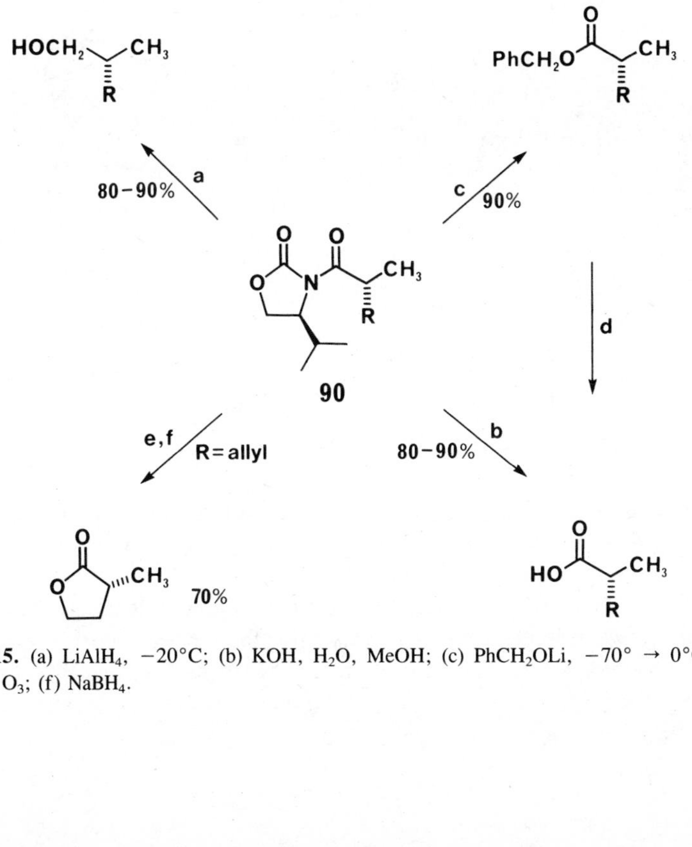

HOCH₂ CH₃ R
PhCH₂O CH₃ R
a 80–90%
c 90%
d
CH₃ R
90
e,f R=allyl
b 80–90%
CH₃ 70%
HO CH₃ R

Scheme 15. (a) $LiAlH_4$, −20°C; (b) KOH, H_2O, MeOH; (c) $PhCH_2OLi$, −70° → 0°C; (d) H_2, Pd/C; (e) O_3; (f) $NaBH_4$.

In an elegant application of this methodology, Smith and Thompson[36] have enantioselectively synthesized talaromycins A (**97a**) and B (**97b**), two toxins from a fungus produced from chicken litter and that grows on animal feed. The key intermediate in the synthesis is bromide **94**, which is prepared in 40% overall yield from **92** (Scheme 16). Reaction of the Grignard reagent derived from **94** with lactone **95** furnishes spiroketal **96** (after acidic hydration), which is subsequently oxidized and manipulated to the desired products **97**.

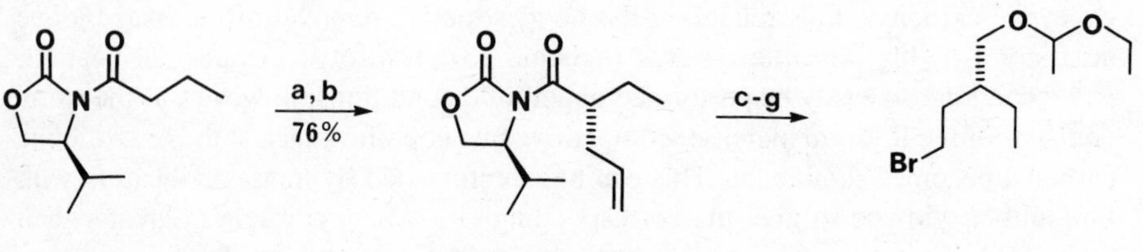

a, b
76%
c-g
Br
92 93 94

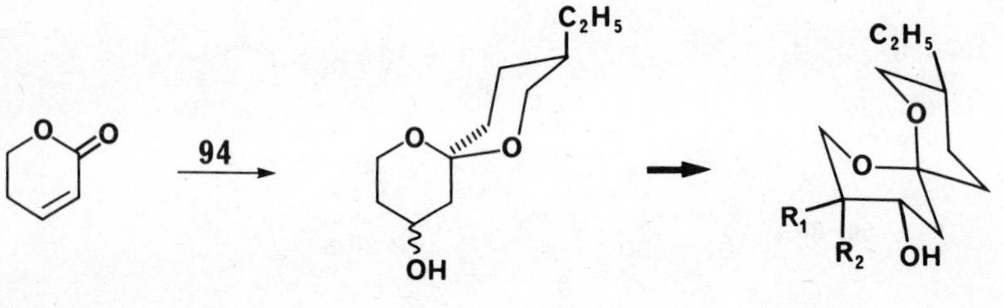

C_2H_5
94
OH
R_1 R_2 OH
95 96
97a R_1= H, R_2= CH_2OH
97b R_1= CH_2OH, R_2= H

Scheme 16. (a) LiHMDS, −78°C; (b) $CH_2{=}CHCH_2Br$, −78° → 0°C; (c) $LiAlH_4$, 0°C; (d) $EtOCH{=}CH_2$, PTSA; (e) O_3, Me_2S; (f) $NaBH_4$; (g) CBr_4, Ph_3P, pyridine.

Acylation Reactions. The lithium enolate derived from **88b** is also subject to acylation by reaction with suitable acylating agents. Cannulation of a solution of the enolate (−78°C) into a solution of an acid chloride at −78°C followed by immediate quench with saturated ammonium chloride diastereoselectively furnishes β-keto imides **98** in high yields.[34] Surprisingly, the new asymmetric center does not easily racemize by way of enolization because of its low kinetic acidity.

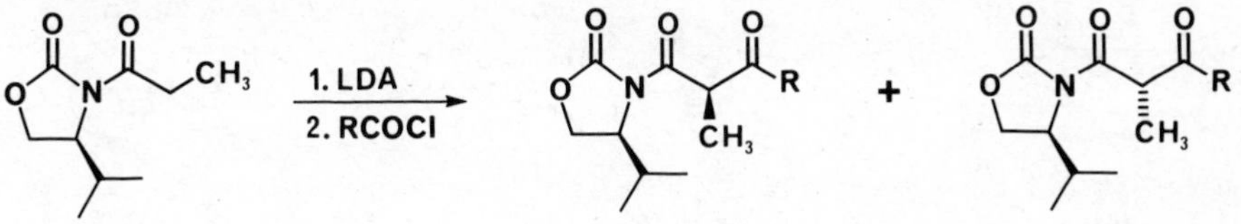

88b → (1. LDA; 2. RCOCl) → 98 + 99 (ratio 96:4)

98 → 100

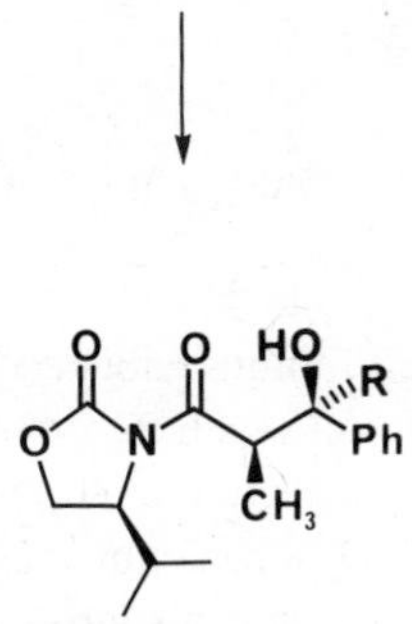

R	98 yield (%)
CH_3	95
C_2H_5	88
C_6H_5	93

The relative stability of **98** renders these derivatives amenable to further chemical manipulations. For example, reduction of **98** (R = C_6H_5) with zinc borohydride (0°C, 30 min) furnishes β-hydroxy imide **100** (R = H) in greater than 95% yield. Likewise, reaction of methylmagnesium bromide (−78°C, 4 hr) cleanly affords **100** (R = CH_3) in more than 90% yield.

(2*S*,3*S*,4*R*)-4-Amino-3-hydroxy-2-methylpentanoic acid (**105**), an important amino acid component in the antitumor antibiotic bleomycin (see Figure 1, Chapter 1) is readily accessible by the aforementioned acylation technique. However, to establish the proper configuration for its three contiguous asymmetric centers, D-valinol must be used to construct the imide **101.**

The lithium enolate of **101,** when allowed to react with Boc–D-alanine anhydride at −78°C, affords **102** in greater than 99% diastereomeric purity. Reduction of **102** with zinc borohydride produces the *erythro*-(2*S*,3*S*,4*R*)-alcohol **103,** again in high diastereoselectivity (>99:1). Transesterification to the methyl ester **104** is accomplished with cold sodium methoxide; then acidic hydrolysis furnishes the desired amino acid **105**[37] (Scheme 17).

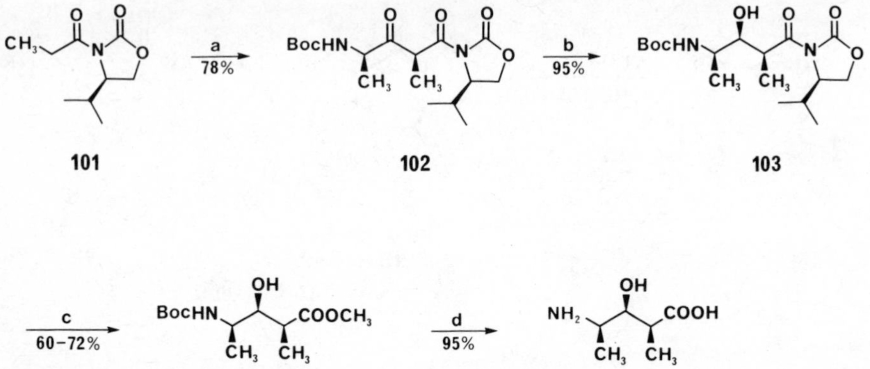

101 102 103

104 105

Scheme 17. (a) LDA, −78°C, $(Boc\text{-}D\text{-}Ala)_2O$; (b) $Zn(BH_4)_2$, −25°C; (c) $NaOCH_3$, 0°C, 5 min; (d) 2 *N* HCl, 55°C.

Aldol Reactions. Although lithium enolates of **88** are remarkably diastereoselective in alkylation and acylation reactions, they surprisingly exhibit low levels of stereoregulation in aldol condensations. The desired *erythro* selectivity, however, can be returned with the use of boron enolates. These are easily formed by treating **88** with di-*n*-butylboryl triflate and diisopropylethylamine at 0°C. After complete enolization occurs (30 min), the aldol reaction is performed at −78°C.[35] As seen in Table 3.1, diastereofacial selection is outstanding.

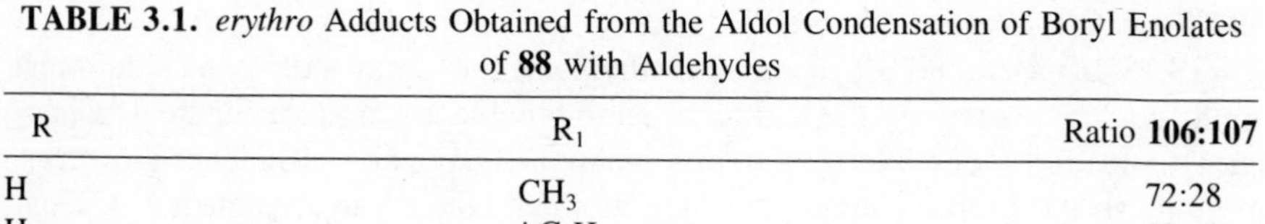

88 106 107

TABLE 3.1. *erythro* Adducts Obtained from the Aldol Condensation of Boryl Enolates of **88** with Aldehydes

R	R_1	Ratio **106:107**
H	CH_3	72:28
H	$i\text{-}C_3H_7$	52:48
CH_3	$i\text{-}C_3H_7$	497:1
CH_3	$n\text{-}C_4H_9$	141:1
CH_3	C_6H_5	>500:1
SCH_3	CH_3	99.6:0.4
SCH_3	$i\text{-}C_3H_7$	98.4:1.6
SCH_3	$n\text{-}C_3H_7$	98.9:1.1
SCH_3	C_6H_5	92.4:7.6

Imides **106** can readily be hydrolyzed to either a β-hydroxy acid (4 equivalents of 2 *N* KOH, CH_3OH, 0°C, 45 min) or β-hydroxy methyl ester (1.1 equivalents of $NaOCH_3$, 0°C, 5 min).

One limiting factor of this method is that R cannot be hydrogen (**88a**). In this case stereoselectivity drops off markedly (see Table 3.1, entries 1 and 2). A viable solution to the problem is the use of **88c.** Its (*Z*)-boryl enolate, on reaction with isobutyraldehyde, affords **108** in nearly 99% diastereomeric purity. Desulfurization to **109** with Raney nickel occurs rapidly (20 min, 60°C, acetone), and base hydrolysis as previously described gives (*S*)-3-hydroxy-4-methylpentanoic acid (**110**) in 80–90% yield.

O O OH
SCH$_3$
RaNi
OH$^-$
HOOC
108 **109** **110**

It is interesting to note that reaction of the boryl enolate of **88b** with a chiral aldehyde (e.g., **111**) cleanly affords the *erythro* anti-Cram adduct **112** (>400:1 diastereoselection).[38] It appears that the "resident enolate chirality largely overrides the intrinsic Cram selectivity imposed by resident chirality in the aldehyde condensation partner."[39]

CH$_3$ + OHC CH$_3$ CH$_3$
Bu_2BOTf
CH$_3$ CH$_3$ CH$_3$
88b (S)-**111** **112**

In an interesting extension of Evans's technology, Lantos and co-workers[40] utilize an aldol condensation of a chiral α-haloimidate (e.g., **113**) to generate bromohydrins **114** or **115** as a precursor for a chirally modified Darzens condensation. Predictably, the boryl enolate of **113** produces a mixture of *erythro*-diastereomers **114** and **115** in a ratio of >20:1. However, with zinc enolates, chirality reversal is observed and diastereomer **115** is formed in excess (ratio **114**:**115** = 1:9) along with minor amounts of *threo*-diastereomers. Treatment of **114** or **115** with lithium benzyloxide at −20°C affords enantiomeric epoxides **116** or **117** in high yields (Scheme 18). Analogous chlorohydrins epimerize to various degrees when subjected to these conditions.

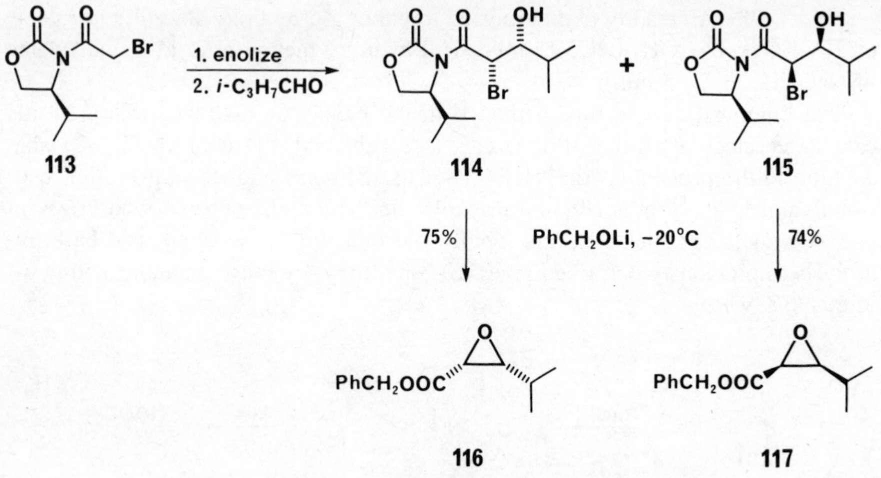

Scheme 18

Chiral crotonate imides such as **118** undergo aldol condensations as easily as in the previously described examples to give adducts **119** in extremely high yields. These adducts have been used to construct chiral synthons for the potential synthesis of thienamycin (Scheme 19) and tylosin (Scheme 20).[41]

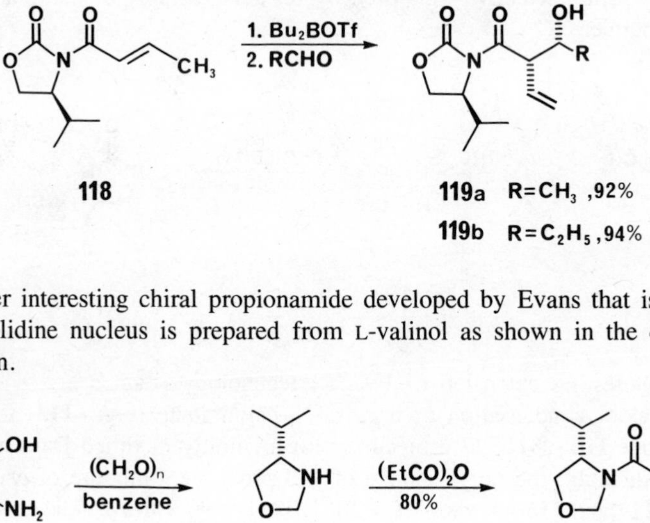

Another interesting chiral propionamide developed by Evans that is based on the oxazolidine nucleus is prepared from L-valinol as shown in the **49** → **120** conversion.

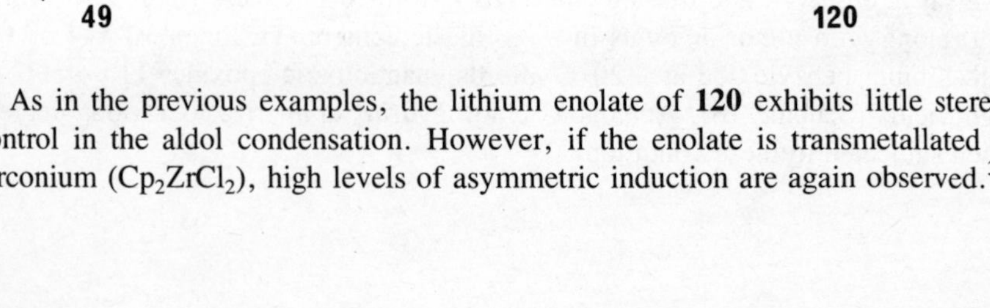

As in the previous examples, the lithium enolate of **120** exhibits little stereocontrol in the aldol condensation. However, if the enolate is transmetallated to zirconium (Cp_2ZrCl_2), high levels of asymmetric induction are again observed.[39]

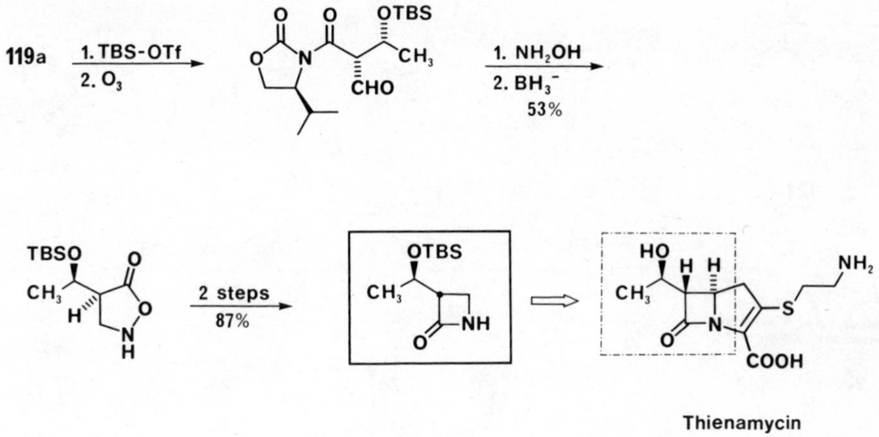

119a — 1. TBS-OTf, 2. O_3 → 1. NH_2OH, 2. BH_3^-, 53%

2 steps, 87%

Thienamycin

Scheme 19

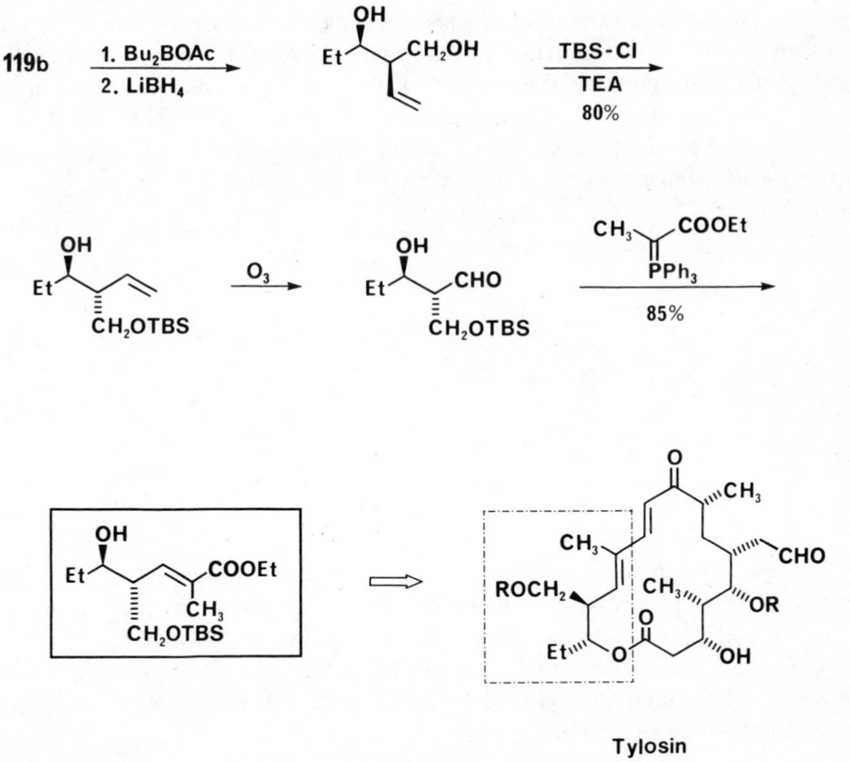

119b — 1. Bu_2BOAc, 2. $LiBH_4$ → TBS-Cl, TEA, 80%

O_3 → $CH_3C(=PPh_3)COOEt$, 85%

Tylosin

Scheme 20

RCHO
−78°C

121 122 + 123

R	120, M[a]	Yield (%)	Ratio 122:123[b]
n-C_4H_9	Li		42:35
	Zr	96	97:1
i-C_3H_7	Li		41:42
	Zr	77	97.5:0.5
C_6H_5	Li		29:41
	Zr	71	94.5:1.5

[a]Zr = $ZrCp_2Cl$.
[b]Remainder of the mixture = *threo*-diastereomers.

In the preceding illustrations of Evans oxazolidones the induction of asymmetry is controlled by the configuration of the isopropyl group that inherits its chirality from valine. If the opposite sense of asymmetric induction is required, one can use a chirally biased oxazolidone **125** based on (1*S*,2*R*)-norephedrine (**124**). Alkylation, acylation, and aldol condensations of imides derived from **125** give the opposite *erythro*-diastereomer with equal facility and diastereoselection.[42]

$COCl_2$

124 125

Alkylation of chiral oxazolidone **87** with phenacyl bromide results in the formation of the 3-phenacyl derivative **126.** When exposed to reducing agents, its prochiral ketone is stereoselectively reduced by means of 1,4-asymmetric induction. The resulting (*S*)-alcohol **127,** the major diastereomer, is obtained in a ratio of 83:17 with sodium borohydride (−20°C) or 71:29 with lithium tri-*t*-butoxyaluminum hydride (−20°C).

At the same temperature, lithium aluminum hydride reduces both carbonyl groups of **126** to give diol **128** in a ratio of 72:28. The treatment of **127** under these conditions also affords **128** in 67% yield.[43]

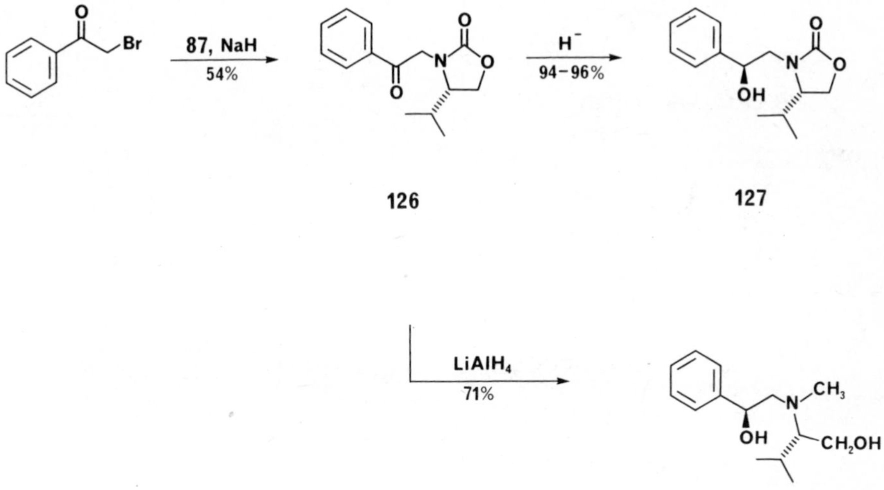

3.3 ADDENDA

Since the original preparation of this manuscript, an enantioselective synthesis of (+)-α-cuparenone (**134**) has been reported[44] that further exemplifies the usefulness of valine-derived imides as auxiliaries for the efficient transfer of chirality. The imide **129,** prepared from **87** and *p*-tolylacetyl chloride, is diastereoselectively methylated (91:9) to give **130** (after chromatography). Conversion to **132** followed by a rhodium-catalyzed intramolecular C–H insertion produces the cyclopentanone **133** with retention of configuration. Standard functional group manipulation furnishes the natural product **134** in 26% yield from **133** (Scheme 21).

Aldol adduct **106,** which is generated in >99% diastereoselectivity from imide **88b** and isobutyraldehyde, is used by Meyers et al.[45] as an intermediate in the synthesis of the depsipeptide unsaturated ester fragment (**137**) of the streptogramin antibiotic madumycin II (**138**) (Scheme 22). Aldehyde **135** is relatively unstable and must be converted directly to the unsaturated ester **136** by using the potassium salt of triethyl phosphonoacetate (45% yield from **106**).

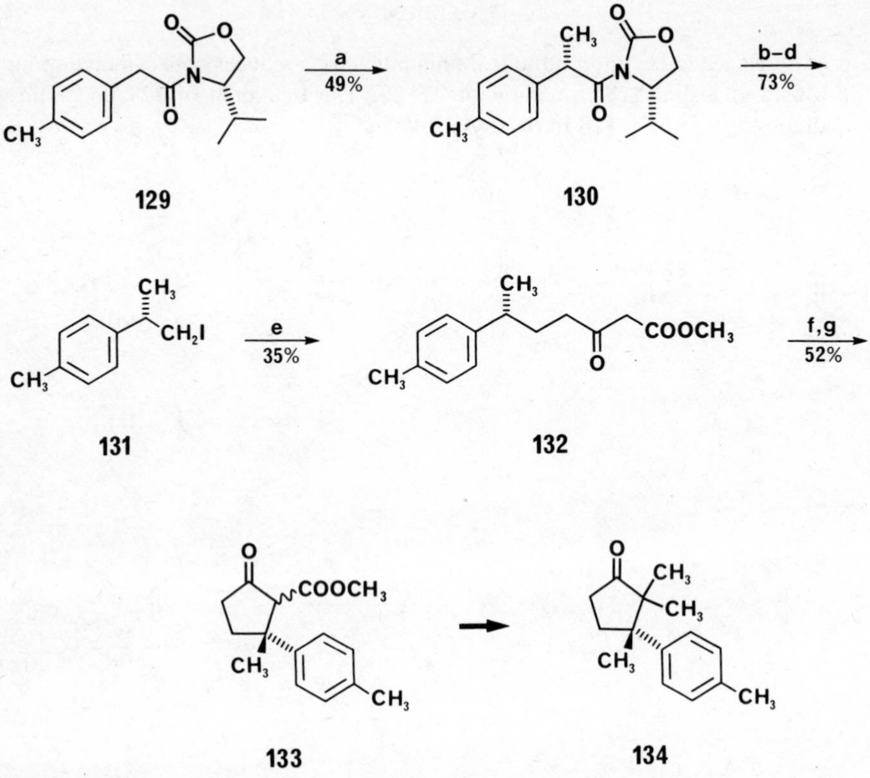

129 → (a, 49%) → 130 → (b–d, 73%)

131 → (e, 35%) → 132 → (f,g, 52%)

133 → 134

Scheme 21. (a) LDA, CH_3I; (b) $LiAlH_4$ (97%); (c) TsCl, pyridine; (d) NaI, acetone (76%); (e) methyl acetoacetate dianion; (f) TsN_3, TEA; (g) Rh_2OAc_4.

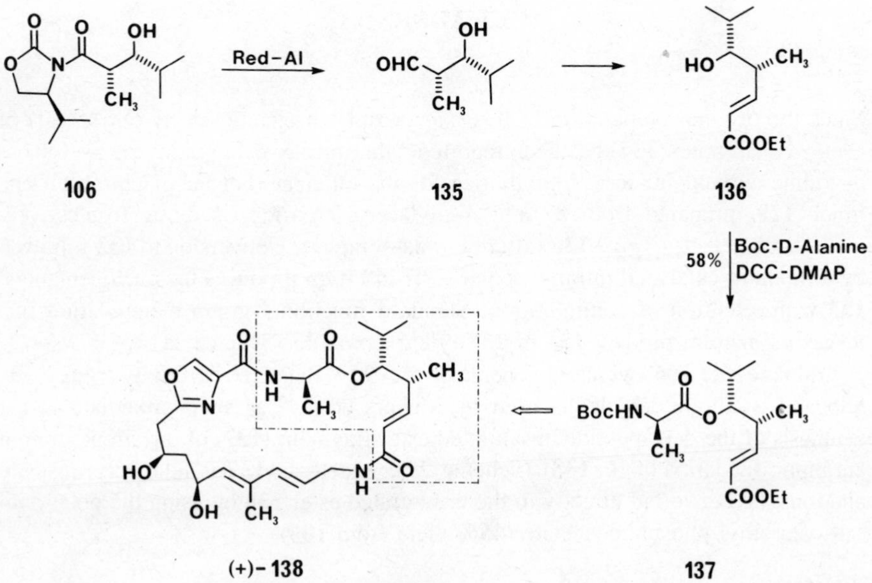

106 → (Red-Al) → 135 → 136 → (Boc-D-Alanine, DCC-DMAP, 58%) → 137 ⇐ (+)-138

Scheme 22

REFERENCES

1. T. Scott and M. Brewer, *Concise Encyclopedia of Biochemistry,* Walter de Gruyter, New York, 1983, p. 495.
2. U. Schöllkopf, U. Groth, and C. Deng, *Angew. Chem., Int. Ed.,* **20,** 798 (1981).
3. (a) U. Schöllkopf, U. Groth, K. Westphalen, and C. Deng, *Synthesis,* 969 (1981); (b) U. Schöllkopf, *Tetrahedron,* **39,** 2085 (1983).
4. J. Nozulak and U. Schöllkopf, *Synthesis,* 866 (1982).
5. U. Schöllkopf, J. Nozulak, and U. Groth, *Synthesis,* 868 (1982).
6. U. Schöllkopf, U. Groth, M. Gull, and J. Nozulak, *Liebigs Ann. Chem.,* 1133 (1983).
7. U. Groth and U. Schöllkopf, *Synthesis,* 673 (1983).
8. U. Schöllkopf, J. Nozulak, and U. Groth, *Tetrahedron,* **40,** 1409 (1984).
9. U. Groth, Y. Chiang, and U. Schöllkopf, *Liebigs Ann. Chem.,* 1756 (1982).
10. U. Schöllkopf and R. Lonsky, *Synthesis,* 675 (1983).
11. U. Groth and U. Schöllkopf, *Synthesis,* 37 (1983).
12. U. Groth and U. Schöllkopf, and Y. Chiang, *Synthesis,* 864 (1982).
13. V. Caplar, A. Lisini, F. Kajfez, D. Kolbah, and V. Sunjic, *J. Org. Chem.,* **43,** 1355 (1978).
14. A. K. Mukerjee and A. K. Singh, *Tetrahedron,* **34,** 1731 (1978).
15. I. Ojima and S. Inaba, *Tetrahedron Lett.,* 2081 (1980).
16. K. Tomioka, K. Ando, Y. Takemasa, and K. Koga, *J. Am. Chem. Soc.,* **106,** 2718 (1984).
17. K. Tomioka, K. Ando, Y. Takemasa, and K. Koga, *Tetrahedron Lett.,* 5677 (1984).
18. R. M. Kellogg, *Angew. Chem., Int. Ed.,* **23,** 782 (1984).
19. J. G. deVries and R. M. Kellogg, *J. Am. Chem. Soc.,* **101,** 2759 (1979).
20. P. Jouin, C. B. Troostwijk, and R. M. Kellogg, *J. Am. Chem. Soc.,* **103,** 2091 (1981).
21. P. Karrer, P. Portmann, and M. Suter, *Helv. Chim. Acta.,* **32,** 1156 (1949).
22. G. A. Smith and R. E. Gawley, *Organic Synthesis,* **63,** 136 (1984).
23. C. F. Lane, U.S. Patent 3,935,280 (1976); *Chem. Abstr.,* **84,** 135101p (1976).
24. A. I. Meyers, L. M. Fuentes, and Y. Kubota, *Tetrahedron,* **40,** 1361 (1984).
25. A. I. Meyers, M. Boes, and D. A. Dickman, *Angew. Chem., Int. Ed.,* **23,** 458 (1984).
26. H. Takahashi and Y. Suzuki, *Chem. Pharm. Bull.,* **31,** 4295 (1983).
27. H. Takahashi, Y. Suzuki, and T. Kametani, *Heterocycles,* **20,** 607 (1983).
28. H. Takahashi, Y. Chida, T. Suzuki, H. Onishi, and S. Yanaura, *Chem. Pharm. Bull.,* **32,** 2714 (1984).
29. H. Takahashi, K. Tanahashi, and K. Higashiyama, *Chem. Pharm. Bull.,* **32,** 4323 (1984).
30. A. I. Meyers, M. Harre, and R. Garland, *J. Am. Chem. Soc.,* **106,** 1146 (1984).
31. A. I. Meyers and K. T. Wanner, *Tetrahedron Lett.,* 2047 (1985).
32. M. J. Kurth, O. H. W. Decker, H. Hope, and M. D. Yanuck, *J. Am. Chem. Soc.,* **107,** 443 (1985).
33. D. A. Evans, M. D. Ennis, and D. J. Mathre, *J. Am. Chem. Soc.,* **104,** 1737 (1982).
34. D. A. Evans, M. D. Ennis, and T. Le, *J. Am. Chem. Soc.,* **106,** 1154 (1984).

35. D. A. Evans, J. Bartroli, and T. L. Shih, *J. Am. Chem. Soc.*, **103,** 2127 (1981).
36. A. B. Smith, III and A. S. Thompson, *J. Org. Chem.*, **49,** 1469 (1984).
37. R. M. DiPardo and M. G. Bock, *Tetrahedron Lett.*, 4805 (1983).
38. D. A. Evans and J. Bartroli, *Tetrahedron Lett.*, 807 (1982).
39. D. A. Evans and L. R. McGee, *J. Am. Chem. Soc.*, **103,** 2876 (1981).
40. A. Abdel-Magid, I. Lantos, and L. N. Pridgen, *Tetrahedron Lett.*, 3273 (1984).
41. D. A. Evans, *Aldrichimica Acta,* **15,** 23 (1982).
42. See Refs. 33, 34, 35, and 41.
43. H. Takahashi, N. Yamada, K. Higashiyama, and K. Kawi, *Chem. Pharm. Bull.*, **33,** 84 (1985).
44. D. F. Taber, E. H. Petty, and K. Raman, *J. Am. Chem. Soc.*, **107,** 196 (1985).
45. A. I. Meyers, R. F. Spohn, and R. J. Linderman, *J. Org. Chem.*, **50,** 3633 (1985).

CHAPTER FOUR

THE LEUCINE FAMILY

All three of the possible isomerically branched butyl (C_4H_9)-substituted α-amino acids comprise the leucine family. L-Leucine, which is particularly abundant in serum albumins and globulins, is both essential in the diet and ketogenic. L-Isoleucine, found in relatively large amounts in hemoglobin and sugar beet molasses, is an essential amino acid that is both glucoplastic (degradation via propionic acid) and ketoplastic (formation of acetate).[1] *tert*-Leucine is a synthetic amino acid classically prepared from pinacolone and resolved into its optical enantiomorphs.[2]

4.1 LEUCINE

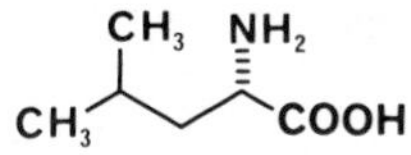

CH3 NH2
CH3 COOH

(*S*)-2-Amino-4-methylpentanoic acid

The efficient formation of a peptidic amide bond requires the activation of the carboxylic acid moiety of an amino acid prior to coupling. L-Leucine (**1**) can be activated by two methods (Scheme 1): (1) treatment of **1** with phosgene in THF at 45°C forms an internally protected and activated *N*-carboxyanhydride **2** in high yield;[3] and (2) reaction of **1** with 5-nitrosalicylaldehyde in the presence of DDC affords benzoxazepinone **3**.[4] Both these derivatives, when allowed to react with an amino acid ester, produce dipeptides in high yield.

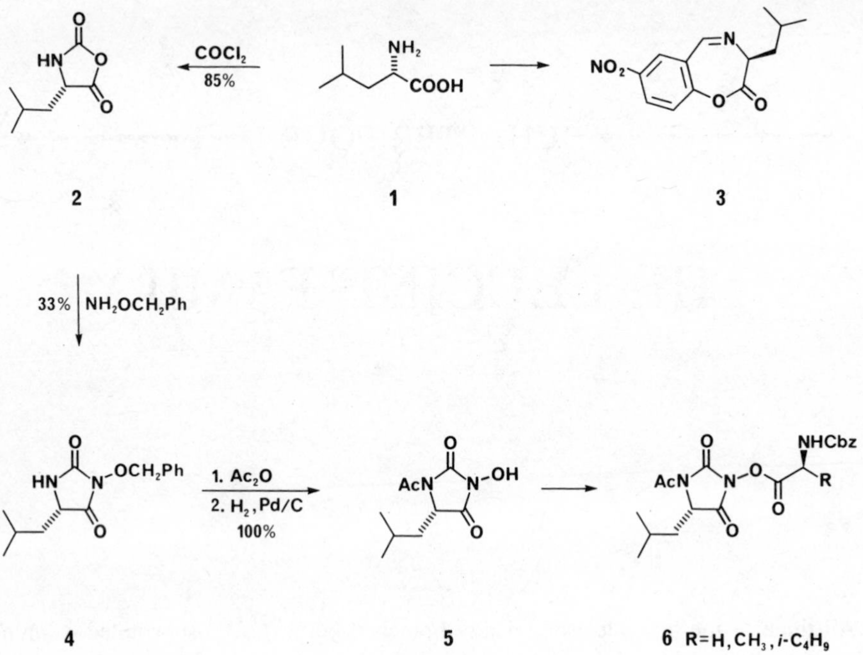

$COCl_2$ 85%

NH_2 COOH

NO_2

2 1 3

33% NH_2OCH_2Ph

HN N OCH_2Ph 1. Ac_2O 2. H_2, Pd/C 100% AcN N–OH AcN N–O NHCbz R

4 5 6 R=H, CH_3, i-C_4H_9

Scheme 1

Anhydride **2** can also be converted to the optically active 3-hydroxyhydantoin **5** by treatment with benzyloxyamine, acetylation with acetic anhydride, and hydrogenolytic debenzylation.[5] This hydantoin is also a chirally biased acyl activating reagent for peptide synthesis. When reacted with a protected amino acid in the presence of DCC, the hydantoin ester **6** is obtained. On reaction with racemic D,L-amino acid esters (e.g., alanine ethyl ester), it selectively forms the peptide bond with the L-isomer, thus providing optically active dipeptides in greater than 91% ee.

α-Alkylleucine methyl esters (**11**) are readily prepared by alkylation of the bis-lactim ether (**9**) of cyclo-(L-Leu–L-Leu) as shown in Scheme 2. Since similar transformations have been described in depth in the previous chapters, the detailed synthetic process is not presented again here. Alkylation of **9** does not produce consistently high stereoselectivities as with bis-lactim ethers of other amino acids (see Chapters 1–3). Here, the diastereomeric excess ranges from 85 to >95%. In addition, the higher steric demands of the isobutyl group at C-6 of the alkylated derivative **10** causes the acid hydrolysis to proceed sluggishly. Where R is bulky (e.g., 2-naphthylmethyl), the hydrolysis practically fails and the product is isolated in only 5% yield.[6]

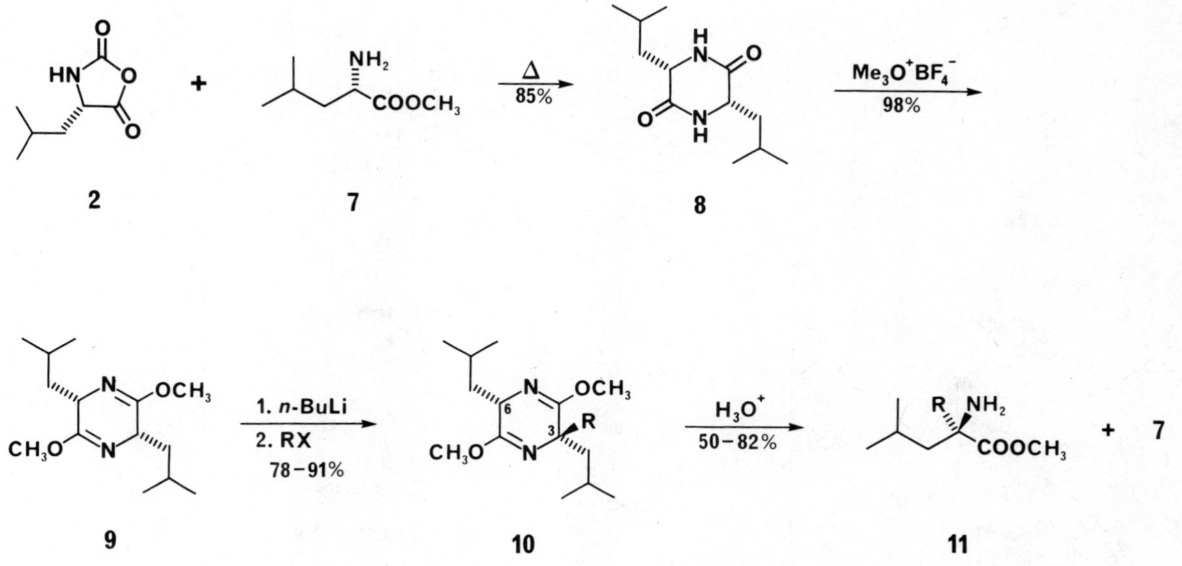

2 + 7 $\xrightarrow[85\%]{\Delta}$ 8 $\xrightarrow[98\%]{Me_3O^+BF_4^-}$

9 $\xrightarrow[78-91\%]{1.\ n\text{-BuLi};\ 2.\ RX}$ 10 $\xrightarrow[50-82\%]{H_3O^+}$ 11 + 7

R= CH_3, n-C_4H_9, $CH_2CH{=}CH_2$, $CH_2C{\equiv}CH$, CH_2Ar, CH_2SCH_3

Scheme 2

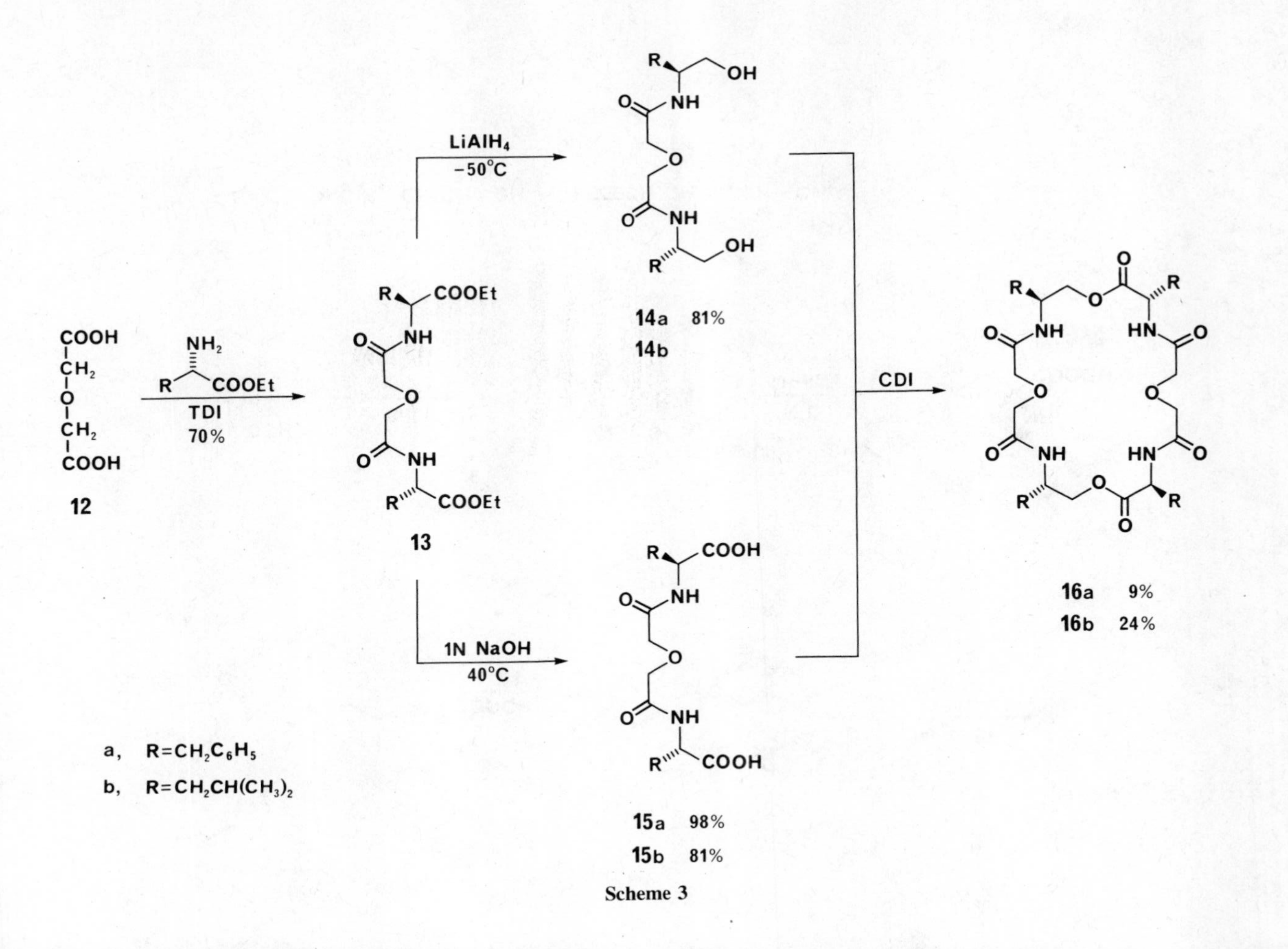
COOH
CH2
O
CH2
COOH
12
NH2
R COOEt
TDI
70%
R COOEt
NH
O
NH
R COOEt
13
LiAlH4
−50°C
R OH
NH
O
NH
R OH
14a 81%
14b
1N NaOH
40°C
R COOH
NH
O
NH
R COOH
15a 98%
15b 81%
CDI
R NH O HN R
O O
16a 9%
16b 24%
a, R=CH2C6H5
b, R=CH2CH(CH3)2

Scheme 3

The area of guest–host complexation of crown ethers has spawned intense interest in the preparation of chiral macrocycles containing both oxygen and nitrogen heteroatoms. Recently, the synthesis of two chiral 24-membered macrocycles was accomplished by using L-phenylalanine or L-leucine as the source of asymmetry.[7] As shown in Scheme 3, the synthesis begins with the amidation of diglycolic acid (**12**) with the amino acid ester in the presence of *N,N'*-thionyldiimidazole. In both cases **13** is formed in approximately 70% yield. Half of the diesters **13** are reduced to diols **14,** while the remaining half is hydrolyzed to the diacids **15.** Macrolactonization of **14** and **15** gives the desired macrocycles **16** in low yield.

Amine–borane complexes are effective reducing agents for a variety of carbonyl-containing substrates. If the complex is chirally biased by incorporation of an asymmetric center adjacent to the nitrogen, reduction of a prochiral ketone should furnish an excess of one enantiomeric alcohol.

Chiral amine–borane complex **17** is prepared by mixing stoichiometric amounts of L-leucine methyl ester (**7**) and diborane in THF. When ketones **18** are reduced with **17** in the presence of boron trifluoride etherate, an excess of the (*S*)-alcohol **19** is obtained in 14.7–22.5% optical yield.[8] Although these reductions are far from being useful in modern asymmetric syntheses, they hopefully can be improved on with further research.

NH_2, $COOCH_3$ — **7** —(B_2H_6)→ $NH_2 \cdot BH_3$, $COOCH_3$ — **17**

$R-C(=O)-CH_3$ — **18** —(**17**, $BF_3 \cdot Et_2O$)→ OH, R, CH_3 — **19**

$R = C_6H_5$, $t\text{-}C_4H_9$, $n\text{-}C_5H_{11}$

An interesting constituent of flue-cured tobacco contains the backbone of L-leucine, where its amino group is incorporated into a pyrrole ring. This derivative, **23,** may have desirable flavoring properties.[9]

Chan and Lee[10] have devised a mild and efficient method for the conversion of primary amines into *N*-substituted pyrroles, and it is an integral feature in the synthesis of **23.** Thus the treatment of a mixture of **7** and 1,4-dichloro-1,4-dimethoxybutane (**20**) in the presence of the weakly basic ion-exchange resin Amberlyst A-21 directly affords pyrrole **21.** Formylation to **22** followed by hydrolysis with either sodium hydroxide or iodotrimethylsilane (55% yield) furnishes the desired product **23.**

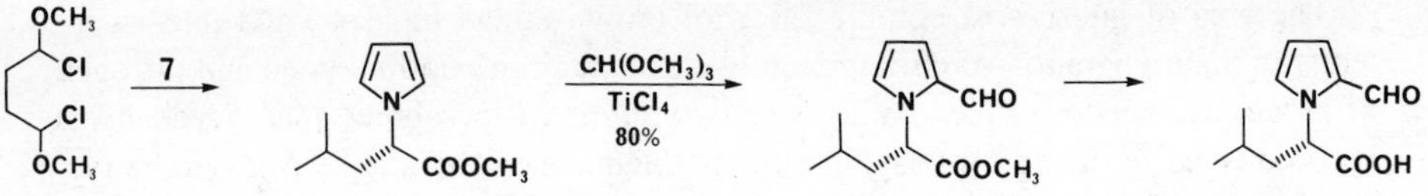

Both the amine and carboxylic acid groups of L-leucine can be manipulated to more suitable functionalities in the quest for synthetically useful chiral intermediates. To convert the acid group into a more reactive species, the amino group must first be protected. One method involves a mild phthaloylation with *N*-carboethoxy phthalimide in the presence of Na_2CO_3 to give *N*-phthaloyl-L-leucine (**24**) in 92% yield.[11] The carboxylic acid can then be converted to an acid chloride (**25**) with thionyl chloride in benzene.

N-Phthalylleucyl chloride (**25**) is a useful intermediate in the synthesis of (3*S*,4*S*)-4-amino-3-hydroxy-6-methylheptanoic acid (statine) (**27**), an amine component of pepstatin. Its basic carbon framework is assembled by reacting **25** with diethyl magnesiomalonate in refluxing ether. The keto group of the resulting diester **26** is then reduced to an alcohol with sodium borohydride, and the product is monodecarboxylated with aqueous HCl. As expected, **27** is isolated as a mixture of (3*R*,3*S*)-diastereomers. These are readily separated by column chromatography on Dowex 50 ion-exchange resin.[12]

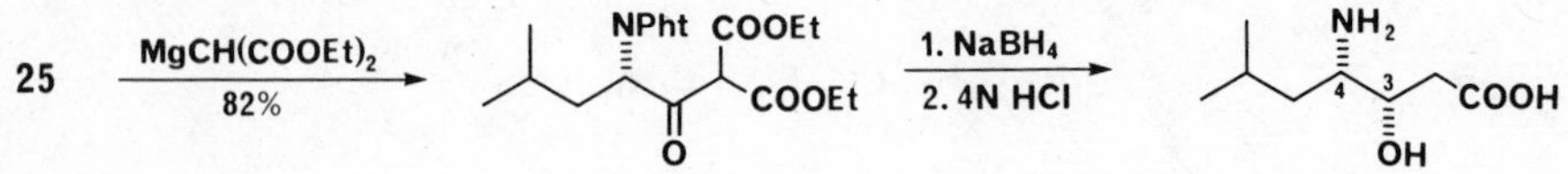

Chiral α-amino aldehydes, derived from amino acids, have been shown in the preceding chapters to be useful synthetic intermediates. The acid group of L-leucine can also be converted to an aldehyde either by partial reduction of Boc-L-leucine methyl ester with diisobutyl aluminum hydride[13] or by the more versatile **28** → **30** transformation.[14] The latter procedure affords Boc–L-leucinal (**30**) in high yield without the need for chromatographic purification, which is known to cause racemization. In this manner, *N*-Boc aldehydes have also been prepared from alanine, phenylalanine, valine, isoleucine, and threonine.

NHBoc COOH — CH_3ONHCH_3, TEA, BOP, 94% → NHBoc CH_3 N OCH_3 O — $LiAlH_4$, 96% → NHBoc CHO

28 29 30

Aldehyde **30** undergoes a Wittig reaction with phosphonium salt **31** to give solely the *trans*-enyne **32.** Treatment of **32** with dicyclohexylborane followed by alkaline hydrogen peroxide efficiently converts the TMS-acetylenic function to the β,γ-unsaturated acid **33.**[15] Removal of the Boc protecting group is effected with 4 *N* HCl and the resulting (5*S*)-amino-7-methyl-3-(*E*)-octenoic acid, when incorporated into certain peptides, exhibits renin inhibitory activity against both hog kidney renin and human amniotic renin.

Br^- $Ph_3P^+CH_2C{\equiv}CTMS$ — 30, *n*-BuLi, 53% → NHBoc TMS — 98% → NHBoc COOH

31 32 33

Recently, a totally asymmetric synthesis of statine (**27**) has been accomplished with high diastereoselectivity (99.9:0.1) by utilizing the highly *erythro*-selective aldol condensation of the boron enolate of a chiral imide with **30** under conditions developed by Evans (see Chapter 3). This single reaction immediately generates the desired 3*S*,4*S* configuration of the functional groups, and, after a series of deprotections, chiral statine is obtained in less than 24% overall yield.[16]

BocHN SCH_3 4 3 N O OH O O — a-d → 27

(a) Raney Ni; (b) NaOEt; (c) saponification; (d) CF_3COOH.

Leucine or its esters is reduced to L-leucinol (**34**) with lithium aluminum hydride,[17] sodium borohydride,[18] or borane–methylsulfide complex.[19] Each of these three methods furnishes optically pure **34** (determined by spectroscopic techniques) even though optical rotations for the product can vary from 50 to 200% (caused by trace impurities).[18]

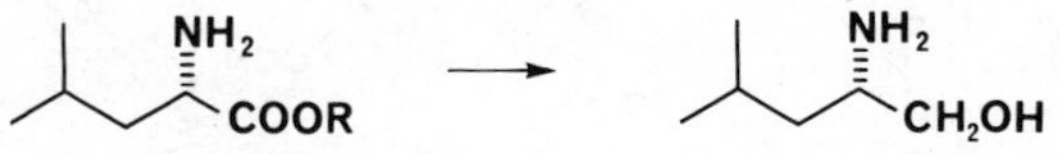

NH_2 COOR → NH_2 CH_2OH

34

L-Leucinol enantioselectively catalyzes the addition of diethylzinc to benzaldehyde.[20] The reaction is performed at room temperature in the presence of 2 mol % of **34,** and after 43 hr, (*R*)-1-phenyl-1-propanol (**36**) is isolated in nearly quantitative yield (48.8% ee).

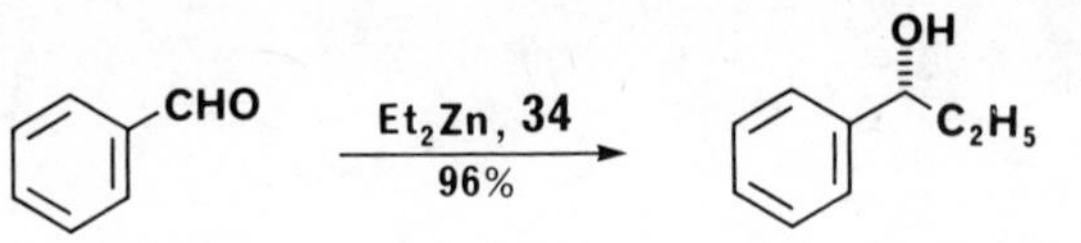

CHO —Et_2Zn, 34 / 96%→ OH C_2H_5

35 **36**

In the presence of nitrous acid, the amino group of L-leucine is subject to hydrolytic deamination, the result of which affords (*S*)-(−)-leucic acid (**37**) with net retention of configuration. Monoprotected glycol (*S*)-**38** is available from **37** by a three-step sequence involving esterification with ethanol and HCl, reaction of the hydroxyl group with dihydropyran, and reduction of the ester with lithium aluminum hydride.

1 —HNO_2 / 57%→ OH COOH —84%→ OTHP CH_2OH

37 **38**

Mori[21,22] employs **38** as a key chiral intermediate in a synthesis of (−)-ipsenol (**39**), which is one of the aggregation pheromones of a bark beetle. It is estimated that the optical purity of **39** is at least 94% (Scheme 4).

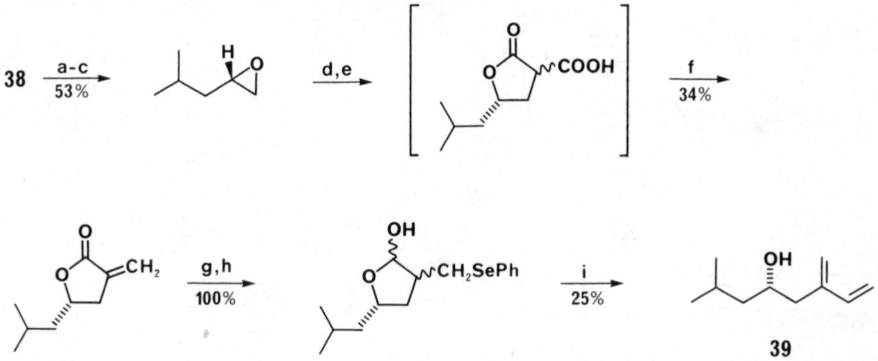

Scheme 4. (a) TsCl, pyridine; (b) HOAc; (c) KOH, H_2O; (d) $CH_2(COOEt)_2$, NaOEt; (e) KOH; (f) CH_2O, Et_2NH, 80–90°C; (g) PhSeNa; (h) DIBAH, −60°C; (i) $Ph_3P{=}CH_2$.

The compound (*R*)-**40**, the enantiomer of (*S*)-**38,** is prepared from D-leucine in the same manner as **38.** By the series of manipulations illustrated in Scheme 5, **40** is transformed to the *cis*-vinyl iodide **41** in 73% overall yield.[23]

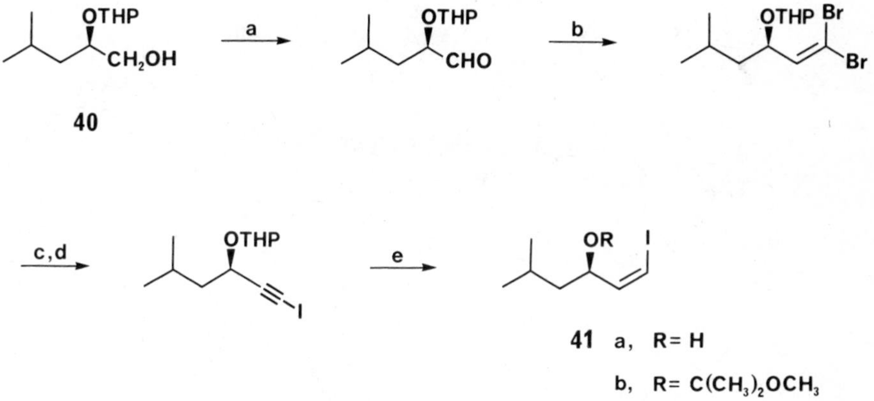

Scheme 5. (a) DMSO, $(COCl)_2$; (b) PPh_3, CBr_4, Zn (0°C); (c) 2 equivalents *n*-BuLi, −78°C; (d) I_2; (e) KOOCN=NCOOK, HOAc (0°C).

Conversion of **41b** to a cuprate[24] [(1) *n*-BuLi, −70°C; (2) CuIP(n-Bu)$_3$] results in the formation of a highly reactive, internally chelated, chirally biased species **44.** This has been used in a stereoselective synthesis of an optically active dihydroindanone **45,** which is a chiral steroid CD-ring synthon. The entire molecule can be formed in essentially one step by a tandem Michael addition of cuprate **44** to 2-methyl-2-cyclopenten-1-one (**43**) and then 2-(trimethylsilyl)-1-buten-3-one (**42**), followed by aldolization during workup. The reaction produces a mixture of diastereomers **45** and **46** in a ratio of 13:1.

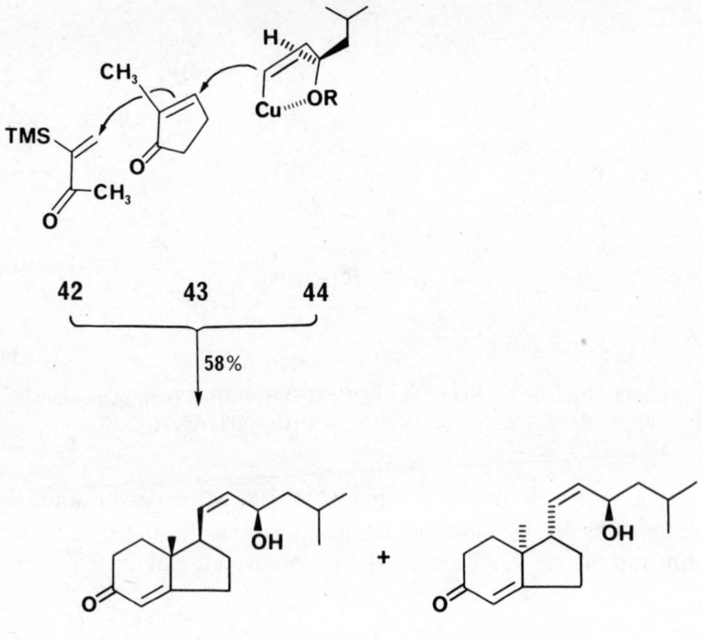

42 43 44

58%

45 + 46

4.2 ISOLEUCINE

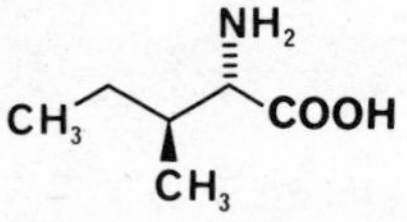

(2*S*,3*S*)-2-Amino-3-methylpentanoic acid (**47**)

Isoleucine is the only member of the leucine family with two asymmetric centers. It is, therefore, capable of transferring both these centers to a new synthetic molecule.

Photochlorination of L-isoleucine (**47**) in concentrated HCl results in the formation of diastereomeric γ-chloro amino acids **48.** Subsequent hydrolysis then produces a diastereomeric mixture of α-amino-γ-lactones **49** and **50** (40% yield), which are isolated as their hydrochloride salts.[25] These lactones are completely separable by thin-layer chromatography (TLC). Compound **50** is the lactone form of L-γ-hydroxyisoleucine, a component of the cyclic octapeptide β-amanitin, one of the poisonous principles of mushrooms of the genus *Amanita.*[26]

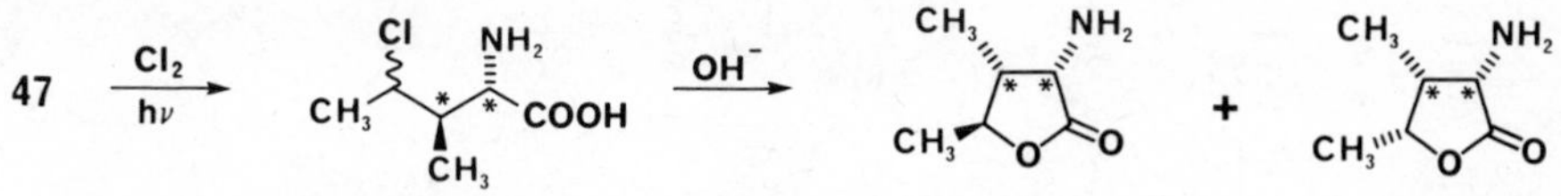

47 $\xrightarrow[h\nu]{Cl_2}$ 48 $\xrightarrow{OH^-}$ 49 + 50

Oxidative decarboxylation of L-isoleucine with silver(II) picolinate affords optically pure (*S*)-2-methylbutanal (**51**) in 71% yield. As shown in Scheme 6, **51** can be converted to either (+)-(*E*)-4,6-dimethyl-4-octen-3-one [(*S*)-manicone, **52**] or (+)-(*E*)-2,4-dimethyl-2-hexenoic acid (**53**). Both of these are mandibular gland secretions of *Manica* and *Componotus* ants.[27] The optical purity of the synthetic (*S*)-manicone (**52**) is estimated at 85% and the (*S*)-acid, at 83%.

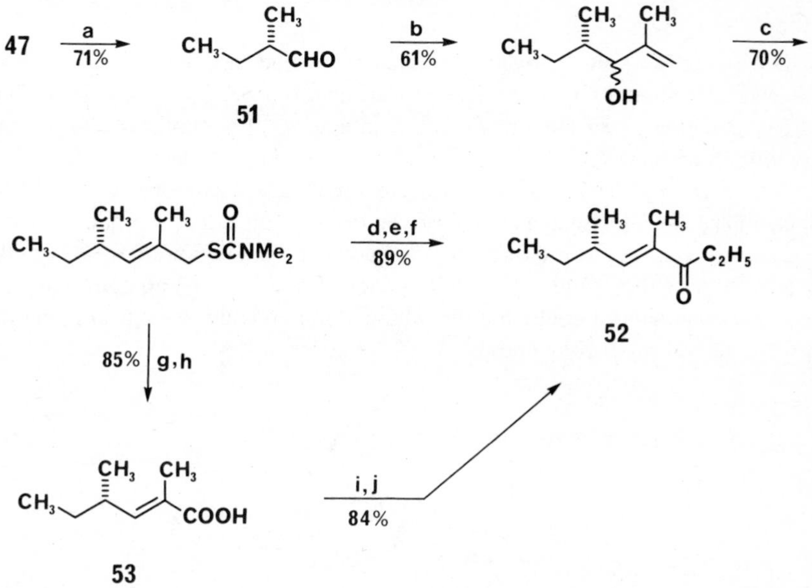

47 $\xrightarrow[71\%]{a}$ 51 $\xrightarrow[61\%]{b}$ $\xrightarrow[70\%]{c}$ $\xrightarrow[89\%]{d,e,f}$ 52

$\xrightarrow[85\%]{g,h}$ 53 $\xrightarrow[84\%]{i,j}$ 52

Scheme 6. (a) $Ag(II)Pic_2$, H_2O; (b) $CH_2{=}C(CH_3)MgBr$; (c) NaH, $ClCSNMe_2$; (d) LDA, CH_3SSCH_3, $-78°C$; (e) LDA, C_2H_5I, $-78°C$; (f) HgO, $BF_3 \cdot Et_2O$; (g) LDA, CH_3SSCH_3 (2 equivalents), $-78°C$; (h) $HgCl_2$, CH_3CN, H_2O; (i) $SOCl_2$; (j) $(C_2H_5)_2CuLi$ (excess), ether, $-78°C$.

Diazotization of an amino acid is a useful method for the conversion of its amine function to other reactive groups. In the presence of water, (*S*)-2-hydroxy acids **54** are produced (Scheme 7). Likewise, in the presence of hydrochloric acid, α-chloroacids **55** are formed. Both these reactions occur with retention of configuration.

Acid **54** is converted in several steps to the (*S*)-alkyloxirane **56,** which is then further used in the synthesis of (2*S*,5*RS*)-chalcogran (**57**), the principal aggregation pheromone of *Pityogenes chalcographus.*[28]

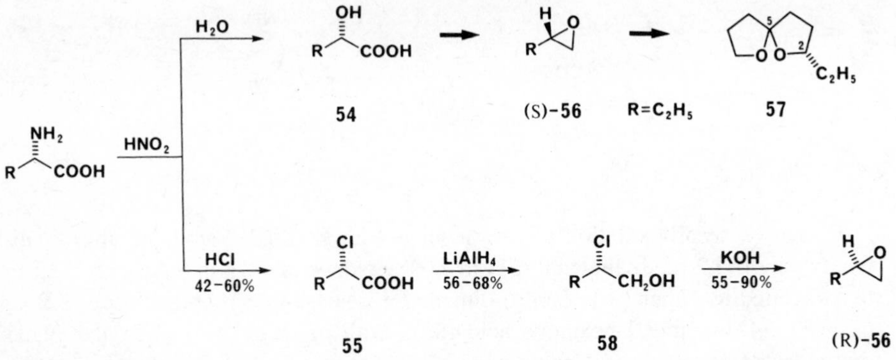

$R = CH_3$, i-C_3H_7, (S)-*sec*-C_4H_9

Scheme 7

(*R*)-Alkyloxiranes **56,** derived from alanine, valine, and isoleucine, are easily produced in a short two-step sequence starting from **55.**[29] A lithium aluminum hydride reduction of **55** (1 : 1 molar ratio) in ether at 0°C furnishes the (*S*)-chlorohydrins **58** within 5 min. If excess of hydride is used in conjunction with extended reaction times, C–Cl hydrogenolysis occurs, giving a dechlorinated (*S*)-alcohol. For example, (*S*)-3-methylpentanol (derived from L-isoleucine) is produced in 57% yield after 3 days. Exposure of **58** to an aqueous 50% solution of KOH results in the formation of the desired (*R*)-alkyloxiranes. The **58** → (*R*)-**56** conversion proceeds with inversion of configuration. These chiral epoxides have also found their way into natural product syntheses.[30,31]

4.3 *tert*-LEUCINE

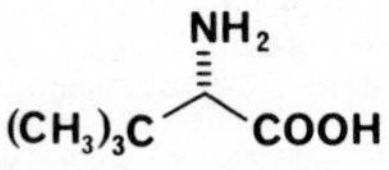

(*S*)-2-Amino-3,3-dimethylbutanoic acid

As a reagent for asymmetric synthesis, *tert*-leucine (**59**) enjoys the massive bulk of a *t*-butyl group adjacent to its chiral center. The steric demands imposed by this group is often superior in asymmetric transformations to other amino acids

For example, in Chapter 3 we described the enantioselective synthesis of (*R*)-α-amino acid esters by alkylation of the lithiated bis-lactim ether of the valine-derived cyclo-(L-Val–Gly). The asymmetric induction of the alkylation step ranged from 60 to >95% depending on the nature of the alkyl halide.

In comparison, if the bis-lactim ether **61** of cyclo-(L-*tert*–Leu–Gly) (**60**) is used in the alkylation, only the (3*S*,6*R*)-diastereomer **62** is formed. On acidic hydrolysis, this translates to an enantiomeric excess of greater than 95% in the resulting (*R*)-α-amino acid methyl esters **63** (Scheme 8).[32]

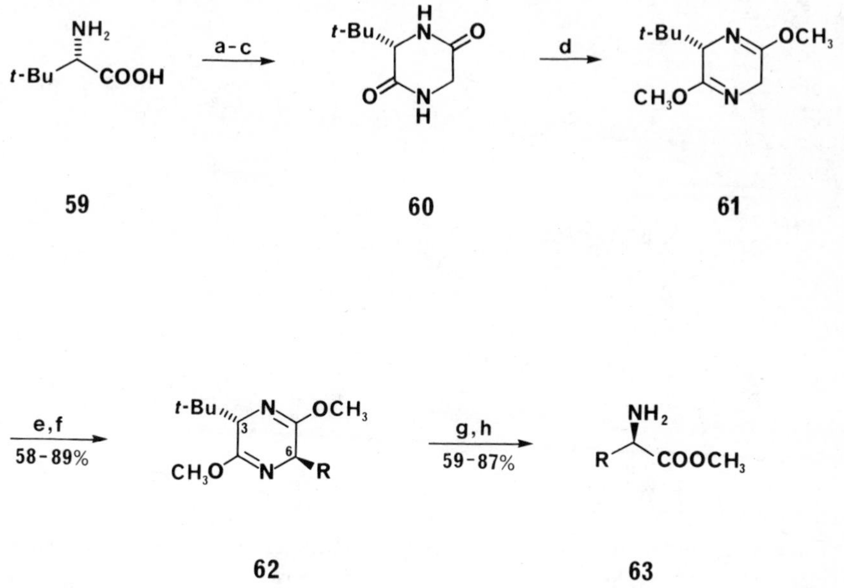

59 **60** **61**

62 **63**

R= CH_2Ph, *n*-C_7H_{15}, allyl, $CH_2C{\equiv}CH$, CH_2COO*t*-Bu

Scheme 8. (a) $COCl_2$ (100%); (b) $H_2NCH_2COOCH_3$ (71%); (c) xylene, reflux; (d) $Me_3O^+BF_4^-$ (94%); (e) *n*-BuLi, −70°C; (f) RBr, −70°C; (g) 0.25 *N* HCl; (h) NH_4OH.

Chiral 2-alkylcycloalkanones have synthetic potential in the construction of biologically active compounds and natural products. Hashimoto and Koga[33,34] have developed a method for synthesizing these intermediates by the use of chiral chelated lithioenamines. The starting imine **65** is easily prepared in 90% yield by refluxing cyclohexanone and *tert*-leucine *t*-butyl ester (**64**) in benzene in a Dean–Stark apparatus. Metallation of **65** in THF with LDA at −78°C produces a yellow solution of the lithioenamine **66** (Scheme 9). Alkylation at this temperature followed by hydrolysis with citric acid furnishes the alkylated cyclohexanones **67** in 56–75% yields (73–98% optical yields). The chiral auxiliary **64** is also recovered without any racemization. Use of valine *t*-butyl ester as the chiral source in this reaction results in lower optical yields of **67**.

Analogous reactions with 2-phenylcycloalkanones afford the 2-methyl-2-phenylcycloalkanones **69** again in high optical yields.

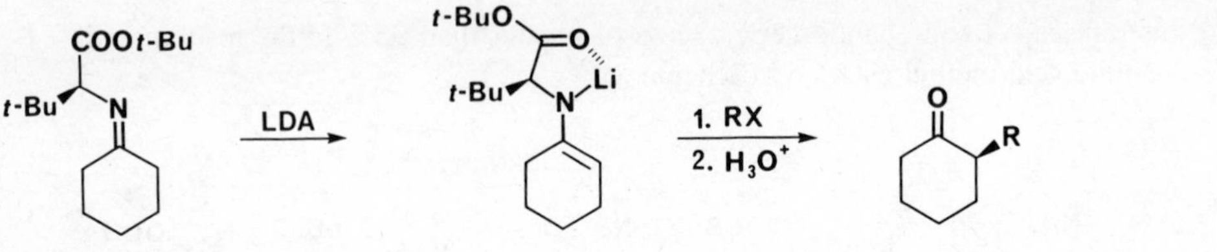

65 $\xrightarrow{\text{LDA}}$ 66 $\xrightarrow{\text{1. RX; 2. } H_3O^+}$ 67 R = CH_3, *n*-Pr, allyl

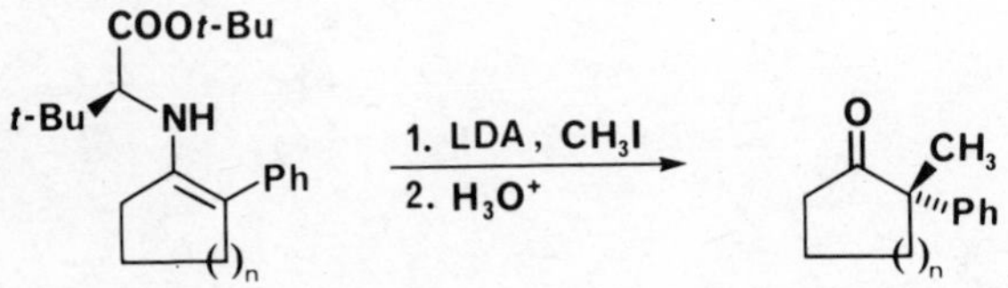

68 $\xrightarrow{\text{1. LDA, } CH_3I\text{; 2. } H_3O^+}$ (S)-69

n = 1 62% (94% opt)
n = 2 40% (96% opt)

Scheme 9

The two examples illustrated in Schemes 8 and 9 clearly demonstrate the superiority of the *t*-butyl group of *tert*-leucine over the isopropyl group of valine in the induction of asymmetry in alkylation reactions.

Two further manipulations of chiral Schiff bases of aldehydes and **64** involve reaction at the imine carbon or, in the case of α,β-unsaturated aldehydes, 1,4-addition reactions.

The first type of transformation, reaction at the imine, has been used in the synthesis of D-phenylglycine (**72**), an important constituent of penicillin and cephalosporin antibiotics. The introduction of the group that will eventually become the acid function of **72** is accomplished by hydrocyanation of **70** in hexane at $-23°C$. Hydrolysis of the nitrile **71** followed by a regiospecific fission of the C–N bond gives **72** in nearly 97% optical purity.[35]

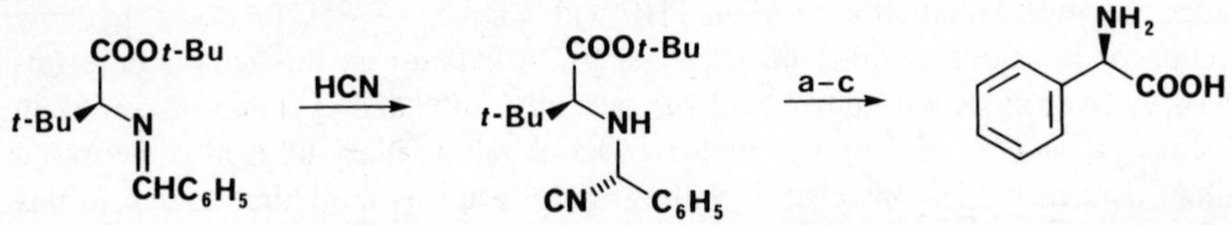

70 $\xrightarrow{\text{HCN}}$ 71 $\xrightarrow{\text{a–c}}$ 72

(a) HCl, dioxane; (b) *t*-BuOCl, Cbz–Cl; (c) 5 *N* HCl.

The second mode of reaction, the conjugate addition to α,β-unsaturated aldimines, is illustrated in Scheme 10. Treatment of the chiral Schiff base **73** of crotonaldehyde with Grignard reagents smoothly affords optically active aldehydes **76** in 40–53% chemical yield and 91–98% optical yield. The high degree of asymmetric induction can be attributed to the formation of a rigid chelated intermediate **74,** which can internally deliver the R group from the less hindered side. Once more, a gain of approximately 30% in optical purity is realized by using *tert*-leucine *t*-butyl ester (**64**) instead of a valine ester as the chiral auxiliary.[36,37]

It is interesting to note that the Michael addition of diethyl malonate to **73** occurs with reversed stereoselectivity to give (*S*)-**77** in 86% optical yield. This reversal occurs as a result of the adoption of an *s-trans* conformer in the transition state as opposed to the *s-cis* conformation in **74.**[38,39]

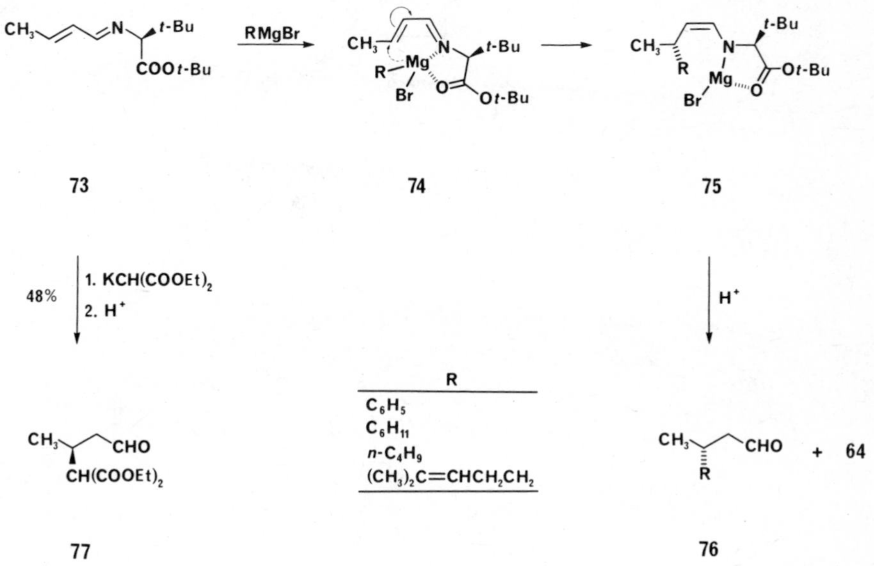

R
C_6H_5
C_6H_{11}
n-C_4H_9
$(CH_3)_2C{=}CHCH_2CH_2$

Scheme 10

By extending the scope of the reaction to include cycloalkenecarboxaldehydes, one can gain access to chiral 2-substituted (**80**) and 1,2-disubstituted cycloalkanecarboxaldehydes (**81**) in 82–93% ee[40,41] (Scheme 11). *In situ* alkylation of magnesioenamine **79** furnishes the *trans*-isomer **81** as a result of the overriding steric effect of the *t*-butyl group in the rigid (*Z*) configuration of **79.** In the absence of the chiral auxiliary, alkylation of **80** occurs from the side opposite the R group to give *cis*-**81.**

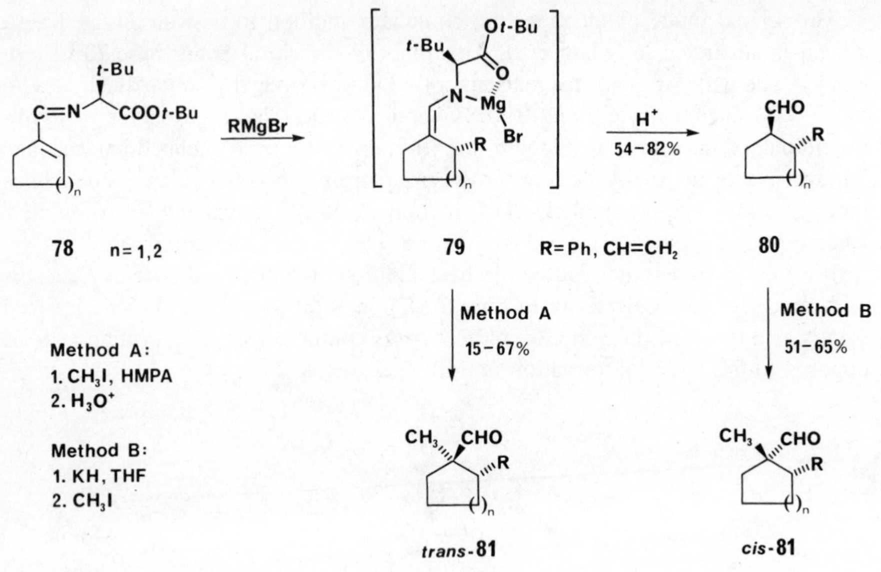

Scheme 11

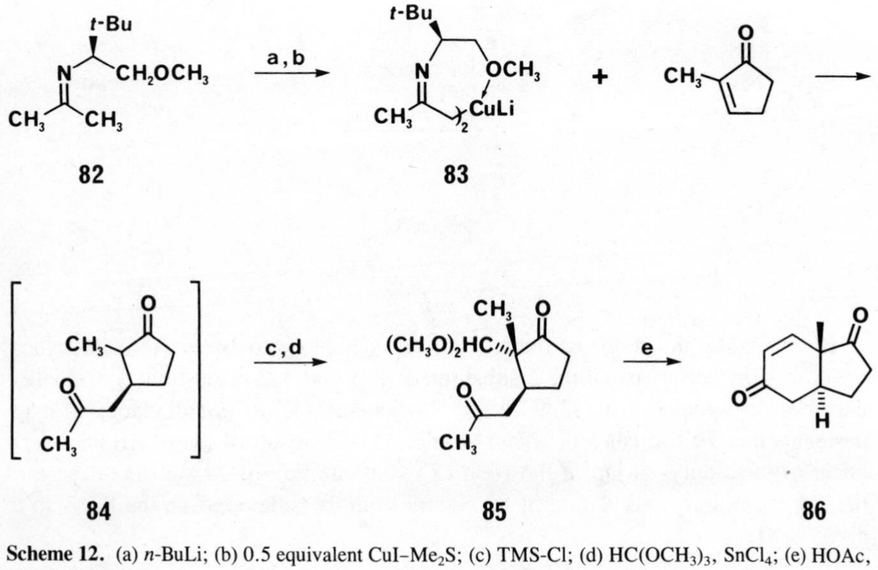

Scheme 12. (a) *n*-BuLi; (b) 0.5 equivalent CuI–Me_2S; (c) TMS-Cl; (d) $HC(OCH_3)_3$, $SnCl_4$; (e) HOAc, $PhCH_2CH_2NH_2$.

tert-Leucine also plays an important role in the construction of chirally biased homocuprates such as **83.** The substrate **82** in this case is formed from (*S*)-*tert*-leucinol methyl ether and acetone. Cuprate **83** is used in the synthesis of the important indandione **86** (Scheme 12).[42] The step that fixes the absolute stereochemistry of the molecule is a conjugate addition of **83** to 2-methylcyclopentenone. The resulting 3-acetonylcyclopentanone **84** is not actually isolated, but is trapped as its enol silyl ether. Introduction of a protected formyl group in the 2 position (directed by the steric influence of the 3-acetonyl group) affords the penultimate intermediate **85,** which furnishes **86** in 24% overall yield (60% ee) on acidic aldolization.

REFERENCES

1. T. Scott and M. Brewer, *Concise Encyclopedia of Biochemistry*, Walter de Gruyter, New York, 1983, p. 233.
2. N. Isumiya, S. J. Fu, S. M. Birnbaum, and J. P. Greenstein, *J. Biol. Chem.*, **205,** 221 (1953).
3. J. L. Bailey, *J. Chem. Soc.*, 3461 (1950).
4. M. Bodanszky, U.S. Patent 3,704,246 (1972); *Chem. Abstr.*, **78,** 58801p (1973).
5. T. Teramoto, T. Kurosaki, and M. Okawara, *Tetrahedron Lett.*, 1523 (1977).
6. U. Schöllkopf, U. Busse, R. Kilger, and P. Lehr, *Synthesis*, 271 (1984).
7. T. Katagi and H. Kuriyama, *Heterocycles*, **19,** 1681 (1982).
8. M. F. Grundon, D. G. McCleery, and J. W. Wilson, *Tetrahedron Lett.*, 295 (1976).
9. R. A. Lloyd, C. W. Miller, D. L. Roberts, J. A. Giles, J. P. Dickerson, N. H. Nelson, C. E. Rix, and P. H. Ayers, *Tob. Sci.*, **20,** 43 (1976).
10. T. H. Chan and S. D. Lee, *J. Org. Chem.*, **48,** 3059 (1983).
11. G. H. L. Nefkens, G. I. Tesser, and R. J. F. Nivard, *Rec. Trav. Chim. Pays-Bas*, **79,** 688 (1960).
12. H. Morishima, T. Takita, and H. Umezawa, *J. Antibiot.*, **26,** 115 (1973).
13. D. H. Rich, E. T. Sun, and A. S. Boparai, *J. Org. Chem.*, **43,** 3624 (1978).
14. J. Fehrentz and B. Castro, *Synthesis*, 676 (1983).
15. R. L. Johnson, *J. Med. Chem.*, **27,** 1351 (1984).
16. P. W. K. Woo, *Tetrahedron Lett.*, 2973 (1985).
17. P. Karrer, P. Portmann, and M. Suter, *Helv. Chim. Acta.*, **31,** 1617 (1948).
18. G. S. Poindexter and A. I. Meyers, *Tetrahedron Lett.*, 3527 (1977).
19. C. F. Lane, U.S. Patent 3,935,280 (1976); *Chem. Abstr.*, **84,** 135101p (1976).
20. N. Oguni and T. Omi, *Tetrahedron Lett.*, 2823 (1984).
21. K. Mori, *Tetrahedron Lett.*, 2187 (1975).
22. K. Mori, *Tetrahedron*, **32,** 1101 (1976).
23. T. Takahashi, H. Okumoto, J. Tsuji, and N. Harada, *J. Org. Chem.*, **49,** 948 (1984).
24. T. Takahashi, Y. Naito, and J. Tsuji, *J. Am. Chem. Soc.*, **103,** 5261 (1981).
25. H. Faulstich, J. Dölling, K. Michl, and T. Wieland, *Liebigs Ann. Chem.*, 560 (1973).

26. T. Wieland, *Science*, **159,** 946 (1968).
27. T. Mimura, Y. Kimura, and T. Nakai, *Chem. Lett.*, 1361 (1979) and references cited therein.
28. K. Mori, M. Sasaki, S. Tamada, T. Suguro, and S. Masuda, *Tetrahedron*, **35,** 1601 (1979).
29. B. Koppenhoefer, R. Weber, and V. Schurig, *Synthesis*, 316 (1982).
30. B. D. Johnston and K. N. Slessor, *Can. J. Chem.*, **57,** 233 (1979).
31. B. Seuring and D. Seebach, *Helv. Chim. Acta*, **60,** 1175 (1977).
32. U. Schöllkopf and H. Neubauer, *Synthesis*, 861 (1982).
33. S. Hashimoto and K. Koga, *Tetrahedron Lett.*, 573 (1978).
34. S. Hashimoto and K. Koga, *Chem. Pharm. Bull.*, **27,** 2760 (1979).
35. S. Yamada and S. Hashimoto, *Chem. Lett.*, 921 (1976).
36. S. Hashimoto, S. Yamada, and K. Koga, *Chem. Pharm. Bull.*, **27,** 771 (1979).
37. S. Hashimoto, S. Yamada, and K. Koga, *J. Am. Chem. Soc.*, **98,** 7450 (1976).
38. S. Hashimoto, N. Komeshima, S. Yamada, and K. Koga, *Chem. Pharm. Bull.*, **27,** 2437 (1979).
39. S. Hashimoto, N. Komeshima, S. Yamada, and K. Koga, *Tetrahedron Lett.*, 2907 (1977).
40. S. Hashimoto, H. Kogen, K. Tomioka, and K. Koga, *Tetrahedron Lett.*, 3009 (1979).
41. H. Kogen, K. Tomioka, S. Hashimoto, and K. Koga, *Tetrahedron Lett.*, 4005 (1980).
42. K. Yamamoto, M. Iijima, Y. Ogimura, and J. Tsuji, *Tetrahedron Lett.*, 2813 (1984).

CHAPTER FIVE

HYDROXY AMINO ACIDS

The hydroxy group is probably one of the most biologically important functionalities in nature. It is, therefore, not surprising to find several naturally occurring amino acids with an OH substituent attached to the group in the α position. Tyrosine, which was discussed in Chapter 2, contains a phenolic hydroxyl that can be exploited to manipulate its aromatic ring.

Serine and threonine, also members of the common α-amino acids, are alkyl amino acids that possess a β-hydroxyl moiety as their common feature. This hydroxy gives the chemist an additional handle to work with when mapping a synthetic course toward a target molecule containing one or more asymmetric centers.

5.1 SERINE

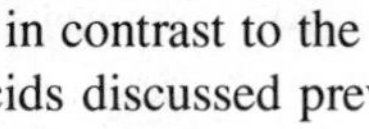

NH_2
HO COOH
(*S*)-2-Amino-3-hydroxypropanoic acid

L-Serine (**1**), the simplest β-hydroxy amino acid, is proteogenic and glucogenic and is a major component of silk fibroin.[1] On reaction of **1** with phosgene, carbonyl insertion occurs between its amine and hydroxyl functionalities to form the (*S*)-4-carboxy-2-oxazolone **2**.[2] This is in contrast to the formation of an *N*-carboxyanhydride as observed with amino acids discussed previously.

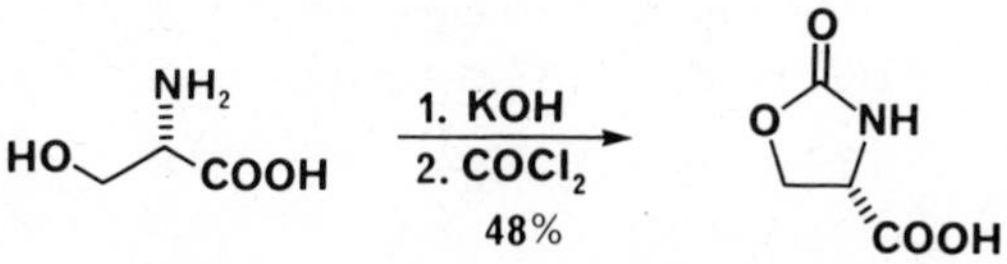

NH_2 HO COOH **1** → 1. KOH, 2. $COCl_2$, 48% → O, O, NH, COOH **2**

Methyl esters of L-serine (**3a**) or L-threonine (**3b**) react with *t*-butyl isocyanide in the presence of 10 mol % of $PdCl_2$ to give optically active 2-oxazolines **4** in high yields. Only the isonitrile carbon atom is incorporated in the heterocycle. The remainder of the reagent is lost as *t*-butylamine.[3]

NH_2 HO COOCH$_3$ R —— *t*-BuN=C, $PdCl_2$ ——→ O N R $COOCH_3$

3a R=H **4a** 91%

3b R=CH_3 **4b** 98%

When heated with pivaldehyde–triethylamine under azeotropic conditions, **3a** is converted to a mixture of diastereomeric oxazolidines **5a** and **6a** in a ratio of approximately 1:1. Formylation of this mixture with the mixed anhydride of acetic and formic acids gives a 95:5 mixture of the *N*-formyloxazolidines **5b** and **6b.** The major diastereomer (2*R*,4*S*)-2-*t*-butyl-4-carbomethoxy-3-formyloxazolidine (**5b**) is isolated by crystallization in 68% yield.[4]

3a —— *t*-BuCHO, TEA ——→ *t*-Bu O NR $COOCH_3$ + *t*-Bu O NR $COOCH_3$

5 a, R=H **6**

b, R=CHO

The lithium enolate of **5b** is easily generated with LDA in THF at −78°C. Reactions with electrophiles (alkyl halides, acetone, benzaldehyde) occur preferentially from the *re* face of the donor center with greater than 95% diastereoselectivity (1,3-asymmetric induction). In the case of alkyl halides (e.g., methyl iodide), yields are rather mediocre (46%) in the absence of a cosolvent; however, the addition of HMPA or DMPU raises the yield to 68%. Hydrolysis of the oxazolidine **7** with refluxing 6 *N* HCl produces (*S*)-α-methylserine, a component of the antibiotic amicetine.[5]

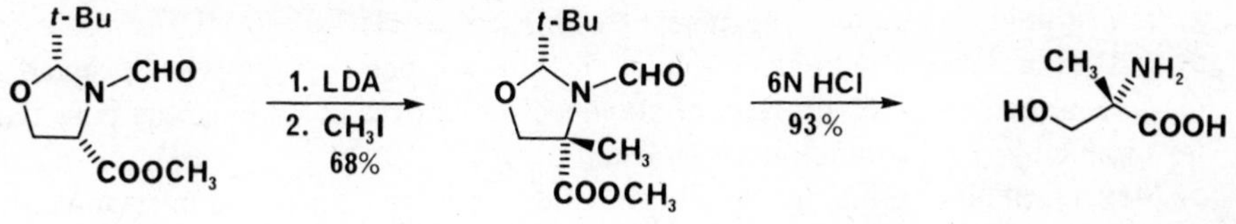

t-Bu; CHO; COOCH$_3$ — 1. LDA, 2. CH$_3$I, 68% → t-Bu; CHO; CH$_3$; COOCH$_3$ — 6N HCl, 93% → CH$_3$, NH$_2$, HO, COOH

5b 7 8

Similar chemistry can be performed on the 1,3-dioxolane nucleus. The requisite heterocycle **10** is prepared from L-serine by diazotization to (*S*)-glyceric acid[6] (**9**) followed by acetalization with pivaldehyde dimethylacetal. In this case, the crystalline *trans*-isomer of (2*R*,4*S*) configuration predominates in the mixture (*cis* : *trans* ratio 1 : 3). Although the enolate of **10** can be generated, alkylation cannot be effected under any conditions. Consequently, this is circumvented by converting **10** to either *t*-butylester **11** or *t*-butylthioester **12.** The thioester **12** is preferred because of its higher acidity.

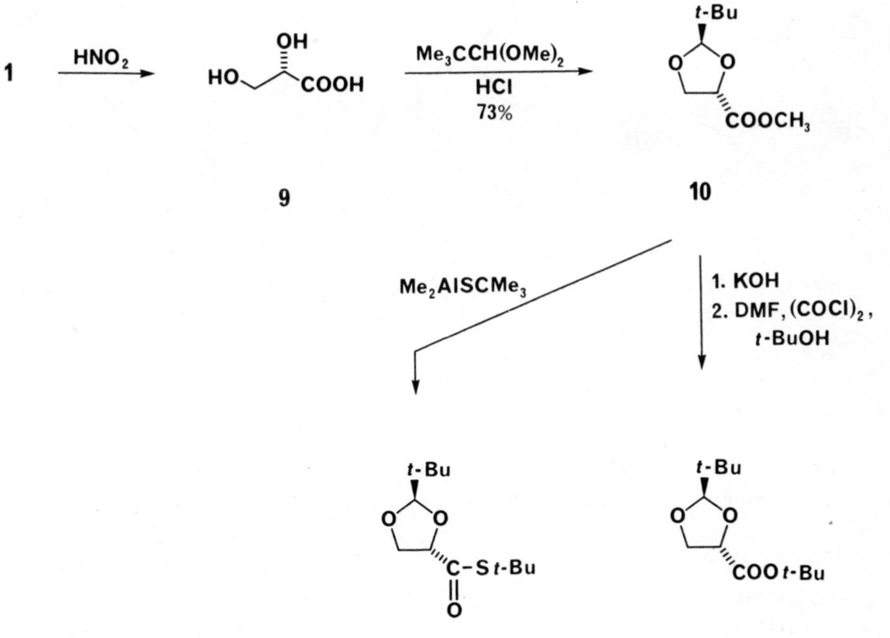

1 — HNO$_2$ → HO, OH, COOH (9) — Me$_3$CCH(OMe)$_2$, HCl, 73% → t-Bu, COOCH$_3$ (10)

10 — Me$_2$AlSCMe$_3$ → 12 (t-Bu; C(=O)–S t-Bu)

10 — 1. KOH, 2. DMF, (COCl)$_2$, t-BuOH → 11 (t-Bu; COO t-Bu)

12 11

The lithium enolate of **12** (generated with LDA in THF at −75°C) is stable up to −20°C and reacts with electrophiles (*cis* to the *t*-butyl group) with acceptable diastereoselectivites.[7] Hydrolysis of alkylated derivatives of **12** have the potential for producing α-substituted glyceric acids.

Meyers, in his continuing study of the utility of chiral oxazolines in asymmetric synthesis, uses the ethyl ester of L-serine (**13**) to construct the heterocycle **14** by treatment with ethyl acetimidate. After conversion to the desired derivative **16,** a 1,4-addition of *n*-butyllithium to the styryl group followed by hydrolytic unmasking of the carboxyl function furnishes (*S*)-(+)-3-phenylheptanoic acid (**17**) in >90% ee[8] (Scheme 1).

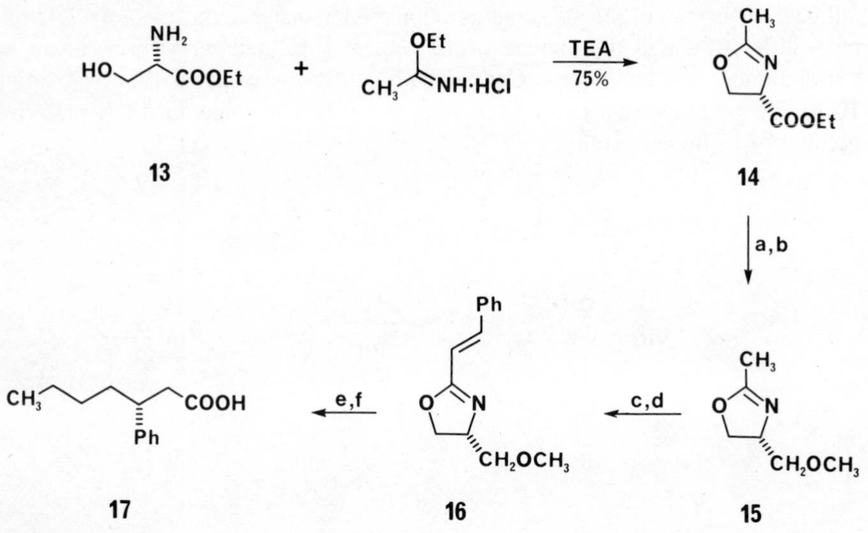

Scheme 1. (a) bis-(2-methoxyethoxy)aluminum hydride (73%); (b) NaH, CH_3I (68%); (c) *n*-BuLi, PhCHO, −78°C (97%); (d) CF_3COOH, 80°C (51%); (e) *n*-BuLi, −78°C (92%); (f) 3M H_2SO_4 (52%).

The technique of using imidates to construct chiral oxazolines is also used by Mori and Funaki[9] in their synthesis of the fruiting-inducing cerebroside in a basidiomycete *Schizophyllum commune*. They cleverly employ **3a** as the sole source of chirality in the synthesis of their target molecule. The optically active sphingadienine portion **21** is assembled (Scheme 2) beginning with the reaction of **3a** with ethyl phenylimidate. The resulting oxazoline **18** is reduced with diisobutylaluminum hydride to give a rather unstable aldehyde **19,** which, when reacted immediately with alkenylalane **20,** gives the product as an almost statistical mixture of diastereomers. These isomers are easily separated by column chromatography, and the desired crystalline diastereomer **21** is isolated in 23% yield. Other ceramide and cerebroside derivatives are also available with the use of this synthetic scheme, simply by varying the alane that reacts with **19.**[10]

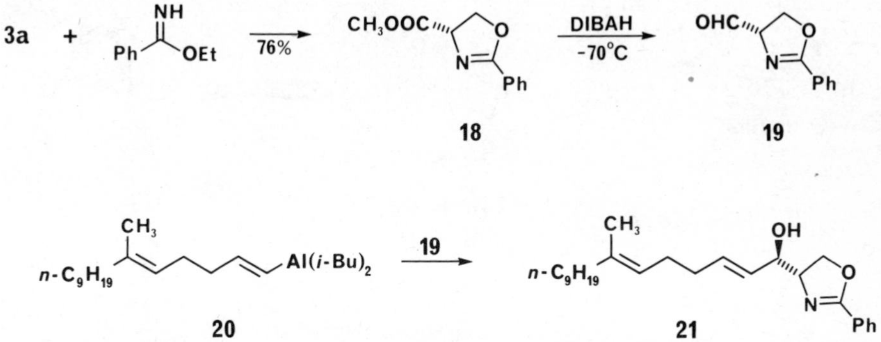

Scheme 2

The ester function of **18** can be transformed not only into an aldehyde carbonyl (**19**), but also a ketone moiety. Low-temperature addition of a lithiodithiane (or even *n*-butyllithium) to the ester of **18** results in the formation of ketones **22** whose optical purity exceeds 95%. The net result of this manipulation is the conversion of serine to a protected keto amino alcohol. This reaction cannot be performed directly on Cbz–*O*-benzylserine methyl ester because of preferential β-elimination of benzyl alcohol.[11] The ability to form these chiral ketones may have synthetic implications in the biomimetic synthesis of the antitumor antibiotic mitomycin C.

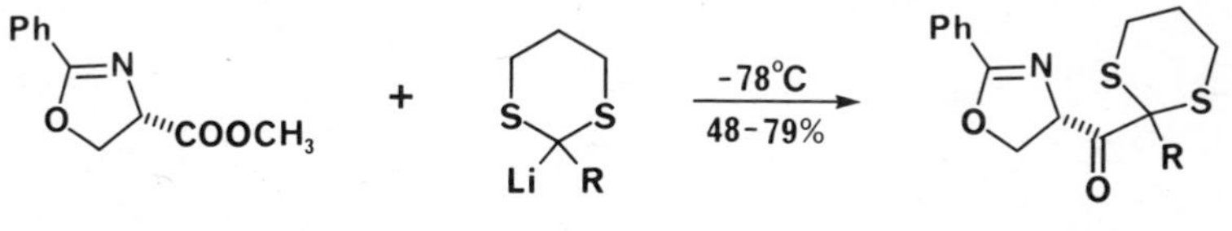

R= H, $CH_2CH_2OCH_2Ph$, $CH_2CH(OEt)_2$

threo-β-Hydroxyl-L-glutamic acid (**27**), a component of an antibiotic peptide, is a unique amino acid that has been used synthetically in the preparation of L-tricholomic acid.[12] Its synthesis, outlined in Scheme 3, begins with the condensation of Boc–D-serine methyl ester (**23**) with DMP. The resulting oxazolidine ester **24a,** when reduced with DIBAH, produces aldehyde **24b** in high yield. This aldehyde is configurationly stable even on exposure to silica gel. A zinc chloride-mediated cyclocondensation of Danishefsky's diene to the penaldic acid equivalent **24b** produces the predominantly *threo*-pyranone (5βH) (**25**). Functional group manipulation leads to the product **27** which is isolated as its hydrochloride salt.[13]

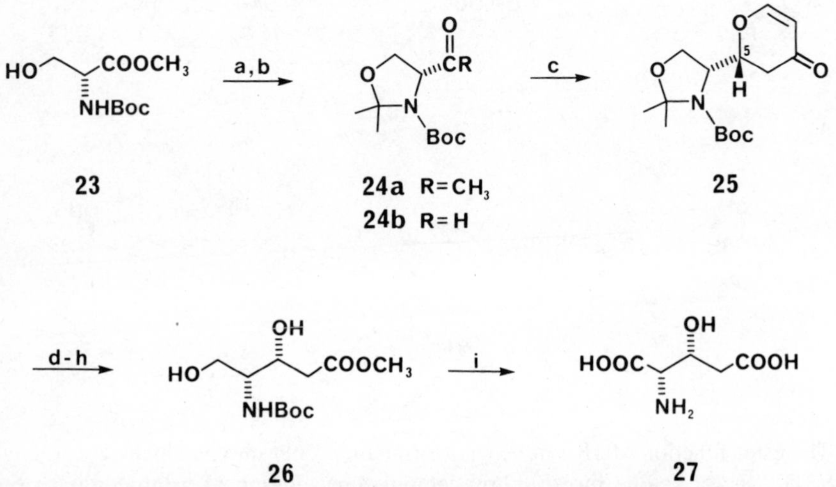

Scheme 3. (a) 2,2-Dimethoxypropane, PTSA (60%); (b) DIBAH, −78°C (80%); (c) Danishefsky's diene, $ZnCl_2/CH_2Cl_2$ (70%); (d) $NaIO_4$ + catalytic RuO_4; (e) NaOH; (f) CH_2N_2; (g) Et_2NTMS, flash chromatography; (h) CH_3OH, PTSA; (i) $KMnO_4$, NaOH.

An interesting chiral aldehyde that has found prominence in the asymmetric synthesis of natural products is (2*S*)-3-acetoxy-2-phthalimidopropanal (**30**). Its preparation is easily accomplished from L-serine according to the four-step process outlined in Scheme 4.[14,15] It is relatively unstable and should be used immediately following its isolation.[16]

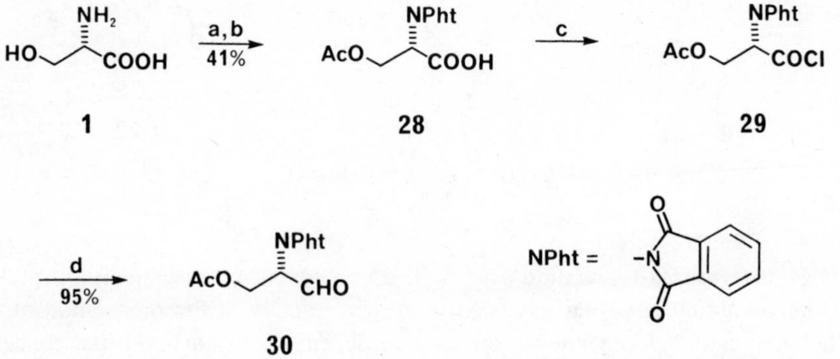

Scheme 4. (a) PhtNCOOEt, Na_2CO_3; (b) Ac_2O; (c) $SOCl_2$, benzene; (d) H_2, benzene, 40°C.

Addition of organometallic reagents to the aldehyde group of **30** is *erythro*-selective, and it is this stereochemistry that is present in the family of sphingolipid bases. D-*erythro*-Sphingosine (**32a**), the most widely occurring member of this class of compounds, is obtained by treating **30** with *trans*-pentadecenyldiisobutylalane at 5–10°C. The intermediate *O*-acetyl-*N*-phthaloylsphinogosine (**31a**) is produced in approximately 15% yield as a mixture of isomers (*erythro*: *threo* ratio 79:21). The pure *erythro*-isomer **31a** is readily separated by partition chromatography. Removal of the protecting groups furnishes the natural product in 96% yield.[14]

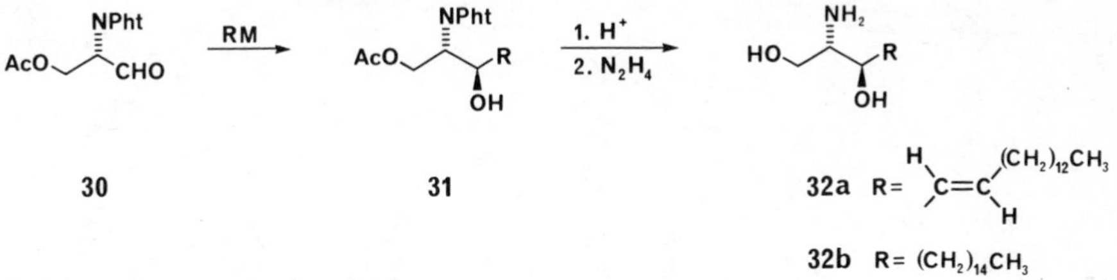

32a R = $H(C=C)(CH_2)_{12}CH_3$ (trans, H)

32b R = $(CH_2)_{14}CH_3$

Dihydrosphinogosine (**32b**) is similarly prepared by treating **30** with pentadecylmagnesium bromide in ether between −88 and −80°C. Chromatographic separation of the major *erythro*-isomer followed by hydrazinolysis of the phthaloyl group affords the desired product **32b** in roughly the same yield as **32a.**[17]

Since the minor *threo*-isomer, D-*threo*-sphinganine (**36**), is isolated only in less than 1% yield from the **30** → **32b** transformation, this method is unacceptable if research quantities of the material are needed. Interestingly, **36** can be synthesized stereospecifically also from L-serine by a modification in the route (see Scheme 5). In this instance the chiral building block, acid chloride **29,** is converted to diazo ketone **33,** which is further elaborated to the 3-dehydrosphinganine derivative **34** (22% overall yield from acid **28**). A stereospecific reduction of the keto group of **34** with tri-*t*-butoxyaluminum hydride produces the desired *threo* configuration in alcohol **35** (≈100% yield). Acidic methanolysis of the acetoxy group, followed by hydrazinolysis of the phthaloyl moiety, gives D-*threo*-sphinganine **36.**[18] Removal of the protecting groups at an earlier stage (e.g., **34**) usually results in racemization.[19]

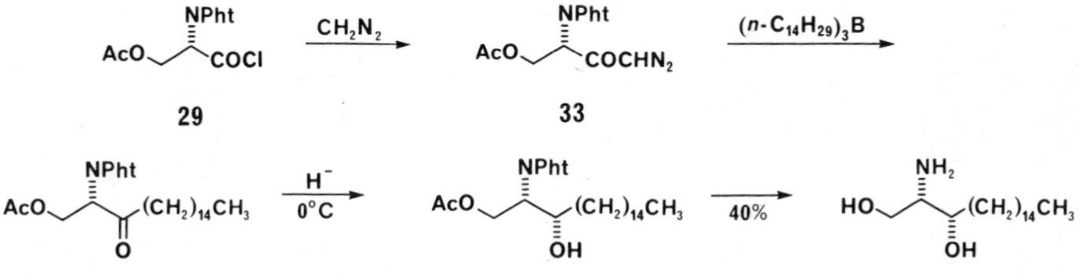

Scheme 5

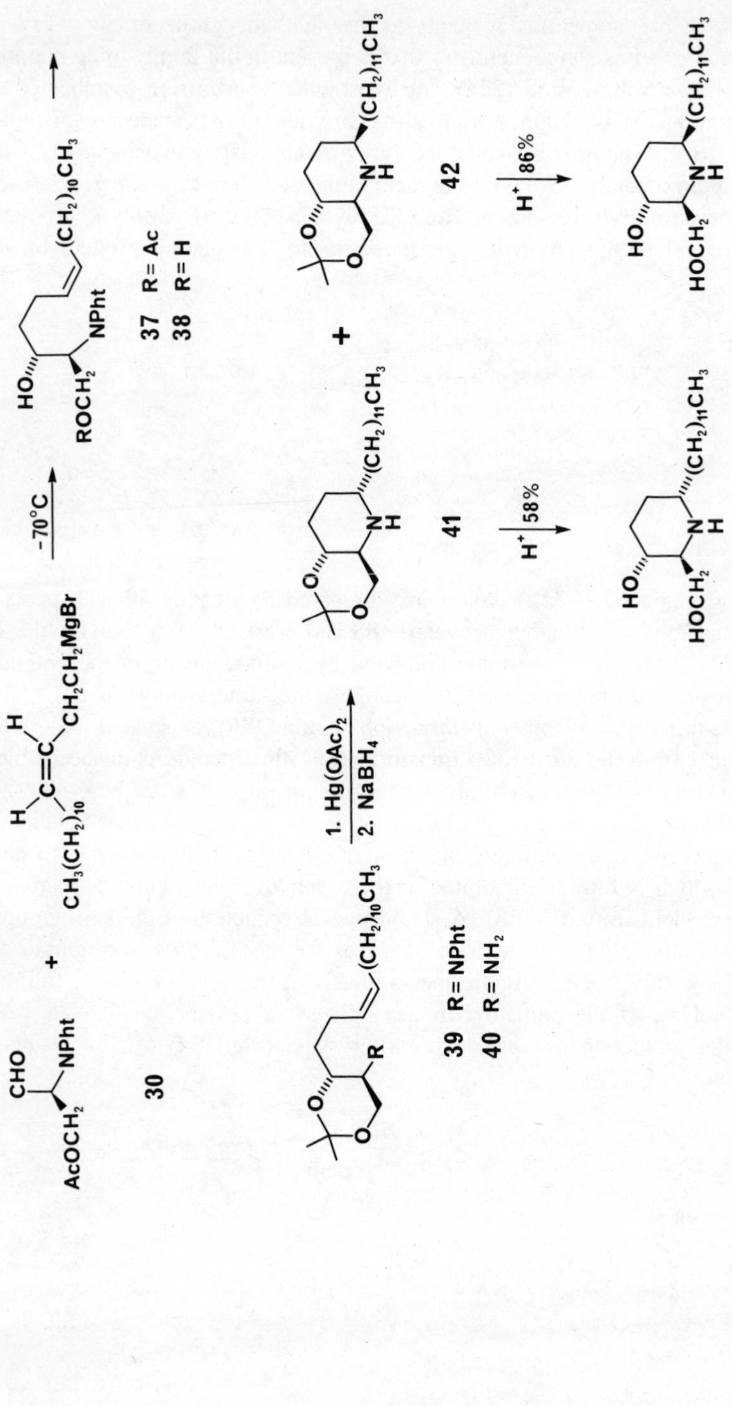

AcOCH$_2$ CHO NPht **30** + CH$_3$(CH$_2$)$_{10}$(H)C=C(H)CH$_2$CH$_2$MgBr $\xrightarrow{-70^\circ C}$ HO ROCH$_2$ NPht (CH$_2$)$_{10}$CH$_3$ $\longrightarrow$

37 R = Ac
38 R = H

39 R = NPht
40 R = NH$_2$

1. Hg(OAc)$_2$
2. NaBH$_4$

(CH$_2$)$_{10}$CH$_3$ → **41** N H (CH$_2$)$_{11}$CH$_3$ + **42** N H (CH$_2$)$_{11}$CH$_3$

41 $\xrightarrow{H^+,\ 58\%}$ **43** HO HOCH$_2$ N H (CH$_2$)$_{11}$CH$_3$

42 $\xrightarrow{H^+,\ 86\%}$ **44** HO HOCH$_2$ N H (CH$_2$)$_{11}$CH$_3$

Scheme 6

Two members of the Prosopis alkaloids, whose common feature is a chiral 2,3,6-trisubstituted piperidine heterocycle, have succumbed to total synthesis with the use of L-serine *via* aldehyde **30** (Scheme 6). The synthetic sequence begins in much the same manner as previously described for the sphingolipid bases.

A stereoselective Grignard reaction of **30** with (Z)-3-pentadecenylmagnesium bromide produces an inseparable mixture of *erythro*- (**37**) and *threo*- (not shown) diol monoacetates (32% total yield). Separation is achieved on acid hydrolysis by column chromatography to give *erythro*-diol **38** (17.6% yield from **30**) and *threo*-diol (2.6% yield from **30**). Isomer **38** is converted to acetonide **39** with DMP in the presence of PTSA (97% yield), and the phthaloyl group is then quantitatively removed with hydrazine.

The resulting amine **40,** when treated with mercuric acetate, undergoes an intramolecular aminomercuration cyclization to form a mixture of two diastereomeric piperidine acetonides **41** and **42** in 3.3 and 76% yields, respectively. After chromatographic separation, acid hydrolysis affords (−)-deoxoprosopinine (**43**) and (−)-deoxoprosophylline (**44**) in overall yields of 0.12 and 4.3%, respectively, from L-serine.[16,20]

Since serine contains three different functional groups, differentiated protection is necessary for successful manipulation of any one of these groups in the presence of the others. An interesting example of this is the conversion of L-serine to amino diol **45** (Scheme 7), which is used in the synthesis of enzyme-inhibitory phospholipid analogs.[21]

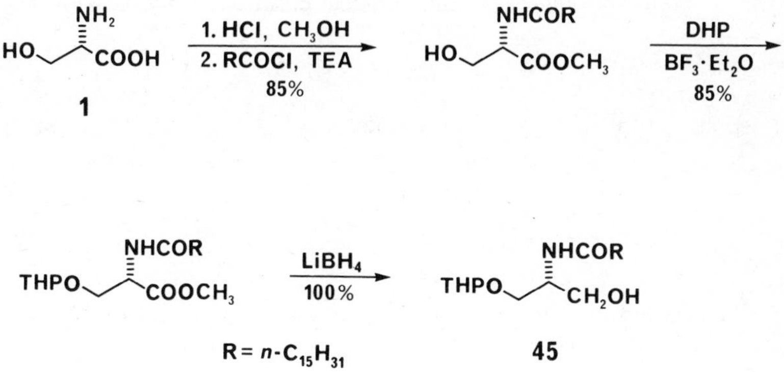

NH_2
HO COOH
1
1. HCl, CH_3OH
2. RCOCl, TEA
85%
NHCOR
HO $COOCH_3$
DHP
$BF_3\cdot Et_2O$
85%

NHCOR
THPO $COOCH_3$
$LiBH_4$
100%
NHCOR
THPO CH_2OH
45
R = n-$C_{15}H_{31}$

Scheme 7

The hydroxyl group of L-serine can be utilized to transform this common amino acid into uncommon L- or D-amino acids. Thus Boc–L-serine methyl ester (**46**) is easily converted to its tosylate **47** under standard conditions.[22] Displacement of the tosyl group by an organocuprate affords the L-amino acid esters **48** with high optical purity. It is imperitive to perform the reaction in diethyl ether in order to minimize β-elimination to a dehydroalanine derivative.[23]

HO–CH₂–CH(NHBoc)–COOCH₃ (46) → TsCl, pyridine → TsO–CH₂–CH(NHBoc)–COOCH₃ (47) → R_2CuLi, -60°C → R–CH₂–CH(NHBoc)–COOCH₃ (48)

R	Yield **48** (%)	Percent ee
CH_3	42	>95
n-C_3H_7	72	85
n-C_4H_9	75	88

D-Amino acids are usually rare and generally considerably more expensive than their L-antipodes. Rapoport and co-workers[24] have devised an ingenious method of converting inexpensive L-serine into a variety of D-amino acids. *N*-(Phenylsulfonyl)-L-serine (**49**) is obtainable from **1** with sodium carbonate and phenylsulfonyl chloride (70% yield). The key carbon–carbon bond forming step is accomplished by an aminoacylation of an organolithium or Grignard reagent with **49.** The resulting α-amino ketones **50** are optically pure with complete retention of configuration at the chiral center.

The keto group of **50** is reduced to **51** by a two-step process that involves thioketal formation and subsequent desulfurization with Raney nickel. Oxidation of the primary alcohol of **50** is accomplished catalytically with oxygen in the presence of platinum. Deblocking to the free amino acid is performed with either 48% HBr or sodium in liquid ammonia. Optical purities are in excess of 99%.

HO–CH₂–CH(NHSO₂Ph)–COOH (49) → RM, 55–83% → HO–CH₂–CH(NHSO₂Ph)–C(=O)R (50) [OXIDIZE; REDUCE] → 55–73% → HO–C(=O)–CH(NHSO₂Ph)–CH₂R (D-51)

M = Li or MgBr R = CH_3; n-C_3H_7; $(CH_2)_4OTMS$; 3,4-dimethoxyphenyl

Enterobactin (**60**), a cyclic serine trimer, is an iron-binding ionophore of enteric bacteria such as *Escherichia coli.* Because it is a trimer of serine, its synthesis is accomplished by a series of coupling reactions on differentially protected and activated L-serine derivatives (Scheme 8).

Commercially available Cbz–L-serine is converted to the *p*-bromophenacyl ester **52** in 98% yield by treatment with *p*-bromophenacyl bromide and potassium bicarbonate. A THP protecting group is introduced by reacting **52** with dihydropyran in the presence of a catalytic amount of PTSA (yield **53** = 93%). Free acid **54** is then obtained by reductive cleavage of the phenacyl group.

Activation of the acid group of **54** is effected by forming the imidazolyl thioester **55,** which is simultaneously coupled with another molecule of **52** to give a diserine. After reductive removal of the phenacyl group (leading to **56**), the process is repeated, giving the penultimate triserine derivative **57.** Cyclization, followed by deprotection and then acylation with 2,3-dihydroxybenzoyl chloride, furnishes the natural product **60.**[25] Its antipode, enantioenterobactin, is synthesized conceptually in the same fashion from D-serine.[26]

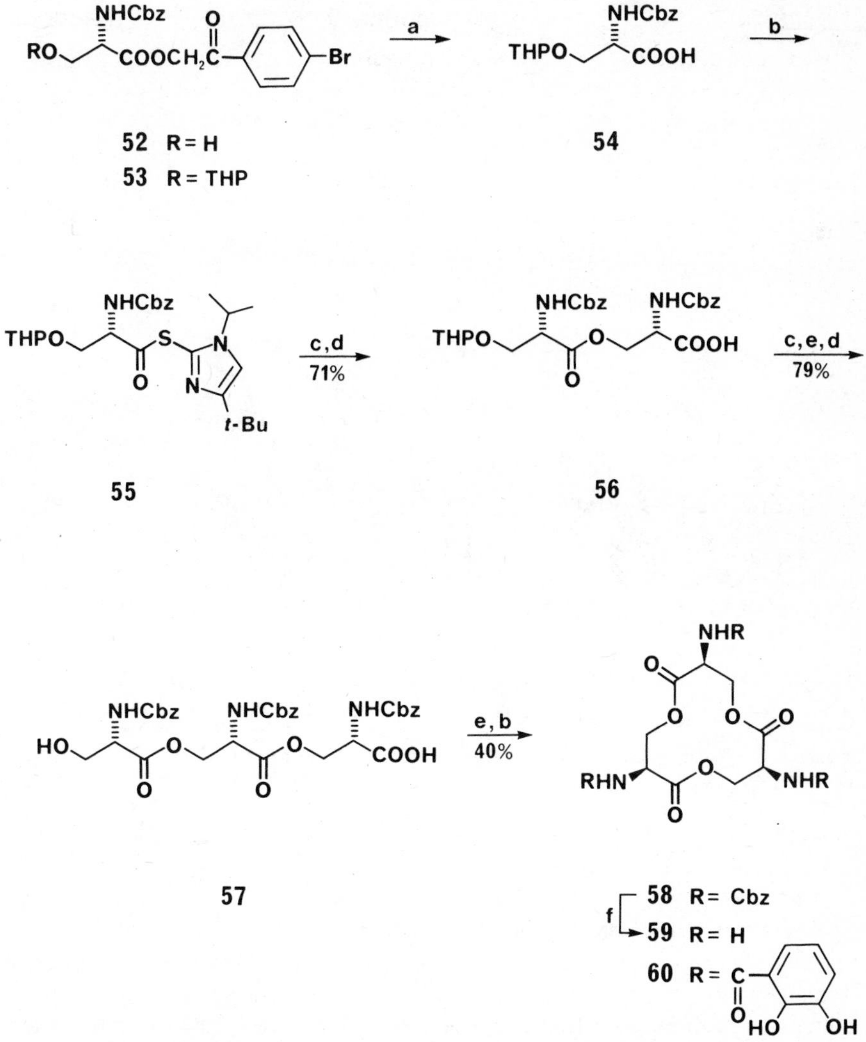

52 R = H
53 R = THP
54
55
56
57
58 R = Cbz
59 R = H
60 R = C(=O)-C₆H₃(OH)₂

Scheme 8. (a) Zn, HOAc (88%); (b) 2,2′-(4-*t*-butyl-1-isopropylimidazolyl)disulfide, Ph_3P; (c) **52;** (d) Zn, HOAc; (e) HOAc–MeOH–THF (4 : 1 : 1) (90%); (f) H_2, Pd/C, 4 equivalents HCl.

The chiral center of serine can sometimes be introduced into a molecule in a less obvious manner. Take, for example, the natural product (*S*)-(−)-wybutine (**66**); its asymmetric side chain is attached to the heterocycle by means of a Wittig reaction of aldehyde **65** with phosphonium salt **64** (obtained from L-serine benzyl ester in four steps; see Scheme 9). Reduction of the olefinic linkage furnishes the saturated side chain.[27]

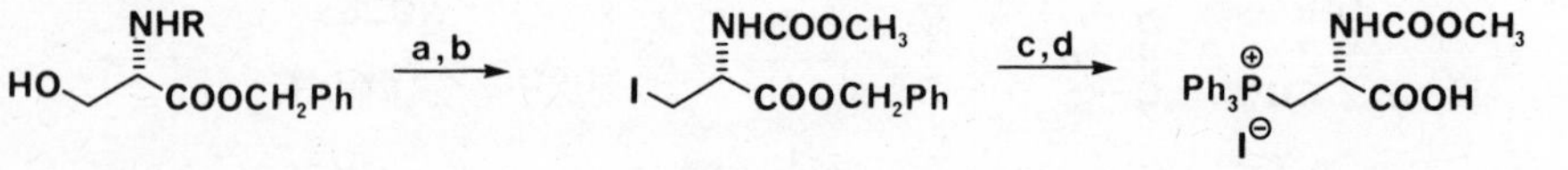

NHR
HO COOCH₂Ph
a,b
NHCOOCH₃
I COOCH₂Ph
c,d
NHCOOCH₃
Ph₃P⊕ COOH
I⊖

61 R=H[28]
62 R=COOCH₃
63
64

OHC O CH₂Ph
CH₃ N N N N N CH₃
e,f,g
CH₃OOC NHCOOCH₃
O H
CH₃ N N N N N CH₃

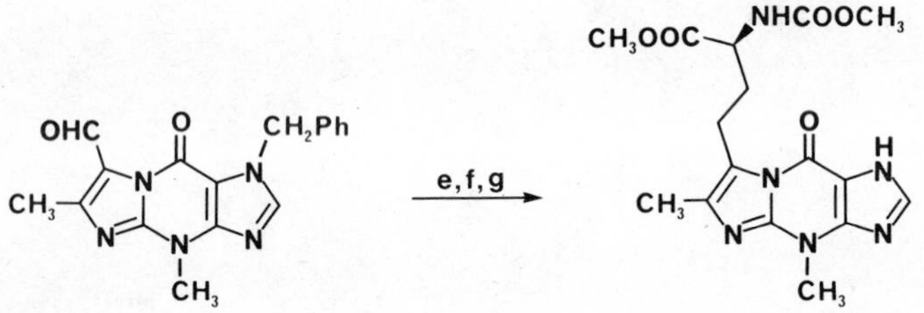

65
66

Scheme 9. (a) TsCl; (b) NaI; (c) Ph_3P; (d) H_2, Pd/C (96%); (e) **64**, *n*-BuLi (5%); (f) Me_3SiCHN_2; (g) H_2, Pd/C (74%)

Helquist and co-workers[29] have taken advantage of the chirality of D-serine to construct the sulfur-containing nine-membered lactone fragment (**68**) of the streptogramin antibiotic griseoviridin (**69**) (Scheme 10). The iodo serine derivative **67** is prepared in three steps from D-serine by protection with Cbz–Cl, esterification with Ph_2CN_2, and iodination (CH_3SO_2Cl, TEA; NaI, acetone).

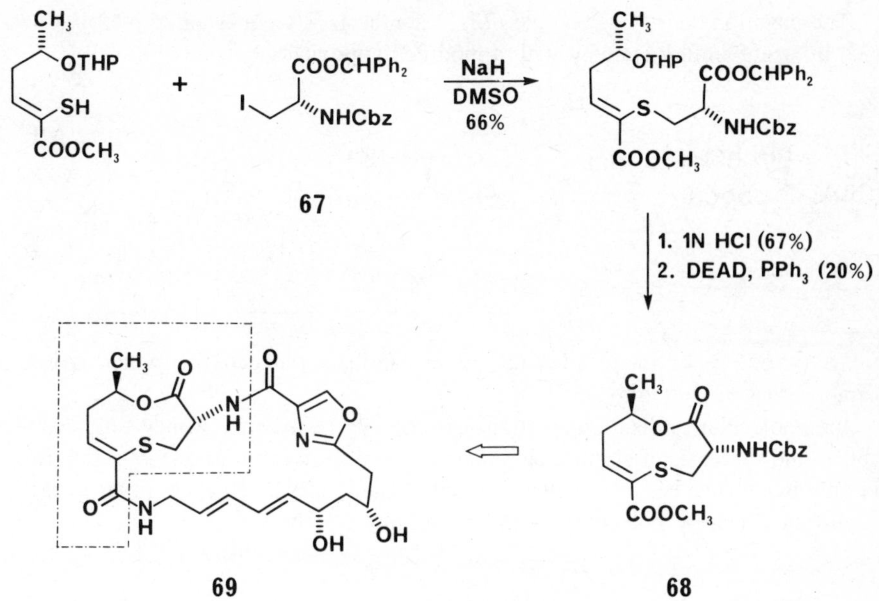

67

NaH, DMSO, 66%

1. 1N HCl (67%)
2. DEAD, PPh_3 (20%)

69

68

Scheme 10

One final mode, and perhaps the most important transformation of serine, is the ability to undergo intramolecular cyclization reactions with its β substituent. These reactions are normally stereospecific with no loss of integrity at the chiral center. An interesting approach to each enantiomer of 4-amino-3-isoxazolidinone (cycloserine) uses such a cyclization as the key step in the formation of the heterocycle.

L-Cycloserine (**72**) is prepared from Cbz–L-serine (**70**) according to the following sequence of reactions. Initial treatment of **70** with benzyloxyamine and DCC furnishes hydroxamic ester **71a** in 93% yield. Tosylation of **71a** is accomplished with *p*-toluenesulfonyl chloride in pyridine to give **71b** in 73% yield. Catalytic hydrogenolysis simultaneously removes the *O*-benzyl and Cbz protecting groups to afford a free hydroxamic acid that spontaneously undergoes cyclization by displacement of the tosylate to form **72**.

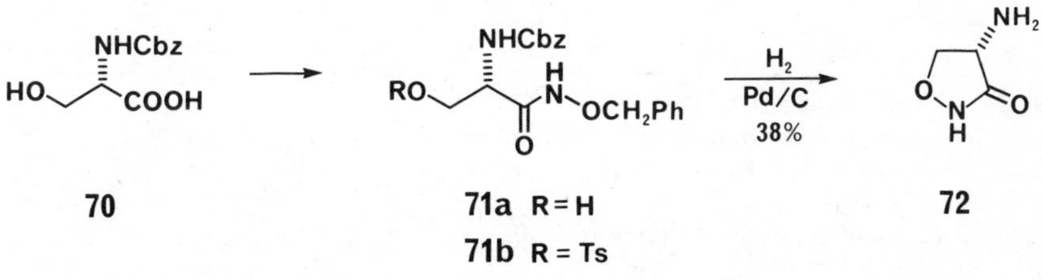

H_2, Pd/C, 38%

70

71a R = H
71b R = Ts

72

The enantiomeric D-cycloserine (**75**) is synthesized from D-serine methyl ester (**73**) by using similar strategy with a modified approach, shown as follows:[30]

$$\underset{\mathbf{73}}{HOCH_2CH(NH_2\cdot HCl)COOCH_3} \xrightarrow[89\%]{PCl_5} \underset{\mathbf{74}}{ClCH_2CH(NH_2\cdot HCl)COOCH_3} \xrightarrow[NaOH,\ 40\%]{NH_2OH} \mathbf{75}$$

D-Cycloserine is of interest because of its antibacterial activity, which closely parallels that of streptomycin.[31]

Intramolecular cyclization reactions of serine derivatives are wonderfully adaptable to the synthesis of chiral 2-azetidinones. The ability to form the β-lactam ring in an efficient manner is highly important because this heterocycle is an integral feature in a host of molecules exhibiting antibiotic activity.

TABLE 5.1. Formation of Chiral β-Lactams from Cyclization of L-Serine Derivatives

$$R_1HN\text{-}CH(CH_2OH)\text{-}COOH \xrightarrow{R_2NH_2} \underset{\mathbf{76}}{R_1HN\text{-}CH(CH_2X)\text{-}CONHR_2} \xrightarrow{\text{METHOD}} \mathbf{77}$$

β-Lactam	R_1	X	R_2	Method[a]	Yield (%)	Reference
a	$PhCH_2CO$	OMs	OCH_2Ph	A	75	34
		OH		C	34[b]	35
b	Cbz	Cl	OCH_2Ph	B	74–86	36, 37
		OH		C	83	36, 37
c	Boc	Cl	OCH_2Ph	B	75–88	36, 37
		OH		C	62	36, 37
		OH		D	86	36, 37
d	Cbz	OH	$OCOCH_3$	D	66	38
e	Cbz	OH	$4\text{-}CH_3C_6H_4$	C	53	39
f	Cbz	OH	$4\text{-}OCH_3C_6H_4$	E	75	40
g	Boc	OH	Tetrazole	F	71	41, 42

[a] Method A: Potassium *t*-butoxide, DMF, −23°C
B: Sodium hydride, DMF/CH_2Cl_2, 50°C, 12 hr
C: DEAD, Ph_3P, room temperature
D: TEA, Ph_3P
E: (a) Sulfonylbisimidazole; (b) NaH, DMF, −40°C
F: *i*-PrOOCN=NCOO*i*-Pr, Ph_3P

[b] Low yield due to competitive oxazoline formation.[34]

The strategic approach to these heterocycles involves the conversion or activation of the hydroxyl to a suitable leaving group (**76**) followed by an intramolecular displacement of that group to form the β-lactam **77.** Table 5.1 illustrates the variety of approaches used to effect the cyclization. All the methods listed give good yields of the azetidinones.

Some of these derivatives (**77**) have been transformed further into a variety of synthetically useful intermediates. Consequently, if **77c** is hydrogenolyzed over 10% Pd/C, a quantitative yield of the 1-hydroxy-2-azetidinone **78** is realized.[32] α-Aminoxyacetate derivatives of **78** have been prepared as potential antibacterial agents.[33]

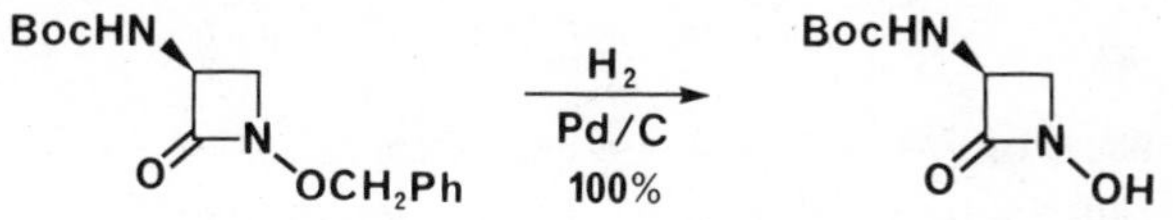

BocHN, O, N, OCH_2Ph — H_2, Pd/C, 100% → BocHN, O, N, OH

77c **78**

More importantly, **78** is used as a precursor to L-quisqualic acid (**81**), which is an exceptionally potent agonist of the neurotransmitter L-glutamate. Its β-lactam ring is isomerized cleanly to the isoxazolidin-5-one (**79**) by treatment with a catalytic amount of lithium ethanethiolate. Exposure of **79** to ethoxycarbonyl isocyanate gives the urea derivative **80,** which on immediate ring opening with sodium hydride followed by deprotection, furnishes **81** in 89% yield from **79**[43] (Scheme 11).

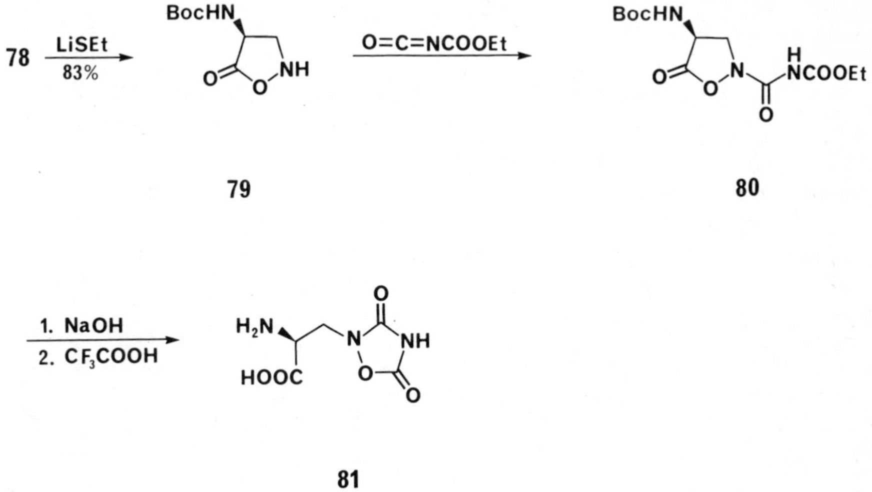

78 —LiSEt, 83%→ BocHN, O, O, NH (**79**) —O=C=NCOOEt→ BocHN, O, O, N, H, NCOOEt, O (**80**)

—1. NaOH, 2. CF_3COOH→ H_2N, HOOC, N, NH, O, O, O (**81**)

Scheme 11

The reduction of **77a** to **82** allows one to gain immediate access to a novel class of totally synthetic monocyclic β-lactam antiobiotics. The active monosulfalactam **83** is simply prepared by treatment of **82** with pyridine–sulfur trioxide complex. Interestingly, intermediate **82** possesses no antibiotic activity.[35]

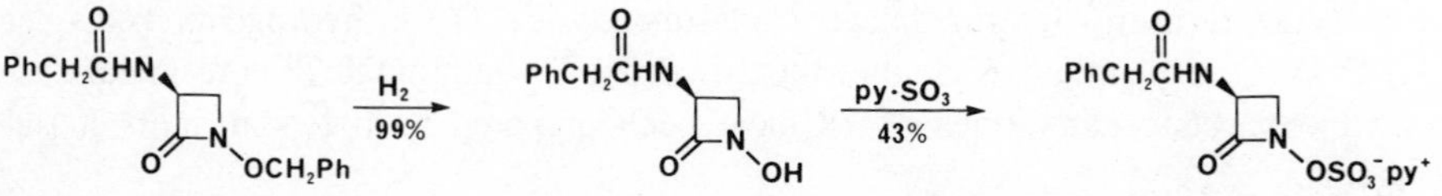

H_2, 99%; py·SO_3, 43%

77a **82** **83**

A related class of *N*-oxy-3-amino-2-azetidinones called oxamazins (**84**) also exhibit significant antibacterial activity. Their β-lactam moiety of the parent compound (R = H) is derived from L-serine by use of the previously described methodology, whereas the methyl analog is derived from L-threonine by a parallel sequence of reactions.[44,45]

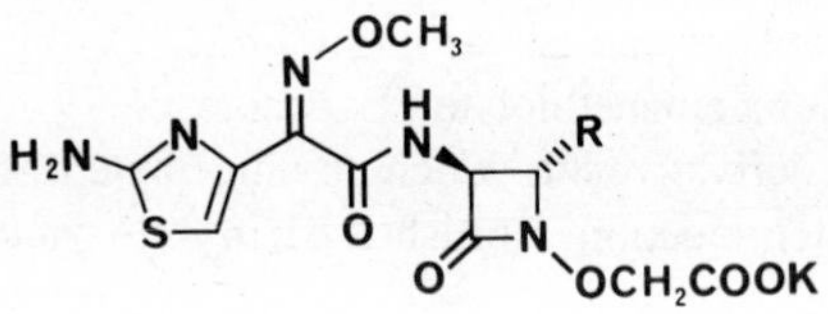

84 R = H, CH_3

Pyrimidoblamic acid (**88b**), one of the five building blocks used in the construction of the antitumor antibiotic bleomycin (Figure 1, Chapter 1), contains the chiral segment (*S*)-2,3-diaminopropionamide (L-DAPA) (outlined portion). The entire skeleton of this fragment is readily obtained from azetidinone **77c** in high yield by ring opening with methanolic ammonia to give intermediate **85.** Hydrogenolytic cleavage of the benzyloxy group furnishes Boc–L-DAPA (**86**) also in high yield.[46]

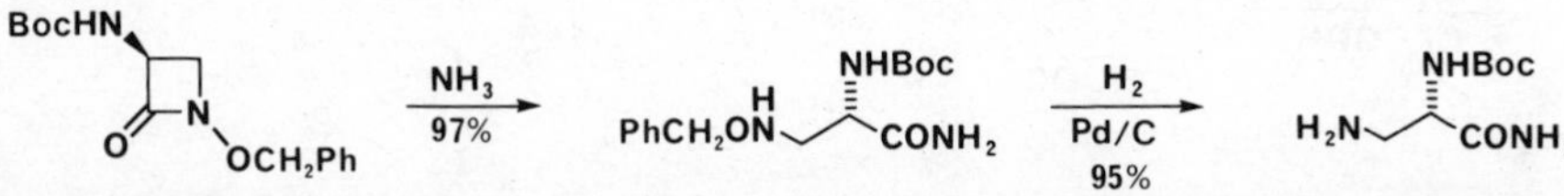

NH_3, 97%; H_2, Pd/C, 95%

77c **85** **86**

The protected Boc–pyrimidoblamic acid (**88a**) is then prepared by formation of the Schiff base **87** followed by addition of an amide anion equivalent to the carbon–nitrogen double bond of the imine.[47,48]

86, 3Å seives

"$H_2NOCCH_2^{\ominus}$"

87

88 a, R = Boc
b, R = H

An alternate method for the synthesis of **86** proceeds directly from the serine derivative **46** without the need to resort to β-lactam formation. The β-amino substituent is readily introduced by means of a Mitsunobu reaction with hydrazoic acid. The resulting azide **89** is formed in surprisingly high yield when one considers the propensity of the hydroxy β to the carbonyl toward elimination. Hydrogenolysis of **89** followed by protection of the free amino group quantitatively furnishes the differentially protected **90.**[49] The ester group is converted to a primary amide in 73% yield, and selective deprotection of the Cbz group (99% yield) gives Boc–L-DAPA (**86**).

DEAD, HN_3, PPh_3, 73%

1. H_2, Pd
2. CbzCl
100%

46 **89** **90**

Nuclear *N*-unsubstituted-2-azetidinones are also key intermediates for the synthesis of β-lactam antibiotics. Protected versions such as **91** are available from several previously described β-lactams by cleavage of the function group attached to the ring nitrogen. Conditions for this cleavage must be compatible with the 3-amino protecting group as well as the base-sensitive chiral site.

Mattingly and Miller[32] have developed an extremely mild method for reductive cleavage of the *N–O* bond of *N*-hydroxy-2-azetidinones. Exposure of **78** to buffered titanium trichloride cleanly reduces the desired bond to give the 3-Boc-amino-2-azetidinone (**91a**) in good yield.

Since **78** is produced by catalytic hydrogenolysis of **77c,** this method is not directly applicable to the corresponding 3-Cbz–amino derivative because of the incompatibility of the Cbz group to hydrogenation. Hanessian et al.[40] approach the problem by using the *N*-aryl-2-azetidinone **77f.** The *p*-methoxyphenyl group is removed under oxidative conditions with ceric ammonium nitrate to provide 3-Cbz–amino-2-azetidinone (**91b**) in good yield. The enantiomers of **91** have also been prepared starting from D-serine.[50]

BocHN ... N–OH → ($TiCl_3$, 67%) → RHN ... NH ← (CAN, 70%) ← CbzHN ... N–(p-C$_6$H$_4$)–OCH_3

78 **91a** R = Boc **77f**

91b R = Cbz

Carbenoid insertion of **92** into the N–H bond of azetidinone **91a** provides facile access to the protected form of 3-aminonocardicinic acid (**93**). The reaction produces a mixture of diastereomers (ratio **93**:**94** = 1:1.2) from which the desired isomer **93** is isolated by chromatography. The undesired isomer **94** can be epimerized with triethylamine to a new diastereomeric mixture of **93** and **94** (ratio 4:3) from which more of **93** is isolated.[51] The entire process from Boc-L-serine → **77c** → **78** → **91a** → **93** is very efficient, with an overall yield of 45%.

RO–(p-C$_6$H$_4$)–C̈COOR → (**91a**, 67%) → BocHN ... N–CH(COOR)–(p-C$_6$H$_4$)–OR + BocHN ... N–CH(COOR)–(p-C$_6$H$_4$)–OR

(1:1.2)

92 R = CH_2Ph **93** **94**

94 → (TEA) → **93**

Townsend et al.[52] have found that cyclization to the azetidinone can be accomplished later in the synthetic sequence (Scheme 12). Their intermediate, serine amide **96,** is prepared by condensation of *N*-phthaloyl-L-serine[53] (**95**) and (*R*)-methyl (*p*-methoxyphenyl)glycinate under peptide coupling conditions (DCC). A Mitsunobu-type cyclization affords a 2:1 mixture of diastereomers **97a** and **98a.**

Hydrogenation of this mixture gives **97b** and **98b,** which, on crystallization, furnishes the desired pure diastereomer **97b** in 43% yield. Sequential deprotection of **97b** leads to (−)-3-aminonocardicinic acid (**99**).

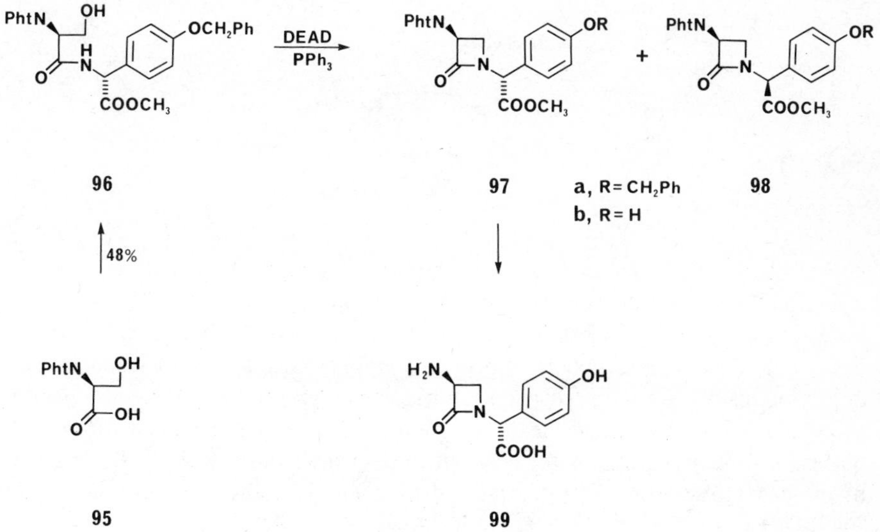

Scheme 12

5.2 THREONINE

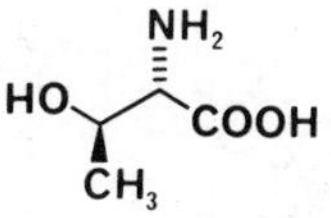

(2*S*,3*R*)-2-Amino-3-hydroxybutanoic acid (**100**)

L-Threonine (**100**) is a proteogenic, essential amino acid.[54] Besides L-isoleucine, it is the only other member of the common amino acids to contain two asymmetric carbon atoms. Synthetically, threonine is capable of contributing both chiral centers to a target molecule. In addition, its three functional groups provide a variety of handles with which chemical manipulations can be performed.

Monobactams are monocyclic β-lactam antibiotics produced by various gram-negative bacteria and whose common feature is the 2-azetidinone-1-sulfonic acid moiety[50,55] (**101**). As we have seen in the previous section, β-lactams can be assembled from a variety of L-serine derivatives. However, in certain instances the presence of a 4α-methyl substituent on the monobactam nucleus (**101,** R = CH_3) contributes substantially to its biological properties.[56]

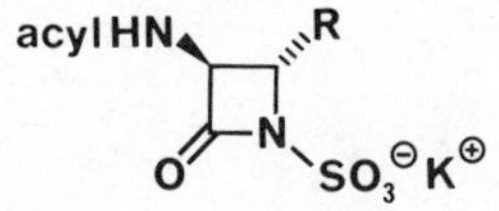

acylHN R N $SO_3^{\ominus}K^{\oplus}$ O

101

An analysis of the azetidinone-forming reaction reveals that the desired 4-methyl group can be introduced into the β-lactam ring simply by starting with L-threonine instead of L-serine. Thus Boc–L-threonine (**102**) is converted in one step to the *O*-benzylhydroxamate **103** by normal peptide coupling procedures. A Mitsunobu-type cylization results in the stereospecific formation of azetidinone **104** with retention of configuration at C-3 and clean inversion at C-4.[36] This inversion fortuitously places the 4-methyl group on the correct α face of the molecule. Reductive removal of the benzyloxy group is accomplished by catalytic hydrogenation followed by $TiCl_3$–NH_4OAc.[50]

It should be noted that *N*-arylamide derivatives of threonine tend to give aziridines under the Mitsunobu conditions by preferential intramolecular displacement of the activated hydroxyl with the α-amino group even if it carries Boc or Cbz protection. This is circumvented by protecting the amino group with a phthaloyl moiety.[39]

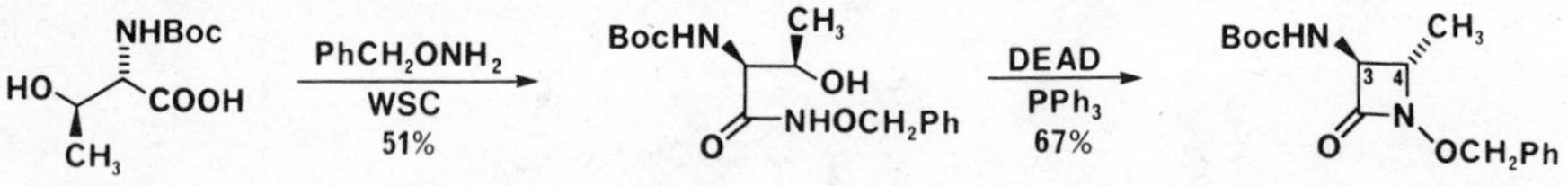

NHBoc HO COOH CH_3 —($PhCH_2ONH_2$, WSC, 51%)→ BocHN CH_3 OH NHOCH$_2$Ph O —(DEAD, PPh_3, 67%)→ BocHN CH_3 3 4 N OCH_2Ph O

102 **103** **104**

Alternatively, the monobactam nucleus is accessible by direct cyclization of an acylsulfamate such as **108** (Scheme 13). The synthesis proceeds as follows: L-threonine is converted to its methyl ester ($SOCl_2$, CH_3OH, 93% yield) and then to its amide **105** (NH_4OH, 5°C). Without purification, **105** is protected with a Boc group (**106**) and mesylated under standard conditions. Treatment of **107** with picoline–sulfur trioxide complex gives the acylsulfamate **108,** which, on treatment with potassium bicarbonate, cyclizes to the azetidinone **109.** The overall yield of the sequence is approximately 50%.[50,57]

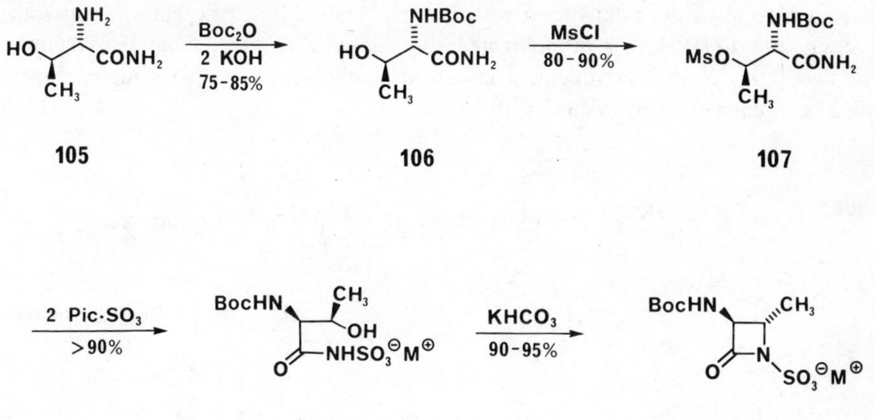

Boc_2O, 2 KOH, 75–85%; MsCl, 80–90%

105 106 107

2 Pic·SO_3, >90%; $KHCO_3$, 90–95%

108 109

Scheme 13

The carbapenems, another interesting class of antibiotics possessing the β-lactam ring, are characterized by the bicyclic nucleus **110.** Several research groups have successfully developed stereospecific syntheses of chiral azetidinone precursors for such systems. In each instance, L-threonine has been the starting material of choice to generate an intermediate chiral epoxide, which is then intramolecularly opened in a stereospecific fashion to form the β-lactam ring.

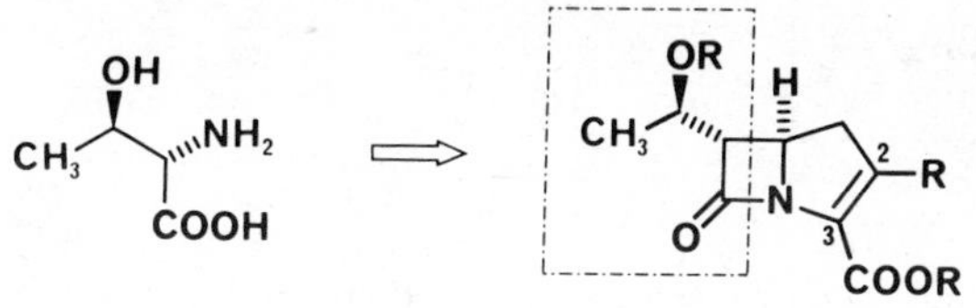

100 110

This technique is particularly suited for the synthesis of thienamycin (**117**), the most potent antibiotic of the class. The 2-azetidinone portion of the molecule is constructed according to the sequence of reactions outlined in Scheme 14.

The action of nitrosyl bromide on **100** produces (2*S*,3*R*)-2-bromo-3-hydroxybutanoic acid (**111**) with retention of configuration at C-2.[58] Condensation of **111** with *N*-2,4-dimethoxybenzylglycinate in the presence of DCC gives amide **112** in high yield. Sequential treatment of **112** with LiHMDS, first at 0°C and then at 20°C, initially produces epoxide **113,** which subsequently undergoes a stereospecific ring opening by the ester enolate to give azetidinone **114.**[59,60] The **112** → **114** transformation proceeds with a double inversion of configuration at the bromine-bearing carbon. Compound **114** can either be converted to alcohol **115** (by protection, reduction of the ester group with $LiAlH_4$, and dedimethoxybenzylation with $K_2S_2O_8$ and K_2HPO_4) or homologated to **116** by way of a diazomethylketone followed by a Wolff rearrangement in benzyl alcohol. Each of these intermediates have been carried on to thienamycin.[61,62]

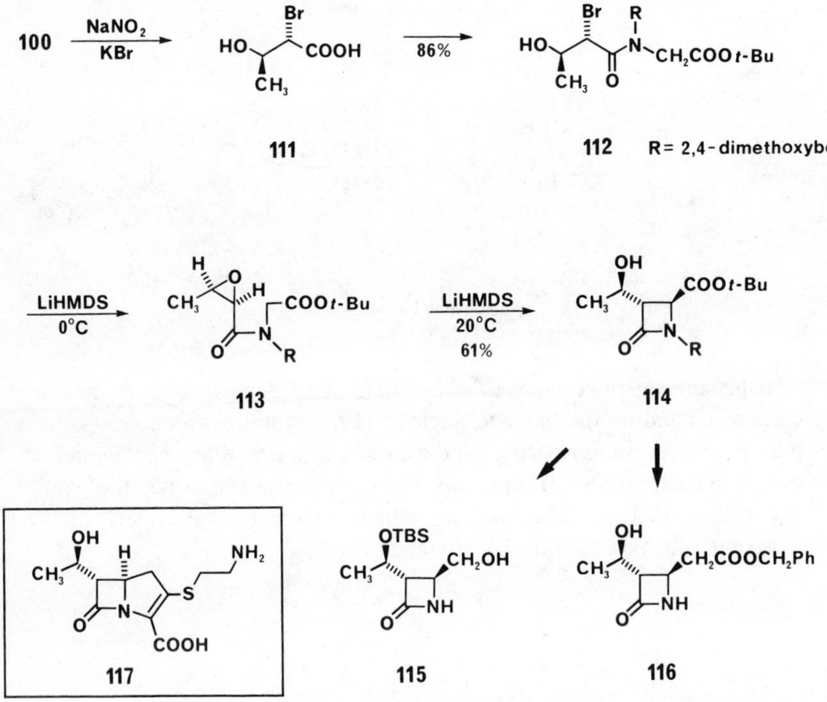

Scheme 14

By a parallel sequence of reactions, **111** is converted to azetidinone **118** in good overall yield (Scheme 15). A key intermediate **119,** required for carbapenem production, is obtained from **118** by sequential protection of the hydroxyl function with a TBS group, hydration of the acetylenic bond (46% overall yield), and removal of the *N*-*p*-methoxybenzyl group with CAN (66% yield).[63] Conversion of **119** to a trialkoxyphosphorane (**120**) followed by an intramolecular Wittig reaction furnishes the carbapenem nucleus **121.**[64]

The synthesis of the (4*R*)-phenylsulfonyl-2-azetidinone derivative **126,** a crucial intermediate for both carbapenems and penem antibiotics,[65] is accomplished in conceptually the same manner (Scheme 16) as the previous two examples but with different experimental techniques. Specifically, the formation of epoxide **123,** carried out under phase-transfer conditions instead of with LiHMDS, is achieved prior to alkylation of the amide functionality. Azetidinone formation is then effected with *n*-butyllithium in THF : HMPA (6 : 1). The HMPA is required for diastereospecificity (structure **125**). Without it, a 3 : 1 mixture of **125** and its 4α-diastereomer is produced.

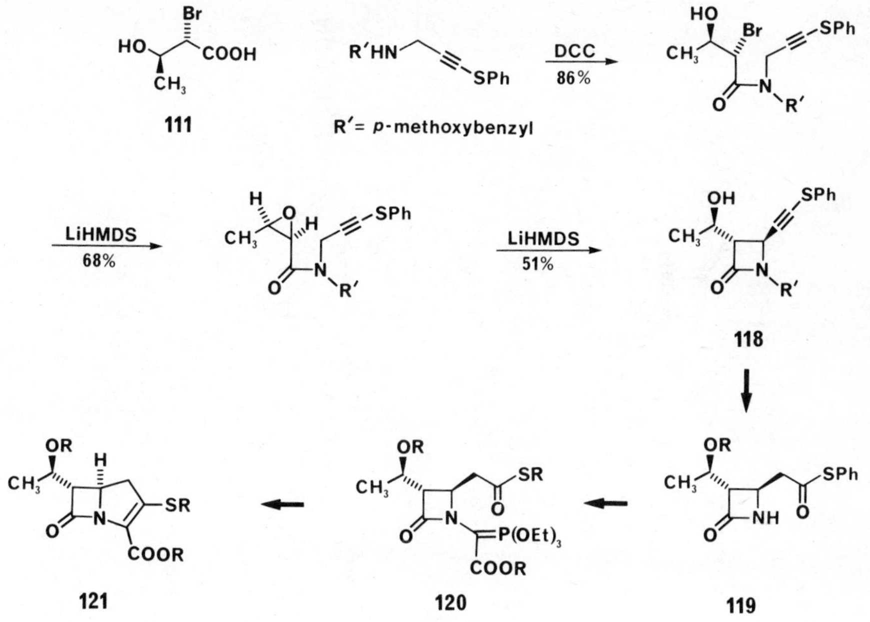

Scheme 15

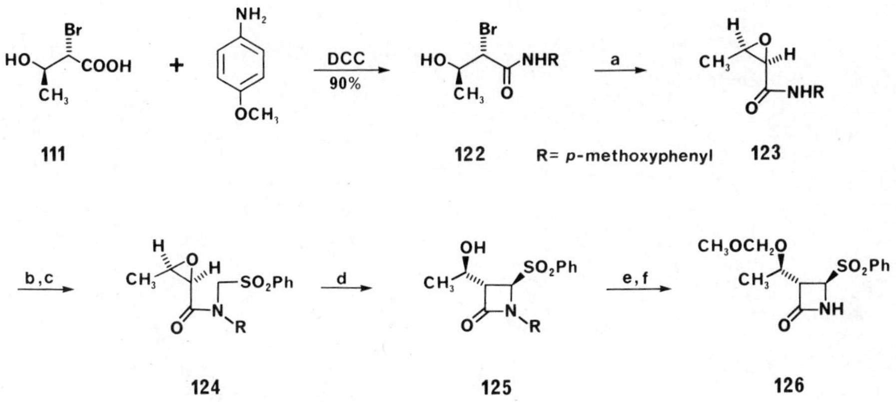

Scheme 16. (a) 50% NaOH, benzyltriethylammonium iodide (100%); (b) $ClCH_2SPh$, phase transfer (87%); (c) MCPBA (80%); (d) *n*-BuLi, −50°C (82%); (e) $ClCH_2OCH_3$, *N*,*N*-diethylaniline (95%); (f) O_3, sodium thiosulfate.

A somewhat similar approach is used by Hanessian et al.[66] to synthesize the biologically active penem derivative FCE22101 (**131**) (Scheme 17). They convert L-threonine via **111** to an epoxyacid in a one-pot procedure, and then to the epoxyamide **127** in good overall yield. Treatment of **127** with potassium carbonate in DMF readily forms the azetidinone **128.** After a Baeyer–Villiger oxidation to **129,** the molecule is elaborated to the desired product **131.**

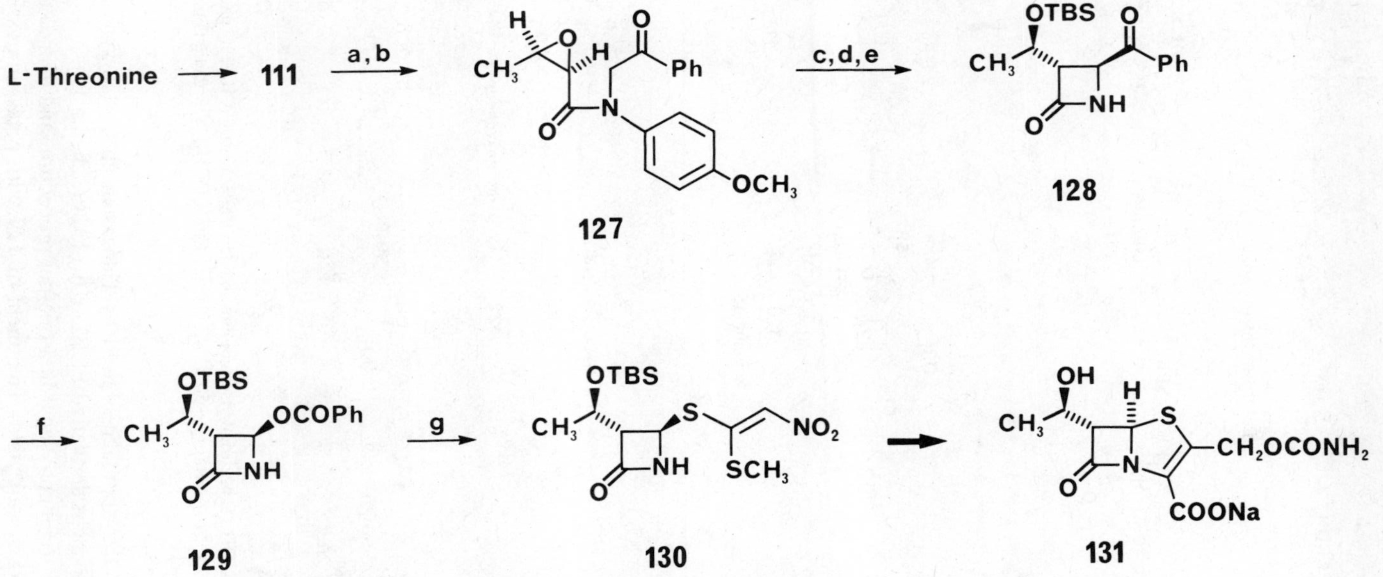

L-Threonine → 111 —a, b→ 127 —c, d, e→ 128

—f→ 129 —g→ 130 → 131

Scheme 17. (a) NaOH; (b) p-$CH_3OC_6H_4NHCH_2COPh$, $ClCOO$*i*-Bu, *N*-methylmorpholine, 4-Å molecular sieves, THF, −20°C, then 25°C (70%); (c) K_2CO_3, DMF, 60°C (75%); (d) TBS–Cl, imidazole, DMF (80%); (e) CAN, CH_3CN, −10°C (87%); (f) monoperphthalic acid (89%); (g) $KS(KS)C{=}CHNO_2$, then Me_2SO_4 (76%).

Diazotative deamination of threonine also plays an important role in the synthesis of chiral amino sugars, which are components of biologically active glycosides. The action of nitrous acid on L-threonine results in the formation of (2*S*,3*R*)-dihydroxybutyric acid (**132**). This is then converted to the chiral dioxolane aldehyde **136** in four steps according to the reaction shown in Scheme 18. This key aldehyde controls the stereochemistry obtained in all subsequent reactions.

Consequently, conversion of **136** to the α,β-unsaturated ester **137,** followed by a stereospecific addition of ammonia to the double bond and acidic hydrolysis of the dioxolane ring, gives rise to D-*xylo*-lactone **138.** Hydride reduction of the lactone cabonyl furnishes *N*-benzoyl-2,3,6-trideoxy-3-amino-D-*xylo*-hexopyranose (**139**).[67]

L-Daunosamine and L-acosamine are the amino sugar constituents of the antitumor glycosides daunomycin and adriamycin. Classically, chiral syntheses of these compounds are carbohydrate-based[68] and involve lengthy multistep sequences from D-sugars or from the rare deoxy sugar L-rhamnose.

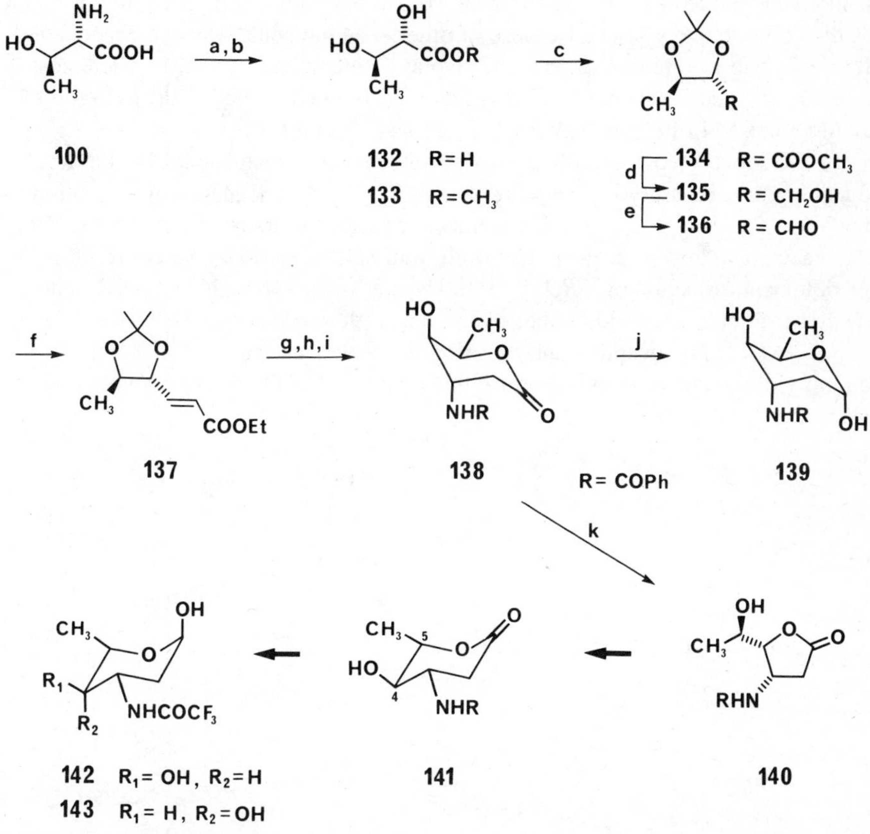

100

132 R=H
133 R=CH_3

134 R=$COOCH_3$
135 R=CH_2OH
136 R=CHO

137 138 139

R= COPh

142 R_1= OH, R_2=H
143 R_1= H, R_2=OH

141 140

Scheme 18. (a) HNO_2; (b) CH_3OH, HCl; (c) 2,2-dimethoxypropane, PTSA (45% overall yield from **100**); (d) $LiAlH_4$ (85%); (e) PCC; (f) $Ph_3P{=}CHCOOEt$; (g) NH_3; (h) 2 *N* HCl; (i) PhCOCl, pyridine/CH_2Cl_2; (j) DIBAH, −50°C; (k) HCl, CH_3OH (100%).

Their syntheses can now be performed in a more expiditious manner starting with L-threonine by an extension of the reactions illustrated in Scheme 18. The lactone **138,** on exposure to methanolic HCl, isomerizes to the γ-lactone **140** in nearly quantitative yield.[69] At this stage it is necessary to convert the hydroxy group from the *xylo* to *arabino* configuration by mesylation and displacement with acetate. After reisomerization to the six-membered lactone **141** and conversion to a trifluoroacetyl derivative, DIBAH reduction of the lactone carbonyl furnishes *N*-trifluoroacetyl-L-acosamine (**142**).

N-Trifluoroacetyl-L-daunosamine (**143**) is also available by this methodology; however, an inversion of configuration of the OH at C-4 of lactone **141** from *arabino* to *lyxo* is essential prior to completion of the synthesis.

This inversion of configuration, which is a major drawback in the synthesis of **143,** can be circumvented by using a chiral dioxolane analog with the correct stereochemistry already built in. This, coincidentally, is accessible from the enantiomeric amino acid D-threonine.

Let us digress momentarily and consider D-threonine. Economically, even though D-threonine is the unnatural form of the common amino acid L-threonine, it is not as disproportionately priced as those of other D-amino acids we have encountered. It is only two to three times as expensive as L-threonine. However, where cost is a factor, D-threonine (**147**) can by synthesized in good overall yield from the bis-lactim ether **144** of cyclo-(L-Val-Gly) (see also Chapter 3).

In the initial step, the addition of acetaldehyde to metalated **144,** the key to obtaining the necessary high *threo*:*erythro* ratio, is transmetalation from lithium to titanium. The adoption of the six-membered pericyclic transition state **145** allows the reactants to approach each other with minimal interaction. The net result is the predominant formation of (3*R*,1′*S*)-**146.** Hydrolysis of **146** with 0.25 *N* HCl affords methyl D-threoninate, and further hydrolysis with concentrated HCl gives D-threonine (**147**).[70] The overall transformation is an elegant example of the use of the chirality of a common amino acid (in this case L-valine) to produce another amino acid of greater rarity.

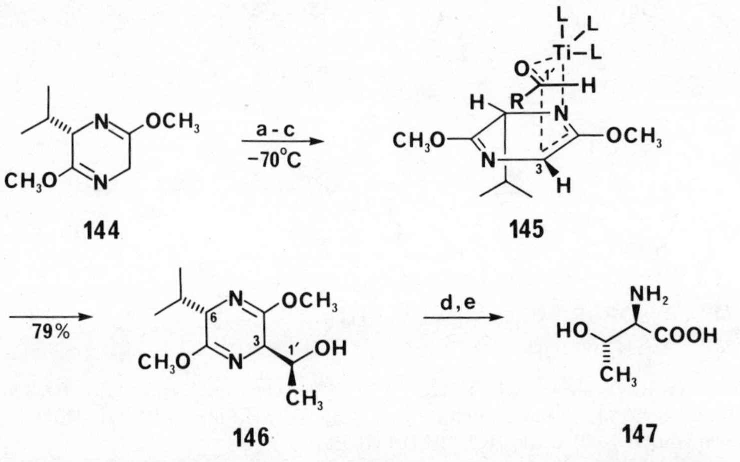

(a) *n*-BuLi; (b) $ClTi[N(CH_3)_2]_3$; (c) CH_3CHO; (d) 0.25 *N* HCl; (e) concentrated HCl.

The requisite dioxolane derivative (**149**) needed for the synthesis of daunosamine is prepared from **147** by the same protocol used for the preparation of **134.** In this instance, the ester group is converted directly to the aldehyde by a controlled reduction with Vitride (50% overall yield).

Addition of allylmagnesium bromide (at −78°C) to the aldehyde carbonyl of **149** produces an 8:2 mixture of inseparable epimeric alcohols **150** with the *threo*-isomer predominating. Lowering the temperature to −120°C does not appreciably affect the isomer ratio. This mixture is carried along the synthetic pathway to the *N*-benzoyl intermediate **152,** where isomer separation is accomplished by crystallization. Ozonolysis of the desired isomer affords *N*-benzoyl-L-daunosamine (**153**)[71] (Scheme 19).

Dioxolane **149** also plays an important role in the introduction of chirality in the asymmetric synthesis of *exo*-brevicomin (**158**), the aggregation pheromone of the western pine beetle. Addition of ethylmagnesium bromide to **149** produces a separable 8:2 mixture of alcohols with the *threo*-isomer **154** as the major component. *O*-Benzylation followed by acetal hydrolysis and periodate glycol cleavage furnishes the key enantiomeric (2*R*)-aldehyde **155.** Treatment of this aldehyde with Grignard reagent **156** gives a 6:4 mixture of adducts **157,** which, when hydrolyzed and debenzylated, affords pure (1*R*,7*R*)-*exo*-brevicomin (**158**) after preparative gas chromatographic separation[72] (Scheme 20).

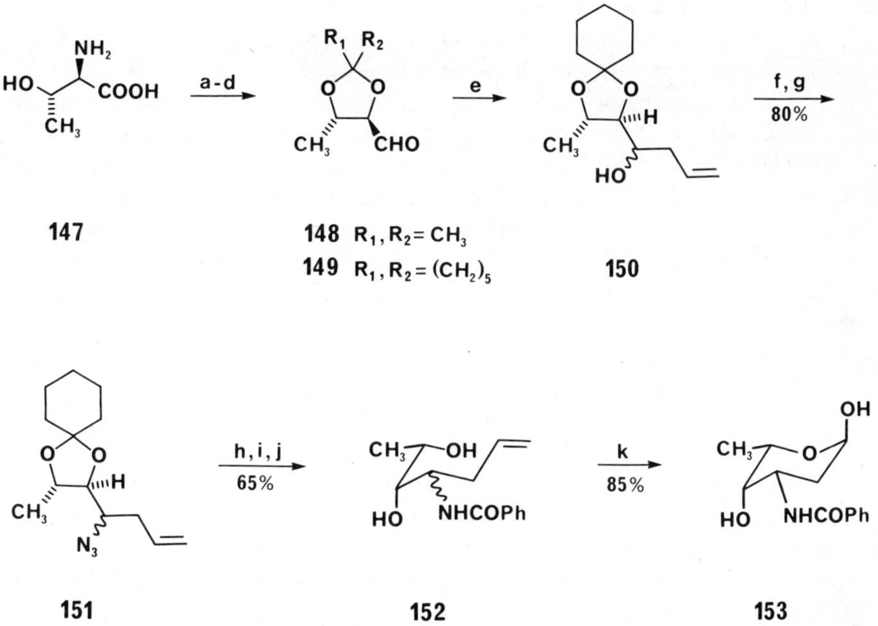

147 $\xrightarrow{\text{a-d}}$ **148** $R_1, R_2 = CH_3$; **149** $R_1, R_2 = (CH_2)_5$ $\xrightarrow{\text{e}}$ **150** $\xrightarrow[\text{80\%}]{\text{f,g}}$

151 $\xrightarrow[\text{65\%}]{\text{h,i,j}}$ **152** $\xrightarrow[\text{85\%}]{\text{k}}$ **153**

Scheme 19. (a) HNO_2; (b) CH_3OH, H^+; (c) $R_1{-}\overset{O}{\overset{\|}{C}}{-}R_2$, PTSA; (d) $NaAlH_2(OCH_2CH_2OCH_3)_2$, −50°C; (e) $CH_2{=}CHCH_2MgBr$, −78°C (75%); (f) PTSCl, pyridine; (g) NaN_3, NH_4Cl, DMF, 100°C; (h) $LiAlH_4$; (i) CH_3COOH, H_2O; (j) PhCOCl, K_2CO_3; (k) O_3, CH_3OH, −20°C, then Me_2S.

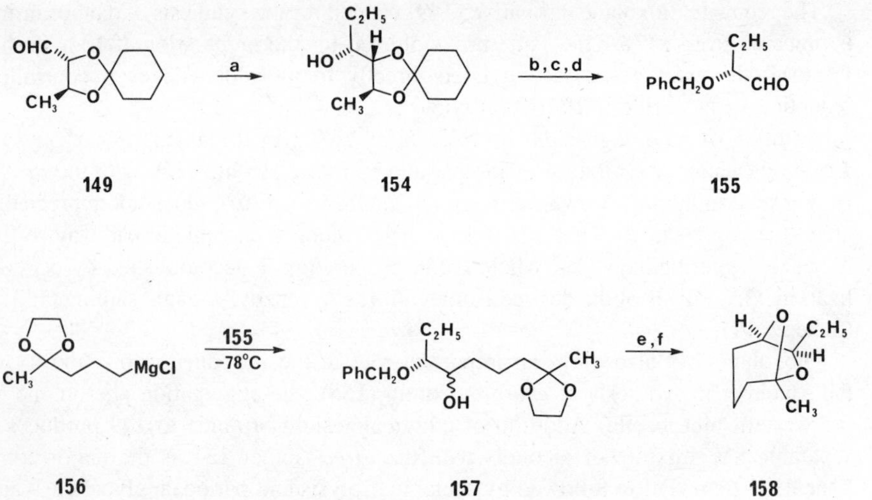

149 154 155

156 157 158

Scheme 20. (a) EtMgBr, $-120°C$; (b) $PhCH_2Cl$, NaH; (c) CH_3COOH, H_2O; (d) HIO_4; (e) H^+; (f) H_2, Pd/C.

Until now, the β-lactam-forming reactions we have explored produced 2-azetidinones with a *trans* relationship between the C-3 and C-4 substituents. Another, and perhaps more classical, approach to the β-lactam ring system is the cycloaddition of a ketene to a Schiff base. Depending on reaction conditions, *cis*-3,4-disubstituted azetidinones are formed. If the Schiff base is chirally biased, asymmetric induction occurs to give diastereomerically enhanced β-lactams. For placement of the peripheral substituents on the β-lactam nucleus in the biologically preferred geometry, D-threonine is required as the source of chirality.

Consequently, the Schiff base **160,** prepared from D-threonine benzyl ester (**159**) and cinnamaldehyde, when treated with azidoacetyl chloride in the presence of triethylamine affords a 1:1 mixture of diastereomeric β-lactams **163** and **164.**[73] The lack of diastereoselectivity in the cycloaddition can be attributed to the Schiff base adopting a nearly planar conformation as a result of strong intramolecular hydrogen bonding between the free hydroxyl of **160** and the ester carbonyl. The reactant can approach this planar structure with equal facility from either the *re* or *si* face, thereby giving both isomers **163** and **164** in equal proportions.

Functionalization of the hydroxy group of **160** eliminates the hydrogen bonding, thus causing the two faces of the Schiff base to become unequal. Therefore, when the *t*-butyldimethylsilyl ether derivative **161** is used in place of **160,** asymmetric induction occurs and the mixture of diastereomeric β-lactams is enhanced strongly in favor of **163** (**163**:**164** ratio = 9:1).[74] Even a higher induction is observed with the triphenylsilyl ether **162.** In this case **163** accounts for more than 95% of the mixture.[75]

Further manipulations of **163** allow one to gain access to the *cis*-monocyclic β-lactam **166,**[75] a precursor for an important intermediate in the synthesis of isocephalosporins,[76] or to the *O*-2-isocephem class of antibiotics **169**[74] (Scheme 21).

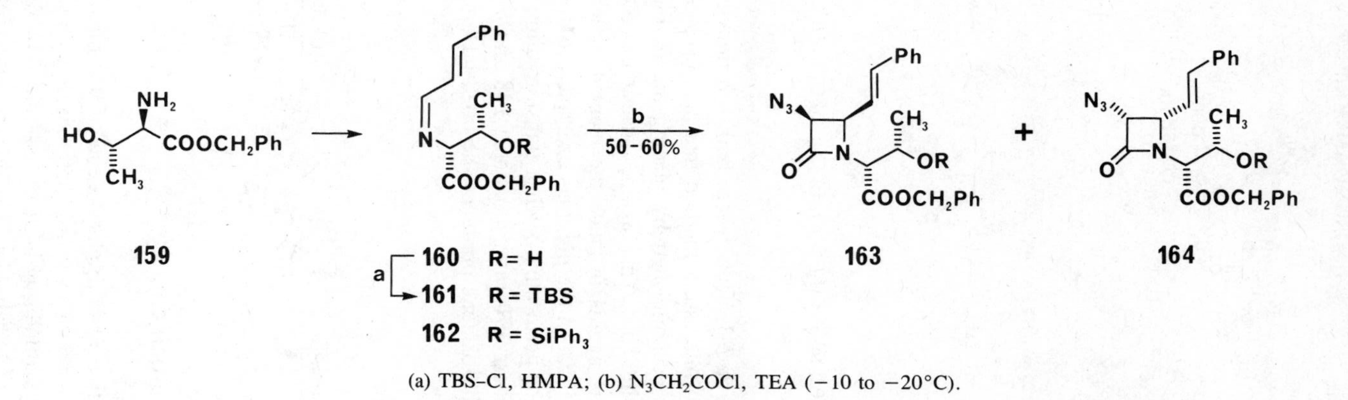

(a) TBS–Cl, HMPA; (b) N_3CH_2COCl, TEA (−10 to −20°C).

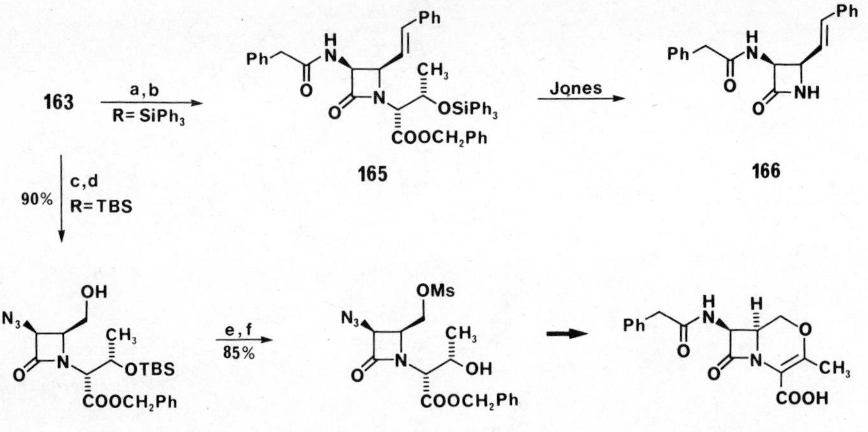

Scheme 21. (a) H_2, Pd; (b) $PhCH_2COOH$, EEDQ; (c) O_3 (−70°C); (d) $NaBH_4$; (e) mesylation; (f) CF_3COOH.

The enantiomeric β-lactam, where both substitutents are on its α face, can be synthesized by an interesting method that also uses D-threonine as the source of chirality. This is accomplished by placing the asymmetric center at a different location on the Schiff base. This, again, causes the two faces of the imine to become unequal; however, in this instance the reagent molecule must approach from a direction opposite that in the previous example (**161** → **163**).

The strategy for performing such a transformation involves the preparation of the chiral Schiff base **170** from dioxolane **148** and *p*-anisidine. Cycloaddition of this imine with diazoketene produces the *cis*-β-lactam **171** with complete diastereospecificity.[75]

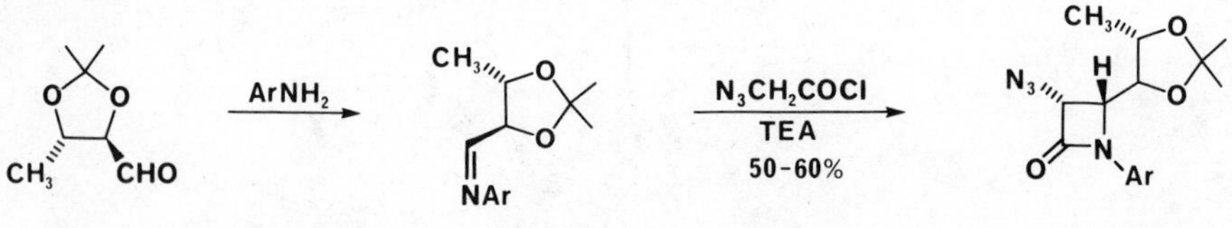

Ar = *p*-$OCH_3C_6H_4$

The incorporation of both chiral centers of L-threonine into an oxazoline ring provides the chemist with a versatile heterocycle with which to perform asymmetric syntheses. The oxazoline ring is easily assembled by the reaction of an L-threonine ester with an alkyl benzimidate. In this manner, either the methyl[77] or ethyl[78] oxazoline ester **172** is obtained in good yield with retention of configuration (4*S*,5*R*) at the two asymmetric carbon atoms. The (4*S*,5*S*)-isomer **173** can also be prepared from L-threonine by treating its *N*-benzoyl derivative with thionyl chloride. The nature of the cyclization results in a clean inversion of configuration at the β-carbon to produce the observed *cis* relationship of the substituents.[79]

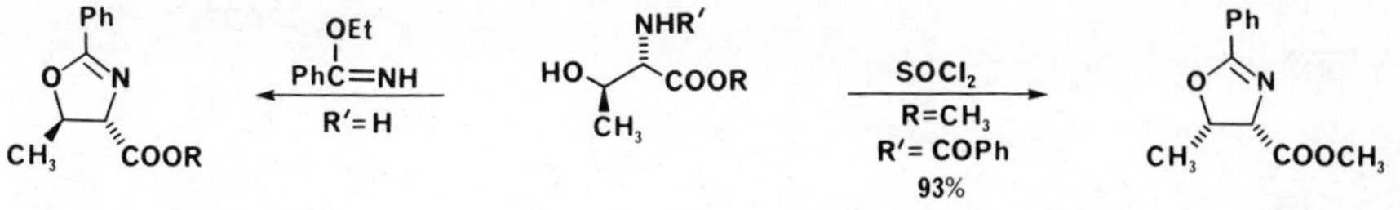

Ph; O, N; CH_3, COOR ← (OEt, PhC=NH; R′=H) — HO, NHR′, COOR, CH_3 — ($SOCl_2$; R=CH_3, R′= COPh; 93%) → Ph; O, N; CH_3, $COOCH_3$

172a R=CH_3 (71%)
172b R= Et (70%)

173

Both oxazolines **172a** and **173** are readily deprotonated with LDA in THF at −75°C. The solutions of the enolates are stable up to −35°C, where β-elimination begins. These enolates are readily alkylated with carbon electrophiles such as primary and secondary alkyl halides, aldehydes, and ketones to give products with diastereoselectivities exceeding 95%. With less reactive alkylating agents, the addition of HMPA or DMPU significantly improves the yields.[77]

On treatment with lithium aluminum hydride, ester **172b** cleanly reduces to alcohol **174.** Tosylation under standard conditions gives the primary tosylate **175.** It is this common intermediate that is responsible for the asymmetry produced in the synthesis of L-dihydroelaiomycin[78] (**176**) (Scheme 22) and (+)-actinobolin[80] (**179**) (Scheme 23).

The interesting feature in the actinobolin synthesis is an intramolecular Diels–Alder reaction of the chiral (*Z*)-diene **177.** The adduct **178,** formed in high yield with an isomer ratio of over 20 : 1, illustrates that the cycloaddition proceeds through a single diastereomeric transition state.

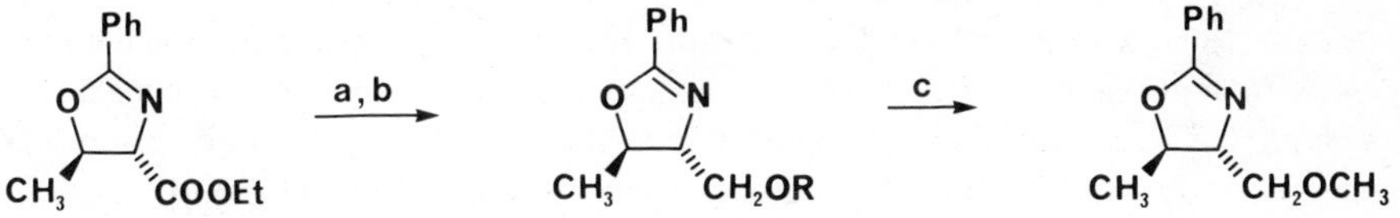

Ph; O, N; CH_3, COOEt — a, b → Ph; O, N; CH_3, CH_2OR — c → Ph; O, N; CH_3, CH_2OCH_3

172b

174 R= H
175 R= Ts

— d, e; 48% → CH_3OCH_2, CH_3, OH; H–N–COOEt → CH_3OCH_2, CH_3, OH; O←N=N; n-C_8H_{17}

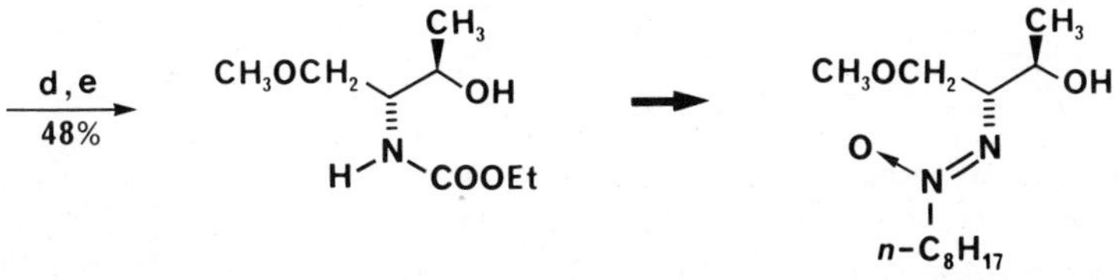

176

12 steps (8% overall yield) from L-Threonine

Scheme 22. (a) $LiAlH_4$ (98%); (b) PTSCl, pyridine; (c) $NaOCH_3$ (70%); (d) 6 *N* HCl; (e) ClCOOEt.

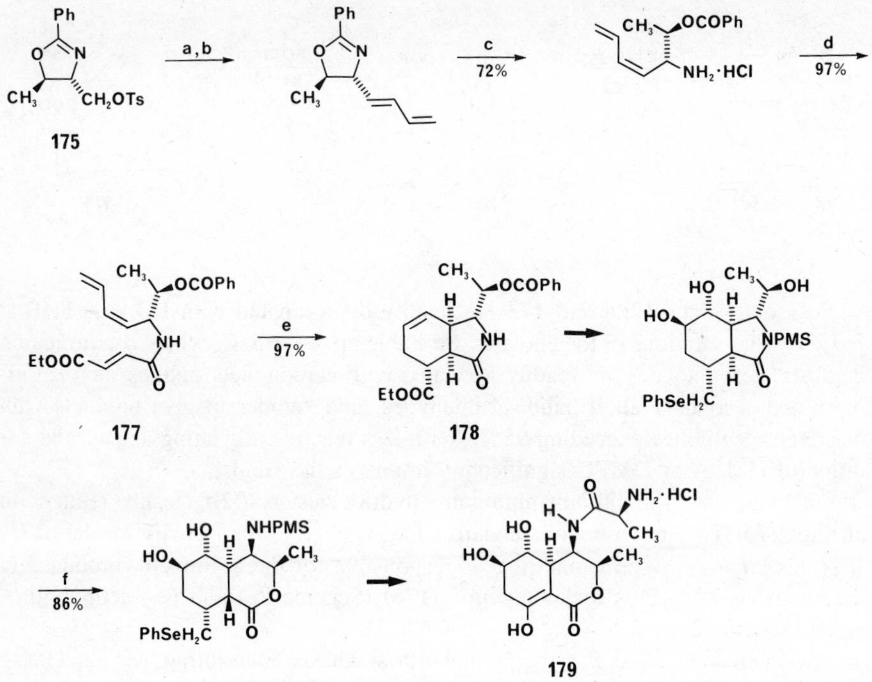

29 steps (20-25% overall yield)

Scheme 23. (a) PPh_3, 130°C (85%); (b) *n*-BuLi, acrolein, −78°C; (c) 1 *N* HCl (95%); (d) ethyl (*E*)-3-(chloroformyl)acrylate, *N*-methylimidazole; (e) 180°C, benzene; (f) $NaOCH_3$, CH_3OH.

2-Substituted furans of general structure **180** are viable precursors for the synthesis of 6-amino-6,8-dideoxyoctoses. Examination of the molecule reveals that without the furan ring, the remaining arrangement of atoms closely resembles threonine, with the C-1 hydroxy coming from its carboxylic acid group.

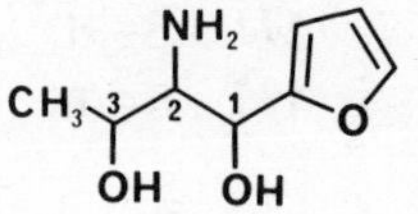

180

Oxazoline **181,** a protected version of L-threonine, contains all the required elements present in the amino diol side chain of **180.** Attachment of the furan ring to **181** is accomplished with 2-furyllithium to give ketone **182.** Reduction of the keto group of **182** ($LiAlH_4$, 0°C) occurs in a stereoselective fashion to produce a 3:1 mixture of alcohols **183** and **184.** After chromatographic separation of the diastereomers, acid hydrolysis of the oxazoline ring of each isomer gives

(1*R*,2*S*,3*R*)-**185** and (1*S*,2*S*,3*R*)-**186**, respectively, in greater than 93% yield[81] (Scheme 24).

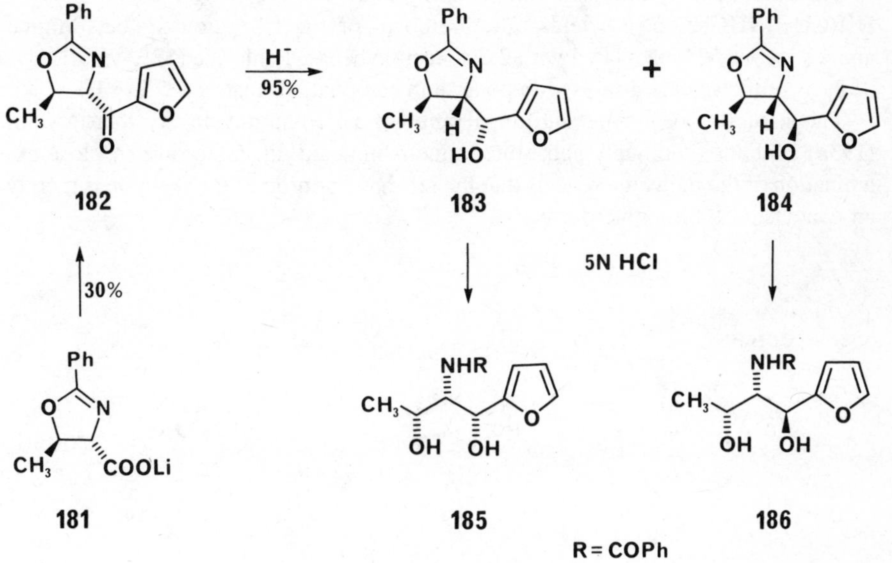

R = COPh

Scheme 24

Parabactin (**187**), a catecholamide spermidine siderophore isolated from iron-starved cultures of *Paracoccus denitrificans*,[82] contains a 2-aryloxazoline heterocycle that by now is highly recognizable as being derived from threonine.

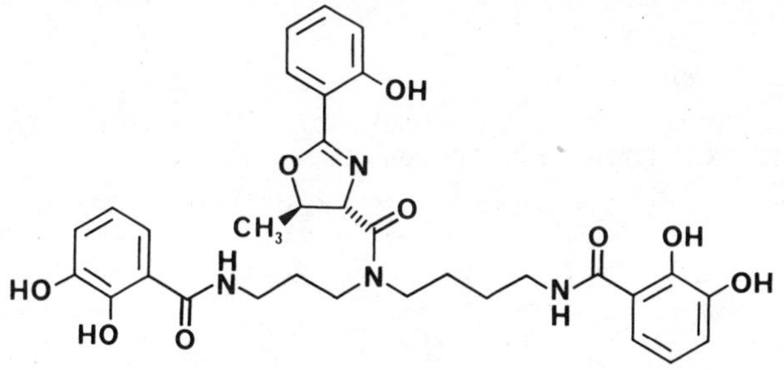

187

The oxazoline ring, with its (4*S*,5*R*) configuration, would be expected to derive from L-threonine. Interestingly, the architects of this heterocycle[83] use D-threonine (**147**) for its preparation (Scheme 25). The oxazoline is built in a stepwise manner by condensing **147** with the thiazolidine derivative **188.** Cyclization of **190** with thionyl chloride occurs with complete stereoinversion at C-5 to afford (4*S*,5*S*)-oxazoline **191.** Epimerization at C-4 with sodium ethoxide produces the desired stereochemistry in **192.**

The parabactin synthesis is completed by condensing **192** with CbzHN$(CH_2)_3$-NH$(CH_2)_4$NHCbz (55% yield); after removal of the Cbz groups, the terminal amines are N^1,N^{10}-diacylated with 2,3-diacetoxybenzoyl chloride (52% yield). Hydrolysis of the acetate groups with potassium carbonate furnishes **187** in 83% yield.

The naturally occurring antifungal antibiotic (+)-antimycin A_3 (blastmycin) (**193a**) contains a uniquely substituted nine-membered dilactone ring. A close examination of the molecule reveals that the left-hand portion of the skeleton is merely an *N*-acylated L-threonine derivative.

188 **189** **190**

191 **192**

Scheme 25. (a) D-Threonine, TEA; (b) CH_2N_2 (96%); (c) H_2, Pd/C (76%); (d) $SOCl_2$; (e) Na_2CO_3, $CHCl_3$; (f) NaOEt, EtOH; (g) H_2O, reflux; (h) Amberlite IR-120.

193 a, R= $\overset{O}{\overset{\|}{C}}CH_2CH(CH_3)_2$

b, R= H

The macrolide framework is assembled by esterification of the threonine derivative **194** with chiral alcohol **195.** The *O*-protected threonine **194** is obtained in 64% yield from Cbz–L-threonine by silylation with *t*-butyldimethylsilyl chloride and imidazole in DMF. After removal of the silyl protecting group of **196** with fluoride (64% yield), the alcohol **197** is subjected to dye-sensitized photooxygenation. The resulting activated triamide **198** is then lactonized under buffered acid catalysis to give the desired dilactone **199** in 20% yield from **197**[84] (Scheme 26). This intermediate has been carried on to the natural product **193a** by a separate group of researchers.[85,86] The dilactone moiety of deisovalerylblastmycin (**193b**), another naturally occurring member of this family, is synthesized in a strategically similar fashion.[87]

Chiral bisphosphines form cationic rhodium (I) complexes, which are excellent catalysts for the homogenous asymmetric hydrogenation of prochiral olefins. Bisphosphine **204,** a new addition to this family of ligands, is prepared from Boc–L-threonine methyl ester according to the sequence of reactions outlined in Scheme 27.[88]

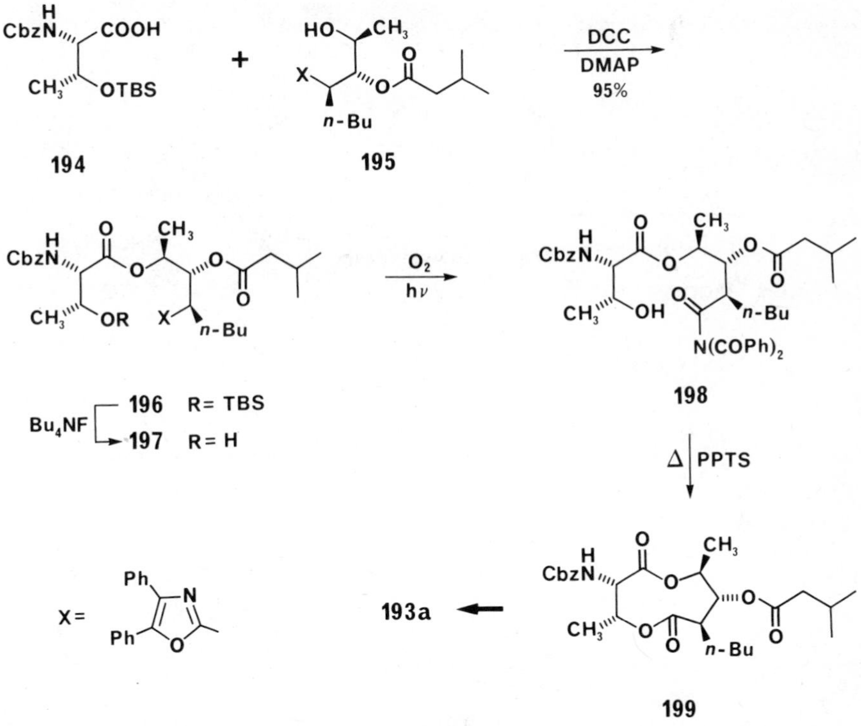

Scheme 26

When complexed with rhodium norbornadiene perchlorate, **204** forms a useful catalyst for the asymmetric hydrogenation of dehydro-α-amino acids (**205**). These reductions proceed under 1 atm of hydrogen over a period of 18–22 hr to furnish (*S*)-*N*-acylamino acids **206** in high chemical and optical yields (see Table 5.2).

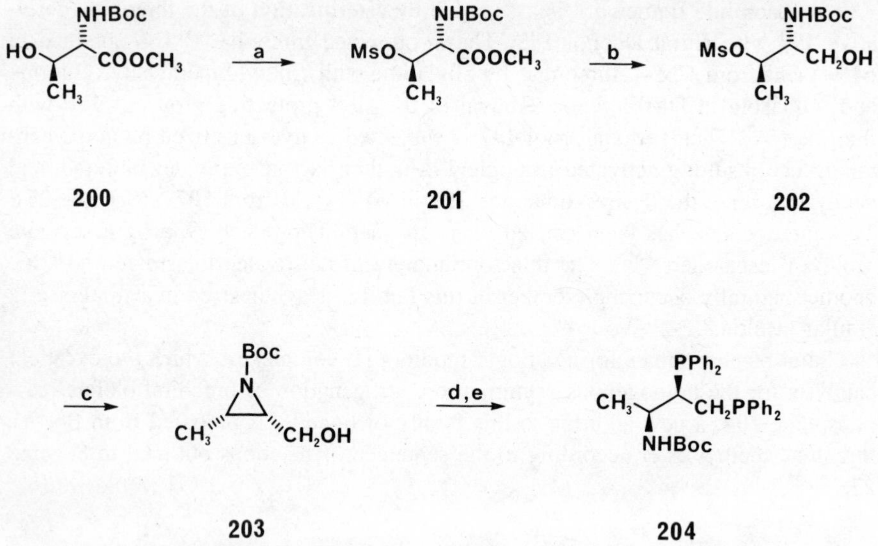

Scheme 27. (a) MsCl, diisopropylethylamine; (b) $NaBH_4$, 0°C (68%); (c) K_2CO_3, CH_3CN, 75°C (61%); (d) MsCl, TEA (89%); (e) $NaPPh_2$, THF, −40 → 0°C (60%).

TABLE 5.2. Asymmetric Hydrogenation of *N*-Acyldehydro Amino Acids

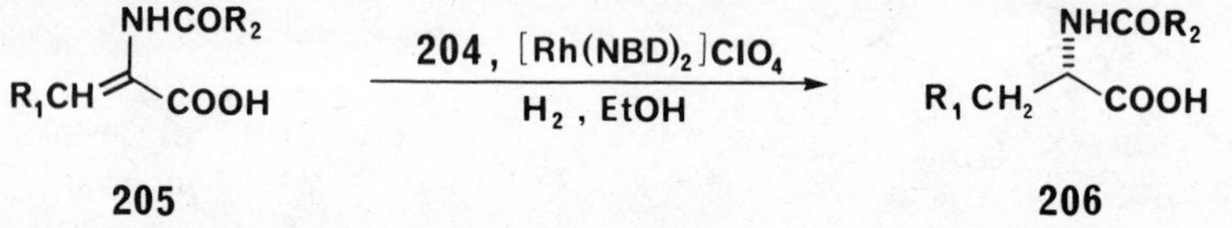

205 → 206: **204**, $[Rh(NBD)_2]ClO_4$, H_2, EtOH

R_1	R_2	Yield (%)	% ee
H	CH_3	99	86
$(CH_3)_2CH$	CH_3	90	89
C_6H_5	CH_3	92	88
C_6H_5	C_6H_5	93	91
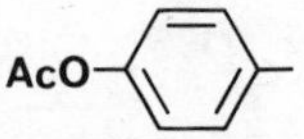 4-AcO-C_6H_4	CH_3	99	84
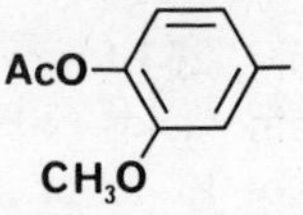 4-AcO-3-CH_3O-C_6H_3	CH_3	93	85

Acetonide **209,** obtained in three steps from Boc–D–threonine (**207**),[89] is an interesting diprotected form of (2*S*,3*S*)-2-aminobutane-1,3-diol. Although it has not been applied for synthetics, the multifunctionality inherent in the molecule should make it an attractive candidate for asymmetric syntheses.

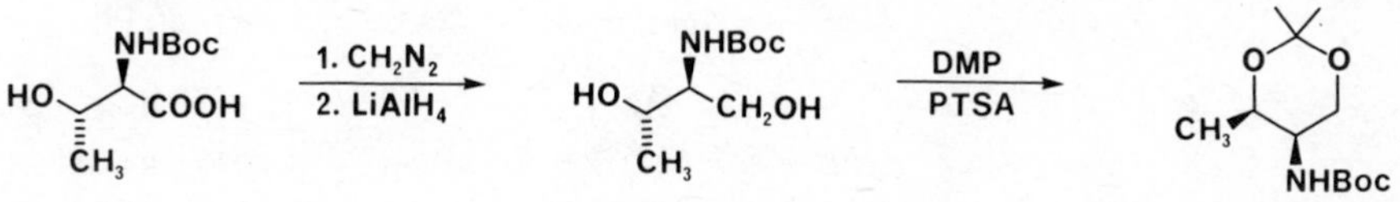

1. CH_2N_2 2. $LiAlH_4$ DMP PTSA

207 **208** **209**

5.2.1 *allo*-Threonine

Inversion of configuration at C-3 of the threonine nucleus leads to its corresponding *allo* form. Thus L-*allo*-threonine possesses the (2*S*,3*S*) configuration and D-*allo*-threonine, the (2*R*,3*R*) configuration. Although both isomers are commercially available, they are extremely expensive and, hence, impractical for large-scale syntheses. Each of the *allo*-isomers, however, is available in enantiomerically pure form by a simple manipulation of either L- or D-threonine.

Treatment of *N*-benzoyl-L-threonine methyl ester with thionyl chloride results in the formation of oxazoline ester **173** with total inversion at C-3. Acid hydrolysis of the oxazoline ring with 6 *N* HCl affords L-*allo*-threonine (**210**) in 96% yield.[79] Likewise, D-threonine can be converted also in high yield to D-*allo*-threonine (**222**) by a similar sequence of reactions.

Sibriosamine (**218**), an amino sugar derived from the potent antitumor antibiotic sibiromycin, is stereochemically equitable at C-4 and C-5 to L-*allo*-threonine.

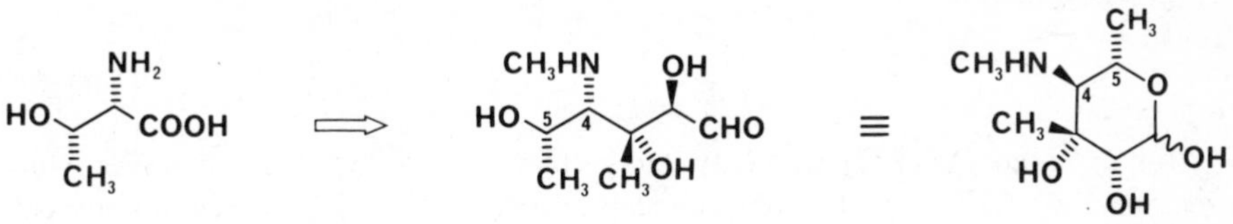

210 acyclic form **218**

Rapoport and co-workers[90] have ingeniously used an interesting transformation of **210** to introduce the remaining stereocenters at C-2 and C-3 (Scheme 28).

L-*allo*-Threonine is initially protected as its *N*-phenylsulfonyl derivative **211** is then converted to methyl ketone **212** by treating the lithium salt of **211** with methylmagnesium iodide. Reaction of **212** with vinylmagnesium bromide provides olefin **213,** which represents the complete carbon skeleton of sibirosamine. Catalytic osmylation of **214** produces a 4 : 1 mixture of tetraol **215** and its C-2 diastereomer. After chromatographic separation of the major isomer, a platinum-catalyzed oxidation of **215** gives lactone **216** in high yield. Functional group manipulation leads to the methyl glycoside **217**, which can be hydrolyzed to the unstable amino sugar **218**.

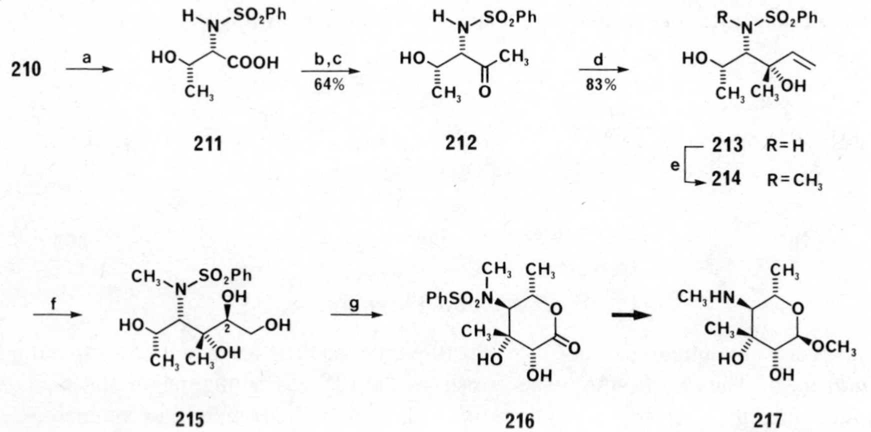

Scheme 28. (a) $PhSO_2Cl$, Na_2CO_3 (67%); (b) CH_3Li, $-78 \rightarrow 0°C$; (c) CH_3MgI, 35°C; (d) $CH_2{=}CHMgBr$, $-10 \rightarrow 23°C$; (e) K_2CO_3, CH_3I, *i*-propanol, 65°C (83%); (f) *N*-methylmorpholine *N*-oxide, OsO_4 (64%); (g) O_2, Pt (91%).

(*S*)-2-Methyloxetane (**221**) is accessible from L-*allo*-threonine by way of (*S*)-1,3-butanediol (**220**). The **210** → **220** conversion proceeds in 18% overall yield.[91] The enantiomeric (*R*)-2-methyloxetane is available from a parallel sequence of reactions starting from L-threonine.

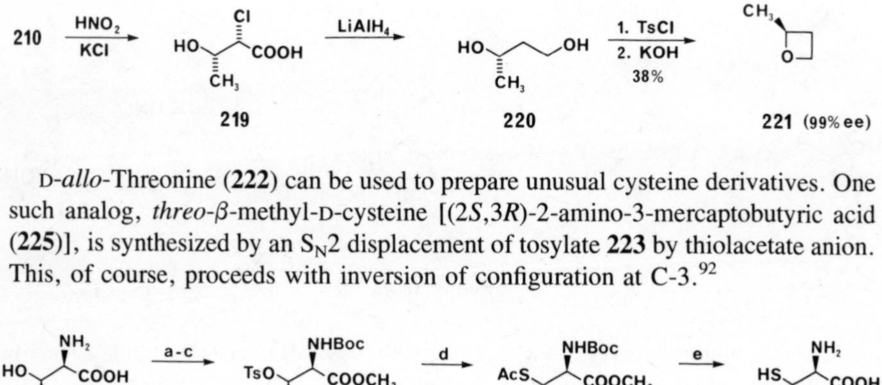

D-*allo*-Threonine (**222**) can be used to prepare unusual cysteine derivatives. One such analog, *threo*-β-methyl-D-cysteine [(2*S*,3*R*)-2-amino-3-mercaptobutyric acid (**225**)], is synthesized by an S_N2 displacement of tosylate **223** by thiolacetate anion. This, of course, proceeds with inversion of configuration at C-3.[92]

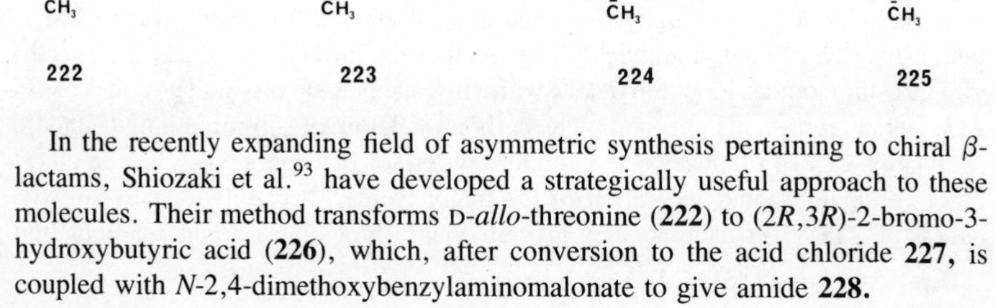

In the recently expanding field of asymmetric synthesis pertaining to chiral β-lactams, Shiozaki et al.[93] have developed a strategically useful approach to these molecules. Their method transforms D-*allo*-threonine (**222**) to (2*R*,3*R*)-2-bromo-3-hydroxybutyric acid (**226**), which, after conversion to the acid chloride **227**, is coupled with *N*-2,4-dimethoxybenzylaminomalonate to give amide **228**.

Treatment of the desilylated amide **229** with 2 equivalents of DBU produces a three-component mixture consisting of bicyclic lactone **230** (75%) and its epimer **231** (12%) and a hydroxy-γ-lactam (7%, not shown). Alternatively, cyclization of **228** with DBU gives azetidinone **232,** which, after desilyation and further treatment with DBU, provides only **230** in high yield (Scheme 29). In both cases the transformation proceeds with complete inversion of configuration at the bromine-bearing carbon atom.

Processing L-threonine (**100**) through azetidinone **234** using the same protocol as **222** → **232** gives access to synthetically useful quantities of the minor isomer **231.**

This methodology is quite useful for the synthesis of a series of chiral intermediates (**243, 244, 245**) for the penems and carbapenems, including thienamycin. Strategically, the assemblage of the key azetidinone **237** follows the same protocol as for the synthesis of **232.**[59]

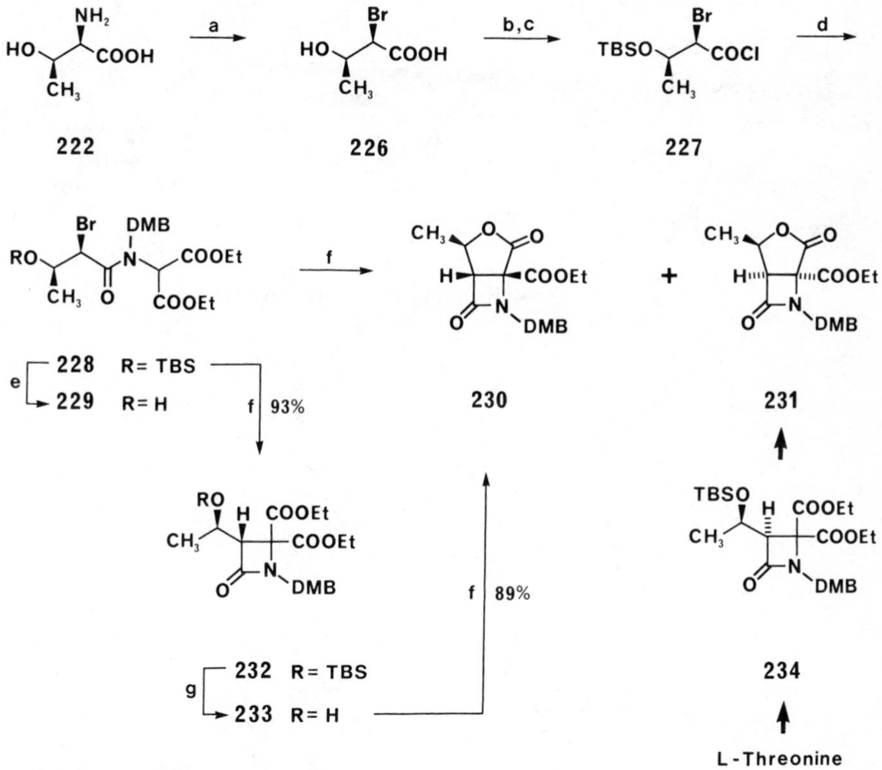

DMB = 2,4-dimethoxybenzyl

Scheme 29. (a) $NaNO_2$, KBr, H_2SO_4; (b) TBS–Cl; (c) $(COCl)_2$, THF, 20°C; (d) *N*-2,4-dimethoxybenzylaminomalonate, TEA, THF (100%); (e) EtOH, H_2O, HCl (9:2:1) (100%); (f) DBU, THF, 25°C, 4 hr; (g) EtOH, H_2O, HCl (9:2:1) (92%).

The ester on the less hindered β face of **237** is stereospecifically saponified to give **238.** Thermal decarboxylation of **238** stereoselectively produces a 4:1 mixture of *cis-* and *trans-***239.** Following base hydrolysis of the ester, exposure of the *cis-*

carboxylic acid to concentrated HCl leads to the formation of lactone **240.** Grignard reaction of **240** with methylmagnesium bromide followed by silylation furnishes the 4-acetyl azetidinone **241** in high yield. A Baeyer–Villiger oxidation of the debenzylated derivative **242** gives the 4-acetoxy-2-azetidinone **243** in nearly quantitative yield.

Compound **243** can be further elaborated to **244** or **245** by reaction with the TMS enol ether of benzyl acetate or sodium benzenesulfinate, respectively.[94] The route outlined in Scheme 30 is an attractive complement to the methods previously presented in this chapter for the construction of key β-lactam intermediates.

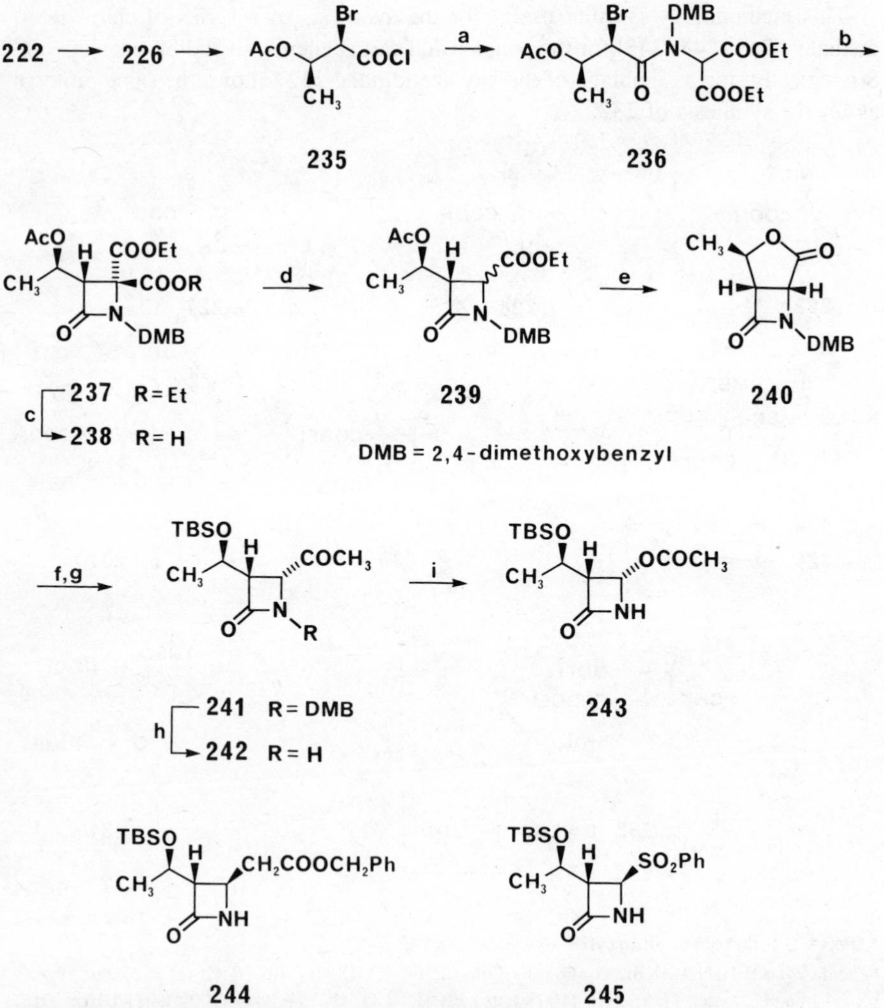

DMB = 2,4-dimethoxybenzyl

Scheme 30. (a) *N*-2,4-Dimethoxybenzylaminomalonate, TEA, THF (95%); (b) DBU, benzene (96%); (c) 1 equivalent 1 N NaOH (62%); (d) 2,4,6-collidine, 160°C (75%); (e) catalytic concentrated HCl (64%); (f) 1.8 equivalent CH_3MgBr (98%); (g) TBS–Cl, DMAP, DMF (88%); (h) $K_2S_2O_8$, K_2HPO_4; (i) MCPBA (96%).

REFERENCES

1. T. Scott and M. Brewer, *Concise Encyclopedia of Biochemistry*, Walter de Gruyter, New York, 1983, p. 427.
2. M. Fujino, T. Fukuda, and C. Hatanaka. Ger. Patent 2,408,324 (1973); *Chem. Abstr.*, **81,** 169852n (1974).
3. Y. Ito, I. Ito, T. Hirao, and T. Saegusa, *Syn. Commun.*, **4,** 97 (1974).
4. D. Seebach and J. D. Aebi, *Tetrahedron Lett.*, 2545 (1984).
5. E. H. Flynn, J. W. Hinman, E. L. Caron, and D. O. Woolf, Jr., *J. Am. Chem. Soc.*, **75,** 5867 (1953).
6. C. M. Lok, J. P. Ward, and D. A. van Dorp, *Chem. Phys. Lipids*, **16,** 115 (1976).
7. D. Seebach and M. Coquoz, *Chimia*, **39,** 20 (1985).
8. A. I. Meyers and C. E. Whitten, *Heterocycles*, **4,** 1687 (1976).
9. K. Mori and Y. Funaki, *Tetrahedron*, **41,** 2379 (1985).
10. P. Tkaczuk and E. R. Thornton, *J. Org. Chem.*, **46,** 4393 (1981).
11. S. J. Blarer, *Tetrahedron Lett.*, 4055 (1985).
12. T. Kamiya, *Chem. Pharm. Bull.*, **14,** 1307 (1966).
13. P. Garner, *Tetrahedron Lett.*, 5855 (1984).
14. H. Newman, *J. Am. Chem. Soc.*, **95,** 4098 (1973).
15. J. C. Sheehan, M. Goodman, and G. P. Hess, *J. Am. Chem. Soc.*, **78,** 1367 (1956).
16. Y. Saitoh, Y. Moriyama, H. Hirota, T. Takahashi, and Q. Khuong-Huu, *Bull. Chem. Soc. Jpn.*, **54,** 488 (1981).
17. Y. Saitoh, Y. Moriyama, H. Hirota, and T. Takahashi, *Bull. Chem. Soc. Jpn.*, **53,** 1783 (1980).
18. H. Newman, *J. Org. Chem.*, **39,** 100 (1974).
19. H. Newman, *Chem. Phys. Lipids*, **12,** 48 (1974).
20. Y. Saitoh, Y. Moriyama, T. Takahashi, and Q. Khuong-Huu, *Tetrahedron Lett.*, 75 (1980).
21. N. S. Chandrakumar and J. Hajdu, *Tetrahedron Lett.*, 2949 (1981).
22. N. T. Boggs, III, B. Goldsmith, R. E. Gawley, K. A. Koehler, and R. G. Hiskey, *J. Org. Chem.*, **44,** 2262 (1979).
23. J. A. Bajgrowicz, A. El Hallaoui, R. Jacquier, Ch. Pigiere, and Ph. Viallefont, *Tetrahedron Lett.*, 2759 (1984).
24. P. J. Maurer, H. Takahata, and H. Rapoport, *J. Am. Chem. Soc.*, **106,** 1095 (1984).
25. E. J. Corey and S. Bhattacharyya, *Tetrahedron Lett.*, 3919 (1977).
26. W. H. Rastetter, T. J. Erickson, and M. C. Venuti, *J. Org. Chem.*, **45,** 5011 (1980).
27. T. Itaya and A. Mizutani, *Tetrahedron Lett.*, 347 (1985).
28. G. Fölsch, *Acta Chem. Scand.*, **13,** 1407 (1959).
29. J. Butera, J. Rini, and P. Helquist, *J. Org. Chem.*, **50,** 3676 (1985).
30. von Pl. A. Plattner, A. Boller, H. Frick, A. Fürst, B. Hegedüs, H. Kirchensteiner, St. Majnoni, R. Schläpfer, and H. Spiegelberg, *Helv. Chim. Acta.*, **40,** 1531 (1957).
31. W. C. Cutting, *Handbook of Pharmacology*, Meredith, New York, 1969, p. 58.
32. P. G. Mattingly and M. J. Miller, *J. Org. Chem.*, **45,** 410 (1980).

33. F. R. Atherton and R. W. Lambert, *Tetrahedron*, **40,** 1039 (1984).
34. M. A. Krook and M. J. Miller, *J. Org. Chem.*, **50,** 1126 (1985).
35. E. M. Gordon, M. A. Ondetti, J. Pluscec, C. M. Cimarusti, D. P. Bonner, and R. B. Sykes, *J. Am. Chem. Soc.*, **104,** 6053 (1982).
36. M. J. Miller, P. G. Mattingly, M. A. Morrison, and J. F. Kerwin, Jr., *J. Am. Chem. Soc.*, **102,** 7026 (1980).
37. P. G. Mattingly, J. F. Kerwin, Jr., and M. J. Miller, *J. Am. Chem. Soc.*, **101,** 3983 (1979).
38. M. J. Miller, A. Biswas, and M. A. Krook, *Tetrahedron*, **39,** 2571 (1983).
39. A. K. Bose, D. P. Sahu, and M. S. Manhas, *J. Org. Chem.*, **46,** 1229 (1981).
40. S. Hanessian, S. P. Sahoo, C. Couture, and H. Wyss, *Bull. Soc. Chim. Belg.*, **93,** 571 (1984).
41. A. Andrus, B. Partridge, J. V. Heck, and B. G. Christensen, *Tetrahedron Lett.*, 911 (1984).
42. A. Andrus, B. Partridge, J. V. Heck, B. G. Christensen, and J. P. Springer, *Heterocycles*, **22,** 1713 (1984).
43. J. E. Baldwin, R. M. Adlington, and D. J. Birch, *J. Chem. Soc., Chem. Commun.*, 256 (1985).
44. S. R. Woulfe and M. J. Miller, *Tetrahedron Let.*, 3293 (1984).
45. S. R. Woulfe and M. J. Miller, *J. Med. Chem.*, **28,** 1447 (1985).
46. H. Arai, W. K. Hagmann, H. Suguna, and S. M. Hecht, *J. Am. Chem. Soc.*, **102,** 6631 (1980).
47. Y. Umezawa, H. Morishima, S. Saito, T. Takita, H. Umezawa, S. Kobayashi, M. Otsuka, M. Narita, and M. Ohno, *J. Am. Chem. Soc.*, **102,** 6630 (1980).
48. M. Otsuka, M. Narita, M. Yoshida, S. Kobayashi, M. Ohno, Y. Umezawa, H. Morishima, S. Saito, T. Takita, and H. Umezawa, *Chem. Pharm. Bull.*, **33,** 520 (1985).
49. M. Otsuka, A. Kittaka, T. Iimori, H. Yamashita, S. Kobayashi, and M. Ohno, *Chem. Pharm. Bull.*, **33,** 509 (1985).
50. C. M. Cimarusti, D. P. Bonner, H. Breuer, H. W. Chang, A. W. Fritz, D. M. Floyd, T. P. Kissick, W. H. Koster, D. Kronenthal, F. Massa, R. H. Mueller, J. Pluscec, W. A. Slusarchyk, R. B. Sykes, M. Taylor, and E. R. Weaver, *Tetrahedron*, **39,** 2577 (1983).
51. P. G. Mattingly and M. J. Miller, *J. Org. Chem.*, **46,** 1557 (1981).
52. C. A. Townsend, A. M. Brown, and L. T. Nguyen, *J. Am. Chem. Soc.*, **105,** 919 (1983).
53. R. S. Hodges and R. B. Merrifield, *J. Org. Chem.*, **39,** 1870 (1974).
54. T. Scott and M. Brewer, *Concise Encyclopedia of Biochemistry*, Walter de Gruyter, New York, 1983, p. 460.
55. R. B. Sykes, C. M. Cimarusti, D. P. Bonner, K. Bush, D. M. Floyd, N. H. Georgopapadakou, W. H. Koster, W. C. Liu, W. L. Parker, P. A. Principe, M. L. Rathnum, W. A. Slusarchyk, W. H. Trejo, and J. S. Well, *Nature*, **291,** 489 (1981).
56. See Ref. 36 and references cited therein.
57. D. M. Floyd, A. W. Fritz, and C. M. Cimarusti, *J. Org. Chem.*, **47,** 176 (1982).

58. Y. Shimohigashi, M. Waki, and N. Izumiya, *Bull. Chem. Soc. Jpn.*, **52,** 949 (1979).
59. M. Shiozaki, N. Ishida, T. Hiraoka, and H. Yanagisawa, *Tetrahedron Lett.*, 5205 (1981).
60. M. Shiozaki, N. Ishida, T. Hiraoka, and H. Maruyama, *Tetrahedron*, **40,** 1795 (1984).
61. T. N. Salzmann, R. W. Ratcliffe, B. G. Christensen, and F. A. Bouffard, *J. Am. Chem. Soc.*, **102,** 6161 (1980).
62. D. G. Melillo, T. Liu, K. Ryan, M. Sletzinger, and I. Shinkai, *Tetrahedron Lett.*, 913 (1981).
63. H. Maruyama, M. Shiozaki, S. Oida, and T. Hiraoka, *Tetrahedron Lett.*, 4521 (1985).
64. A. Yoshida, Y. Tajima, N. Takeda, and S. Oida, *Tetrahedron Lett.*, 2793 (1984).
65. H. Yanagisawa, A. Ando, M. Shiozaki, and T. Hiraoka, *Tetrahedron Lett.*, 1037 (1983) and references cited therein.
66. S. Hanessian, A. Bedeschi, C. Battistini, and N. Mongelli, *J. Am. Chem. Soc.*, **107,** 1438 (1985).
67. G. Fronza, C. Funganti, P. Grasselli, and G. Marinoni, *Tetrahedron Lett.*, 3883 (1979).
68. I. Dyong and R. Wiemann, *Angew. Chem., Int. Ed.*, **17,** 682 (1978).
69. G. Fronza, C. Fuganti, and P. Grasselli, *J. Chem. Soc., Chem. Commun.*, 442 (1980).
70. U. Schollkopf, J. Nozulak, and M. Grauert, *Synthesis*, 55 (1985).
71. C. Fuganti, P. Grasselli, and G. Pedrocchi-Fantoni, *Tetrahedron Lett.*, 4017 (1981).
72. R. Bernardi, C. Fuganti, and P. Grasselli, *Tetrahedron Lett.*, 4021 (1981).
73. A. K. Bose, M. S. Manhas, J. E. Vincent, K. Gala, and I. F. Fernandez, *J. Org. Chem.*, **47,** 4075 (1982).
74. S. M. Tenneson and B. Belleau, *Can. J. Chem.*, **58,** 1605 (1980).
75. A. K. Bose, M. S. Manhas, J. M. van der Veen, S. S. Bari, D. R. Wagle, V. R. Hegde, and L. Krishnan, *Tetrahedron Lett.*, 33 (1985).
76. A. K. Bose, M. S. Manhas, J. M. van der Veen, S. G. Amin, I. F. Fernandez, K. Gala, R. Gruska, J. C. Kapur, M. S. Khajavi, J. Kreder, L. Mukkavilli, B. Ram, M. Sugiura, and J. E. Vincent, *Tetrahedron*, **37,** 2321 (1981).
77. D. Seebach and J. D. Aebi, *Tetrahedron Lett.*, 3311 (1983).
78. R. A. Moss and T. B. K. Lee, *J. Chem. Soc., Perkin Trans. I*, 2778 (1973).
79. D. F. Elliott, *J. Chem. Soc.*, 62 (1950).
80. M. Yoshioka, H. Nakai, and M. Ohno, *J. Am. Chem. Soc.*, **106,** 1133 (1984).
81. B. Szechner, *Tetrahedron*, **37,** 949 (1981).
82. T. Peterson and J. B. Neilands, *Tetrahedron Lett.*, 4805 (1979).
83. Y. Nagao, T. Miyasaka, Y. Hagiwara, and E. Fujita, *J. Chem. Soc., Perkin Trans. I*, 183 (1984).
84. H. H. Wasserman and R. J. Gambale, *J. Am. Chem. Soc.*, **107,** 1423 (1985).
85. M. Kinoshita, S. Aburaki, M. Wada, and S. Umezawa, *Bull. Chem. Soc. Jpn.*, **46,** 1279 (1973).
86. S. Aburaki and M. Kinoshita, *Bull. Chem. Soc. Jpn.*, **52,** 198 (1979).
87. S. Aburaki and M. Kinoshita, *Chem. Lett.*, 701 (1976).
88. K. Saito, S. Saijo, K. Kotera, and T. Date, *Chem. Pharm. Bull.*, **33,** 1342 (1985).

89. Y. Ohfune and N. Kurokawa, *Tetrahedron Lett.*, 1587 (1984).
90. P. J. Maurer, C. G. Knudsen, A. D. Palkowitz, and H. Rapoport, *J. Org. Chem.*, **50,** 325 (1985).
91. K. Hintzer, B. Koppenhoefer, and V. Schurig, *J. Org. Chem.*, **47,** 3850 (1982).
92. J. L. Morell, P. Fleckenstein, and E. Gross, *J. Org. Chem.*, **42,** 355 (1977).
93. M. Shiozaki, N. Ishida, T. Hiraoka, and H. Maruyama, *Chem. Lett.*, 169 (1983).
94. M. Shiozaki, N. Ishida, H. Maruyama, and T. Hiraoka, *Tetrahedron*, **39,** 2399 (1983).

CHAPTER SIX

SULFUR-CONTAINING AMINO ACIDS

6.1 CYSTEINE

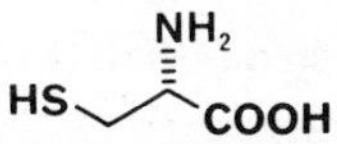

NH2
HS COOH

(*R*)-2-Amino-3-mercaptopropanoic acid (**1**)

L-Cysteine (**1**) is a sulfur-containing proteogenic amino acid that is important for the tertiary structure of proteins and/or enzymatic activity. In solution at neutral or alkaline pH, **1** is easily air-oxidized to L-cystine.[1] Unlike the other naturally occurring L-amino acids, which possess the (*S*) configuration, L-cysteine has the (*R*) configuration as a result of the prioritization of sulfur in the Cahn–Ingold–Prelog (CIP) sequence rules.[2]

As a β-mercaptoalkylamine, **1** provides both the nitrogen and sulfur atoms necessary for the syntheses of a variety of valuable heterocyclic compounds. The inherent chirality of **1** offers the extra advantage of introducing asymmetry into these heterocyclic systems.

Thiazolidines, a somewhat neglected class of heterocyclic compounds, received considerable attention as a result of the importance of these compounds in the chemistry of carbonyl compounds with β-mercaptoalkylamines.[3,4]

The treatment of **1** with 40% formaldehyde at room temperature, followed by the addition of ethanol and pyridine, provies (4*R*)-4-thiazolidinecarboxylic acid (**2**) in 92% yield and with no racemization.[5–7] Aliphatic,[7–10] aromatic,[8–12] and heterocyclic aldehydes[8–10] react readily and in good yields to afford 2-substituted-(4*R*)-4-thiazolidinecarboxylic acids (**3**) (Scheme 1 and Table 6.1). Aldehydic sugars, such as D-(+)-galactose[13] or L-arabinose,[14] react as aldehydes with **1** to afford **3**, in which the 2-substituent is a sugar moiety.

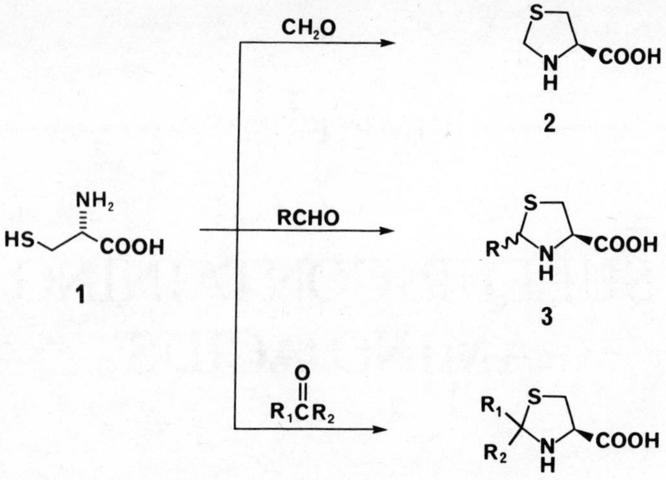

CH_2O

RCHO

R_1COR_2

Scheme 1

TABLE 6.1. 2-Monosubstituted (4*R*)-4-Thiazolidinecarboxylic Acids (3)

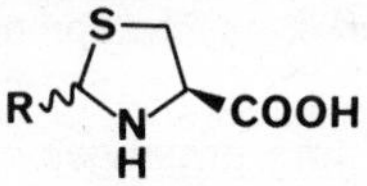

R	Yield **3** (%)	Reference
CH_3	42 (24[7])	9
n-C_3H_7	54	9
n-C_6H_{13}	99 (52.5)[9]	8
C_6H_5	75.5	9
$C_6H_5CH_2$	90	8
4-HOC_6H_4	93	8
2′-Thienyl	94	8
2-Furanyl	55.2	9
4-Pyridyl	47	9
4-ClC_6H_4	70	9
4-$NO_2C_6H_4$	37	9

Aliphatic ketones,[7,9,10,15,16] as well as certain 3-ketosteroids,[17] react with **1** to successfully afford the corresponding 2,2-dialkyl-(4*R*)-4-thiazolidinecarboxylic acids (**4**) (Scheme 1 and Table 6.2). However, some carbonyl compounds such as certain α,β-unsaturated aldehydes and ketones, benzophenone, acetophenone, α-tetralone, and glucose may fail to provide thiazolidines.[4]

TABLE 6.2. 2,2-Dialkyl-(4*R*)-4-thiazolidinecarboxylic Acids (4)

R1 S N H COOH

4

R_1	R_2	Yield (%)	Reference
CH_3	CH_3	96.5	7
CH_3	CH_2CH_3	70 (35.5)[9]	15
CH_3	i-C_3H_7	44	15
CH_2CH_3	CH_2CH_3	35.5	9
$-(CH_2)_4-$		56	15
$-(CH_2)_5-$		47	9
3,17-Androstanedione		67	17
3,20-Pregnanedione		59	17

Antihypertensive agents exhibiting inhibitory activity against angiotensin-converting enzyme (ACE) have been prepared from **2–4** by acylating the basic nitrogen of the thiazolidine with a variety of sulfur-containing acid chlorides.[10,18] Interestingly, the type of product obtained from this acylation reaction with **3** depends on the method used to carry out the acylation. Under aqueous conditions (Na_2CO_3/H_2O), the (2*R*,4*R*)-isomer **5** is obtained in 88% yield with a selectivity of not less than 97%. On the other hand, carrying out the acylation in pyridine results in the isolation of a nearly 1 : 1 mixture of **5** and its (2*S*)-enantiomer **6.**[12] Spectroscopic as well as X-ray structure studies confirm the *cis* geometry with an absolute configuration of (2*R*,4*R*).[19]

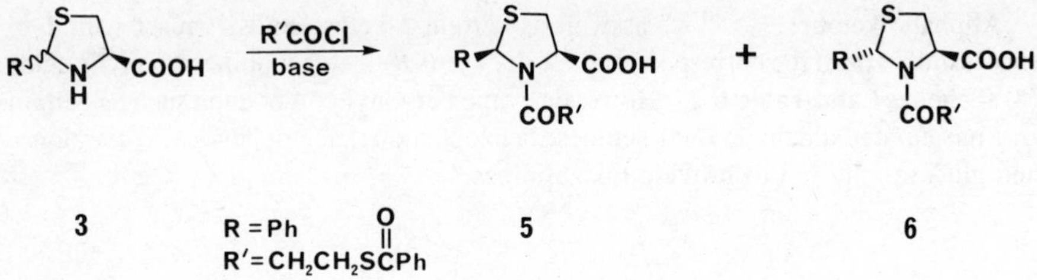

Deprotection of the sulfur followed by oxidation to their symmetrical disulfides affords products that show an inhibition of aldose reductase (AR), the enzyme involved in the initiation of cataract formation.[20]

Nucleoside derivatives such as **11** and **12,** which incorporate 2,2-dimethyl-4(*R*)-carboxythiazolidine (**7**), have been prepared for possible anticancer activity (Scheme 2). It is important to note that when **7** is treated with benzoyl chloride in 200 equivalents of pyridine and methylated with diazomethane, only the (4*R*)-ester **8** is obtained. The use of only 2 equivalents of pyridine followed by esterification with diazomethane affords a racemic mixture.[16]

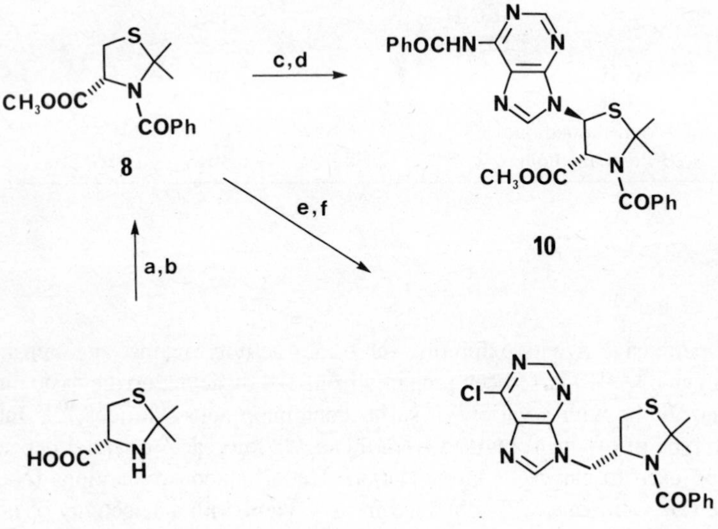

Scheme 2. (a) ClCOPh/pyridine (200 equivalents) (74%); (b) CH_2N_2 (83%); (c) dibenzoyl peroxide, benzene (20%); (d) 6-benzamidopurine (**9**) (38%); (e) $NaBH_4$ (80%); (f) 6-chloropurine (**11**) (62%).

A feature of the thiazolidine ring system that makes it useful synthetically is its inability to undergo a cyclopentanelike pseudorotation. The introduction of the heteroatoms, various substituents, and the sp^2-hybridized nitrogen create a potential barrier restricting any pseudorotation.[19]

Antimicrobial substances such as penicillins and its congeners cephalosporins, clavulanic acid, the norcardicins and thienamycin require the β-lactam moiety for their valuable activity. In Woodward's initial total synthesis of cephalosporin C (**19**), the thiazolidine **13**, easily prepared in two steps from **1**, provides the restricted rotation around C-2–C-3 needed to stereospecifically functionalize α to the sulfur atom. Treatment of the resulting *cis*-amino ester **14** with triisobutylaluminum affords the bicyclic β-lactam **15**, which, when converted to the adduct **17** by condensation with dialdehyde **16**, rearranges to the thiazine **18** with trifluoroacetic acid. Subsequent functionalization of **18** provides **19**, which is identical with the natural material[21] (Scheme 3).

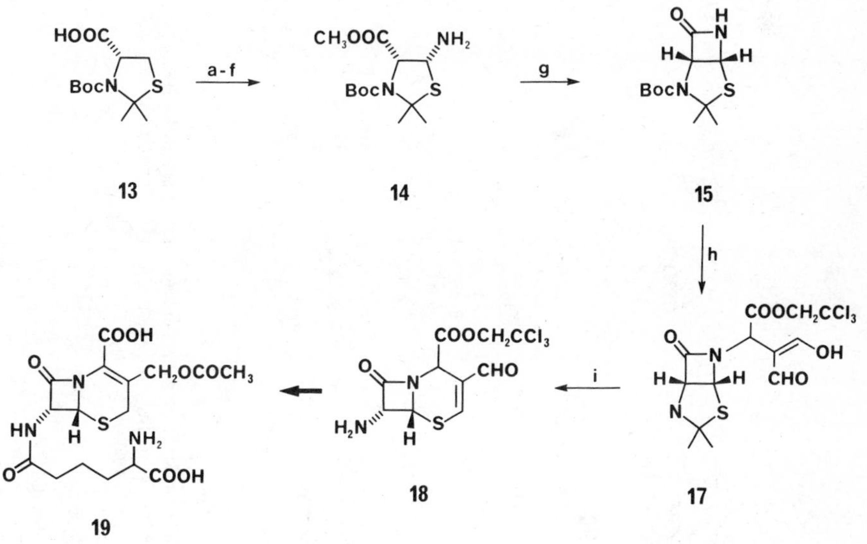

Scheme 3. (a) CH_2N_2; (b) DMAD; (c) $Pb(OAc)_4$, benzene; (d) MsCl/DMF; (e) NaN_3; (f) Al·Hg/CH_3OH; (g) $(i\text{-}C_4H_9)_3Al$; (h) $Cl_3CCH_2OOCCH{=}C(CHO)_2$ (**16**); (i) CF_3COOH.

Approaching the synthesis from a more biogenetic perspective in which the penam and cepham ring systems are derived from a tripeptide, Baldwin[22,23] prepares 6-aminopenicillanic acid (**25**). The thiazolidine **20** allows for the stereospecific functionalization necessary for the preparation of the bicyclic β-lactam **22.** An acid-catalyzed ring opening of **22** affords the ketosulfide **23**, which, after a series of transformations, is converted to the penicillin sulfoxide **24** (overall yield of ~11% from **22**). Smooth deoxygenation yields **25**[23] (Scheme 4).

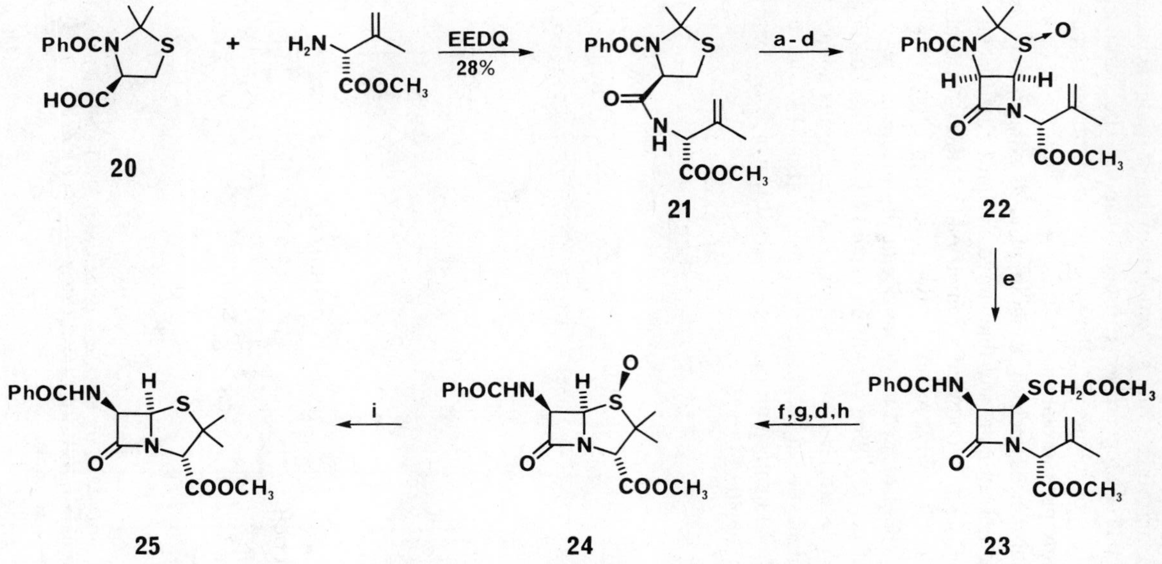

Scheme 4. (a) Benzoyl peroxide (40%); (b) HCl/CH_2Cl_2 (94%); (c) $NaH/DMF/CH_2Cl_2$ (82%); (d) MCPBA (67%); (e) H_2SO_4/benzene/DMA (41%); (f) CH_2N_2 (53%); (g) $BF_3 \cdot Et_2O$; (h) benzene, 80°C (21%); (i) PBr_3/DMF (61%).

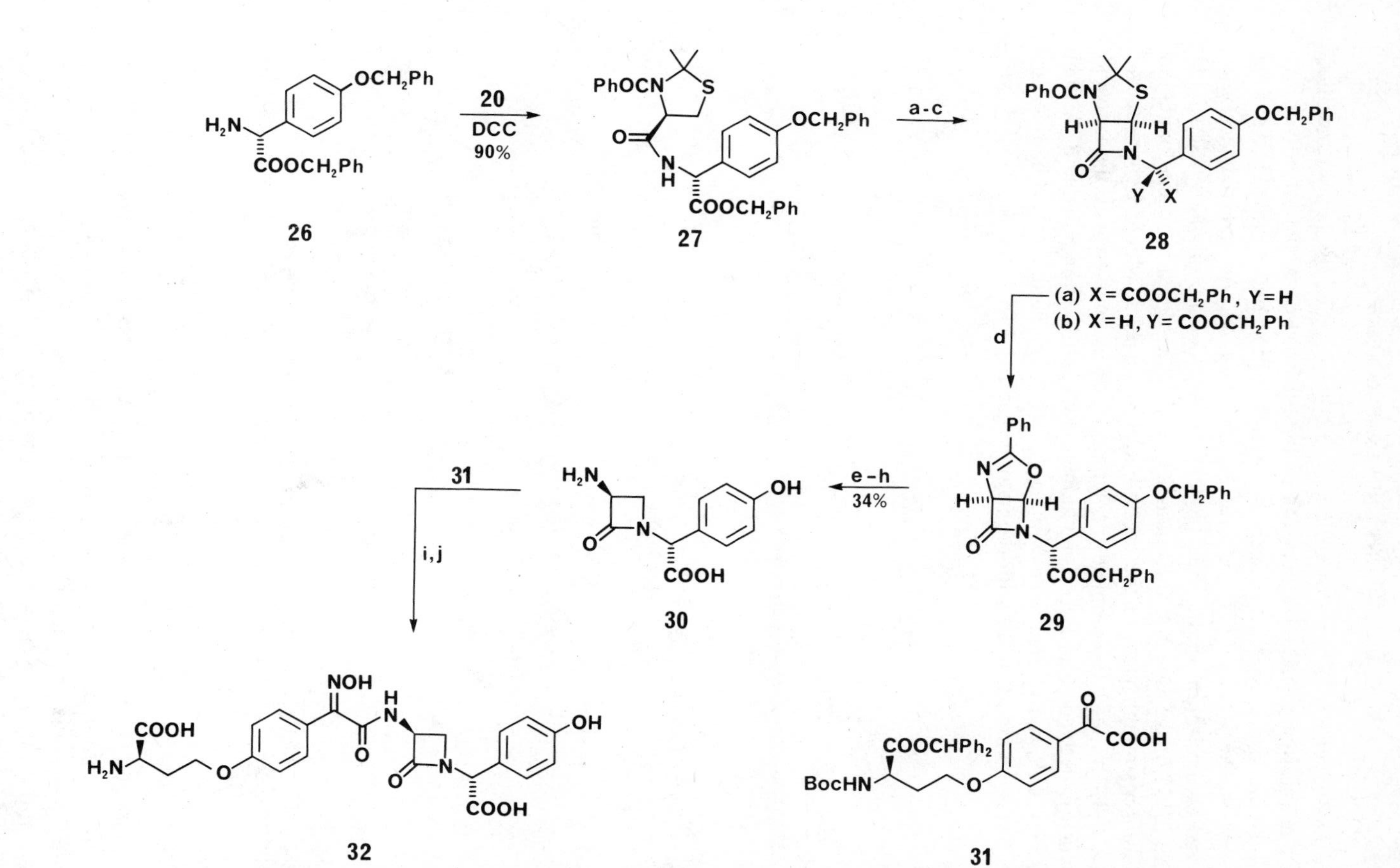

Scheme 5. (a) Benzoyl peroxide (45–55%); (b) HCl/CH_2Cl_2 (100%); (c) NaH (85%); (d) $Hg(OAc)_2$; (e) PCl_5/CH_2Cl_2; (f) n-Bu_3SnH; (g) $PTSA/H_2O$; (h) 5% Pd/C; (i) CF_3COOH; (j) $NH_2OH \cdot HCl$.

The monocyclic β-lactam norcardicin A (**32**) is prepared by using a modified version of this synthesis.[24] The reaction of the dibenzyl derivative of *p*-hydroxyphenylglycine (**26**) with **20** furnishes the protected dipeptide **27** (90%), which is converted stereospecifically to the bicyclic β-lactam **28.** An unavoidable base-catalyzed epimerization during ring closure to the β-lactam provides a separable mixture (3 : 1) of the isomeric β-lactams **28a** and **28b.** Mercuric acetate oxidation of **28a** affords a high yield of the crystalline oxazoline **29,** which is converted to 3-ANA (**30**). In order to complete the synthesis, **30** is coupled with the ketoacid **31,** which is derived from Boc–D-methionine **33**[25] as shown in Scheme 6. A final deblocking followed by treatment with aqueous hydroxylamine hydrochloride furnishes **32,** which is identical with the natural material[24] (Scheme 5).

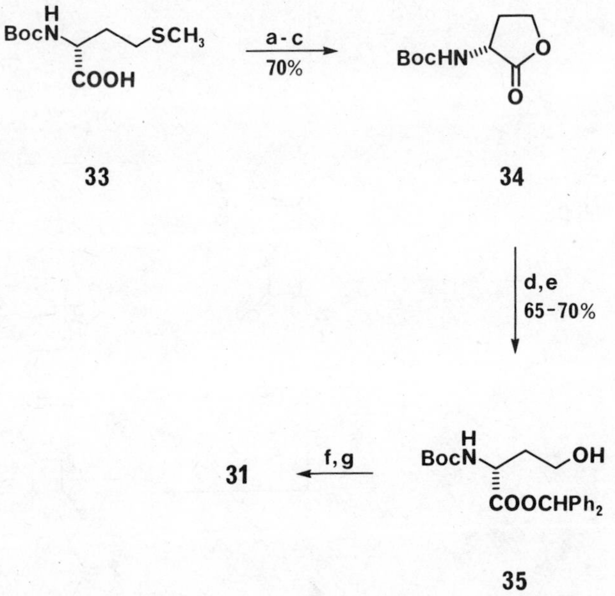

H
BocN SCH₃
COOH
33
a-c
70%
BocHN O
O
34
d,e
65-70%
31 f,g
H
BocN OH
COOCHPh₂
35

Scheme 6. (a) TMSCl, CH_3CN; (b) CH_3I; (c) *t*-BuOK/THF; (d) KOH/H_2O:dioxane (1:1); (e) Ph_2CHBr/18-crown-6; (f) (4-hydroxyphenyl)glyoxylic acid *p*-nitrobenzyl ester (68%); (g) NaOH/dioxane/H_2O (90%).

A shortened stereospecific synthesis of penicillins takes advantage of the stereoelectronic control offered by the bicyclic thiazolidine **36.** Although **36** is prepared in rather poor yield from the reaction of **1** with methyl levulinate, the fixed geometry of the bicyclic system brings the more desirable S–C_5 bond of the tricyclic β-lactam **38** into greater orthogonality, a feature necessary for its preferred cleavage. Consequently, treatment of **38** with *tert*-butyl hypochlorite in wet THF affords the 2β-chloromethylpenam **39,** which quantitatively isomerizes to the 3β-chloro-cepham **40** by standing neat at room temperature or warming to 45°C in DMF[26] (Scheme 7).

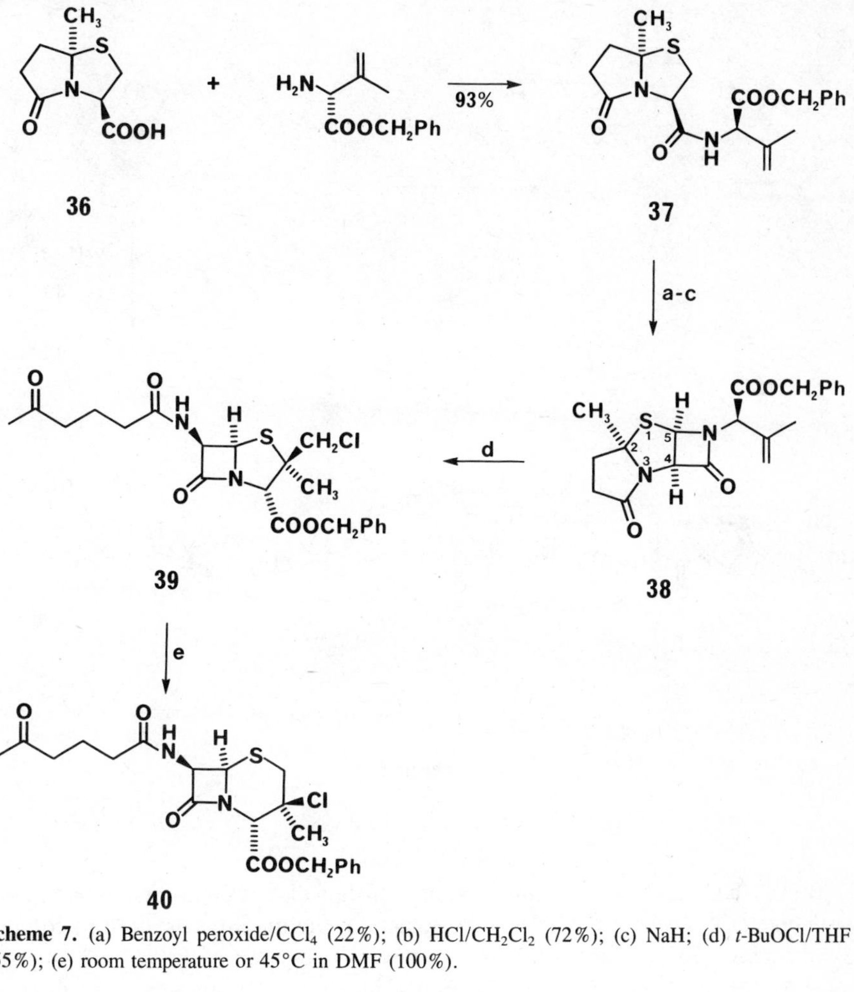

Scheme 7. (a) Benzoyl peroxide/CCl_4 (22%); (b) HCl/CH_2Cl_2 (72%); (c) NaH; (d) *t*-BuOCl/THF (55%); (e) room temperature or 45°C in DMF (100%).

The antibiotic althiomycin (**46**), isolated from *Streptomyces althioticus*,[27] possesses a unique structure where the (*S*) configuration of the thiazoline moiety is derived from D-cysteine. The coupling of the active ester of *N*,*S*-ditrityl-D-cysteine (**41**) with the sodium salt of 4-methoxy-3-pyrrolin-2-one (**42**) followed by a detritylation affords amino thiol **43.** Condensation with ethyl Cbz–aminoacetimidate (**44**) provides the thiazoline **45** (45% overall yield from **41**). Further elaboration of **45** provides the natural product **46**[28a,b] (Scheme 8).

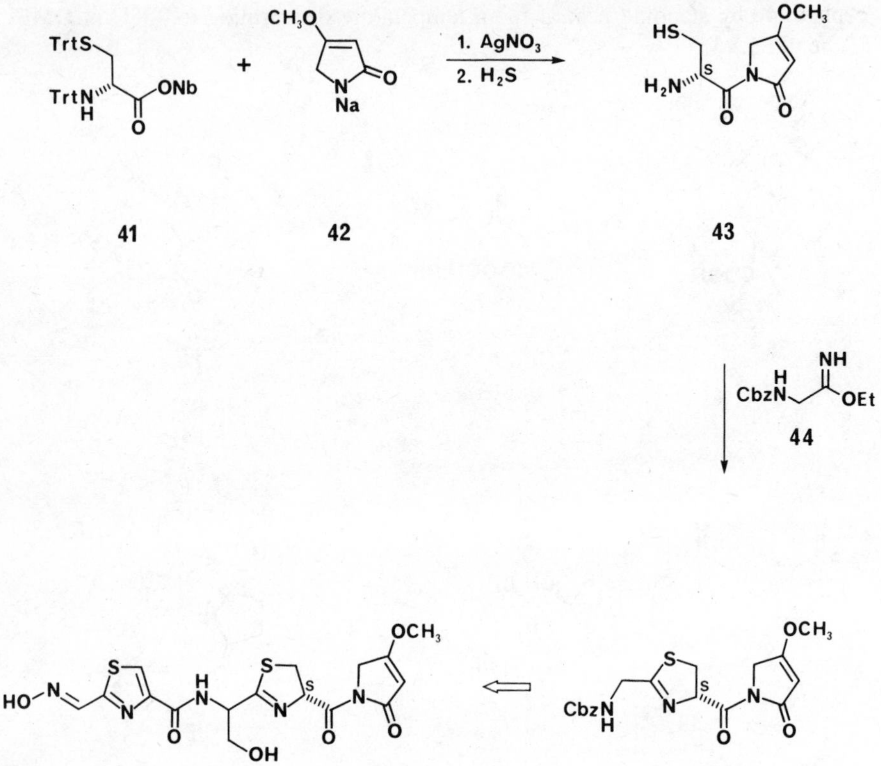

Scheme 8

Biotin, also known as vitamin H, is nutritionally essential and is a growth promotant. Of the eight possible stereoisomers, D-biotin (**53**) is the most important. Synthetically, **53** is accessible from L-cysteine (**1**) by way of the (2*R*,4*R*)-thiazolidine **47** (easily prepared from **1** and benzaldehyde). After conversion of the acid to aldehyde **48** [>90%(*R*) at C-4], a Grignard reaction followed by a Claisen rearrangement furnishes the *trans*-olefin **49** in excellent yield. When treated with pyridinium hydrobromide perbromide, **49** undergoes a dramatic oxidative cyclization–rearrangement without any C-3 epimerization to provide **50.** After equilibration of **50** to **51** in acetic acid, elucidation to D-bisnorbiotin methyl ester (**52**) is accomplished with azide under S_N2 conditions. Chain elongation completes the sequence to give **53**[29,30] (Scheme 9).

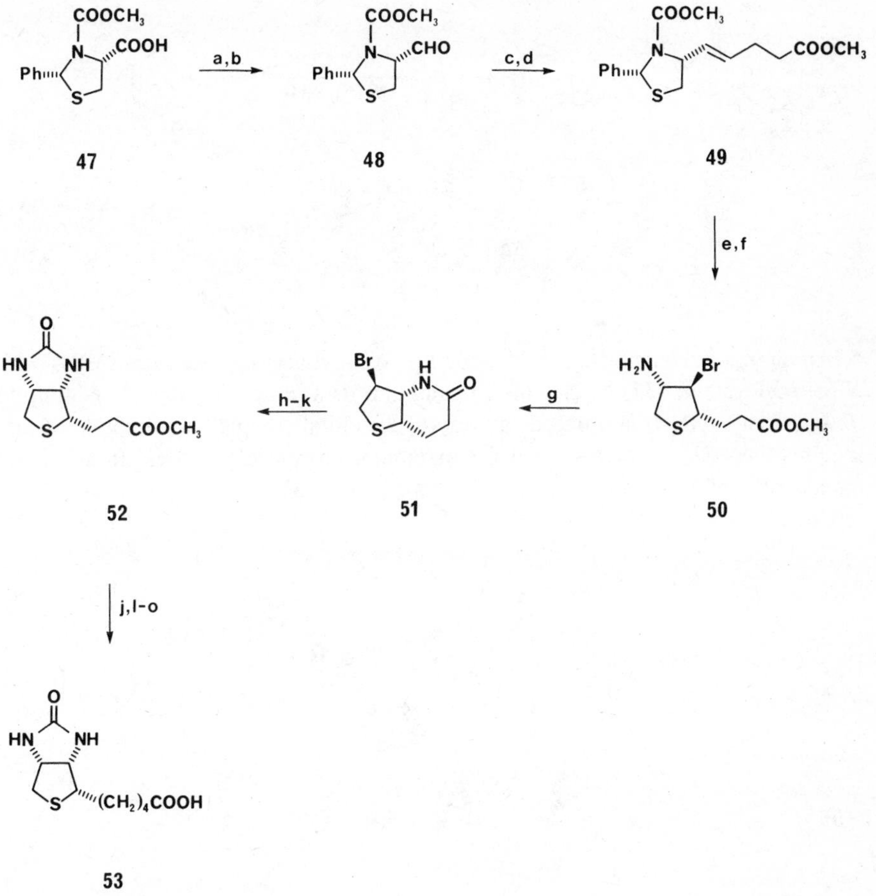

47 →(a,b) 48 →(c,d) 49 →(e,f) 50 →(g) 51 →(h–k) 52 →(j,l–o) 53

Scheme 9. (a) B_2H_6 (99%); (b) CrO_3, py (80%); (c) $CH_2{=}CHMgCl$, CH_2Cl_2 (87%); (d) $CH_3C(OCH_3)_3$, benzene, 92°C (95%); (e) py · Br_2 · HBr (47%); (f) HBr/CH_3COOH (79%); (g) HOAc/NaOAc (96%); (h) LiN_3/DMF (16%); (i) 10% Pd/C, H_2 (68%); (j) $Ba(OH)_2$, H_2O; (k) phosgene (45%); (l) $LiBH_4$/THF (87%); (m) HBr/HOAc (68%); (n) $CH(COOEt)_2$/Na; (o) 190°C.

The chirality possessed by thiazolidines prepared from L-cysteine provides an opportunity for asymmetric induction in the enantioselective homogeneous hydrosilylation of ketones with diphenylsilane.[31,32] With the procatalyst $[Rh(COD)Cl]_2$ and thiazolidine **54,** prepared from 2-pyridine carboxaldehyde and L-cysteine ethyl ester, the reduction of acetophenone proceeds with 97% ee and in quantitative yield. However, in the case of benzyl methyl ketone or *t*-butyl methyl ketone, best results are obtained with thiazolidines other than **54.** This result suggests that the reaction is substrate dependent with respect to the thiazolidine. Interestingly, optically pure thiazolidines as well as those of diastereomeric nonhomogeneity at C-2 give similar inductions in the hydrosilylation reaction.

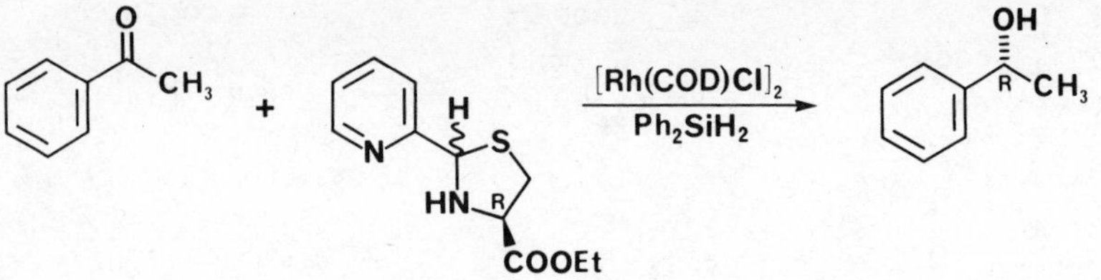

54

An alternative enantioselective reduction of alkyl aryl ketones with $LiBH_4$ with *N*-benzoylcysteine (**55**) as the chiral ligand affords asymmetric alcohols with up to 92% ee. This reduction proceeds by way of the chiral complex borohydride **56** and requires the addition of *t*-butanol for maximum enantioselectivities. In all cases, the alcohols of (*R*) configuration are obtained from (*S*)-**55**.[33]

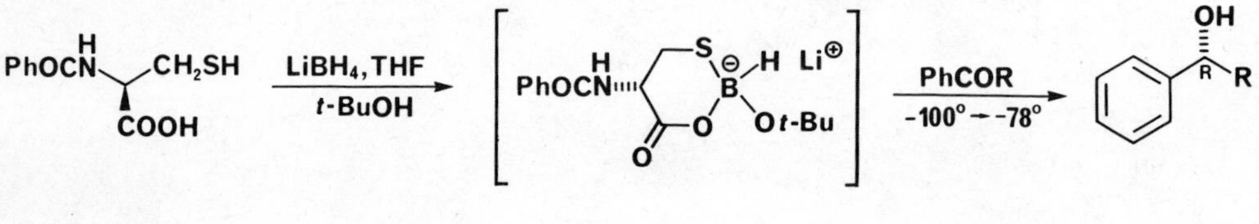

(S)-**55** **56**

R	Yield (%)	Percent ee
CH_3	66	87
C_2H_5	51	89
n-C_3H_7	44	92

L-Cysteine methyl ester, on reaction with carbon disulfide, forms the chiral auxiliary (4*R*)-(methoxycarbonyl)-1,3-thiazolidine-2-thione (**57**) [(4*R*)-MCTT] with complete retention of the (4*R*) configuration. Acylation on nitrogen with a variety of acid chlorides affords 3-acyl-(4*R*)-methoxycarbonyl-1,3-thiazolidine-2-thione derivatives (**58**) [(4*R*)-AMTT]. Aminolysis of **58** with racemic amines results in a chiral recognition leading to optically active amides **59** [(*S*) excess] with 12.4–64.4% ee and amines **60** [(*R*) excess][34] (Scheme 10). This methodology has been successfully used to determine the absolute configuration of chiral amines.[35]

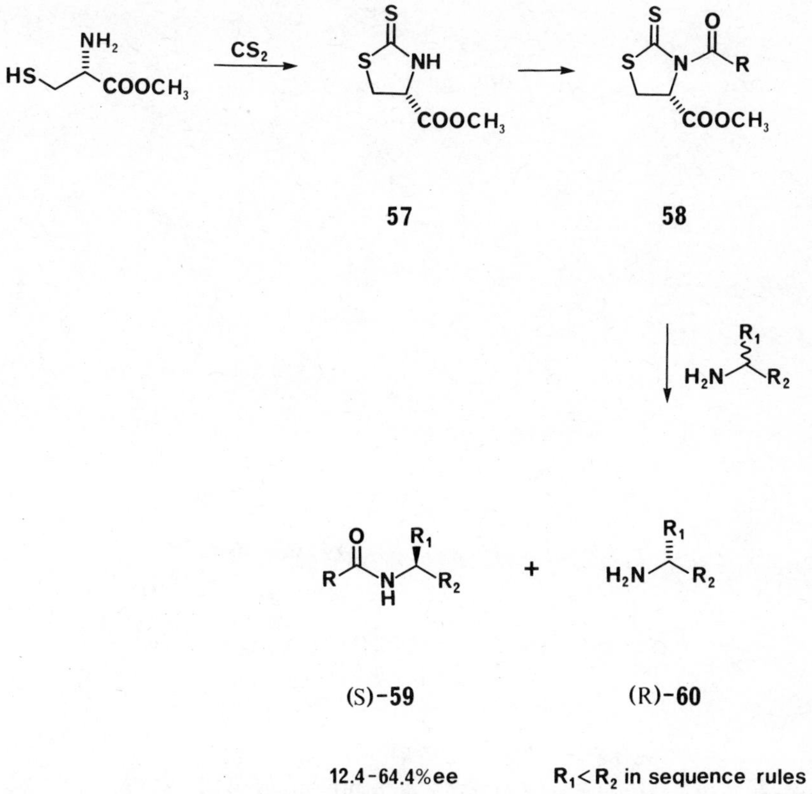

57 58

(S)-59 (R)-60

12.4-64.4%ee $R_1 < R_2$ in sequence rules

Scheme 10

One of the most exciting applications of this methodology is to the highly regioselective differentiation between two identical groups in an acyclic molecule having a prochiral center [e.g., 3-methylgutaric acid (**61**)]. Under the standard chemical conditions, it is very difficult to distinguish the pro-(*S*) ligand ($HOOCCH_2$–) from the pro-(*R*) ligand (–CH_2COOH) of **61.** However, in the diamide **62,** which results when two molecules of **57** react with **61,** the sterical situation between the two groups is sufficiently different that a suitable nucleophile preferentially attacks the amide carbonyl of the pro-(*S*) side from the least hindered α face in the transition state. The best nucleophiles displaying this anticipated regioselectivity are cyclic secondary amines. Treatment of **62** with piperidine provides a separable mixture of diastereomers with component **63** predominating. This is easily converted to (−)-(3*S*)-3-methylvalerolactone (**64**). This novel nonenzymatic asymmetric synthesis holds considerable promise for the differentiation of enantiotopic groups in *meso* compounds[36] (Scheme 11).

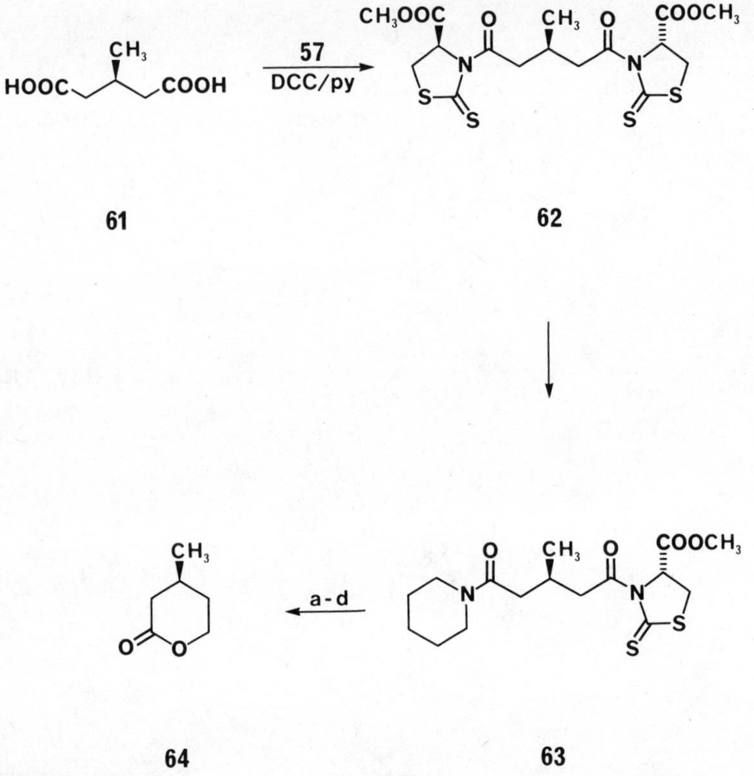

Scheme 11. (a) 4-Bromothiophenol/NaH (97%); (b) $NaBH_4$/THF (84%); (c) 6 *N* HCl; (d) benzene, reflux (65%).

The ability of **1** to readily form various heterocyclic systems is also useful for the α-hydroxyalkylation of **1** without racemization and without the use of a chiral auxiliary.[37,38] The reaction of **1** with formaldehyde and pivalaldehyde readily provides (2*R*,5*R*)-2-*tert*-butyl-1-aza-3-oxa-7-thiabicyclo[3.3.0]octan-4-one (**65**) as a single stereoisomer. While the enolate cannot be generated directly because of a facile β elimination of the thiolate, addition of **65** to a combined solution of LDA and a nonenolizable aldehyde (such as benzaldehyde) affords the adduct **66.** Of the four possible diastereomeric products, only one is formed with >90% selectivity.[37] More importantly, **66** can undergo various possible transformations to monocyclic or acyclic amino acid derivatives such as the oxazolidinones **67** and **71** (cleavage of the sulfur-containing hetero ring), thiazolidines **70, 73** and **74** (cleavage of the cyclic *N*,*O*-acetal), and the α-branched cysteine **68** as well as the phenylserines **69** and **72** (cleavage of both rings)[38] (Scheme 12).

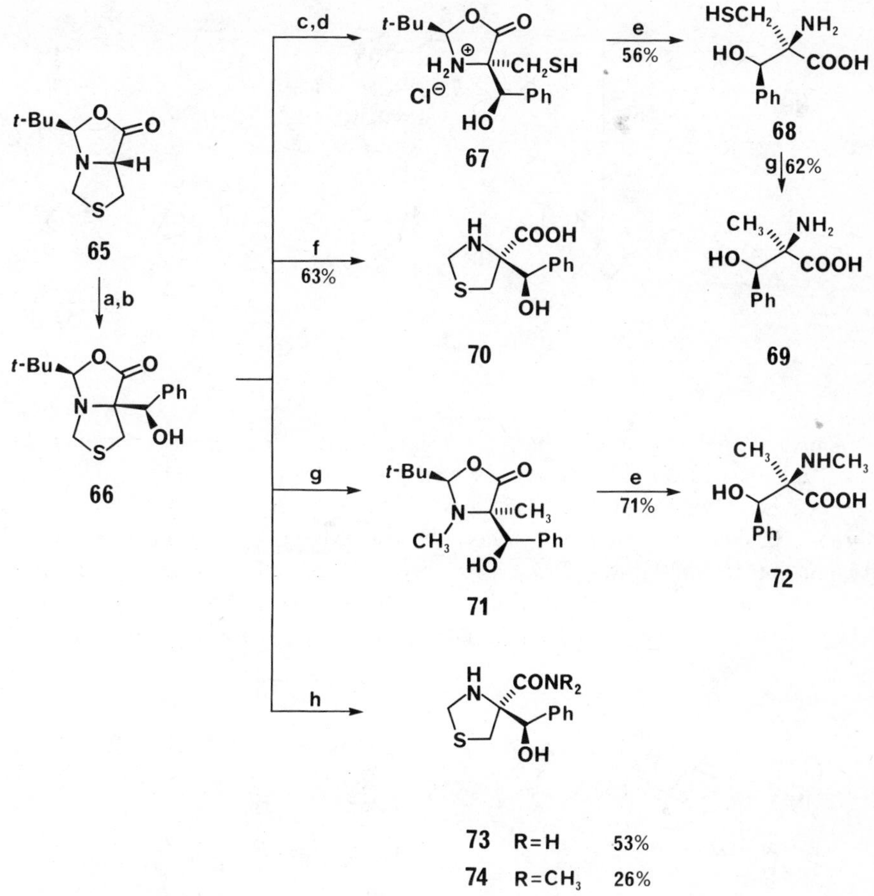

Scheme 12. (a) Benzaldehyde/LDA, $-100°C$; (b) H_3O^+; (c) $HgCl_2$/EtOH; (d) H_2S; (e) 6 *N* HCl; (f) 48% HBr; (g) Raney Ni; (h) $LiNR_2$.

The synthesis of chiral thiirancarboxylates from **1** illustrates another valuable synthetic application of this versatile amino acid. Optically pure (*S*)-thiirancarboxylate esters **75** are prepared by diazotization of the corresponding L-cysteinate esters. Hydrolysis of these esters with sodium hydroxide occurs without significant racemization to provide (*S*)-thiirancarboxylic acid (**76**).[39,40] L-Cysteine (**1**) also undergoes the deaminative cyclization but affords instead the (*R*)-thiirancarboxylic acid (**78**) with 53% optical purity. Mechanistically, two competing pathways provide this observed result: (1) the direct S_N2 displacement of nitrogen by the thiol group (with inversion) to give **75** (less preferred route) and (2) a displacement of nitrogen by the carboxy group followed by an isomerization of this α-lactone **77** to **78**[40] (Scheme 13).

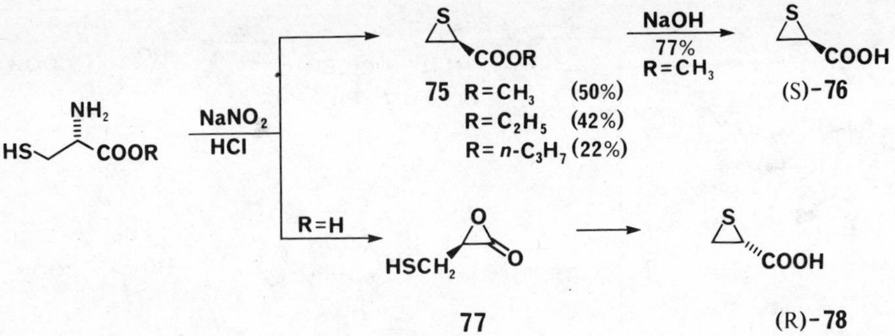

Scheme 13

A natural product that has attracted considerable attention because of its unusual chiral dithioacetal *S*-oxide moiety, as well as its antitumor activity, is (*S*)-(+)-sparsomycin (**79**), a metabolite of *Streptomyces sparsogenes*[41] and *S. curpidosporus*.[42] Besides its antitumor activity, **79** also exhibits activity against various bacteria, fungi, and viruses. From a retrosynthetic analysis, **79** can be viewed as an amide consisting of β-(6-methyluracil)acrylic acid (**80**) and the chiral amine **81**, which is ultimately available from either D-cysteine or L-serine[43] (Scheme 14).

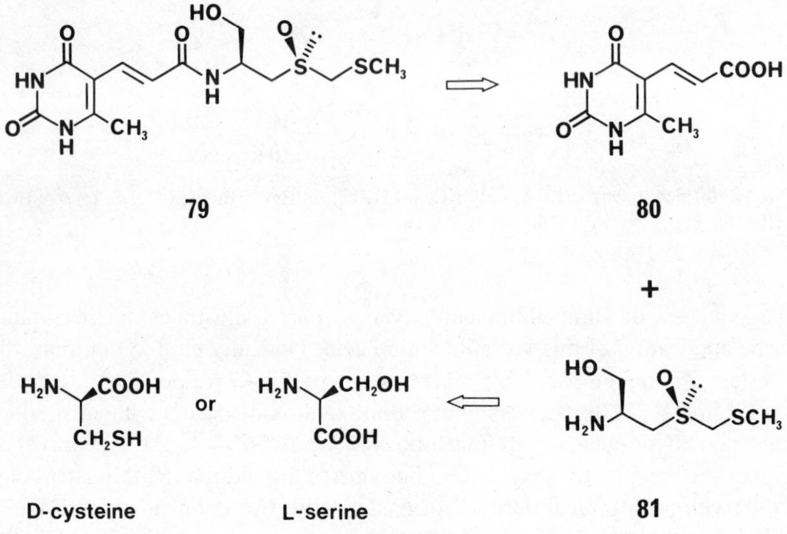

Scheme 14

Helquist has successfully prepared the (R_c)-enantiomer of **79** from inexpensive L-cysteine (**1**)[43,44] (Scheme 15).

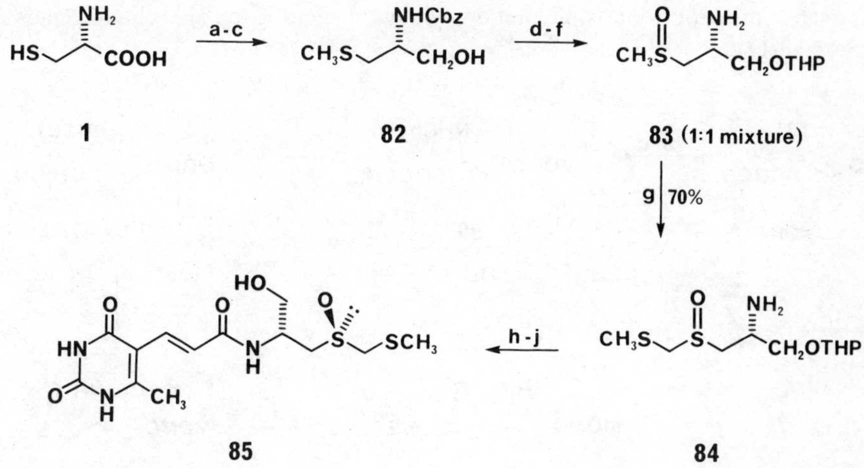

Scheme 15. (a) $NaOCH_3/CH_3I$ (88%); (b) $LiAlH_4$ (70%); (c) Cbz–Cl/NaOH (88%); (d) $NaIO_4/CH_3CN$ (90%); (e) DHP/PTSA (93%); (f) Na/NH_3 (83%); (g) 2 LDA/CH_3SSCH_3; (h) chromatographic separation; (i) **80** (51%); (j) 1 *N* HCl (80%).

For preparation of **79** with the (*S*) configuration, D-cysteine is required. However, because of its significant cost, Helquist approaches the synthesis using L-cysteine (**1**), in which a racemization and a resolution step is introduced a short distance into the synthesis[43] (Scheme 16).

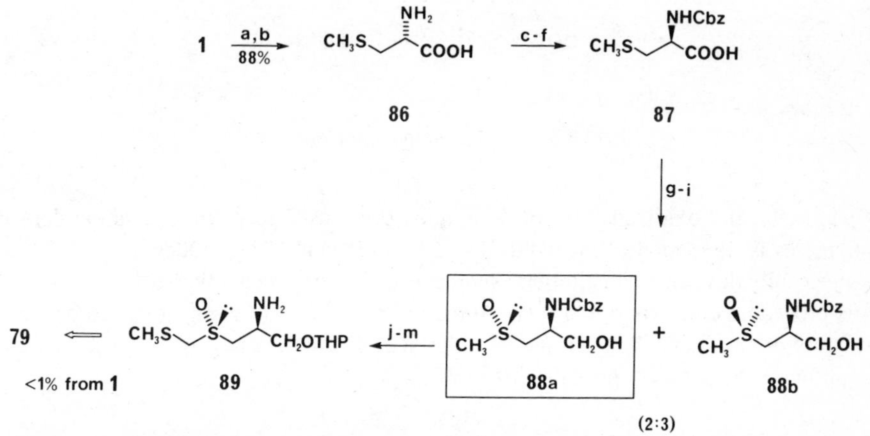

Scheme 16. (a) $NaOEt/CH_3I$; (b) HI; (c) HOAc (racemization); (d) HCl (87%); (e) Cbz–Cl (81%); (f) ephedrine (60–65%); (g) CH_2N_2 (98%); (h) $LiBH_4$ (51%); (i) $NaIO_4$ (90%); (j) chromatographic separation; (k) DHP/PTSA (97%); (l) Na/NH_3 (72%); (m) 2 LDA/CH_3SSCH_3 (40%).

A more attractive approach to the synthesis of **79** involves the use of L-serine. By operating separately on the carboxy and hydroxy groups, one accomplishes a formal ''inversion'' of configuration without operating on the chiral center[43] (Scheme 17).

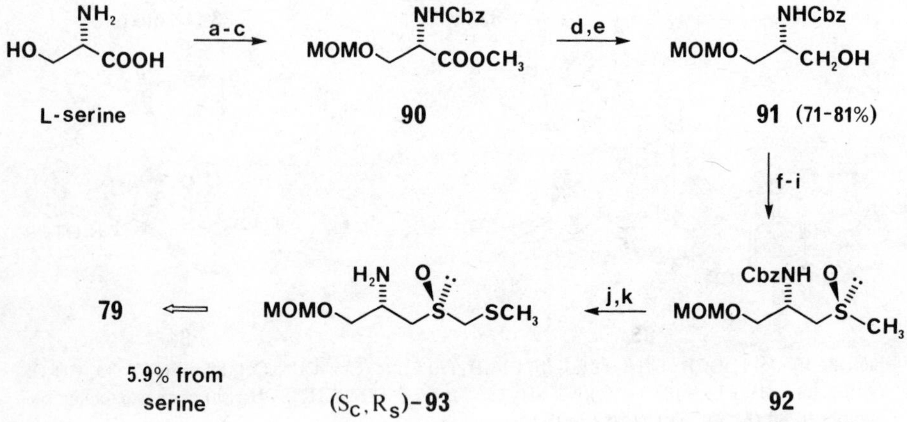

L-serine → (a-c) → 90 → (d,e) → 91 (71-81%) → (f-i) → 92 → (j,k) → (S_C, R_S)-93 ⇒ 79 (5.9% from serine)

Scheme 17. (a) Cbz–Cl (90%); (b) $SOCl_2/CH_3OH$ (95%); (c) $ClCH_2OCH_3$/Huenig's base (91%); (d) DIBAH; (e) $NaBH_4$; (f) TsCl (98%); (g) $NaSCH_3$ (91%); (h) $NaIO_4$ (92%, 1 : 1 mixture); (i) chromatographic separation; (j) Na/NH_3 (45–71%); (k) LDA/CH_3SSCH_3 (40%).

6.2 CYSTINE

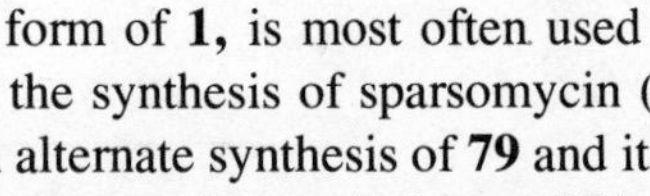

94

(*R*, *R*′)-3,3′-Dithiobis(2-aminopropionic acid)

Cystine **94,** the oxidized form of **1,** is most often used to prepare cysteine derivatives, as is illustrated in the synthesis of sparsomycin (**79**).[47] Ottenheijm[45–48] has successfully developed an alternate synthesis of **79** and its enantiomer from cystine. His procedure converts Boc–D-cystine methyl ester **95** in five steps to the separable mixture of amino alcohols **81** and **96** (14 and 25% yield, respectively). Subsequent coupling of **81** with **80** provides **79** in 33% yield.[47]

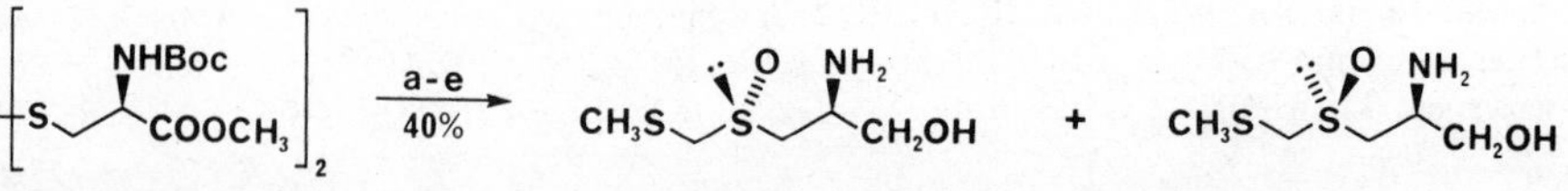

95 → (a-e, 40%) → **81** + **96**

(a) Cl_2/Ac_2O; (b) CH_2N_2; (c) $LiBH_4$, separation; (d) $NaSCH_3$; (e) CF_3COOH.

Another alternative synthesis of **81** involves a nucleophilic ring opening of a cyclic sulfinate **97**, a γ-sultine.[49] Such an intermediate has the sulfur atom activated toward nucleophilic attack and simultaneously provides a protected alcohol function.[48] Prepared from **94** as a 1 : 1 mixture of diastereomers, the undesired **97a** can be converted to **97b** via epimerization by heating at 120–130°C.[50] In this way, an optimal yield of **97b** is possible after chromatographic separation. Moreover, when **97b** is treated with (methylthio)methyl lithium, smooth ring opening occurs with inversion at sulfur to afford **98** in 70% yield. Acidic deprotection affords **81** (Scheme 18).

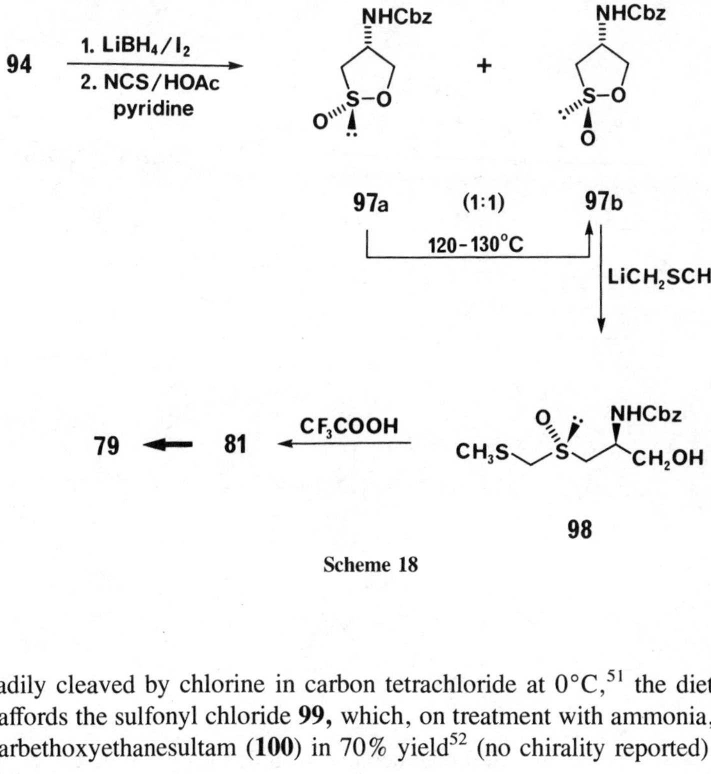

94 —(1. $LiBH_4/I_2$; 2. NCS/HOAc, pyridine)→ 97a + 97b (1:1)

97a —(120–130°C)→ 97b

97b —($LiCH_2SCH_3$)→ 98 (CH_3S–CH_2–S(O)–CH_2–CH(NHCbz)–CH_2OH)

98 —(CF_3COOH)→ 81 → 79

Scheme 18

Readily cleaved by chlorine in carbon tetrachloride at 0°C,[51] the diethyl ester of **94** affords the sulfonyl chloride **99**, which, on treatment with ammonia, cyclizes to 3-carbethoxyethanesultam (**100**) in 70% yield[52] (no chirality reported).

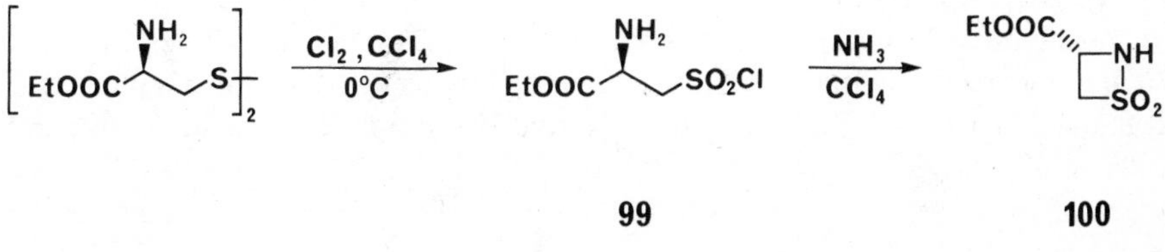

[EtOOC–CH(NH_2)–CH_2–S–$]_2$ —(Cl_2, CCl_4, 0°C)→ EtOOC–CH(NH_2)–CH_2–SO_2Cl (**99**) —(NH_3, CCl_4)→ **100**

In this manner, *N*-Cbz–L-cysteamide ethyl ester (**102**) is prepared in good yield from **94.**[52] This is an important intermediate for the preparation of the heterocyclic compound **105,** which possesses a tetrahedral sulfur atom. It is presumed that **105** would mimic the tetrahedral transition state for the enzymatic conversion of *N*-carbamylaspartate to dihydroorotate and thus act to inhibit this process. Such an inhibition could have implications for the development of anticancer drugs.[53]

Hydrogenation of **102** provides the sulfonamide ester **103,** which, on treatment with 1,1-carbonyldiimidazole, gives the ethyl ester of the 1,2,4-thiadiazine dioxide **104.** Saponification affords 3-oxo-3,4,5,6-tetrahydro-2*H*-1,2,4-thiadiazine-5-carboxylic acid 1,1-dioxide (**105**), but no mention of chirality is reported[53] (Scheme 19).

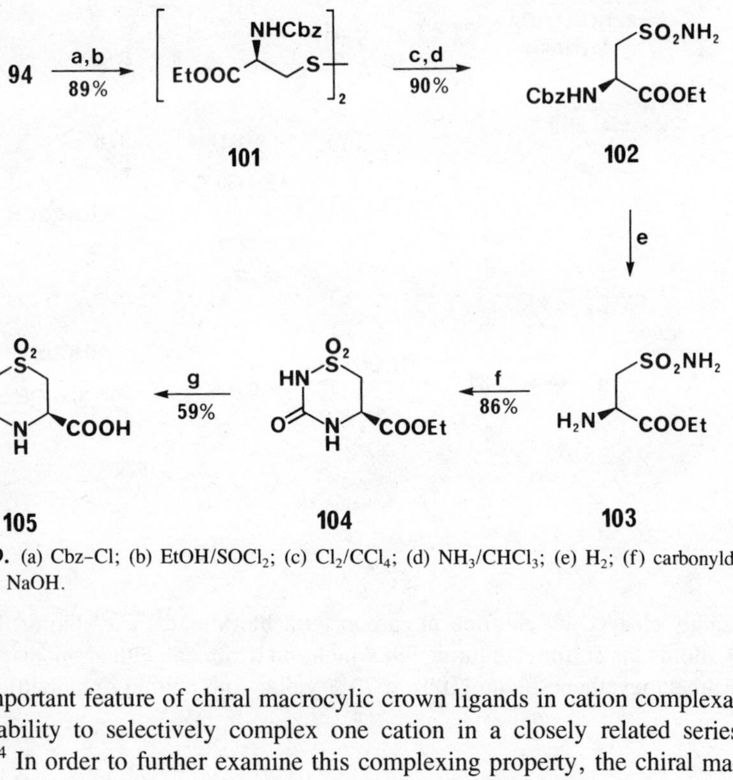

Scheme 19. (a) Cbz-Cl; (b) EtOH/$SOCl_2$; (c) Cl_2/CCl_4; (d) $NH_3/CHCl_3$; (e) H_2; (f) carbonyldiimidazole; (g) NaOH.

An important feature of chiral macrocylic crown ligands in cation complexation is their ability to selectively complex one cation in a closely related series of cations.[54] In order to further examine this complexing property, the chiral macrocyclic dilactones **109a** and **109b** were prepared by a simple method based on a tin "template-driver" process.[55] The conversion of the L-cystine derivatives **106a** and **106b** to the diacid fluorides **107** is accomplished by using Mukaiyama's method.[56] Treatment of **107** with the alkoxytin derivative **108** affords the dilactones **109a** and **109b** in 57 and 48%, respectively.[57,58]

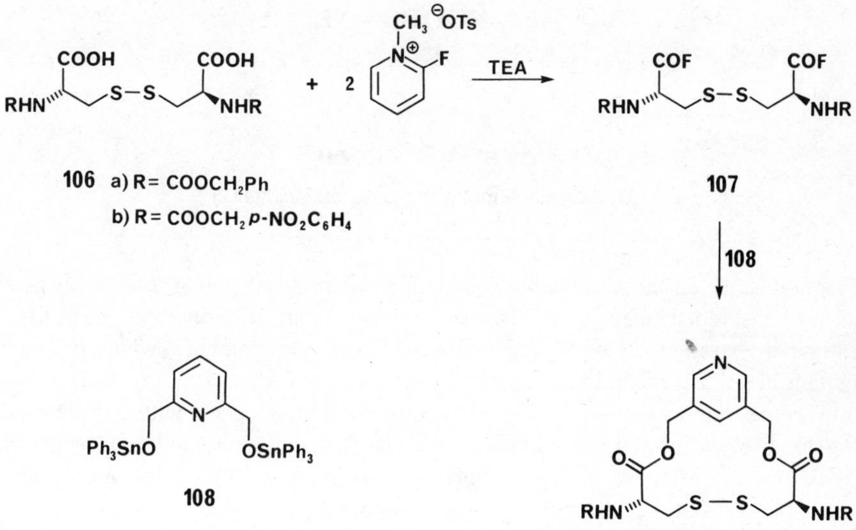

TEA

106 a) R = $COOCH_2Ph$
b) R = $COOCH_2$*p*-$NO_2C_6H_4$

107

108

109a 57%
109b 48%

Cystine can also function as a chiral auxiliary in the asymmetric reduction of prochiral 3-aryl-3-oxo-esters **110** with $LiBH_4$. The partial decomposition of $LiBH_4$ with *N,N'*-dibenzoyl-L-cystine and *tert*-butanol affords a complex borohydride that, on reduction of **110,** provides optically active 3-aryl-3-hydroxyesters (**111**) in good yield and in 80–92% ee. The chiral auxiliary is recovered in over 70% yield. The absolute configuration of **111** is *R* if L-cystine is used and *S* if D-cystine is used.[59]

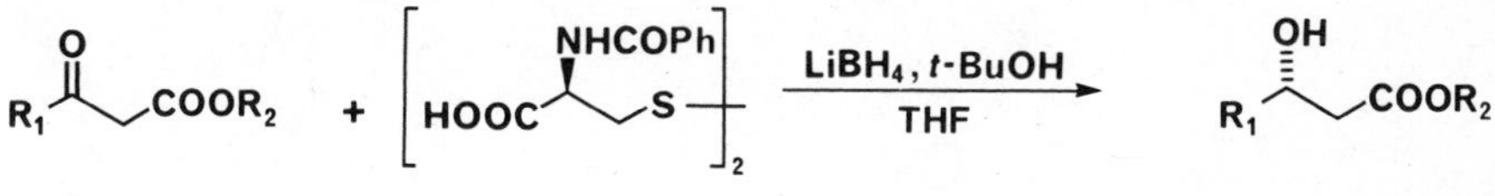

$LiBH_4$, *t*-BuOH, THF

110 (R)-**111**

R_1	R_2	Yield (%)	Percent ee
C_6H_5	CH_3	78	84
C_6H_5	C_2H_5	94	87
C_6H_5	*i*-C_3H_7	83	91
C_6H_5	*t*-C_4H_9	88	90
4-$CH_3C_6H_4$	C_2H_5	88	85
4-$MeOC_6H_4$	C_2H_5	85	84
1-Naphthyl	C_2H_5	90	92

6.3 METHIONINE

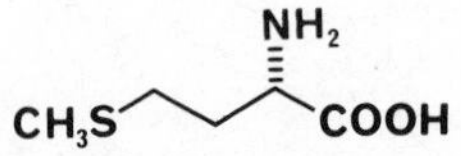

(S)-2-amino-4-(methylthio)butanoic acid (**112**)

Methionine is an essential amino acid that in mammals gives rise to cysteine through the replacement of the hydroxy oxygen atom of serine by a transsulfurization. It also interacts with ATP to form *S*-adenosylmethionine, which is an important biological methylating agent.[60]

Whereas the reaction involving *S*-adenosylmethionine occurs with complete sulfur stereoselectivity,[61] little is known about the absolute configuration of pyramidal trialkylsulfonium ions. A highly stereospecific intramolecular alkylation occurs when (*S*)-*tert*-butyl-2-(1-tosyloxy-4-methylthio)butylcarbamate (**114**) is heated in benzene at 60°C for a few hours. The crude tetrahydrothiophenium salt (**116**), having an optical purity of not less than 80%, is twice recrystallized in acetone to afford optically pure **116** possessing the (*S*) configuration at the sulfur atom. This high diastereoselectivity may result from the cyclic transition state **115,** where the steric requirements of the *S*-methyl and BocNH favor nucleophilic displacement by the pro-*S* lone pair[62] (Scheme 20).

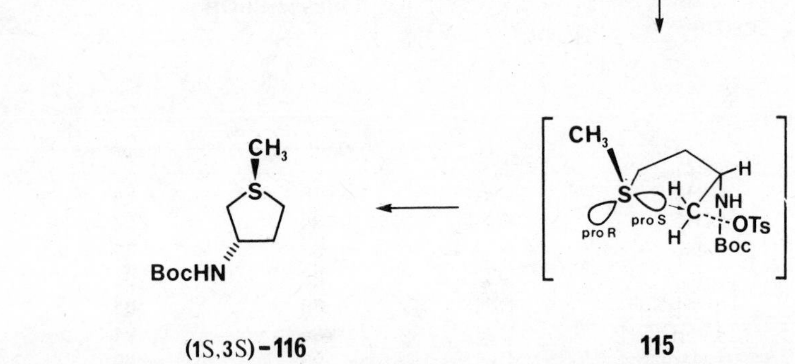

1. $LiAlH_4$ (71%)
2. TsCl/py, -20°C

113 114 115 (1S,3S)-**116**

Scheme 20

L-Methionine is used to prepare optically pure (*S*)-vinylglycine (**122**) through a thermal *syn* elimination. Conversion of **112** to Cbz–L-methionine methyl ester (**117**) proceeds in excellent yield. Oxidation of **117** with sodium metaperiodate provides a quantitative yield of methyl 2-(Cbz–amino)-4-methylsulfinyl butyrate (**118**). Pyrolysis of **118** by Kugelrohr [148°C (3 mm)] gives the protected vinylglycine **119** in 80% isolated yield after simple chromatographic purification. Under basic conditions, **119** is converted almost quantitatively to either **120** or **121.** However, acid hydrolysis of **119** affords **122** in greater than 90% yield with an optical purity of >99%[63] (Scheme 21).

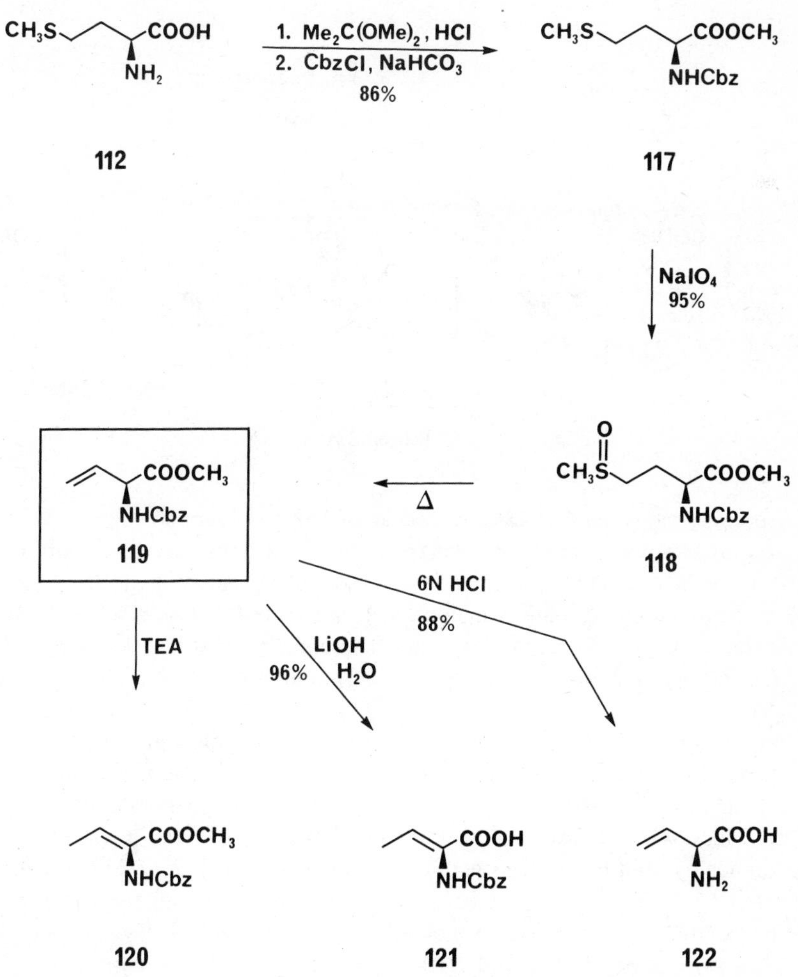

CH_3S COOH NH_2 **112** → (1. $Me_2C(OMe)_2$, HCl; 2. CbzCl, $NaHCO_3$; 86%) → CH_3S $COOCH_3$ NHCbz **117**

117 → ($NaIO_4$, 95%) → $CH_3S(O)$ $COOCH_3$ NHCbz **118**

118 → (Δ) → $COOCH_3$ NHCbz **119**

119 → (TEA) → $COOCH_3$ NHCbz **120**

119 → (LiOH, H_2O, 96%) → COOH NHCbz **121**

119 → (6N HCl, 88%) → COOH NH_2 **122**

Scheme 21

An important application of readily available **122** is to the synthesis of the antitumor agent AT-125 (**126**) [(αS,5S)-α-amino-3-chloro-4,5-dihydro-5-isoxazoleacetic acid]. Cyanogen chloride *N*-oxide, generated *in situ* by the addition of silver nitrate to dichloroformaldoxime (**124**), undergoes a 1,3-dipolar cycloaddition with **123** to afford a diastereomeric mixture of isoxazoles **125a** and **125b** in 52% yield. Chromatographic separation and deprotection of **125a** (unfortunately, the minor product) affords the natural product **126**[64] (Scheme 22).

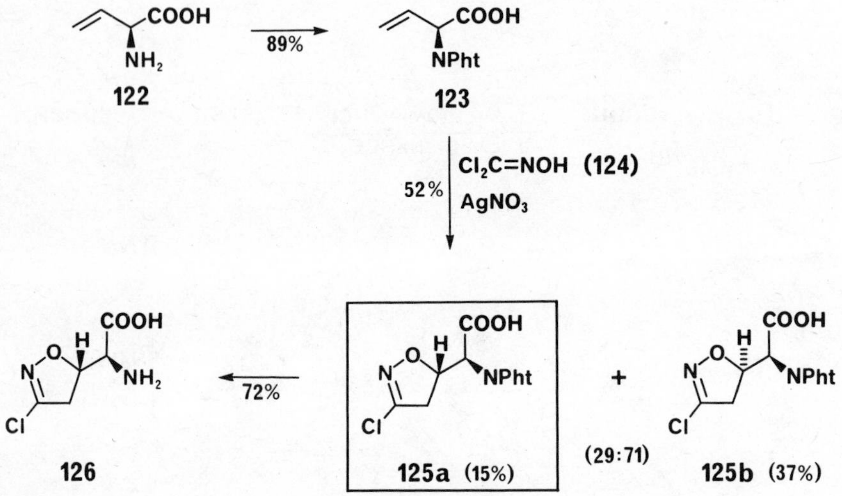

Scheme 22

Aspergillomarasmine A (**134**) is a metal chelating polyamino acid isolated from *Aspergillus fluvas oryzae.*[65] Whereas the amino acid units may be constructed by using the reductive amination method with sodium cyanoborohydride, it is likely that the required chiral 2-aminomalonic half aldehyde (''serine aldehyde'') would racemize, leading to the formation of a diastereomeric mixture. However, by employing the masked D-serine equivalent, (2*R*)-amino-3-butenol (**130**), this can be avoided.

Boc–D-methionine (**127**) is easily converted to amino alcohol **128** in 89% yield without any observed racemization. Oxidation of **128** to the sulfoxide followed by a thermal *syn* elimination provides optically pure (2*R*)-*N*-Boc-amino-3-butenol (**129**), which when deprotected yields **130.** Similarly, one can prepare the (2*S*)-enantiomer by starting from L-methionine (**112**). Ozonolysis of **129** provides an aldehyde that affords the adduct **131** on reductive coupling with **130.** After protection of the imino group, **131** is ozonolyzed and the resulting aldehyde coupled with the L-aspartic moiety **132,** affording the triamino acid **133.** The conversion of **133** to **134** requires the oxidation of both alcohol groups to their corresponding acids. However, this has not yet been accomplished[66] (Scheme 23).

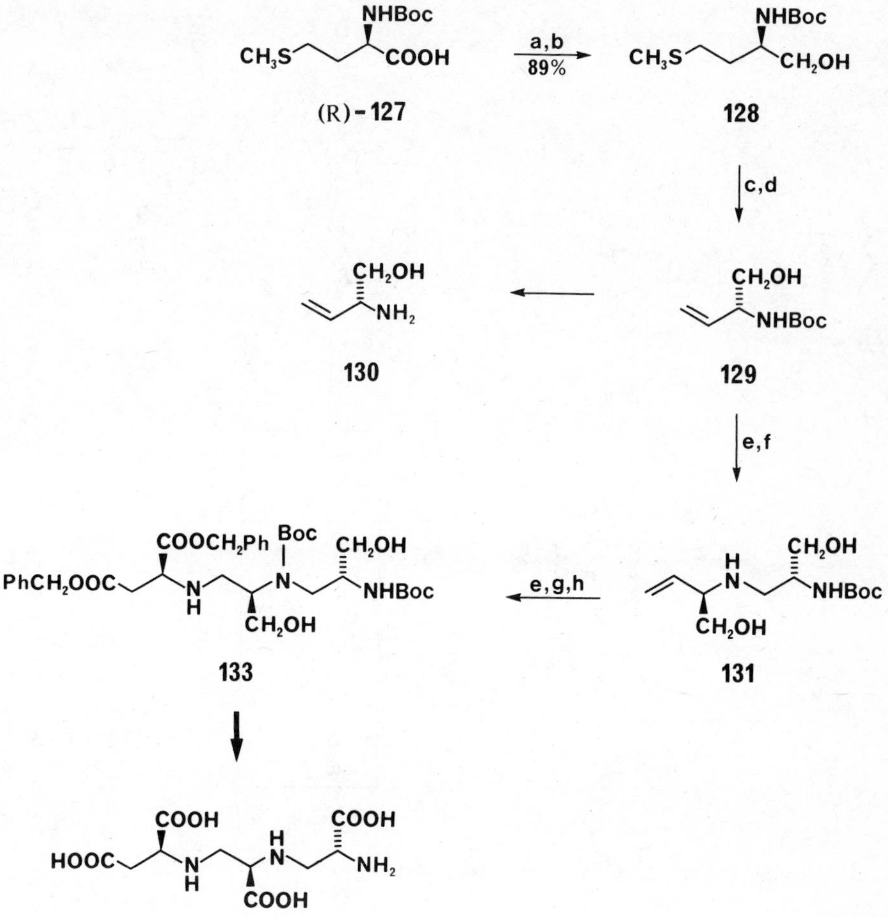

NHBoc
CH_3S COOH → a,b 89% → CH_3S CH_2OH (NHBoc)

(R)-127 128

c,d

CH_2OH NH_2 ← CH_2OH NHBoc

130 129

e,f

Boc
$COOCH_2Ph$ CH_2OH
$PhCH_2OOC$ N H N NHBoc ← e,g,h ← H N CH_2OH NHBoc
CH_2OH CH_2OH

133 131

COOH COOH
HOOC N H H N NH_2
COOH

134

Scheme 23. (a) CH_2N_2/ether; (b) $LiAlH_4$/THF; (c) $NaIO_4$ (91%); (d) 170°C, 2 hr (65%); (e) O_3; (f) **130**/$NaCNBH_3$/CH_3OH; (g) $(Boc)_2O$/TEA; (h) L-aspartic acid dibenzyl ester (**132**) (64%).

Chiral **129** is also a useful synthon for the synthesis of (+)-galantinic acid (**141**), a unique pyran-containing amino acid in the peptide antibiotic galantin I. Epoxidation of **129** affords, in a highly stereoselective manner, the *threo*-3,4-epoxy-2-aminobutanol **135a.** Regiospecific epoxide ring opening with divinylcuprate **136** provides a mixture of adducts that are converted without separation to the diacetate **137.** Removal of the silyl group and oxidation of the allylic alcohol yields the ester **138.** Deprotection of the acetates results in a spontaneous cyclization to afford a 1:1 mixture of **139** and **140.** Chromatographic separation provides **140** in 39% yield. This is then converted to **141,** having the (3*S*,5*S*,6*S*) absolute configuration[67] (Scheme 24).

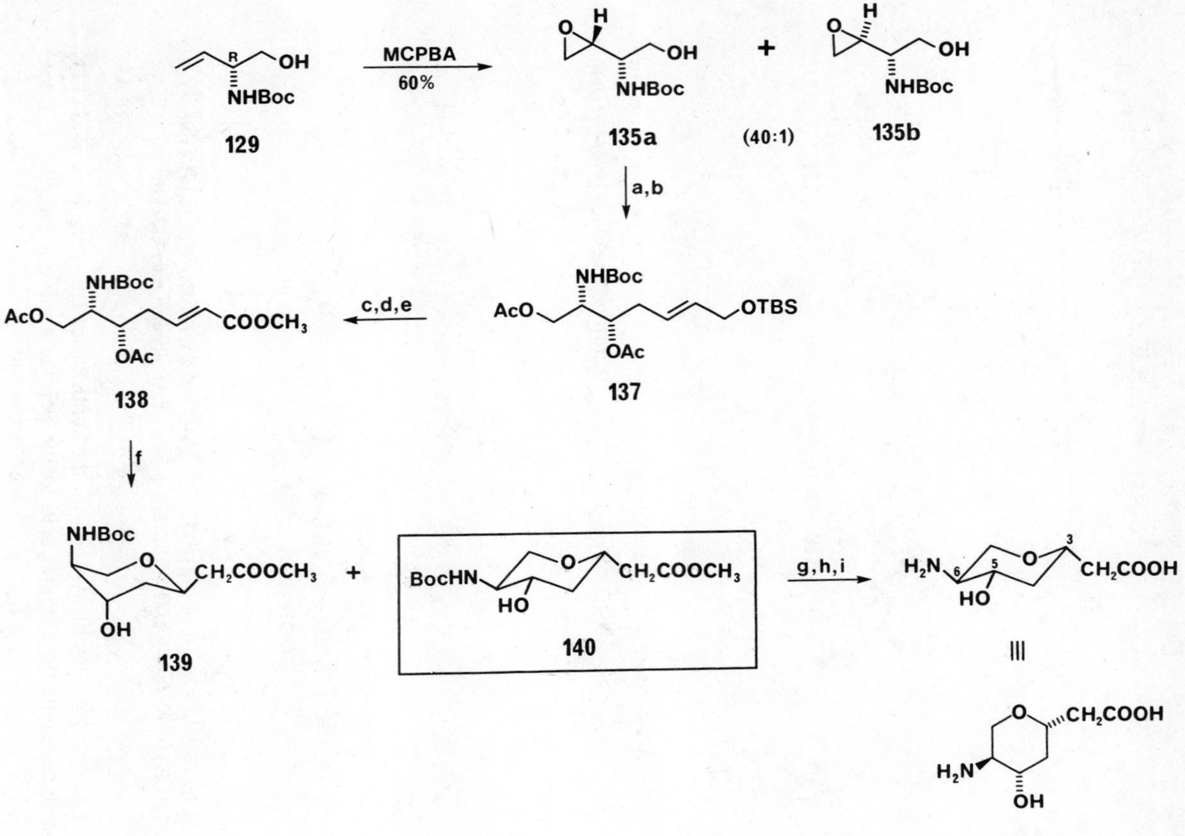

Scheme 24. (a) $Li_2(CN)Cu(CH{=}CHCH_2OTBS)$ (**136**) (51%); (b) Ac_2O/pyridine; (c) PTSA/CH_3OH (88%); (d) PCC; (e) MnO_2/NaCN/HOAc (75%); (f) K_2CO_3/CH_3OH; (g) chromatography; (h) KOH/CH_3OH; (i) Dowex.

Methionine (**112**) is also useful for the preparation of chiral ligands used in the transition-metal-catalyzed carbon–carbon bond formation. Methphos (**143**), prepared enantiomerically pure from **112,** solubilizes catalytic amounts of $NiCl_2$, which mediates the coupling of vinyl bromide with the racemic Grignard reagent of 1-phenyl-1-chloroethane. The coupled product, (*S*)-3-phenyl-1-butene (**144**), is formed in excellent yield and with 65% ee. A variety of other chiral ligands prepared from **112** do not produce such a high ee[68] (Scheme 25).

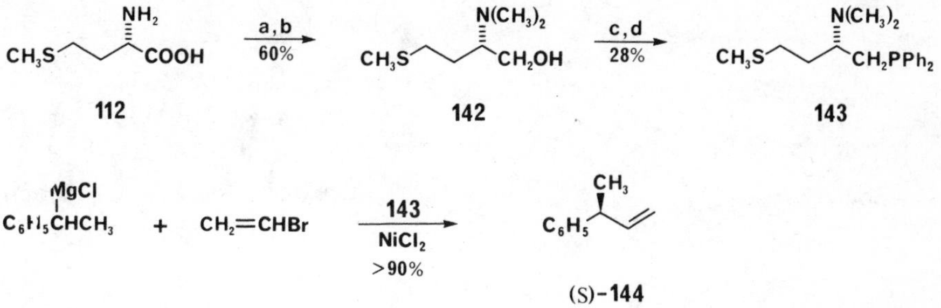

Scheme 25. (a) CH_2O, Pd/C, H_2; (b) $LiAlH_4$; (c) CH_3SO_2Cl, proton sponge, THF; (d) Ph_2PH/*t*-BuOK.

6.4 ADDENDA

Since the completion of this chapter, new research involving both cysteine and methionine for the construction of chiral products has appeared in the literature.

The highly enantioselective differentiation between two identical ligands in the prochiral σ-symmetric dicarboxylic acid 3-methylglutaric acid (**61**) was achieved by utilizing (4*R*)-MCTT (**57**).[36] Further application of **57** to the discrimination between conformational enantiomers of *cis*-4-cyclohexen-1,2-ylenebis (acetic acid) (**145**) has led to a new diastereoselective chiral method for their recognition.[69] When **145** reacts with **57,** the diamide **146** is obtained in 63.6% yield. Subsequent aminolysis of **146** with 1 equivalent of piperidine provides a yellow solid containing a 94 : 6 ratio of products. Recrystallization furnishes the pure major product **147** in 51.2% overall yield. This bifunctional and optically active synthon should be useful for the asymmetric synthesis of biologically active compounds such as prostacarvacyclins and coriolin derivatives.[69]

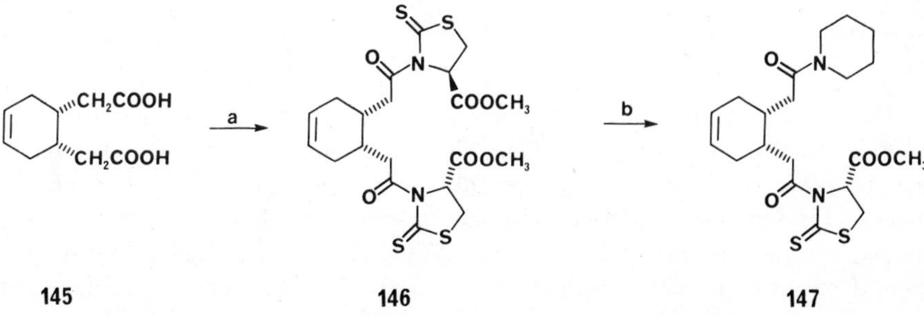

(a) **57,** DCC, pyridine (63.6%); (b) (i) piperidine, −78°C, (ii) recrystallization (51.2%).

The enantiomeric (4*S*)-MCTT (**148**) can be prepared in 84.6% yield by the reaction of D-cysteine methyl ester with carbon disulfide in the presence of triethylamine. When reacted with *meso*-2,4-dimethylglutaric acid anhydride (**149**) in the presence of DCC in pyridine, the diamide **150** is obtained in 67.2% yield. Aminolysis with piperidine and recrystallization of the resulting mixture (97.3 : 2.7) affords pure **151.** This, after conversion to the aldehyde **152,** undergoes an aldol-type condensation with tin enolate **153** to furnish **154** in 74.2% yield (ratio 92 : 8). Pure **154** is smoothly converted to (+)-Prelog–Djerassi lactonic methyl ester **155**[70] (Scheme 26).

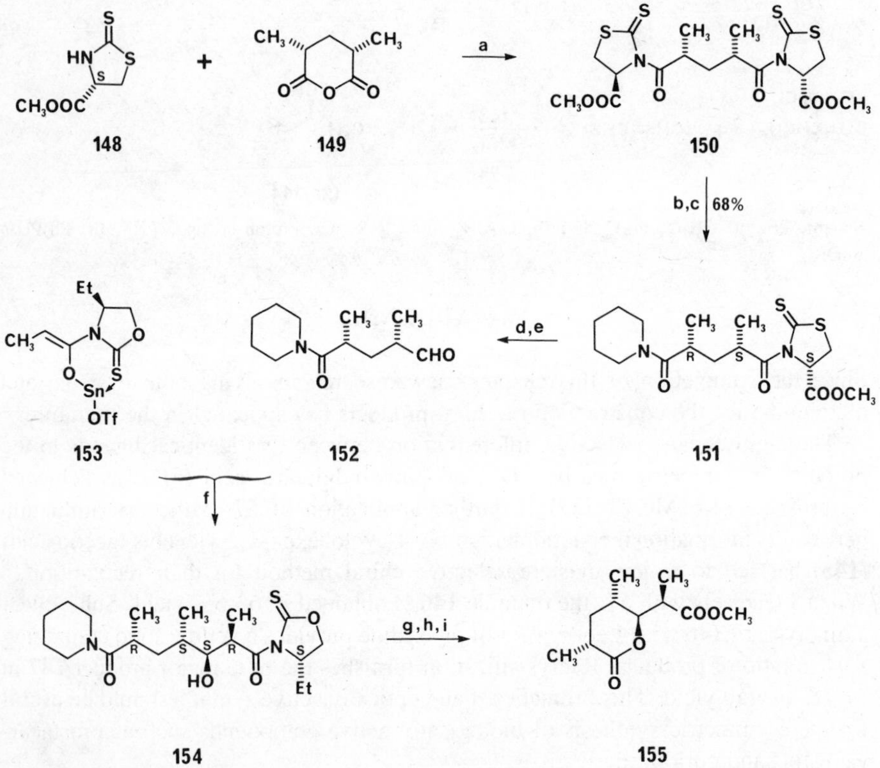

Scheme 26. (a) DCC, pyridine (67.2%); (b) piperidine, CH_2Cl_2, $-20°C$; (c) recrystallization; (d) $NaBH_4$, THF/H_2O (91.1%); (e) pyridine·SO_3/TEA/DMSO (63.5%); (f) CH_2Cl_2, $-78°C$ (74.2%); (g) HCl/benzene (56.7%); (h) LiOH, THF; (i) CH_2N_2, ether (53.3%).

The chiral auxiliary **57** is also compatible with boron enolate chemistry and provides aldol products with significant enantioselectivity. The *N*-acylation of **57** with butyryl chloride provides the optically active amide **156** in 97% yield. Treatment of **156** with bis(*n*-butyl)boron triflate in methylene chloride at 0°C followed by addition of aldehyde **158** at $-78°C$ affords chiral *erythro*-aldol **159** in 77% yield. Conversion of **159** to the hydroxamate **160** with *O*-benzylhydroxylamine and cyclization with triphenylphosphine–diisopropyl azodicarboxylate (Ph_3P–DIAD) gives optically active β-lactam **161.** This is being investigated for the synthesis of PS-5 (**162**), an optically pure carbapenem[71] (Scheme 27).

The same method used for the preparation of L-vinylglycine (**122**) from L-methionine[63] can be used with a slight modification[72] to prepare in 65% yield the D-vinylglycine methyl ester **163** from D-methionine. This, when oxidized with *m*-chloroperbenzoic acid, provides a 4:1 mixture of *syn*-epoxide **164** and *anti*-epoxide **165,** which is separated by medium-pressure chromatography. Treatment of **164** with methanolic HCl affords the ring-opened chlorohydrin **166,** which is transformed to the aziridine **167** in good yield. Homologation and reductive ring closure provides the crystalline aziridinopyrrolidine **168** in 78% yield. Sequential reduction, regiospecific addition of 2,3-dibromo-5-methoxy-6-methylbenzoquinone (**169**), photochemical rearrangement, oxidation, and palladium-catalyzed ring closure afford the (*R,R*)-aziridinomitosene **170** in 58% overall yield. The regioisomeric aziridinoindoloquinone **171** can be obtained directly by the addition of **168** to **169** followed by a copper(II)-catalyzed ring closure[72] (Scheme 28).

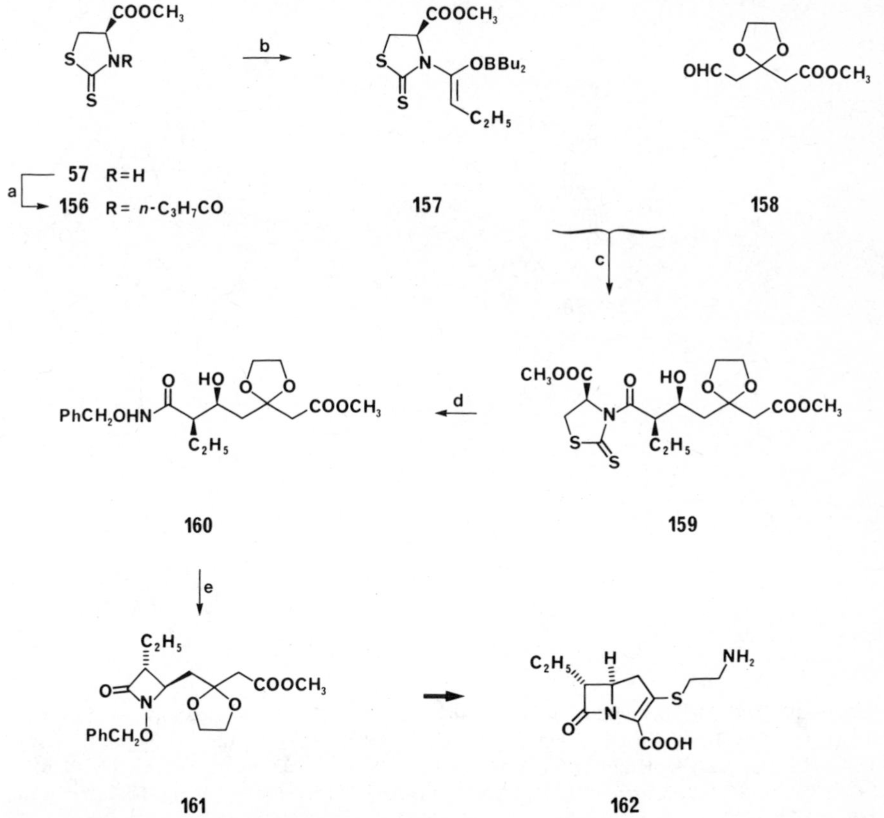

Scheme 27. (a) $CH_3(CH_2)_2COCl$, pyridine, −40°C (97%); (b) $(n\text{-Bu})_2BOTf$, $(i\text{-Pr})_2NEt$; (c) −78°C (77%); (d) H_2NOCH_2Ph, CH_3CN (74%); (e) Ph_3P/DIAD (67%).

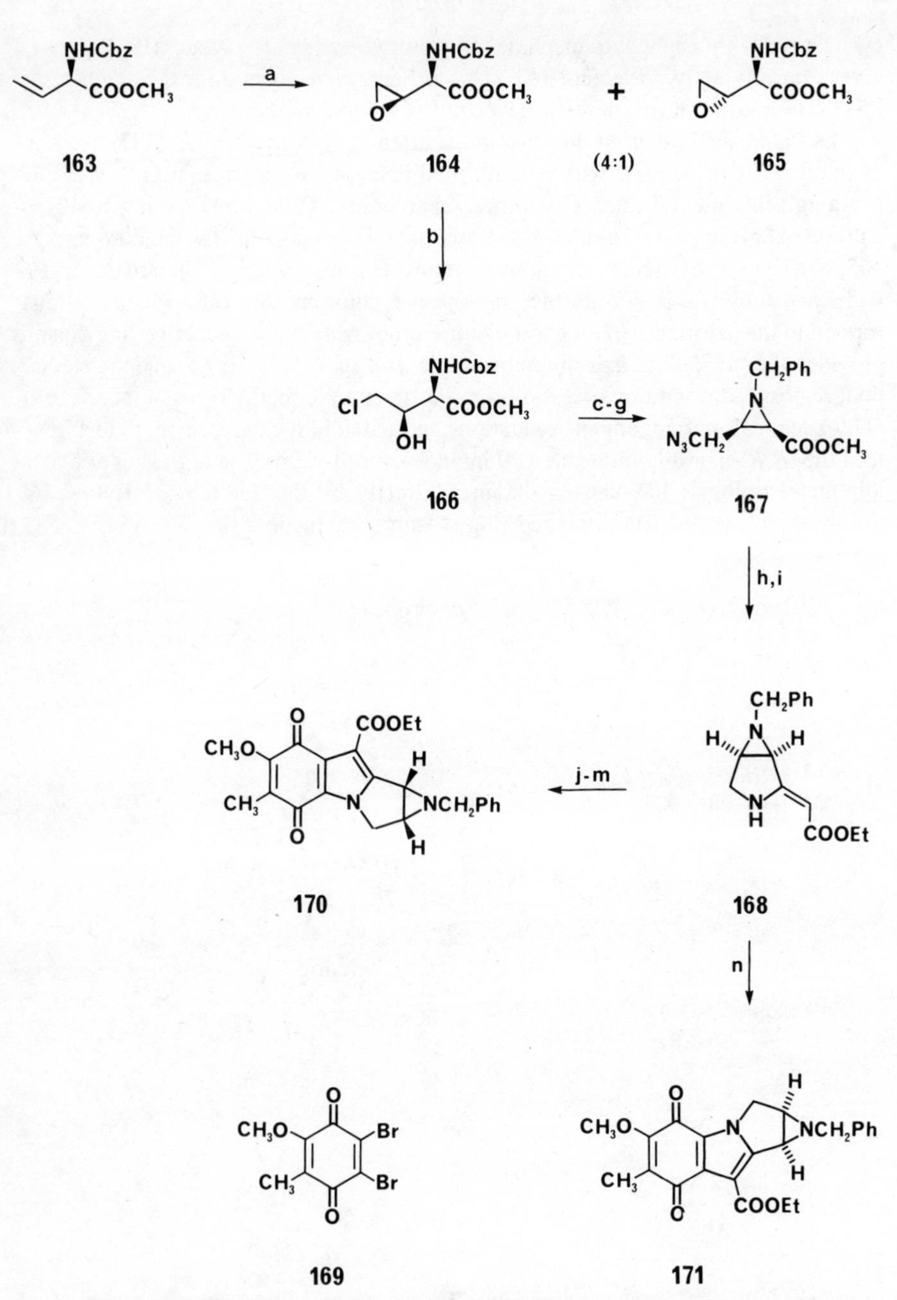

Scheme 28. (a) (i) MCPBA (99%), (ii) MPLC (**164** = 62%); (b) MeOH/HCl (68%); (c) cyclohexene, HOAc, Pd/C (98%); (d) CH_3OH, HOAc, PhCHO, $NaCNBH_3$ (69%); (e) $(Boc)_2O$ (96%); (f) MsCl/TEA (93%); (g) CF_3COOH (70%); (h) LDA, EtOAc (89%); (i) PtO_2/EtOH (76%); (j) $NaCNBH_3$/HOAc (89%); (k) **169,** benzene (93%); (l) (i) *hν,* (ii) Pd/C (61%); (m) $Pd(OAc)_2$, TEA (96%); (n) **169,** CH_3CN (93%).

REFERENCES

1. T. Scott and M. Brewer, *Concise Encyclopedia of Biochemistry,* Walter de Gruyter, New York, 1985, p. 114.
2. Alan Bassindale, *The Third Dimension in Organic Chemistry,* Wiley, New York, 1984, pp. 213–215.
3. A. H. Cook, *Quart. Rev.*, **2,** 231 (1948).
4. R. C. Elderfield, *Heterocyclic Compounds,* Vol. 5, Wiley, New York, 1957, p. 697.
5. M. P. Schubert, *J. Biol. Chem.*, **114,** 341 (1936).
6. S. Ratner and H. T. Clarke, *J. Am. Chem. Soc.*, **59,** 200 (1937).
7. N. J. Lewis, R. L. Inoles, J. Hes, R. H. Matthews, and G. Milo, *J. Med. Chem.*, **21,** 1070 (1978).
8. H. Soloway, F. Kipnis, J. Ornfelt, and P. E. Spoerri, *J. Am. Chem. Soc.*, **70,** 1667 (1948).
9. R. Riemschneider and G.-A. Hoyer, *Nature Forsch.*, **176,** 765 (1962).
10. M. Oya, T. Baba, E. Kato, Y. Kawashima, and T. Watanabe, *Chem. Pharm. Bull.*, **30,** 440 (1982).
11. B. Paul and W. Korytnyk, *J. Med. Chem.*, **19,** 1002 (1976).
12. M. Oya, E. Kato, J.-I. Iwao, and N. Yasuoka, *Chem. Pharm. Bull.*, **30,** 484 (1982).
13. J. Martens and K. Drauz, *Liebigs Ann. Chem.*, 2073 (1983).
14. R. Bognar, Z. Györgydeak, L. Szilagyi, G. Horvath, G. Czira, and L. Radics, *Liebigs Ann. Chem.*, 450 (1976).
15. R. H. Wiley and J. F. Jeffries, *J. Am. Chem. Soc.*, **71,** 1137 (1949).
16. M. Iwakawa, B. M. Pinto, and W. A. Szarek, *Can. J. Chem.*, **56,** 326 (1978).
17. S. Lieberman, P. Brazeau, and L. B. Hariton, *J. Am. Chem. Soc.*, **70,** 3094 (1948).
18. I. Mita, J.-I. Iwao, M. Oya, T. Chiba, and T. Iso, *Chem. Pharm. Bull.*, **26,** 1333 (1978).
19. R. Parthasarathy, B. Paul, and W. Korytnyk, *J. Am. Chem. Soc.*, **98,** 6634 (1976).
20. E. Kato, K. Yamamoto, T. Babb, T. Watanabe, Y. Kawashima, H. Masuda, M. Horiuchi, M. Oya, T. Iso, and J.-I. Iwao, *Chem. Pharm. Bull.*, **33,** 74 (1985).
21. R. B. Woodward, K. Heusler, J. Gosteli, P. Naegeli, W. Oppolzer, R. Ramage, S. Ranganathan, and H. Vorbrüggen, *J. Am. Chem. Soc.*, **88,** 852 (1966).
22. J. E. Baldwin, A. Au, M. Christie, S. B. Haber, and D. Hesson, *J. Am. Chem. Soc.*, **97,** 5957 (1975).
23. J. E. Baldwin, M. A. Christie, S. B. Haber, and L. I. Kruse, *J. Am. Chem. Soc.*, **98,** 3045 (1976).
24. G. A. Koppel, L. McShane, F. Jose, and R. D. G. Cooper, *J. Am. Chem. Soc.*, **100,** 3933 (1978).
25. R. D. G. Cooper, F. Jose, L. McShane, and G. A. Koppel, *Tetrahedron Lett.*, 2243 (1978).
26. J. E. Baldwin and M. A. Christie, *J. Am. Chem. Soc.*, **100,** 4597 (1978).
27. H. Yamaguchi, Y. Nakayama, K. Takeda, K. Tawara, T. Takeuchi, and H. Umezawa, *J. Antibiot.*, **A10,** 195 (1957).

28. (a) K. Inami and T. Shiba, *Tetrahedron Lett.*, 2009 (1984); (b) K. Inami and T. Shiba, *Bull. Chem. Soc. Jpn.*, **58,** 352 (1985).
29. P. N. Confalone, G. Pizzolato, E. G. Baggiolini, D. Lollar, and M. R. Uskokovic, *J. Am. Chem. Soc.*, **97,** 5936 (1975).
30. P. N. Confalone, G. Pizzolato, E. G. Baggiolini, D. Lollar, and M. R. Uskokovic, *J. Am. Chem. Soc.*, **99,** 7020 (1977).
31. H. Brunner, G. Riepl, and H. Weitzer, *Angew. Chem., Int. Ed.*, **22,** 331 (1983).
32. H. Brunner, R. Becker, and G. Riepl, *Organometallics,* **3,** 1354 (1984).
33. K. Soai, T. Yamanoi, and H. Oyamada, *Chem. Lett.*, 251 (1984).
34. Y. Nagao, M. Yagi, T. Ikede, and E. Fujita, *Tetrahedron Lett.*, 201 (1982).
35. Y. Nagao, M. Yagi, T. Ikede, and *E. Fujita, Tetrahedron Lett.*, 205 (1982).
36. Y. Nagao, T. Ikeda, M. Yagi, and E. Fujita, *J. Am. Chem. Soc.*, **104,** 2079 (1982).
37. D. Seebach and T. Weber, *Tetrahedron Lett.*, 3315 (1983).
38. D. Seebach and T. Weber, *Helv. Chim. Acta,* **67,** 1650 (1984).
39. C. D. Maycock and R. J. Stoodley, *J. Chem. Soc., Chem. Commun.*, 234 (1976).
40. C. D. Maycock and R. J. Stoodley, *J. Chem. Soc., Perkin Trans. I,* 1852 (1979).
41. A. D. Argoudelis and R. R. Herr, *Antimicrob. Agents Chemother.*, 780 (1962).
42. E. Higashide, T. Hasegawa, M. Shibata, K. Mizuno, and H. Akaike, *Takeda Kenkyusho Nempo,* **25,** 1 (1966); *Chem. Abstr.*, **66,** 54238 (1967).
43. D.-R. Hwang, P. Helquist, and M. S. Shekhani, *J. Org. Chem.*, **50,** 1264 (1985).
44. P. Helquist and M. S. Shekhani, *J. Am. Chem. Soc.*, **101,** 1057 (1979).
45. H. C. J. Ottenheijm and R. M. J. Liskamp, *Tetrahedron Lett.*, 2437 (1978).
46. H. C. J. Ottenheijm, R. M. J. Liskamp, and M. W. Tijhuis, *Tetrahedron Lett.*, 387 (1979).
47. H. C. J. Ottenheijm, R. M. J. Liskamp, S. P. J. M. Van Nispen, H. A. Boots, and M. W. Tijhuis, *J. Org. Chem.*, **46,** 3273 (1981).
48. H. C. J. Ottenheijm, *Chimia,* **39,** 89 (1985).
49. R. M. J. Liskamp, H. J. M. Zeegers, and H. C. J. Ottenheijm, *J. Org. Chem.*, **46,** 5408 (1981).
50. R. M. J. Liskamp, H. J. Blom, R. J. F. Nivard, and H. C. J. Ottenheijm, *J. Org. Chem.*, **48,** 2733 (1983).
51. H. Baganz and G. Dransch, *Chem. Ber.*, **93,** 782 (1960).
52. H. Baganz and G. Dransch, *Chem. Ber.*, **93,** 784 (1960).
53. C. H. Levenson and R. B. Meyer, Jr., *J. Med. Chem.*, **27,** 228 (1984).
54. S. T. Jolley, J. S. Bradshaw, and R. M. Izatt, *J. Heterocyclic Chem.*, **19,** 3 (1982).
55. A. Shanzer and J. Libman, *J. Chem. Soc., Chem. Commun.*, 846 (1983).
56. T. Mukaiyama and T. Tanaka, *Chem. Lett.*, 303 (1976).
57. P. Tisnes, L. Cazaux, and C. Picard, *J. Chem. Res. (S),* 38 (1984).
58. C. Picard, L. Cazaux, and P. Tisnes, *Tetrahedron Lett.*, 3809 (1984).
59. K. Soai, T. Yamanoi, H. Hikima, and H. Oyamada, *J. Chem. Soc., Chem. Commun.*, 138 (1985).
60. A. L. Lehninger, *Biochemistry,* 2nd ed., Worth Publishers, New York, 1975, p. 699–700.

61. G. De La Haba, G. A. Jamieson, S. H. Mudd, and H. H. J. Richards, *J. Am. Chem. Soc.*, **81,** 3975 (1959).
62. K. Tani, S. Otsuku, M. Kidi, and I. Miura, *J. Am. Chem. Soc.*, **102,** 7394 (1980).
63. A. Afzali-Ardakani and H. Rapoport, *J. Org. Chem.*, **45,** 4817 (1980).
64. P. A. Wade, M. K. Pillayand, and S. M. Singh, *Tetrahedron Lett.*, 4563 (1982).
65. A. L. Haenni, M. Robert, W. Vetter, L. Roux, M. Bailer, and E. Lederer, *Helv. Chim. Acta,* **48,** 729 (1965).
66. Y. Ohfune and N. Kurokawa, *Tetrahedron Lett.*, 1071 (1984).
67. Y. Ohfune and N. Kurokawa, *Tetrahedron Lett.*, 1587 (1984).
68. J. H. Griffin and R. M. Kellogg, *J. Org. Chem.*, **50,** 3261 (1985).
69. Y. Nagao, T. Ikeda, T. Inoue, M. Yagi, M. Shiro, and E. Fujita, *J. Org. Chem.*, **50,** 4072 (1985).
70. Y. Nagao, T. Inoue, K. Hashimoto, Y. Hagiwara, M. Ochiai, and E. Fujita, *J. Chem. Soc., Chem. Commun.*, 1419 (1985).
71. C.-N. Hsiao, S. P. Ashburn, and M. J. Miller, *Tetrahedron Lett.*, 4855 (1985).
72. K. J. Shaw, J. R. Luly, and H. Rapoport, *J. Org. Chem.*, **50,** 4515 (1985).

CHAPTER SEVEN

DIFUNCTIONAL AMINO ACIDS

7.1 ASPARTIC ACID

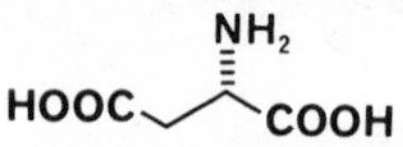

NH_2
HOOC COOH
(*S*)-2-Aminosuccinic Acid (**1**)

Aspartic acid (**1**) is a proteogenic acidic amino acid that is not essential for mammals. It plays an important role in the urea cycle and in purine and pyrimidine biosynthesis.[1]

Alkylation of **1** can occur at both C-α and C-β. The double deprotonation of di-*tert*-butyl (*S*)-*N*-formylaspartate (**2**)[2] with lithium diethylamide proceeds cleanly to give **3,** which undergoes alkylation to furnish a separable mixture of **4** and **5** (2:7 ratio) in total yields of 60–70%. The β-alkylated α-amino acid esters **5** are obtained diastereomerically and enantiomerically pure and have the *erythro* or (2*S*,3*R*) configuration. The α-alkylated α-amino acid derivatives **4** are also optically active to the extent of about 60% ee.[3]

The enantiospecific synthesis of 2-amino-6,7-dihydroxy-1,2,3,4-tetrahydronaphthalene (ADTN) (**11**), a powerful agonist of dopamine, requires as a chiral synthon (*R*)-*N*-(trifluoroacetyl)aspartic acid anhydride (**7**), which is similarly prepared from D-aspartic acid (**6**). The Friedel–Crafts acylation of veratrole with **7**

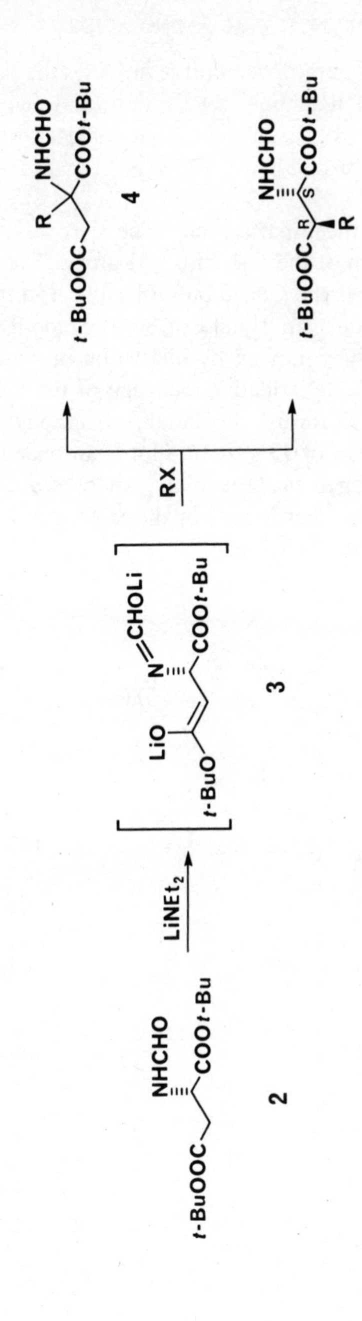

t-BuOOC
NHCHO
COOt-Bu
2
LiNEt2
LiO
t-BuO
N=CHOLi
COOt-Bu
3
RX
R
NHCHO
t-BuOOC
COOt-Bu
4
NHCHO
t-BuOOC
R
S
COOt-Bu
R
5
R = CH3, C2H5, CH2CH=CH2, CH2Ph

in the presence of anhydrous aluminum chloride provides the single isomeric ketone **8** in 55% yield. Subsequent reduction, cyclization, and deoxygenation afford **9,** which, on mild alkaline hydrolysis, yields **10** isolated as its hydrochloride salt. The optical purity of **10** corresponds to $\geq$90% ee. Finally, a didemethylation of **10** provides **11**[4–6] (Scheme 1).

An attractive feature of the aspartic anhydride derivative **12** is its ability to undergo a selective reduction of the C-1 carboxyl group. The regiospecific reduction of **12** with $NaBH_4$ affords the Cbz–δ-butyrolactone **13a** in 91% yield. This is easily converted to **13b** in an overall yield of 80%. Smooth deprotonation to the dianion with LDA at $-78°C$ followed by alkylation provides the favored *trans* product **14a,** which, after facile oxidative cleavage of the vicinal amino alcohol, unmasks the latent aldehyde to furnish the chiral β-dicarbonyl fragment **15.** On the other hand, further conversion of **13b** to **16** affords an oxazoline that, as a result of chelation with the β-nitrogen in the enolate, imposes a diastereofacial bias in the alkylation step opposite to that found for the lactone **13b.** This then provides the enantiomeric fragment **18**[7] (Scheme 2).

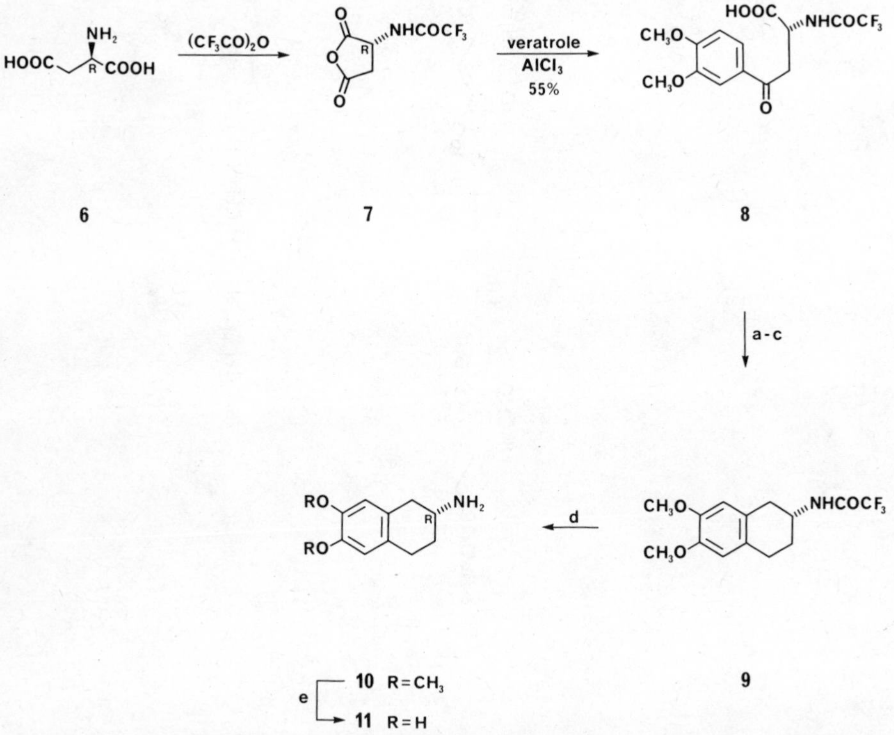

10 R=CH_3
11 R=H

Scheme 1. (a) Et_3SiH/CF_3COOH (72%); (b) PCl_5/CH_2Cl_2, $SnCl_4$ (80%); (c) $Et_3SiH/BF_3 \cdot Et_2O$ (85%); (d) $K_2CO_3/CH_3OH/H_2O$ (46%); (e) HI/Ac_2O (60–70%).

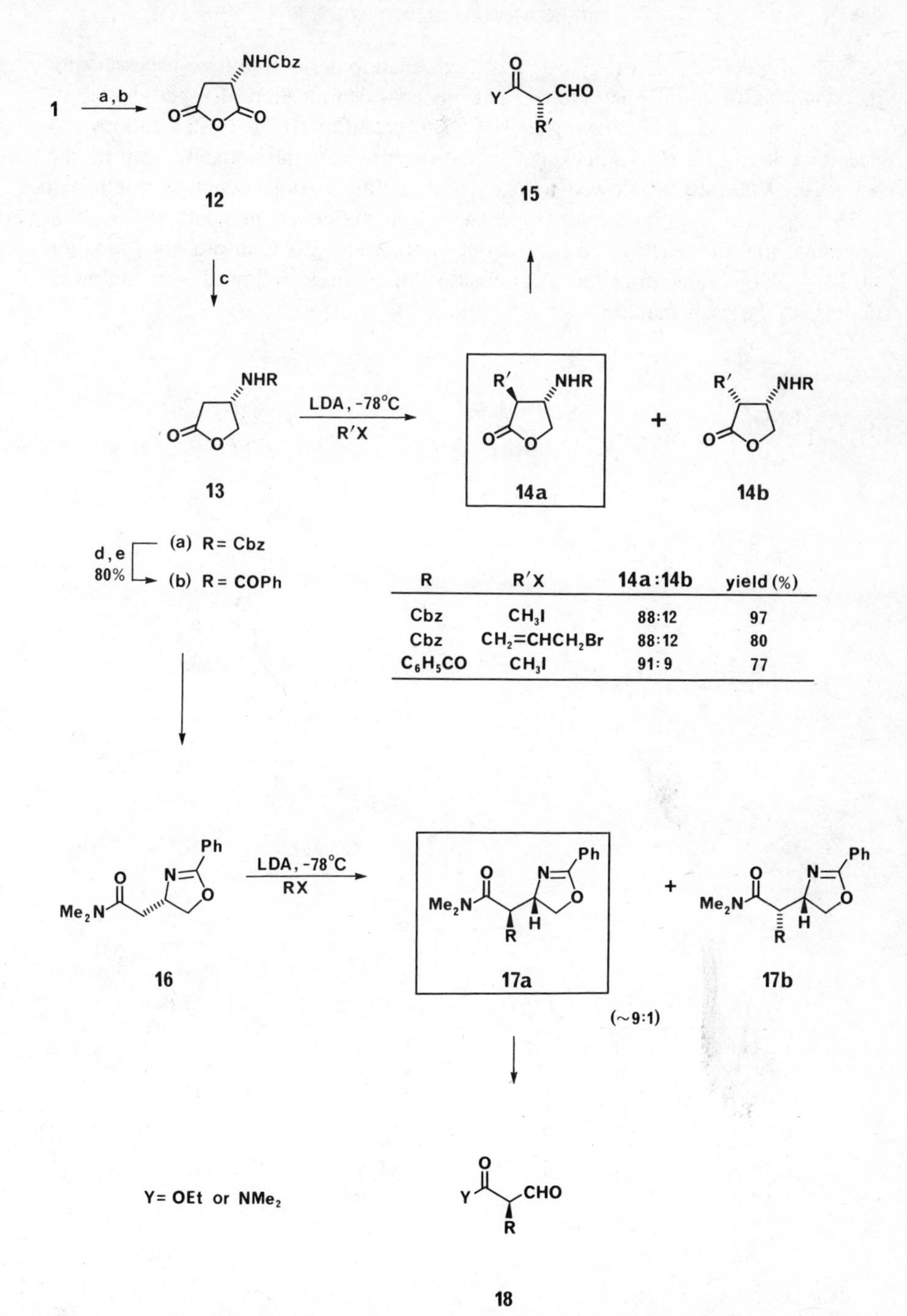

R	R′X	14a:14b	yield (%)
Cbz	CH_3I	88:12	97
Cbz	CH_2=$CHCH_2Br$	88:12	80
C_6H_5CO	CH_3I	91:9	77

Scheme 2. (a) Cbz–Cl/K_2CO_3; (b) Ac_2O/HOAc; (c) $NaBH_4$ (91%); (d) HBr/HOAc; (e) ClCOPh/pyridine.

Optically active α-hydroxy esters of high enantiomeric purity are necessary for the synthesis of chiral pheromones. Nitrous acid deamination of aspartic acid affords the malic acid **19,** having a 94% ee (pure after two recrystallizations). A decarboxylation by Huensdiecker degradation furnishes the optically pure bromo ester **20.** When **20** is allowed to react with a lithium organocuprate in ether at −20°C, optically pure α-hydroxy esters **21** are obtained in good yields. It is important that the reaction be carried out in ether at −20°C to prevent formation of **22,** which results from the β-elimination of bromide followed by addition of cuprate to the ester moiety.[8]

1 $\xrightarrow{HNO_2}$ HOOC–CH(OH)–COOH (**19**) → Br–CH₂–CH(OH)–COOEt (**20**)

R_2CuLi

R–CH₂–CH(OH)–COOEt (**21**) + CH_3–CO–C(R)(R)–OH (**22**)

R	Solvent (°C)	Ratio **21**:**22**	Yield **21** (%)
n-C_3H_7	Ether (−20)	100:0	65
n-C_4H_9	Ether (−20)	100:0	71
n-C_4H_9	Ether (−0)	66:34	35
n-C_4H_9	THF (−20)	21:79	—
s-C_4H_9	Ether (−20)	100:0	51
i-C_5H_{11}	Ether (−20)	100:0	50

The aspartic acid derivative (*S*)-*N*-Cbz–aspartic acid α-*tert*-butyl ester (**23**)[9,10] undergoes a reduction of its mixed anhydride with aqueous $NaBH_4$ to furnish (*S*)-Cbz–homoserine *tert*-butyl ester (**24**) without any racemization. Oxidation with

CrO_3 in pyridine affords an 81% yield of (*S*)-2-Cbz–amino-4-oxo-butanoic acid *tert*-butyl ester (**25**), which is the aldehyde synthon for the synthesis of nicotianamine (**26**).[11,12]

Thienamycin (**33**), a naturally occurring bicyclic β-lactam antibiotic possessing exceptional potency, differs from the classical penicillins and cephalosporins in that its novel carbapenam nucleus has the unusual hydroxyethyl and cysteamine side chains.[13]

COOH HOOC NH$_2$ **1** → COO*t*-Bu HOOC NHCbz **23** —1. ClCOOEt 2. $NaBH_4$ 79%→ COO*t*-Bu HO NHCbz **24** → COO*t*-Bu OHC NHCbz **25** ⇒ COOH N COOH N H COOH NH$_2$ **26**

Dibenzyl aspartate (**27**)[14] is converted without racemization to the silylated iodide **29** in an overall yield of 50% from L-aspartic acid (**1**). Introduction of the hydroxyethyl side chain into the 3 position is accomplished by using an aldol condensation of the enolate with excess acetaldehyde. It proceeds in excellent yield to afford a 1 : 1 mixture of (*R*) and (*S*) alcohol epimers. Subsequent oxidation of the mixture to a ketone followed by a reduction with K-Selectride yields a 9 : 1 mixture in favor of the desired (*R*) epimer, which is obtained pure by a chromatographic separation. The bicyclic system **32** is produced by a highly efficient intramolecular insertion of carbene **31** into the azetidinone N–H bond. This intermediate is then carried on to thienamycin[15] (Scheme 3).

An analog of **33** that incorporates all of the features of **33** into a less strained ring system is (−)-homothienamycin (**36**). When **28** reacts with the dianion of *t*-butylacetoacetate (2 equivalents), the keto ester **34** is obtained in 84% yield. A similar carbene insertion reaction followed by a tosylation with *p*-toluenesulfonic anhydride leads to the stable crystalline enol tosylate **35.** Again, introduction of the hydroxyethyl side chain by way of an aldol reaction affords a separable mixture of alcohol diastereomers. The (*R*) isomer is then elaborated to (−)-homothienamycin (**36**)[13] (Scheme 4).

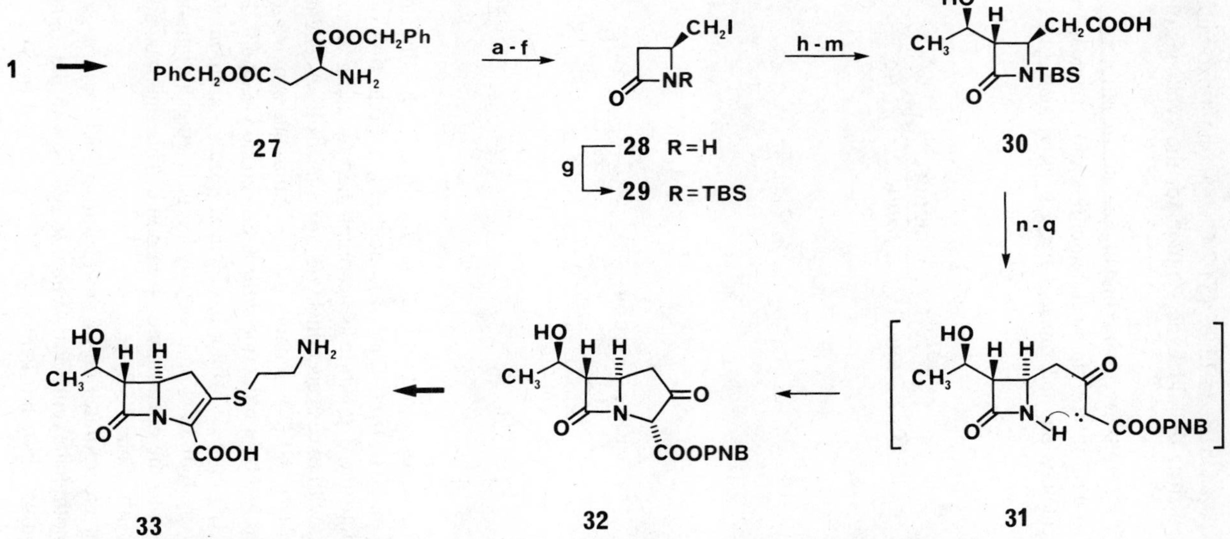

1 → PhCH$_2$OOC–CH$_2$–CH(NH$_2$)–COOCH$_2$Ph (27) —a - f→ 28 R=H; 29 R=TBS (g) —h - m→ 30 —n - q→ [31] → 32 → 33

Scheme 3. (a) TMS–Cl/TEA; (b) *t*-BuMgCl; (c) 2 *N* HCl; (d) $NaBH_4/CH_3OH$; (e) CH_3SO_2Cl/TEA; (f) NaI/acetone; (g) *t*-$BuSiMe_2Cl$/TEA; (h) 2-lithio-2-(trimethylsilyl)-1,3-dithiane; (i) LDA/CH_3CHO (97%); (j) LDA/*N*-acetylimidazole (82%); (k) K-Selectride (87%, 9:1 mixture); (l) $HgCl_2/HgO/CH_3OH$ (93%); (m) H_2O_2/CH_3OH (76%); (n) CDI/THF, $MgOOCCH_2COOPNB$ (86%); (o) HCl/CH_3OH; (p) *p*-carboxybenzenesulfonyl azide/TEA (90%); (q) $Rh_2(OAc)_4$/benzene, 75°C (100%)

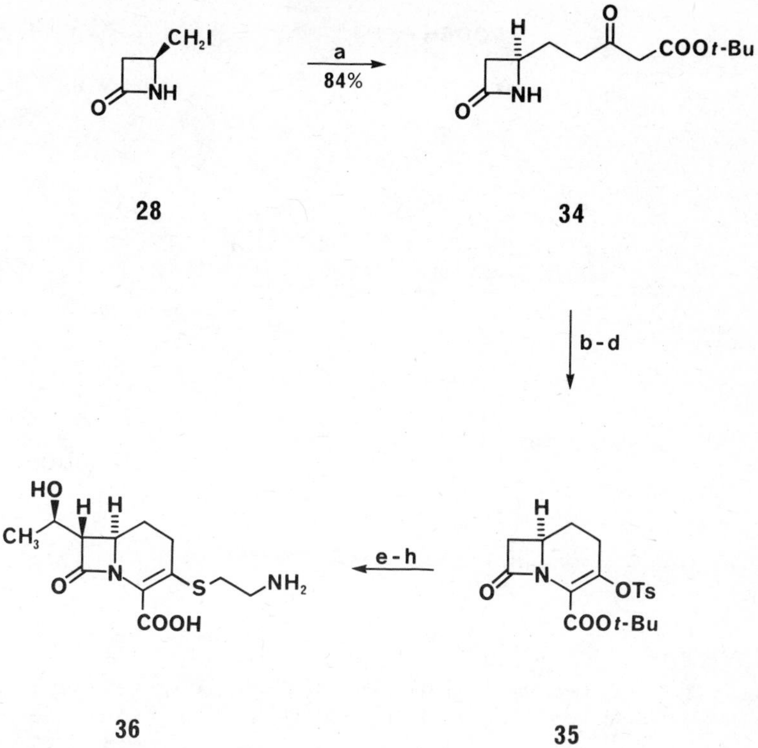

Scheme 4. (a) *t*-Butyl acetoacetate/LDA; (b) *p*-carboxybenzenesulfonyl azide/TEA (93%); (c) $Rh_2(OAc)_4$/benzene, 75°C; (d) Ts_2O/TEA [70% for (c) and (d)]; (e) LDA/CH_3CN (63%); (f) chromatography (35% *trans*-*R*); (g) $HSCH_2CH_2NH_2 \cdot HCl$/DMF; (h) TFA [34% (g), (a)–(d), (h)].

An unpleasant feature of the syntheses of both **33** and **36** is the need to separate the (*R*) and (*S*) alcohol diastereomers that result from the aldol reaction with acetaldehyde. An improvement of this procedure takes advantage of the original chiral center at C-5 (derived from L-aspartic acid) to control the new chirality at the adjacent carbon (C-6) by subsequent formation of the thermodynamically favored *trans*-keto acid **38.** A stereospecific reduction of the keto group of **38** furnishes the desired crystalline (5*S*,6*S*,8*R*)-hydroxyacid **39.** A smooth oxidative decarboxylation introduces an acetoxy group at C-5 (exclusively from the side opposite the hydroxyethyl group at C-6) to afford **40.** This is then elaborated along similar reactions to thienamycin (**33**)[16] (Scheme 5).

Interestingly, **37** can be used to prepare the chiral β-lactam carboxaldehyde **43** without significant racemization, provided it is used immediately on preparation. It is necessary to protect the nitrogen with a *tert*-butyldimethylsilyl group. Reduction of **41** with sodium borohydride affords the alcohol **42** in 86% yield. A Swern oxidation converts **42** almost quantitatively to **43.**[17]

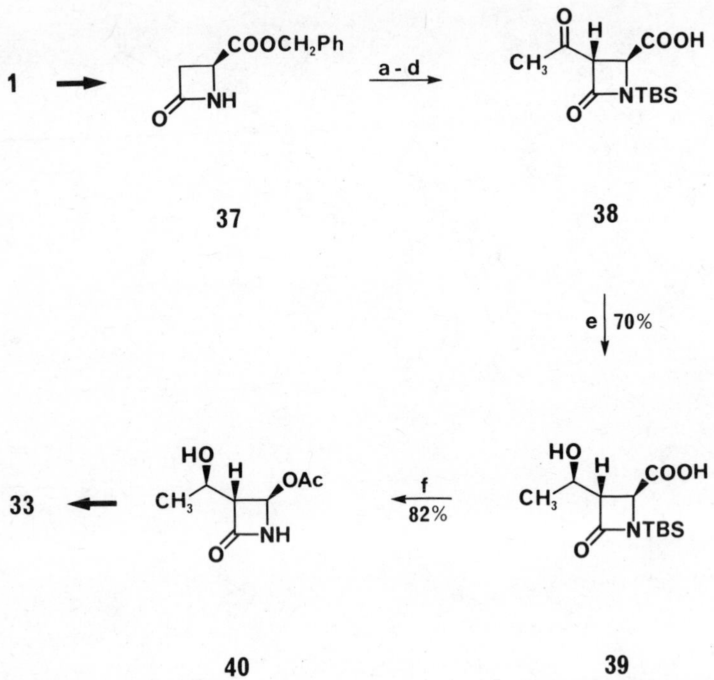

Scheme 5. (a) TBS–Cl/TEA (98%); (b) H_2, Pd/C (92%); (c) LDA, 0°C, CH_3CHO (90%); (d) $Na_2Cr_2O_7/H_2SO_4$ (64%); (e) $(i\text{-}C_3H_7)_2NH \cdot BH_3/Mg(CF_3COO)_2$; (f) $Pb(OAc)_4$/DMF/HOAc.

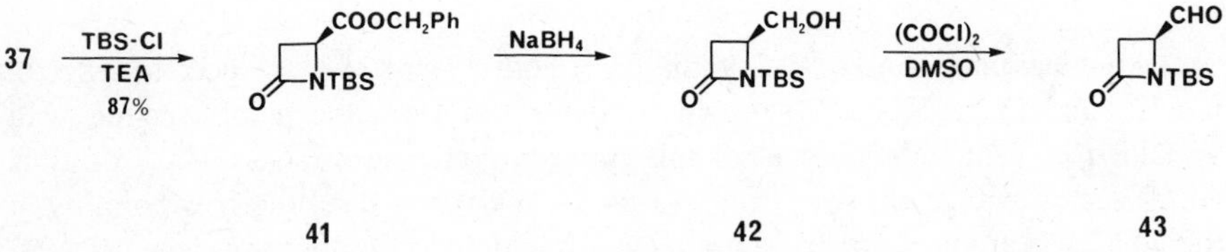

The chiral carbapenam ring system **48** that has the fundamental skeleton of **33** and providing an attractive synthon for the preparation of thienamycin derivatives can also be prepared from L-aspartic acid (**1**)[18,19] as shown in Scheme 6. (*R*)-3-Boc–amino-4-methoxycarbonylbutyrate (**45**) is prepared in 79% yield from aspartate **44**[20] by an Arndt–Eistert synthesis. Acidic cleavage of the Boc group followed by Schiff base formation with benzaldehyde and catalytic hydrogenation affords *N*-benzyl β-amino ester **46** in 74% yield. Cyclization of **44** with thionyl chloride and TEA provides **47,** which is converted along the sequence described for thienamycin (Scheme 3) to the carbapenam **48.**

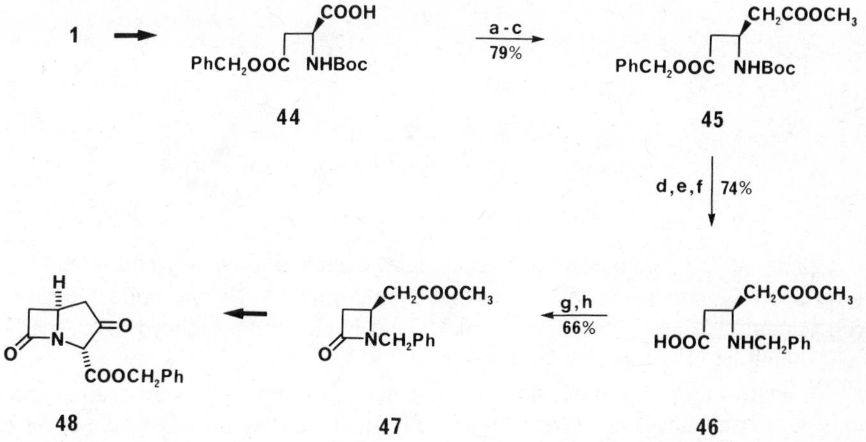

1 → COOH / PhCH$_2$OOC NHBoc **44** —a-c, 79%→ CH$_2$COOCH$_3$ / PhCH$_2$OOC NHBoc **45** —d,e,f 74%→ CH$_2$COOCH$_3$ / HOOC NHCH$_2$Ph **46** —g,h 66%→ CH$_2$COOCH$_3$ / NCH$_2$Ph **47** → H / COOCH$_2$Ph **48**

Scheme 6. (a) ClCOOEt/TMEDA; (b) CH_2N_2; (c) PhCOOAg/TEA; (d) HCl/EtOAc; (e) PhCHO; (f) H_2, Pd/C; (g) $SOCl_2$; (h) TEA/benzene.

An alternate route to **47** takes advantage of the selective reduction of diethyl *N*-Cbz-(*S*)-aspartate (**49**)[21] with sodium borohydride to afford 4-hydroxybutyrate **50** in 60% yield. Protection of **50** as a THP ether followed by reductive removal of the Cbz group then replacing it with a benzyl group affords **51.** Cyclization to the corresponding β-lactam is accomplished in good yield with ethylmagnesium chloride. This is smoothly converted to the nitrile **52,** which provides **47** on treatment with methanolic HCl[17] (Scheme 7).

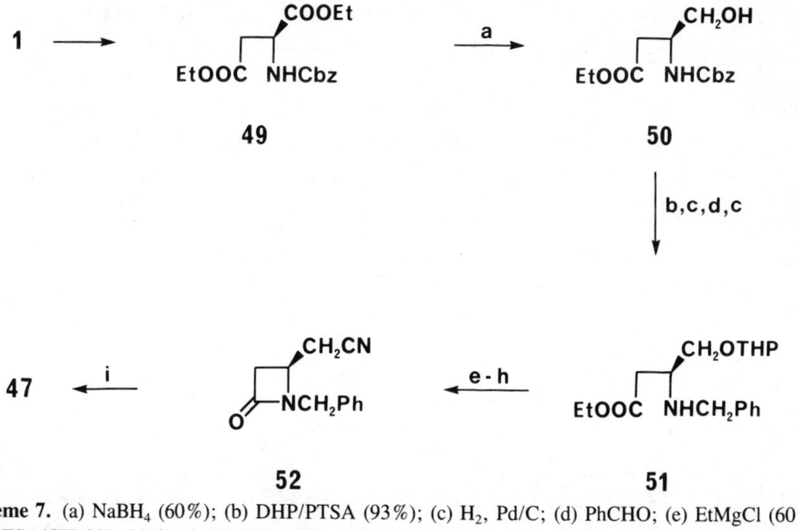

1 → COOEt / EtOOC NHCbz **49** —a→ CH$_2$OH / EtOOC NHCbz **50** —b,c,d,c→ CH$_2$OTHP / EtOOC NHCH$_2$Ph **51** —e-h→ CH$_2$CN / NCH$_2$Ph **52** —i→ **47**

Scheme 7. (a) $NaBH_4$ (60%); (b) DHP/PTSA (93%); (c) H_2, Pd/C; (d) PhCHO; (e) EtMgCl (60%); (f) PTSA/CH_3OH (80%); (g) TsCl/pyridine (90%); (h) KCN, 18-crown-6, CH_3CN (75%); (i) HCl/CH_3OH (40%).

7.1.1 Asparagine

(*S*)-2,4-Diamino-4-oxobutanoic acid (**53**)

L-Asparagine (**53**), a nonessential amino acid, occurs ubiquitously both in the free form and as a component of proteins. It is important in the metabolic control of cell functions in nerve and brain tissues[22] and is structurally related to L-aspartic acid, being its β-half amide derivative.

An interesting reaction occurs when attempts are made to convert L-asparagine to is *N*-carboxyanhydride. When phosgene is passed through a suspension of **53** in dioxane, not only does an *N*-carboxyanhydride form with the α-amino acid portion of the molecule, but also a dehydration occurs at the amide function to provide nitrile **55** directly. The next-higher homologue, L-glutamine (**54**), behaves similarly to give **56** in somewhat lower yield.[23]

$COCl_2$, 60°C

53 n = 1 → **55** n = 1 (60%)

54 n = 2 → **56** n = 2 (42%)

Acetone reacts with the sodium salt of **53** to form the tetrahydropyrimidine-4-carboxylate **57** by way of cyclization with both nitrogens in the molecule.[24]

53 → 1. NaOH, 2. acetone, 90% → **57**

On exposure to a sodium hypohalite, *N*-protected asparagine derivatives **58** and **59** undergo a Hofmann rearrangement to produce the imidazolidin-2-ones **60** and **61**.[25,26] The transformation occurs by initial conversion of the primary amide group to an isocyanate followed by intramolecular cyclization.

H_2N–C(=O)–CH$_2$–CH(NHR)–COOH $\xrightarrow{\text{NaOX}}$ cyclic urea (O=C, N–R, COOH, HN)

58 R = Ac X = Br **60** R = H

59 R = Cbz X = Cl **61** R = Cbz (36%)

If the protecting group on the nitrogen is of sufficient size, the intramolecular formation of the cyclic urea (e.g., **60** and **61**) is suppressed and protected (*S*)-2,3-diaminopropionic acid derivatives **63** are produced.[27,28]

H_2N–C(=O)–CH$_2$–CH(NHR)–COOH $\xrightarrow{\text{NaOBr}}$ H_2N–CH$_2$–CH(NHR)–COOH

62a, R = Ts **63a**, R = Ts (56%)

b, R = Boc **b**, R = Boc

This transformation is employed as the initial step in the synthesis of the chiral portion of pyrimidoblamic acid (**68b**), one of five building blocks used for the construction of the antitumor antibiotic bleomycin (see Figure 1, Chapter 1). Thus Hofmann rearrangement of **62b** followed by treatment with benzyl chloroformate and then diazomethane furnishes the differentially protected 2,3-diaminopropionic acid ester **64** in 12% yield without any purification and the itermediates. Amidation and selective deprotection give the chiral fragment Boc–L-DAPA (**65**).[28]

63b $\xrightarrow[\text{2. } CH_2N_2]{\text{1. Cbz-Cl}}$ CbzHN–CH$_2$–CH(NHBoc)–$COOCH_3$ $\xrightarrow[\text{2. } H_2\text{, Pd (99\%)}]{\text{1. } NH_3 \text{ (73\%)}}$ H_2N–CH$_2$–CH(NHBoc)–$CONH_2$

64 **65**

Boc–pyrimidoblamic acid (**68a**) is obtained from **65** by formation of the Schiff base (**67**) with the 2-formylpyrimidine derivative **66,** followed by addition of an amide anion equivalent to the imine carbon.[29,30]

Pipecolic acid derivatives serve as useful precursors in indole alkaloid syntheses. One derivative in particular that is used in the total synthesis of (+)-apovincamine (**72**) is the (2*S*,3*R*)-3-cyano-3-ethylpipecolate **69.** As shown in Scheme 8, the synthon is obtained from L-asparagine via its Boc derivative **62b.**

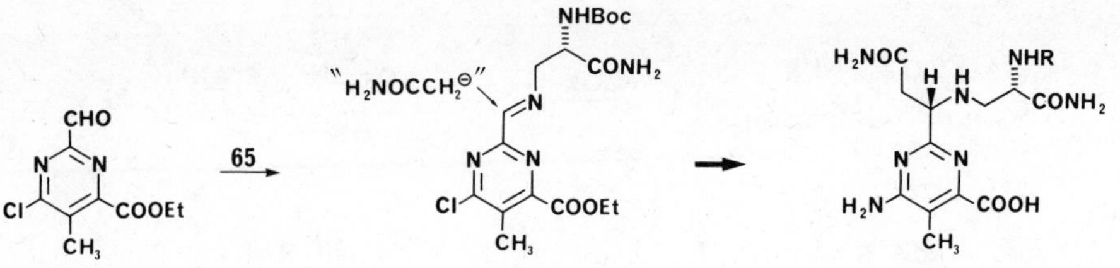

66 67 68a R= Boc
68b R= H

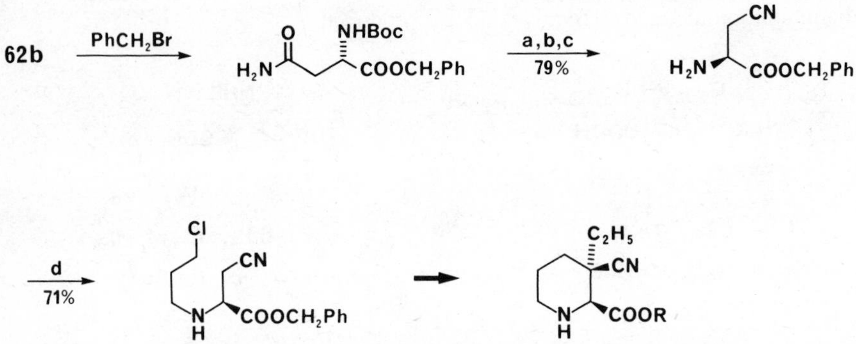

69

Scheme 8. (a) TsCl/pyridine; (b) $NaHCO_3$; (c) CF_3COOH; (d) $CF_3CH_2SO_3(CH_2)_3Cl$.

The alkaloid is assembled by alkylation of **69** with tryptophyl bromide (**70**) and the resulting tryptamine derivative **71** elaborated to the desired product.[31]

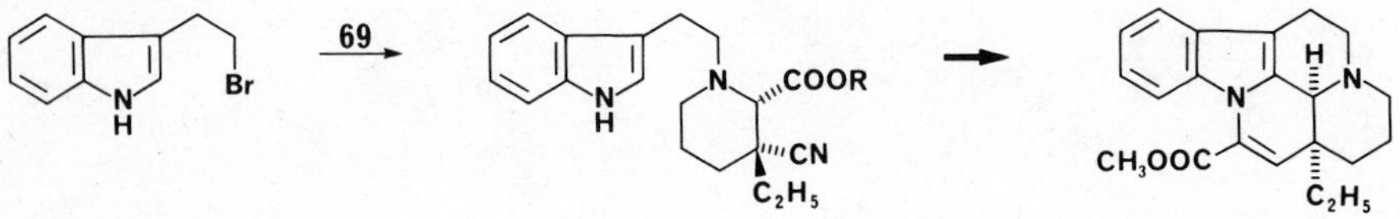

70 71 72

7.2 GLUTAMIC ACID

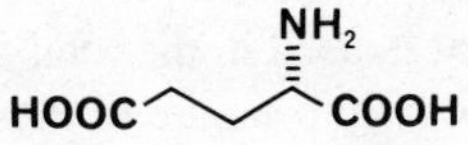

(*S*)-α-Aminoglutaric acid (**73**).

Glutamic acid (**73**) is a proteogenic amino acid with two carboxyl groups of which only one forms a peptide bond and the other gives the polypeptide an acid character. It is involved in the transport of potassium ions in the brain and also detoxifies ammonia in the brain by forming glutamine (**54**), which can cross the blood–brain barrier.[32]

Both the L and the D forms of **73** are commercially available. L-Glutamic acid is the least expensive of all the amino acids. Because of its versatility as a chiral synthon and its availability, **73** is one of the most useful of the amino acids.

The enantioselective α-alkylation of glutamic acid is accomplished through the dilithium enolatocarboxylate **77** of 3-[3-benzoyl-2-(*tert*-butyl)-1-methyl-5-oxo-imidazolidin-4-yl]propionic acid (**76**), which is prepared from commercially available *N*-Cbz–L-glutamic acid (**74**) in an overall yield of about 20%. Recrystallization of crude **75** and **76** provides these products in greater than 93% optical purity.

A double deprotonation of **76** with either lithium diethylamide or LDA affords **77**, which undergoes a regioselective alkylation away from the bulky *tert*-butyl group to furnish **78** with >95% ds. Hydrolysis of the imidazolidinone ring with HBr affords enantiomerically pure α-branched (*S*)-glutamic acids **79** (Scheme 9). This method can also be similarly applied for the enantioselective α-alkylation of aspartic acid.[33]

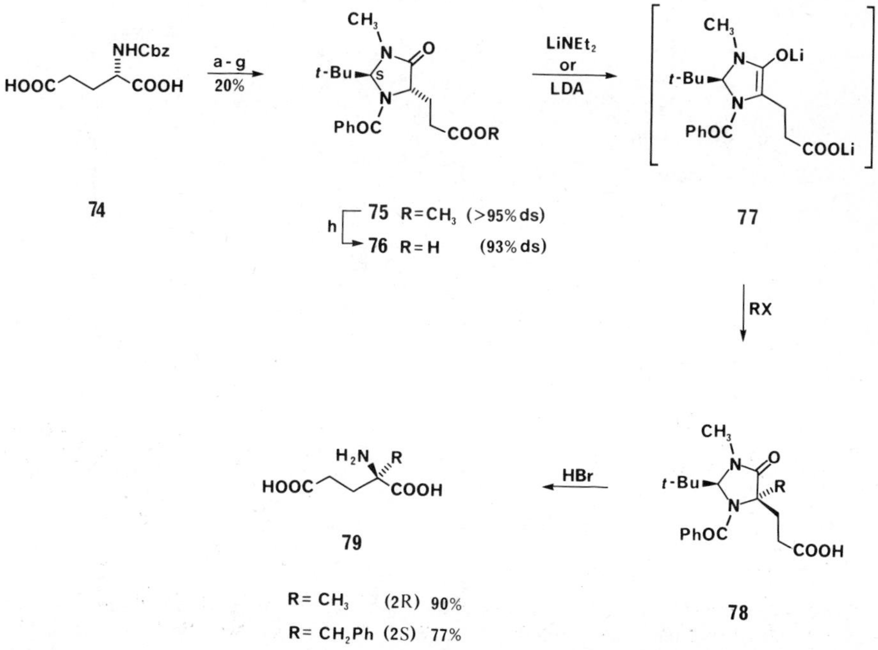

74

a-g 20%

h: 75 R=CH_3 (>95% ds); 76 R=H (93% ds)

$LiNEt_2$ or LDA

77

RX

HBr

78

79

R= CH_3 (2R) 90%

R= CH_2Ph (2S) 77%

Scheme 9. (a) Ac_2O; (b) CH_3NH_2/EtOH; (c) H_2, Pd; (d) HCl/CH_3OH; (e) t-C_4H_9CHO/TEA; (f) HCl/CH_3OH; (g) PhCOCl/TEA; (h) KOH/CH_3OH/H_2O (99%); (i) $LiNEt_2$ or LDA, −78°C; (j) HBr.

The antimetabolite (**α***S*,5*S*)-α-amino-3-chloro-4,5-dihydro-5-isoxazole acetic acid (**83**) is of synthetic interest because of its antitumor and enzyme-inhibitory properties. The key feature in the stereospecific synthesis of **83** is the photochlorination of **73** to afford a 1:1 mixture of L-*threo*- and L-*erythro*-β-chloroglutamic acids **80a** and **80b**, respectively. These diastereomers are separated and obtained pure by ion-exchange chromatography. Isomer **80a** is converted in good yield to the cyclic anhydride **81**, which undergoes attack at the α-carbonyl when treated with the lithium salt of *N*-hydroxyphthalimide. Displacement of the phthalimide with aqueous hydroxylamine followed by cyclization under basic conditions furnishes **82**, which is then elaborated to **83**. This eight-step synthesis (Scheme 10) is accomplished in 17% overall yield from **80a.**[34] A similar synthesis of the (5*R*)-diastereomer **84** proceeds analogously from **80b** in a 15% overall yield.

Siderophores are low-molecular-weight microbial iron chelators that use hydroxamate, thiohydroxamate, or catechol derivatives to chelate iron in an octahedral high-spin complex.[35] One such iron chelator

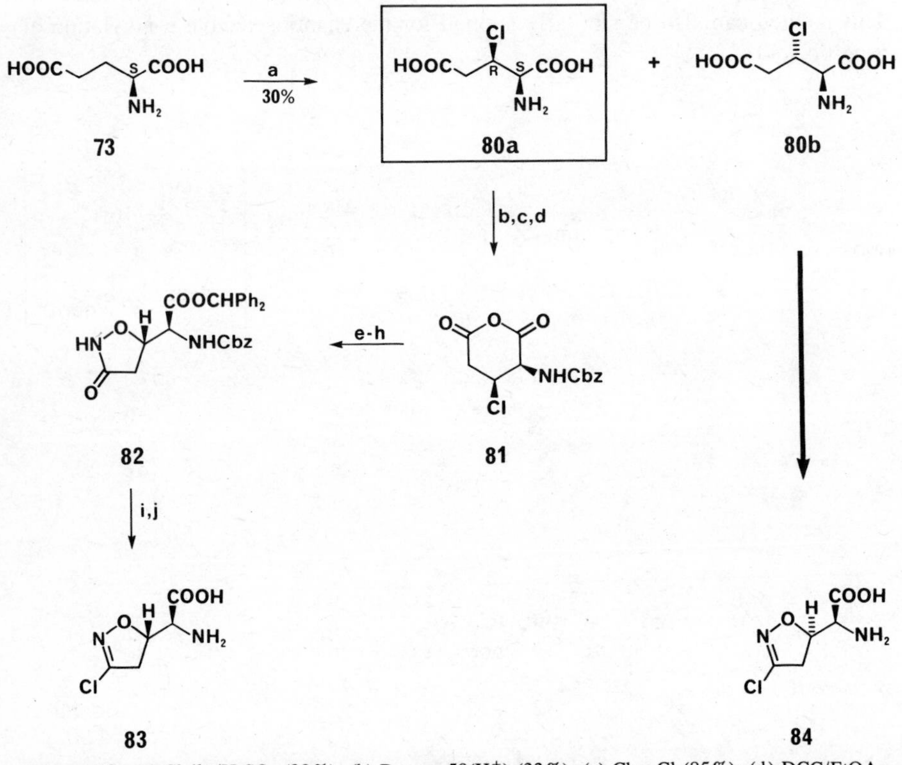

Scheme 10. (a) $Cl_2/h\nu/H_2SO_4$ (30%); (b) Dowex-50(H^+) (33%); (c) Cbz–Cl (85%); (d) DCC/EtOAc (85%); (e) *N*-hydroxyphthalimide; (f) $NH_2OH \cdot HCl$ (66% from **81**); (g) TEA, pH 11; (h) Ph_2CN_2; (i) $Cl_2P(NMe_2)_3$ (54%); (j) CF_3COOH.

is rhodotorulic acid (**90**), a diketopiperazine that is prepared from the protected δ-*N*-hydroxy-L-ornithine derivatives **88** and **89.** These both can be synthesized from commercially available α-*t*-butyl *N*-Boc–L-glutamate (**85**) (Scheme 11). The conversion of **86** to **88** can be accomplished by either the direct reaction of **86a** with protected hydroxylamines under Mitsunobu conditions (DEAD–Ph_3P)[35] or a nucleophilic displacement of the bromo derivative **86b**[36] with *N*-acetyl-*O*-benzylhydroxylamine.

Alternatively, **90** can be prepared from **89** by proceeding through the Leuch anhydride **91.** Subsequent treatment with ethyl *N*-hydroxyacetimidate (**92**) or aziridine affords the benzyl-protected derivative **93** in 20% yield. This is then converted to **90.**[35]

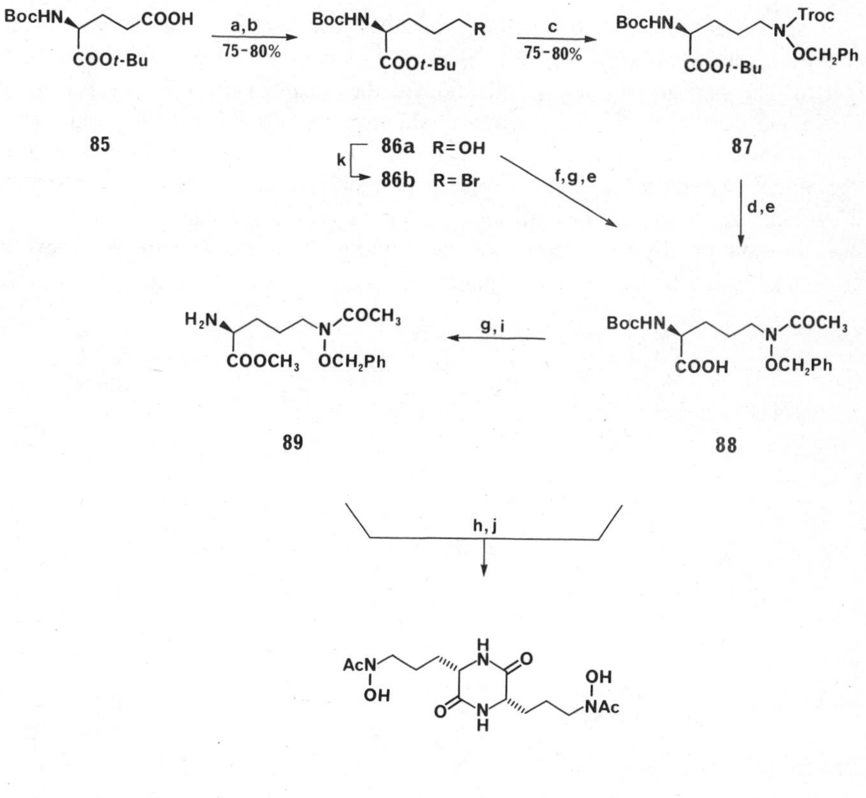

Scheme 11. (a) ClCOOEt/TEA; (b) $NaBH_4$; (c) DEAD/Ph_3P, Troc–$NHOCH_2Ph$; (d) Zn/HOAc/Ac_2O (85%); (e) TFA, $(Boc)_2O$ (75%); (f) CBr_4/Ph_3P (77%); (g) *O*-methyl-*N*,*N*′-dicyclohexylisourea (81%); (h) EEDQ/TEA (67–70%); (i) TFA (42%); (j) H_2, Pd/C (70%); (k) CBr_4/Ph_3P.

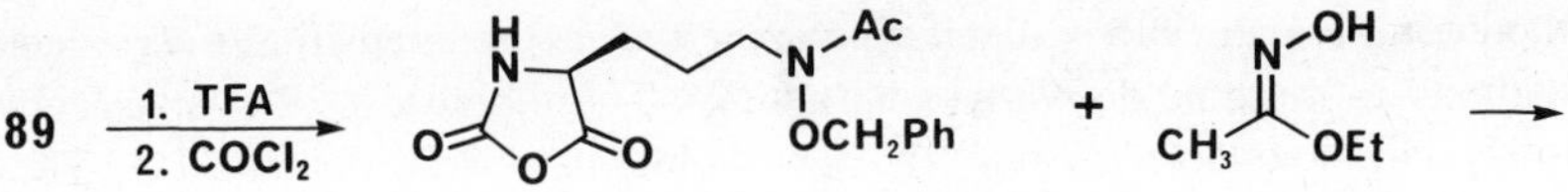

89 →(1. TFA; 2. $COCl_2$) 91 + 92 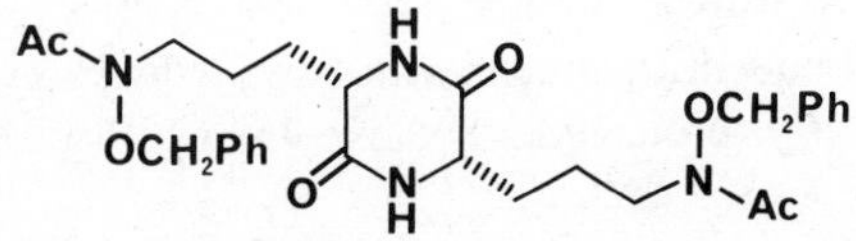 → 93

A second approach to the synthesis of δ-*N*-hydroxy-L-ornithine derivatives incorporates the α-carboxyl and the α-amino group into a protective heterocyclic system that then allows functionalization of the γ-carboxyl group. Treatment of Cbz–L-glutamic acid (**74**) with paraformaldehyde affords **94** in 85% yield. After conversion to the aldehyde, **95** reacts with *O*-benzylhydroxylamine to furnish oxime **96,** which, after reduction and ring opening under basic conditions, provides chiral **97** without any observed racemization[37] (Scheme 12). A similar method, but one that relies on a diallylic protection of the amino group rather than the heterocyclic protection, can also be used to prepare **97.**[38]

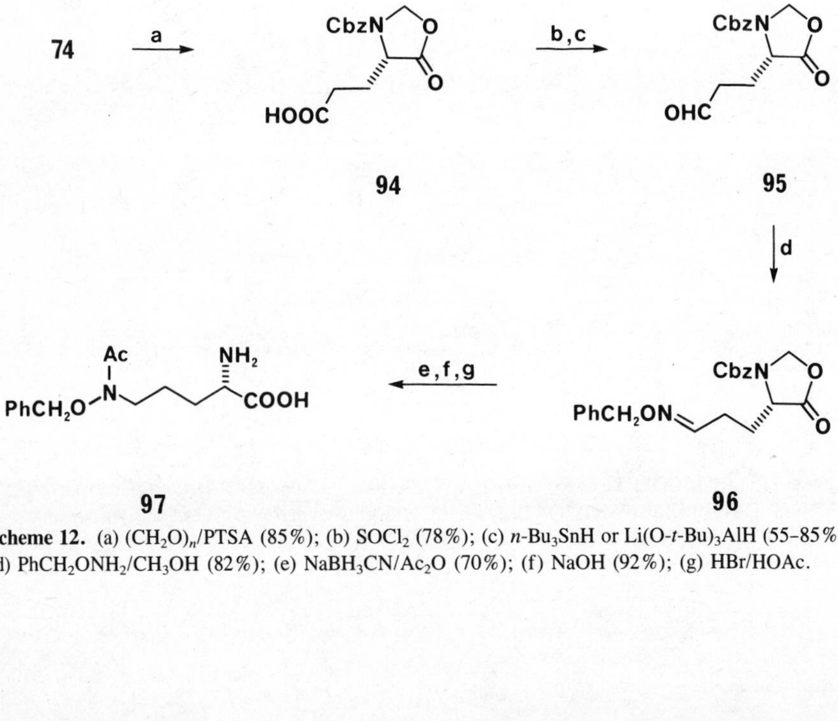

74 →(a) 94 →(b,c) 95 →(d) 96 →(e,f,g) 97

Scheme 12. (a) $(CH_2O)_n$/PTSA (85%); (b) $SOCl_2$ (78%); (c) *n*-Bu_3SnH or Li(O-*t*-Bu)$_3$AlH (55–85%); (d) $PhCH_2ONH_2/CH_3OH$ (82%); (e) $NaBH_3CN/Ac_2O$ (70%); (f) NaOH (92%); (g) HBr/HOAc.

L-Glutamic acid (**73**) provides the correct absolute configuration at C-2 of α-kainic acid (**103**), an algal metabolite possessing potent neuronal excitatory activity. An important requirement in the enantioselective synthesis of **103** is the ability to sterically control the formation of the C-3–C-4 bond by way of an intramolecular thermal type I ene reaction. Accordingly, selective reduction of the α-carboxyl group of **99** with diborane, followed by protection of the resulting primary alcohol with *tert*-butyldimethylsilylchloride and alkylation with 1-bromo-methyl-2-butene, furnishes **100.** Successive selenation, oxidation, and selenoxide elimination smoothly produce **101,** which subsequently undergoes a thermal ene reaction to provide **102** without the formation of any other isomers. Furthermore, the required *cis* geometry of the isopropenyl and ethyl acetate substituents is obtained. The overall yield of optically pure **103** from **99** is 5%[39] (Scheme 13).

Vinylglycine (**106**), a natural β,γ-unsaturated α-amino acid found in mushrooms and implicated as an intermediate in some biochemical processes, was first synthesized optically pure by the pyrolysis of a protected methionine sulfoxide derivative (Chapter 6). Hanessian and Sahoo[40] have developed a simple and practical synthesis of optically pure crystalline **106** by employing a decarboxylative elimination of *N*-Cbz-L-glutamic acid monomethyl ester (**104**) with lead tetraacetate, catalyzed by cupric

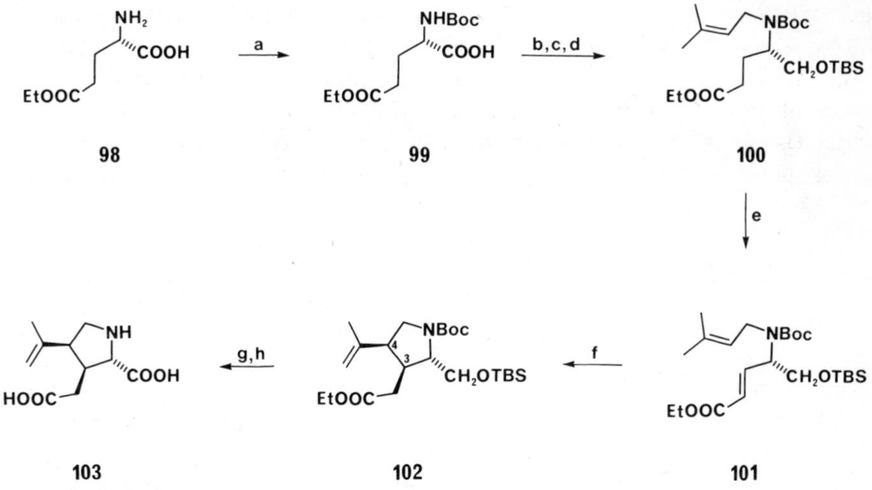

NH2 COOH EtOOC **98** —a→ NHBoc COOH EtOOC **99** —b,c,d→ NBoc CH2OTBS EtOOC **100**

↓ e

NH COOH HOOC **103** ←g,h— NBoc 4 3 CH2OTBS EtOOC **102** ←f— NBoc CH2OTBS EtOOC **101**

Scheme 13. (a) $(Boc)_2O$/TEA/DMF/H_2O (99%); (b) BH_3 (57%); (c) TBS–Cl/TEA (92%); (d) NaH/1-bromo-3-methyl-2-butene/HMPA (77%); (e) LTMP/PhSeCl/H_2O_2 (48%); (f) 130°C, 40 hr (70%); (g) $Bu_4N^+F^-$, Jones oxidation (60%); (h) LiOH/CF_3COOH, ion exchange (56%).

acetate. A high-yielding process for the preparation of **104** proceeds from **74** through the formaldehyde adduct **94.**[41] Moreover, **104** also undergoes the same decarboxylative elimination to provide **107** in about 50% yield, which may be converted to **106** as well (Scheme 14).

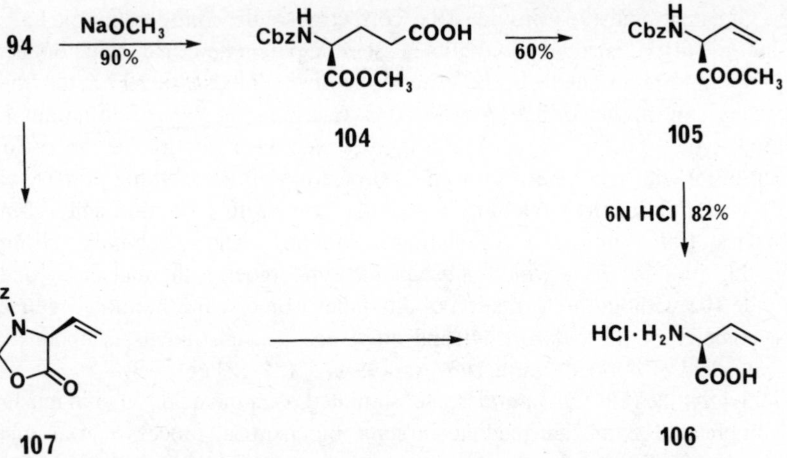

Scheme 14

Barton et al.[42] have synthesized optically pure **106** by two different routes starting from L-glutamic acid. The first involves the free-radical rearrangement of the *O*-ester of selenohydroxamic acid **108** to give the *nor*-alkyl selenide **109**, which readily undergoes an oxidative elimination to provide **106** (after acidic hydrolysis) in about 45% overall yield. The second route uses a modified Huensdiecker reaction on the terminal carboxyl of **104** to prepare the *nor*-bromide **110**, which affords the phenylselenide **111** on treatment with phenylselenide anion. Oxidative elimination followed by acidic deprotection furnishes **106** also in an overall yield of about 45%. Use of the benzyl ester of **104** with Boc protection proceeds with higher yields, but the final acidic deprotection occurs with partial racemization (Scheme 15).

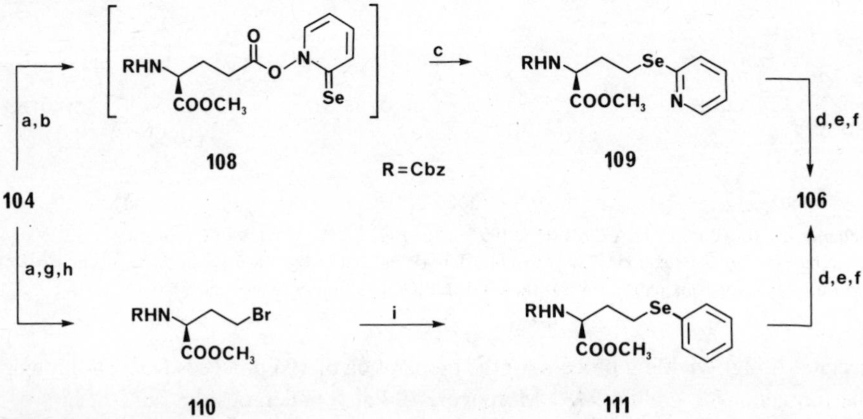

Scheme 15. (a) *N*-Methylmorpholine, −15°C, *i*-C_4H_9OCOCl; (b) *N*-hydroxypyridine-2-selenone/TEA; (c) *hν* (58% from **104**); (d) O_3/CH_2Cl_2, −78°C; (e) CCl_4, 80°C (78% from **108**); (f) 6 *N* HCl; (g) thiohydroxamic acid (77%); (h) CCl_3Br, *hν* (78%); (i) PhSeSePh/$NaBH_4$/EtOH (84%).

The bicyclic dipeptide **115** has the fixed conformation simulating that of the two central amino residues in type II′ β-turn, a common secondary structure of polypeptides intimately involved in protein folding and activity.[43] It can be synthesized from *N*-phthaloyl-L-glutamic anhydride (**112**). The reaction **112** with thiophenol opens the anhydride. Esterification with diazomethane, reduction with Raney nickel, and oxidation with PCC afford the aldehyde **113.** Conversion to the intermediate thiazolidine **114** is effected on treatment with L-cysteine under basic conditions. Without purification, this is thermally cyclized to **115,** in which two of the three chiral centers are derived from the natural amino acids[44] (Scheme 16).

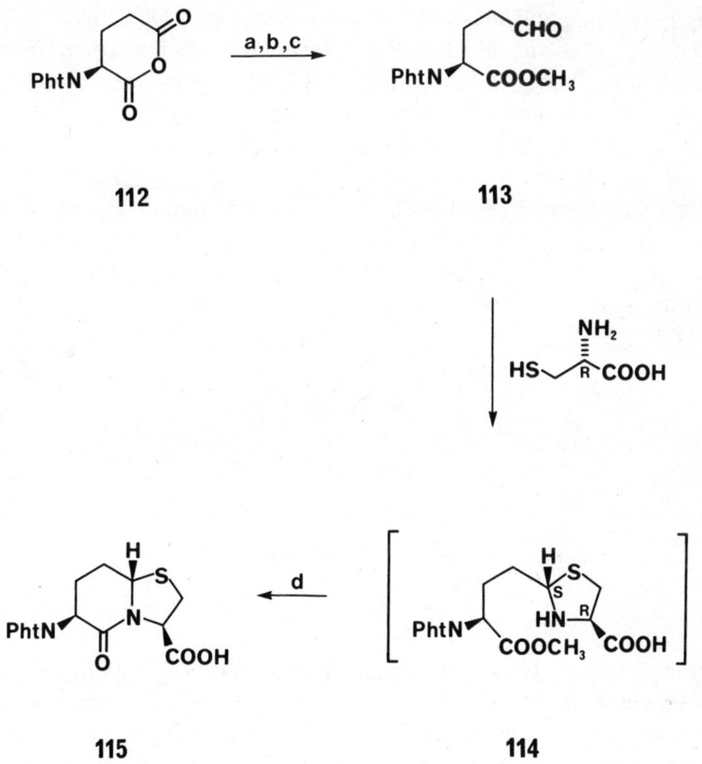

112 **113**

115 **114**

Scheme 16. (a) PhSH (85%); (b) CH_2N_2 (99%); (c) Raney Ni; (d) PCC (34%); (e) 70°C (51%).

7.2.1 Pyroglutamic Acid

L-Pyroglutamic acid (**116**) or (5*S*)-2-oxotetrahydropyrrole-5-carboxylic acid is a versatile chiral synthon. It is easily prepared in one step by the direct dehydration of L-glutamic acid (**73**) in refluxing aqueous unbuffered solution. Heating at 135°C in a sealed tube increases the yield to 95%, but with the occurrence of 4–5% racemization of **116.**[45]

NH_2

HOOC COOH 1. H_2O, 100°C 2. Dowex AG50WX8 86% O N H COOH

73 116

Whereas the solvolytic ring opening of **116** proceeds with difficulty,[46] the *N*-protected derivatives **118(a)–(d)** undergo ring opening with much greater ease. These various *N*-protected derivatives can be prepared through a two-step sequence where the corresponding *N*-protected glutamic acid is first converted to the corresponding L-glutamic anhydride **117** by dehydration with DCC. One may also use PCl_5, $SOCl_2$,[47] or acetic anhydride[48] to effect anhydride formation. When treated with dicyclohexylamine (DCA), followed by acidification with hydrogen chloride in methanol, **117** is converted almost quantitatively to **118.**[46]

Chiral pyrrolidines, important auxiliaries for asymmetric induction, are accessible from **116** and its derivatives. (*R*)-Anilinomethylpyrrolidine (**121**) is a chiral additive for the asymmetric reduction of ketones with lithium aluminum hydride.[49] Heating D-glutamic acid (**119**) in aniline at 195–200°C for 30 min

NHR

HOOC COOH DCC NHR O O O 1. DCA 2. HCl O N R COOH

117 118

(a) R=Cbz, (b) R=$COCH_3$, (c) R=CHO, (d) R=Ts

affords optically pure **120** in 46% yield. Reduction of **120** with lithium aluminum hydride provides (*R*)-**121** in 85% yield, with >95% ee. An attractive feature of this synthesis is the avoidance of protection and deprotection steps[50] (Scheme 17). Alternatively, the thermal dehydration of **119** at 200°C for 1 hr provides a near-quantitative yield of **122** with the (*R*) configuration. When this is reduced with diborane, (*R*)-prolinol (**123**) is obtained in 85% yield. Because of the high cost of D-proline, this method provides an attractive and affordable alternative to this chiral molecule.

Esterification of **122,** followed by treatment with methylamine, affords the (*R*)-amide **124** in excellent overall yield. Subsequent reduction with diborane furnishes (*R*)-2-methylaminomethylpyrrolidine (**125**).[51]

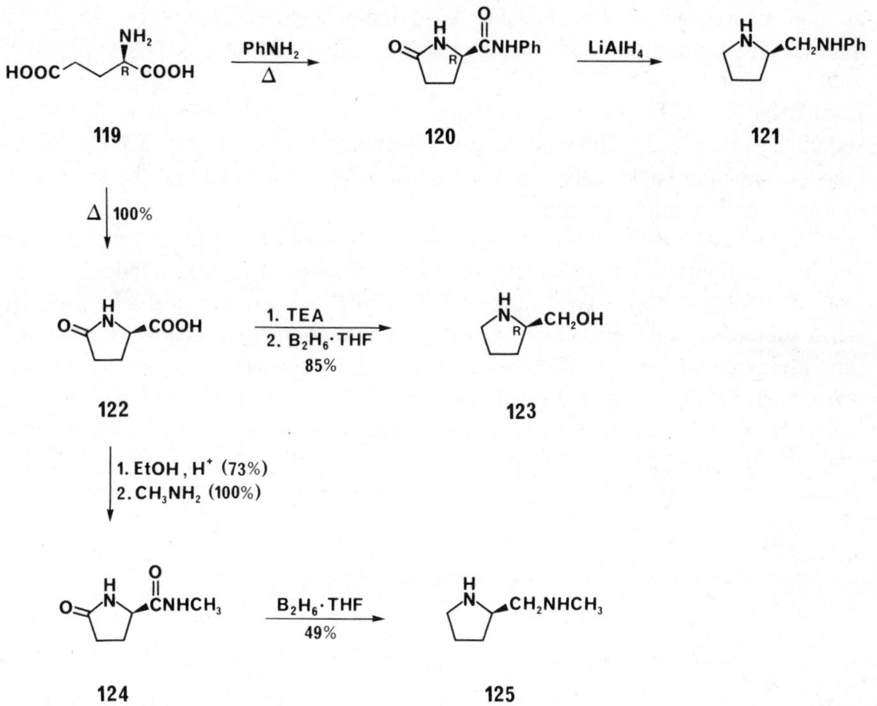

Scheme 17

The (*S*)-enantiomers of the molecules shown in Scheme 17 are similarly prepared when **73** is used in place of **119.**

The esterification of **116** with diazomethane provides **126,** which, when treated with boron trifluoride etherate, affords chiral imino ether **127.** This is converted in two steps to (5*S*)-5-methoxycarbonylpyrroldine-2-acetonitrile (**128**), which is isolated as a 1 : 1 mixture of (*E*) : (*Z*) isomers.[52]

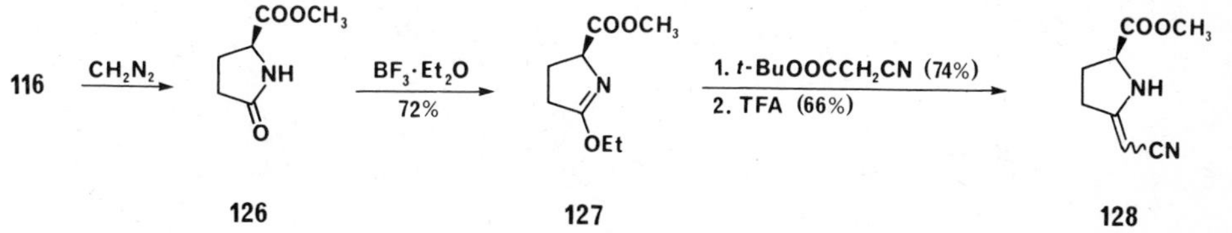

Convulsive seizures can be prevented by inhibition of γ-aminobutyric acid (GABA) transaminase. A new class of such inhibitors, the (*S*)-5-substituted-4-

aminopentanoic acids, are available from the key intermediate (*S*)-5-(hydroxymethyl)-2-pyrrolidinone (**129**). Conversion to either **131** or **132** is accomplished with $(C_6H_5)_3P/CX_4$ (X = Cl or Br). Both fluoro **136** and cyano **137** are available from **132** by a nucleophilic displacement of bromine with either silver fluoride or sodium cyanide. Each correspondingly substituted lactam (**134** or **135**) is hydrolyzed to the amino acid salts in 1 *N* aqueous acid [step (c)] without the occurrence of any racemization[53] (Scheme 18).

Fujimoto and Kishi[54] have used the chirality of nitrile **135** to establish the correct absolute configuration of gephyrotoxin (**143**), a biologically active alkaloid isolated from dendrobatid frogs. Treatment of **135** with P_4S_{10} provides a thiolactam that, when subjected to the Eschenmoser sulfide contraction[55] and deacylation, affords the vinylogous urethane **138.** Hydrogenation of **138** proceeds to give a separable mixture (2.3 : 1) of *cis* and *trans* diastereomers. It is the *cis*-pyrrolidine **139a** that is processed through the cyclic urethane **140** to pyrrolidine **141** (note the potential symmetry). Elaboration of **141** proceeds along established routes to provide **143** possessing the configuration opposite that of natural material (Scheme 19).

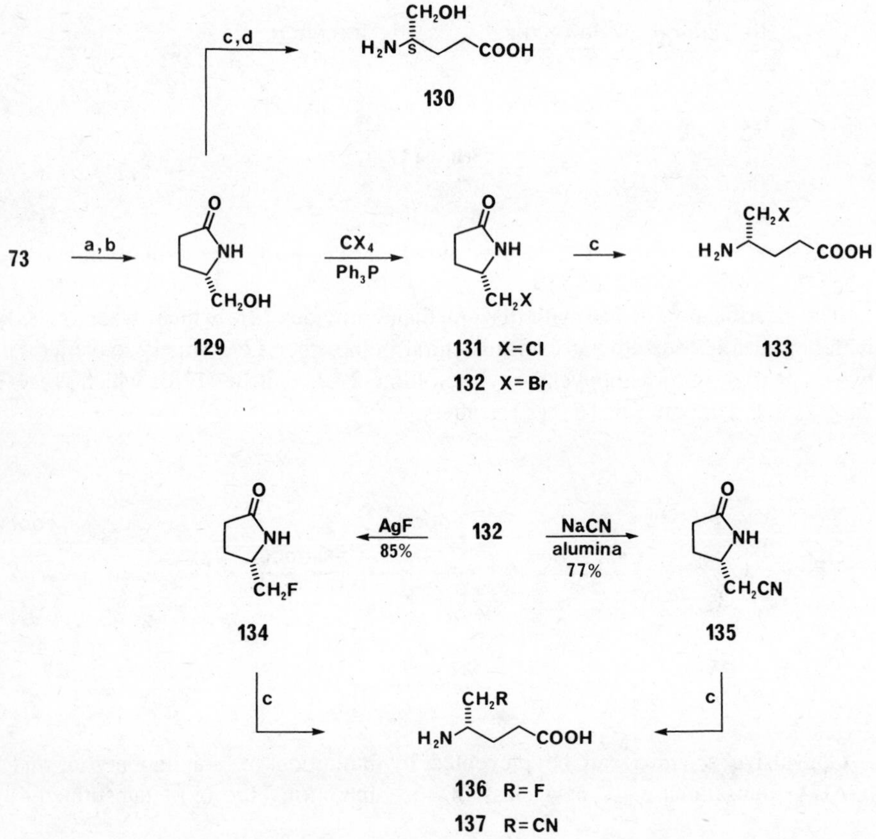

Scheme 18. (a) $SOCl_2$/EtOH (84%); (b) $LiBH_4$ (88%); (c) 1 *N* HCl, reflux; (d) Ag_2O, H_2S.

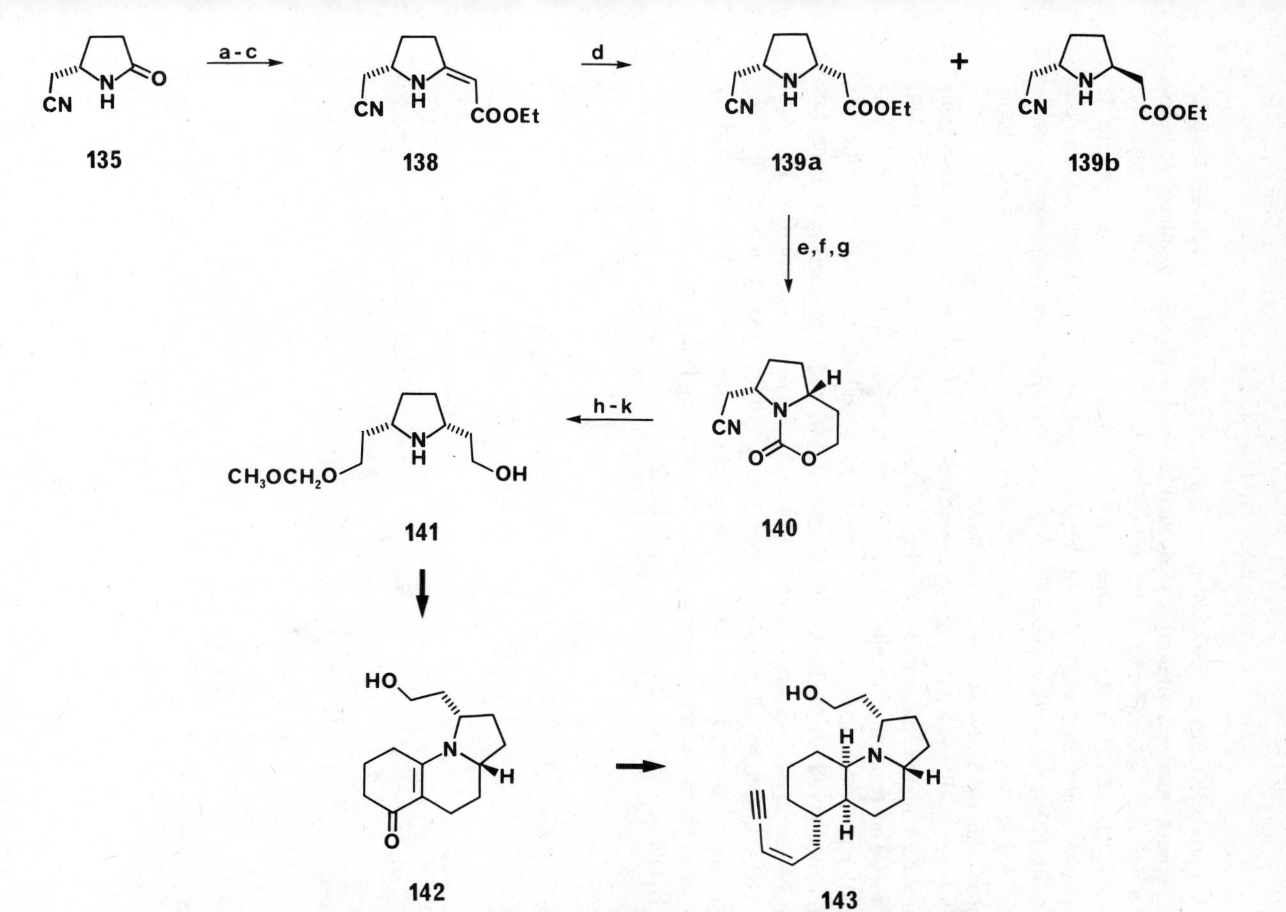

Scheme 19. (a) P_4S_{10}/pyridine, 80°C; (b) $CH_3COCH(Br)COOEt$, $NaHCO_3$; (c) 0.1 *N* KOH, 60°C; (d) H_2, Pt/C, $HClO_4$; (e) PhOCOCl, pyridine; (f) $LiBH_4$; (g) KH; (h) DIBAH, −105°C; (i) $NaBH_4$/DME; (j) CH_3OCH_2Br/TEA; (k) $Ba(OH)_2$.

L-γ-Carboxyglutamate (**147**) is an amino acid present in the calcium-binding sites of prothrombin, a vitamin K-dependent clotting factor. Its synthesis requires the introduction of a carboxyl group into the 4 position of L-glutamic acid while preserving the (*S*) chirality. Commercially available **118a** is first converted to its benzyl ester and then treated with Bredereck's reagent (**144**) to afford the enamine **145** in excellent yield. Exposure of **145** to 2,2,2-trichloroethoxycarbonyl chloride furnishes **146,** which, on treatment with benzyl alcohol followed by a hydrogenolysis, yields optically pure **147**[56] (Scheme 20).

This methodology is adaptable to the synthesis of arogenate (**152**), an important intermediate in the biosynthesis of the aromatic amino acids phenylalanine and tyrosine. The reaction of diphenyl sulfide and tri-*n*-butylphosphine with **145** followed by peracid oxidation of the resulting vinyl sulfide furnishes the sulfoxide diastereomers **148.** A Diels–Alder reaction of **148** with Danishefsky's diene **149** provides the crystalline spirodienone **150** in 57% yield. Reduction of **150** with DIBAH, followed by silica gel chromatography, affords **151.** This is finally converted to **152** by hydrolysis of the benzyl ester with 2 *N* NaOH, followed by removal of the protecting group. Lactam ring opening is accomplished with ethanolic NaOH at 70°C for 20–48 hr[57] (Scheme 21).

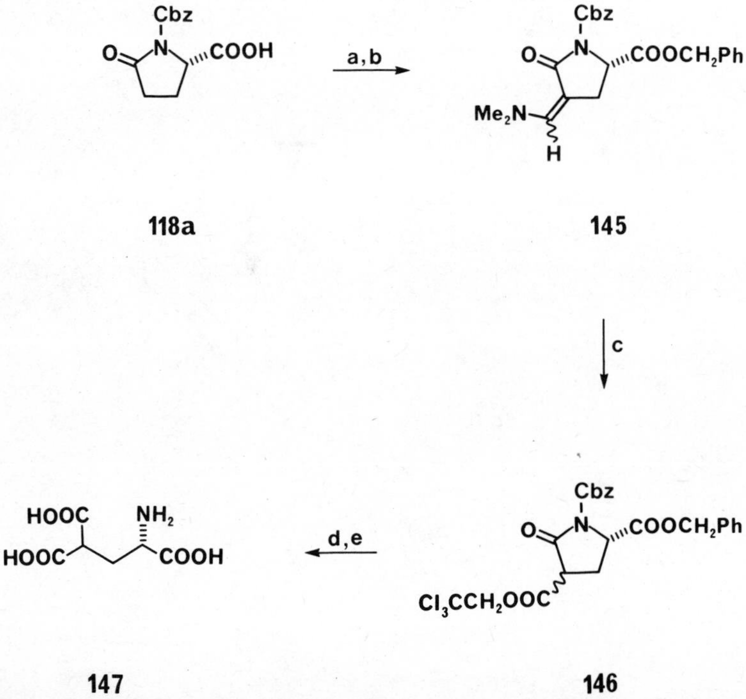

118a **145**

147 **146**

Scheme 20. (a) $PhCH_2Cl$/TEA/acetone (93%); (b) $(Me_2N)_2CHO$*t*-Bu (**144**) (95%); (c) Cl_3CCH_2OCOCl (41%); (d) $PhCH_2OH$/TEA (62%); (e) H_2, Pd/C (93%).

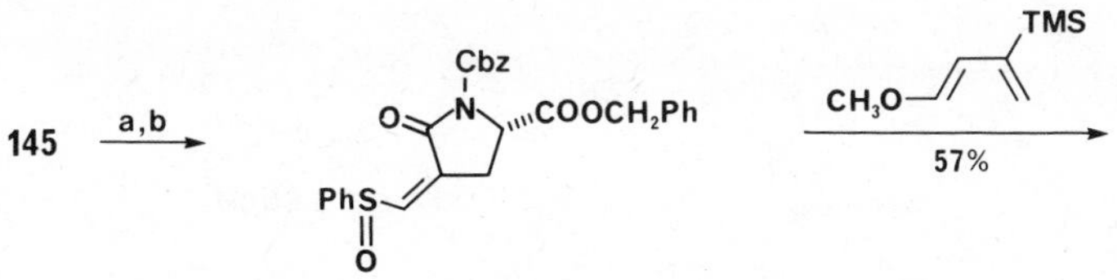

145 →(a,b) 148 →(CH_3O-diene-TMS, 57%) 149

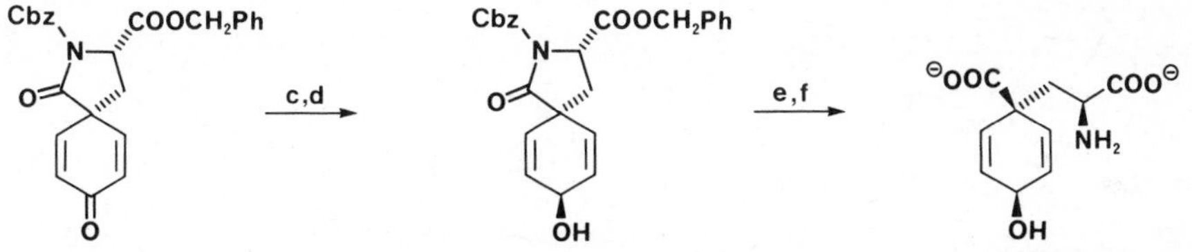

150 →(c,d) 151 →(e,f) 152

Scheme 21. (a) PhSeSePh/Bu_3P (68%); (b) MCPBA (60%); (c) DIBAH (56%, 2.4:1 $\beta:\alpha$); (d) chromatography (24%); (e) NaOH/CH_3OH; (f) NaOH/EtOH, 70°C.

(−)-Domoic acid (**159**), isolated from the red algae *Chondria armata* Okamura (Rhodomelaceae), like (−)-kainic acid (**102**), possess potent neurotransmitting activity. Its structure can be equated to L-pyroglutamic acid, in which the chirality of the C-5 carboxylic acid of **116** controls the stereochemistry of the ring carbons at C-3 and C-4. As shown in Scheme 22, commercially available *N*-Boc–L-pyroglutamic acid (**153**) is converted without racemization to the unsaturated lactam **154** in an overall yield of 70%. A [4 + 2] cycloaddition reaction of 2-(trimethylsilyl)oxy-1,3-pentadiene (**155**) with **154** proceeds stereospecifically to afford the single optically pure adduct **156.** This is converted in four steps to **157,** which is elaborated to **158.** Completion of the synthesis of **159** requires removal of the protecting groups.[58]

The chirospecific synthesis of the neurotoxin (−)-anatoxin a (**168**) can be achieved in 17 steps from L-glutamic acid in an overall yield of 40% and having 98% ee. Central to the synthesis is the initial carbon–carbon bond formation proceeding from the pyroglutamic acid derivative **160** by way of an efficient sulfide-contraction reaction (**161** → **162**). Stereoselective olefin reduction of **162** from the β face affords a *cis*-ester that is converted to the aldehyde **163** in excellent yield. The aldehyde is easily transformed to **165** (1:4 ratio), which is elaborated through a series of standard transformations to chiral **168**[58] (Scheme 23).

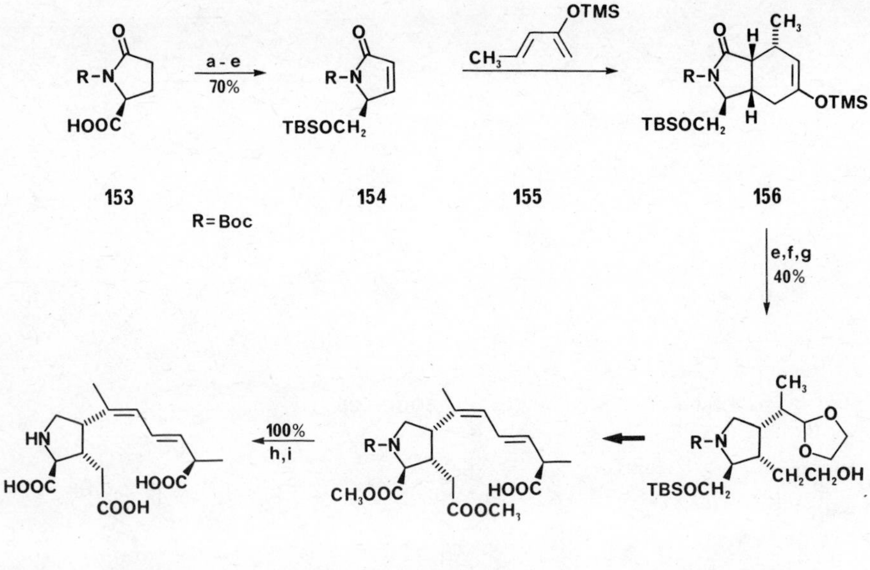

153 154 155 156

R=Boc

159 158 157

Scheme 22. (a) ClCOOEt/TEA, −10°C; (b) $NaBH_4$, −10°C; (c) TBS–Cl/DMF/imidazole; (d) LDA/PhSeCl, −78°C; (e) O_3/CH_2Cl_2, −78°C; (f) CH_2N_2; (g) 2-methyl-2-ethyl-1,3-dioxolane/PTSA; (h) 2.5% KOH; (i) CF_3COOH.

By a more convergent approach, D-glutamic acid (**119**) can be used to prepare (+)-anatoxin a (**174**). The six-carbon side chain is introduced as a single group by way of the sulfide-contraction reaction of **169** with triflate **170,** thus shortening the overall synthesis and improving the yield[59] (Scheme 24).

After the formation of a *cis*-pyrrolidine by way of a sulfide contraction, the newly formed chiral center at C-5 can be used to invert the chirality at C-2, thus forming *trans*-pyrrolidines. This is beautifully exemplified in the synthesis of both enantiomers of *trans*-5-butyl-2-heptylpyrrolidine (**181**), an active and major component in the repellant venom of the ant *Solenopsis fugax*. In this case, the sulfide contraction is most efficiently accomplished by performing the reaction of 0°C in the presence of the base *N*-methylpiperidine. The resulting vinylogous carbamate **176** having 96% ee is stereoselectively reduced to **177** and is then converted in two steps to a kinetic mixture of amino nitriles **178,** which is equilibrated in a silica gel slurry to a 1:9, *cis*:*trans* mixture. Strong mineral acid hydrolysis of **178** and crystallization of the mixture provides >99% *trans*-**179.** Propyllithium addition to **179** followed by acetone quench yields a propyl ketone that is immediately reduced to the alcohol **180.** Bissulfonation followed by removal of the sulfonate with $NaBH_4$ and deprotection with sodium in liquid ammonia or 48% HBr produces *trans*-**181** having 94% ee[61] (Scheme 25). D-Glutamic acid provides the enantiomeric *trans*-**182** when the same synthesis sequence is employed.

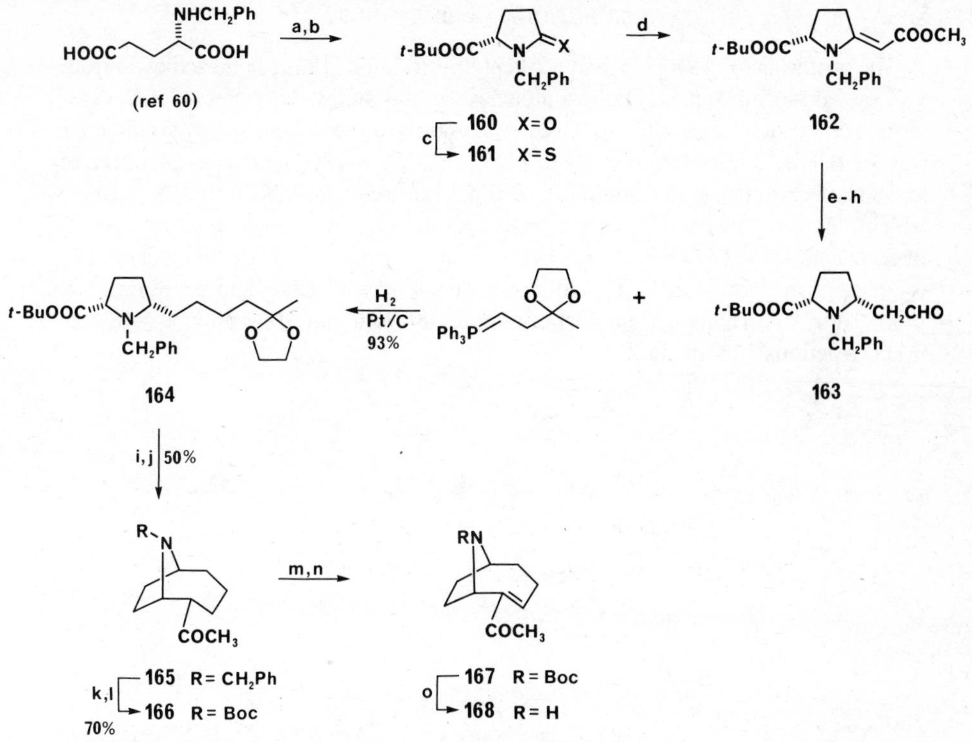

Scheme 23. (a) H_2O, 100°C (80%); (b) $(COCl)_2$/DMF/*t*-BuOH (93%); (c) P_4S_{10} (87%); (d) $BrCH_2COOCH_3/CH_3CN/Ph_3P$ (90%); (e) H_2, Pt/C (89%); (f) K_2CO_3/H_2O (93%); (g) $LiBH_4$ (95%); (h) Swern oxidation (99%); (i) HOAc/*n*-PrOH; (j) $POCl_3$; (k) H_2, Pd/C; (l) $(Boc)_2O$; (m) KH/TMS-Cl/TEA (90%); (n) $Pd(OAc)_2/CH_3CN$ (41%); (o) CF_3COOH (97%).

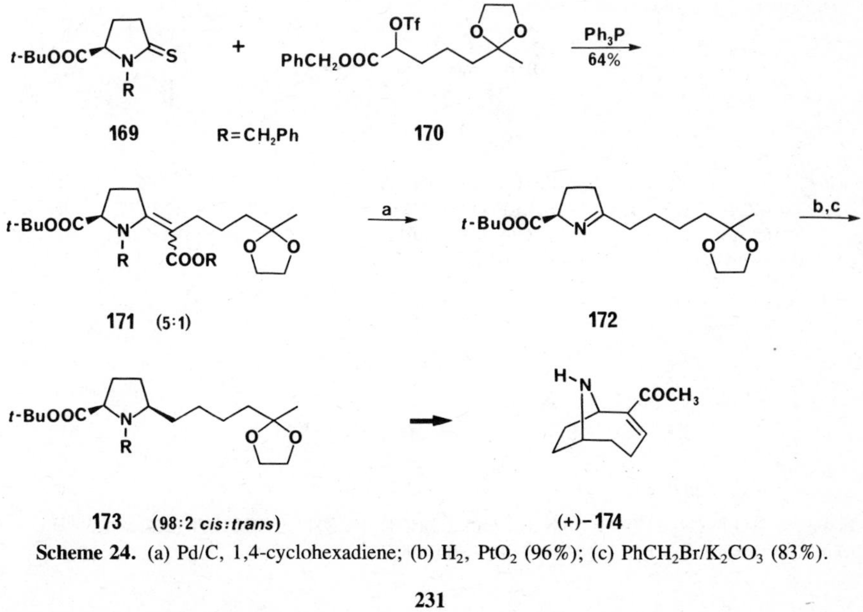

Scheme 24. (a) Pd/C, 1,4-cyclohexadiene; (b) H_2, PtO_2 (96%); (c) $PhCH_2Br/K_2CO_3$ (83%).

The synthesis of (2*S*)-*cis*-5-butyl-2-heptylpyrrolidine (**186**) can be achieved from **177** by taking advantage of the efficiency of the sulfide-contraction process to introduce the butyl side chain at C-2. Preparation of the *cis*-dialkyl pyrrolidine by way of this double-contraction route (**161** → **176** → **185**) represents a reflective transfer of chirality as the original C-2 center is used to establish the C-5 center, which, in turn, reestablishes the original center that had been destroyed at an intermediate stage.[60] Conversion of the carboxyl group of **177** to thiolactam **183** is accomplished by treating the amino nitriles **178** with LDA and excess sulfur, followed by $NaBH_4$ reduction. Elaboration to **186** is achieved by previously established reactions[61] (Scheme 26).

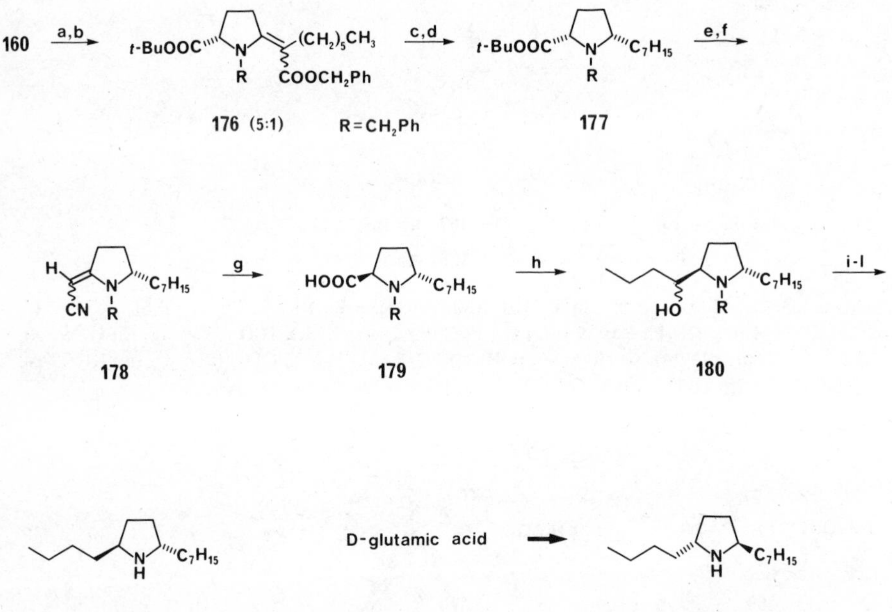

Scheme 25. (a) P_4S_{10}; (b) $CH_3(CH_2)_5CH(OTf)COOCH_2Ph$ (**175**) (70%); (c) $HCOONH_4$, Pd/C (86%); (d) $PhCH_2Br/K_2CO_3$ (97%); (e) HOAc/*n*-PrOH (97%); (f) $POCl_3$/KCN, 100°C (87%); silica gel slurry; (g) HCl (50%); (h) *n*-PrLi, −78°C, then $NaBH_4$ (70%); (i) H_2, Pd/C (95%); (j) $PhSO_2Cl$/NaOH (85%); (k) $NaBH_4$/DMSO (80%); (l) Na/NH_3 (80%).

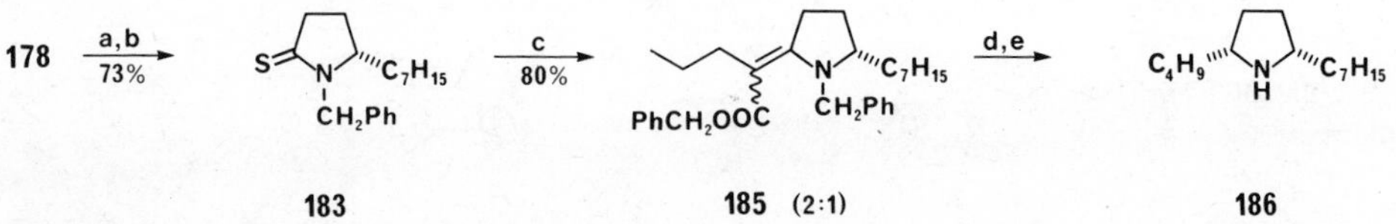

Scheme 26. (a) LDA/S_8, −78°C; (b) $NaBH_4$; (c) $CH_3(CH_2)_2CH(OTf)COOCH_2Ph$ (**184**), Ph_3P/CH_3CN; (d) Pd/C, cyclohexane (70%); (e) $HCOONH_4$, PtO_2 (90%).

An alternate strategy introduces the three-carbon unit directly by the addition of a large excess of *n*-propyllithium to **187.** The resulting 5-butanone group is reduced with $NaBH_4$ to give a mixture of diastereomeric amino alcohols **188** (separable if desired). Deoxygenation of the mixture is accomplished by first hydrogenolyzing the *N*-benzyl group, followed by bissulfonation with (phenylsulfonyl)imidazole activated with trimethyloxonium tetrafluoroborate. Reductive displacement of the resultant sulfonate and deprotection with sodium in liquid ammonia afford **186** having >99% *cis* stereochemistry and >94% optical purity.[61]

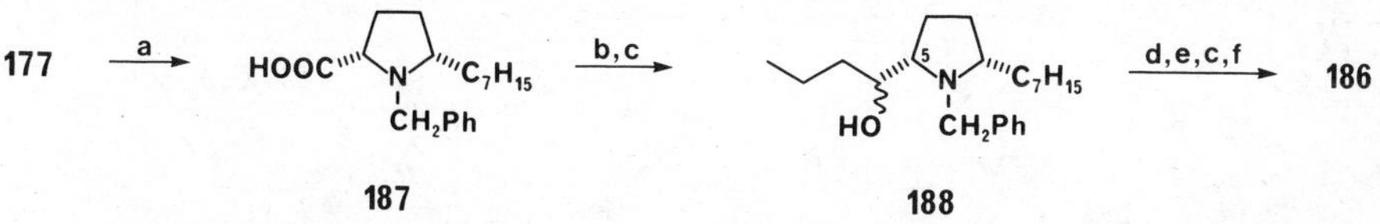

177 —a→ 187 —b,c→ 188 —d,e,c,f→ 186

(a) HOAc/*n*-PrOH (97%); (b) *n*-PrLi, −78°C; (c) $NaBH_4$; (d) H_2, Pd (98%); (e) (phenylsulfonyl)imidazole, $MeO^+BF_4^-$ (93%); (f) Na/NH_3.

The enantiomeric (2*R*)-*cis*-5-butyl-2-heptylpyrrolidine (**192**) is prepared from **160** by introducing the butyl side chain at C-5 through the sulfide-contraction process. *n*-Hexyllithium addition to the acid **190** provides a 2-hexyl ketone that is similarly manipulated to **192**[61] (Scheme 27).

Utilization of the chiral 2-pyrrolidinone derivative **194,** prepared from L-glutamic acid (**73**) by way of **129** as a chiral ketone in the [2 + 2] cycloaddition reaction with imine **195,** affords the *cis*-β-lactam **196** in 62% yield, having 96% optical purity. Interestingly, chiral heterocycles derived from L-(+)-tartaric acid similarly undergo the [2 + 2] cycloaddition reaction to give *trans*-cycloadducts instead[62] (Scheme 28).

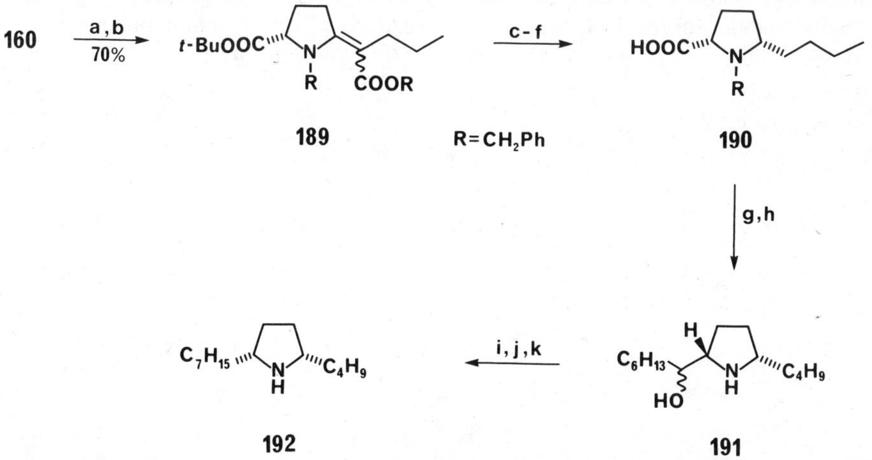

160 —a,b (70%)→ 189 —c-f→ 190 —g,h→ 191 —i,j,k→ 192

R=CH_2Ph

Scheme 27. (a) P_4S_{10}; (b) Ph_3P; (c) Pd/C, cyclohexene (87%); (d) $HCOONH_4$/Pd (96%); (e) $PhCH_2Br$/K_2CO_3 (92%); (f) HOAc/*n*-PrOH (81%); (g) *n*-$C_6H_{13}Li$, $NaBH_4$ (70%); (h) Pd/HOAc (99%); (i) (phenylsulfonyl)imidazole/$MeO_3^+BF_4^-$ (88%); (j) $NaBH_4$ (75%); (k) Na/NH_3 (80%).

Scheme 28

Asymmetric C–C bond formation α to the carbonyl of aldehydes and ketones by use of a chiral version of the hydrazone method[63] requires either (*S*)- or (*R*)-1-amino-2-methoxymethylpyrrolidine (**200**) (SAMP or RAMP) as the chiral auxiliary. An important feature for a practical asymmetric synthesis using stoichiometric amounts of the chiral auxiliary is that this reagent be inexpensive and easily available in large amounts. While SAMP can be prepared from both L-glutamic acid **73** and L-proline (see Chapter 8 for a complete discussion), RAMP **200** is prepared from D-glutamic acid (**119**) that is commercially available and relatively inexpensive. (*R*)-Prolinol (**123**), prepared from **119** in three steps, is converted to the nitrosamine **198** with ethyl- or *tert*-butylnitrite in THF. Treatment with NaH–methyl iodide affords **199**, which is reduced with lithium aluminum hydride to provide **200** in a 35% overall yield from **119**[64] (Scheme 29).

Scheme 29. (a) H_2O (reflux); (b) Dowex ion exchange; (c) CH_2N_2; (d) $LiAlH_4$; (e) $C_2H_5NO_2$; (f) NaH/CH_3I.

By mixing a pure aldehyde or ketone with **200** and stirring at room temperature (aldehydes) or 60°C (ketones), the RAMP–hydrazones **201** are obtained almost quantitatively and in high purity. Metalation with LDA in ether or THF at 0°C generates the lithiated hydrazone **202,** which, as a result of the intramolecular methoxy chelation with the lithium cation, is conformationally rigid. Electrophilic addition occurs predominantly *syn* to the lithium [S_E2 (front) or S_Ei].[65,66]

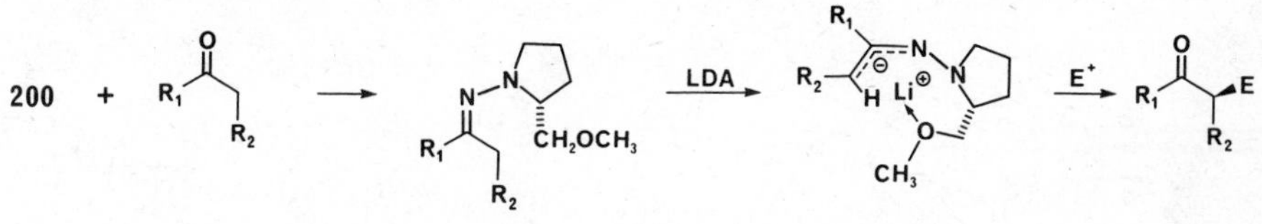

200 + R_1 O R_2 → N N R_1 CH_2OCH_3 R_2 LDA → R_1 N R_2 H Li N O CH_3 E^+ → O R_1 E R_2

201 202 203

A summary of the synthetic scope of the RAMP–hydrazone method is provided in Scheme 30. Since the chemistry of RAMP parallels that of its enantiomer SAMP, the reader is referred to Chapter 8 for a more detailed discussion of the chemistry of some of these reactions.

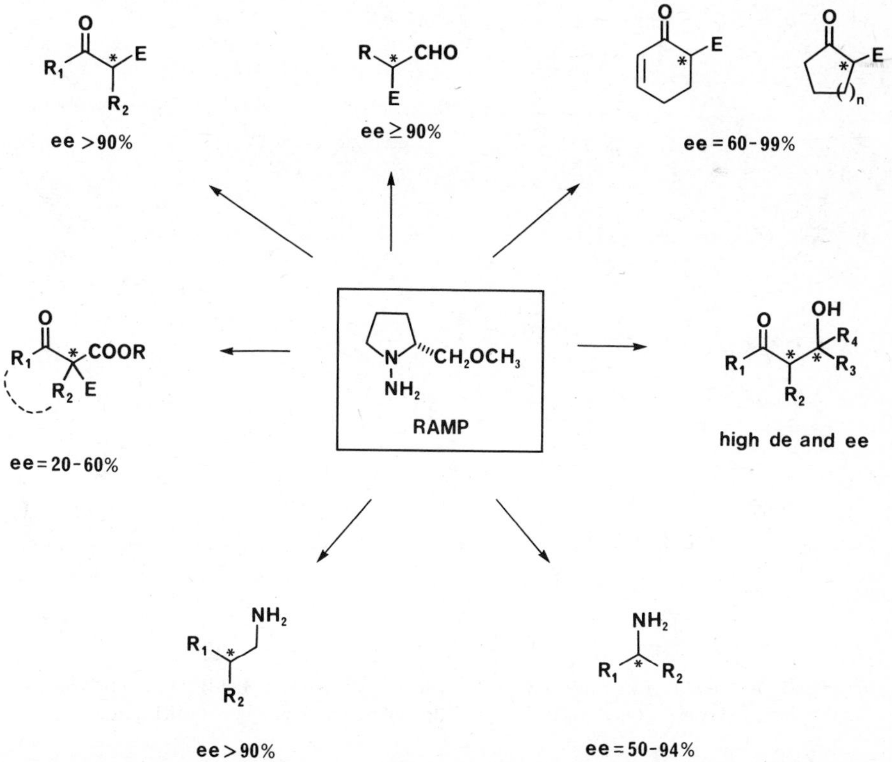

O R_1 * E R_2

ee >90%

R * CHO E

ee ≥ 90%

O * E

O * E ( )$_n$

ee = 60-99%

O R_1 * COOR R_2 E

ee = 20-60%

N CH_2OCH_3 NH_2

RAMP

O OH R_4 R_1 * * R_3 R_2

high de and ee

NH_2 R_1 * R_2

ee > 90%

NH_2 R_1 * R_2

ee = 50-94%

Scheme 30. Reprinted with permission from reference 66. Copyright 1984 Academic Press.)

The chirally specific synthesis of (*S*)-(+)-tylophorine (**208**), a phenanthroindolizidine alkaloid of interest biologically, is achieved by utilizing an intramolecular Friedel–Crafts acylation of the pyroglutamic acid derivative **206** (Scheme 31). The diisopropyl ester of glutamic acid is used to prevent premature pyroglutamate formation.[67]

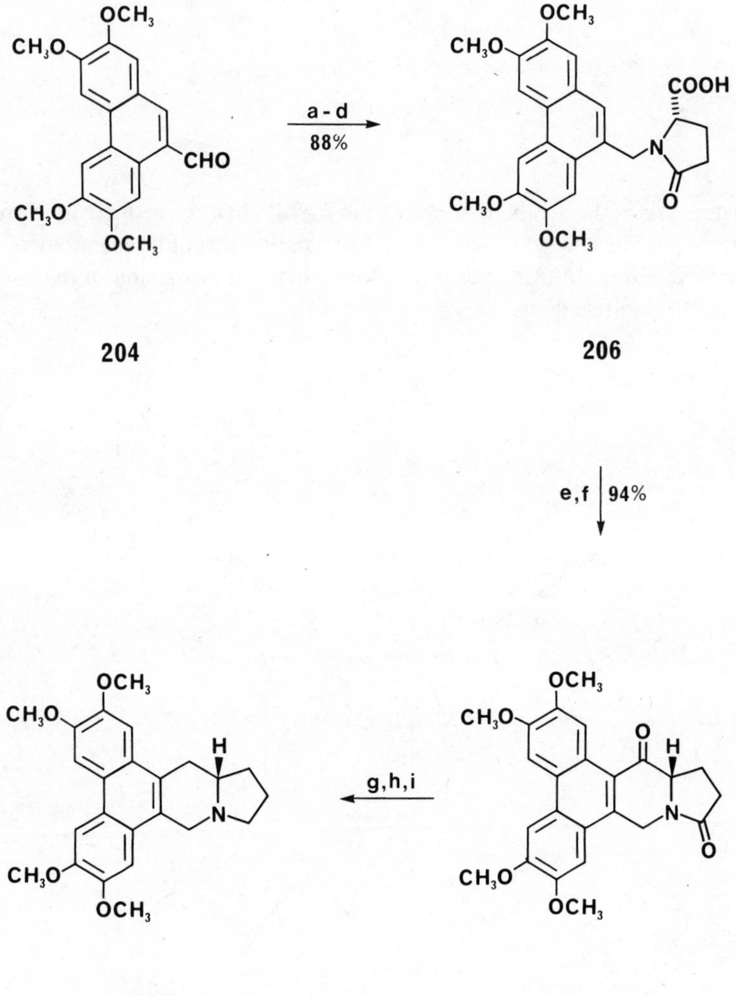

204 206

208 207

Scheme 31. (a) Diisopropylglutamate (**205**); (b) $NaBH_3CN$; (c) $HOAc/CH_3OH$; (d) KOH/dioxane; (e) oxalyl chloride; (f) $SnCl_4$; (g) $Pd(OH)_2/C$ (95%); (h) $SOCl_2$, Pd/C (93%); (i) $LiAlH_4$ (96%).

7.2.2 α-Butyrolactone-α-carboxylic Acid

L-Glutamic acid (**73**) reacts with nitrous acid in an aqueous solution to provide (*S*)-α-butyrolactone-α-carboxylic acid (**209**) with complete retention of configuration due to the participation of the neighboring carboxylate group.[68,69] Using D-glutamic acid (**119**), one obtains an 85% yield of the corresponding enantiomer.[69] The conversion of **209** to the acid chloride **210** proceeds in good to excellent yields with the use of either thionyl chloride[69] or oxalyl chloride.[70]

NH_2 HOOC COOH — HNO_2, 75–85% → O O COOH → O O COCl

73 (S)-**209** **210**

The ability of **210** to undergo displacement of the chloride without loss of optical integrity makes this easily accessible chiral intermediate valuable synthetically. Treatment of **210** with didecylcadmium affords the lactone ketone **211,** which, after reduction with $NaBH_4$ in methanol and chromatographic separation, provides the pure diastereomeric alcohols **212** and **213.** Protection of the alcohol as a THP ether, DIBAH reduction of the lactone to the lactol, and a Wittig reaction with isobutylidene triphenylphosphorane afford, after catalytic reduction of the resulting olefin, compounds **214** and **215.** By the proper sequence of transformations, **214** is converted to *cis*-(7*R*, 8*S*)-dispalure (**216**), the sex attractant emitted by the female gypsy moth, or its enantiomer **217.** Likewise, **215** can be converted to the *trans*-(7*R*, 8*R*)-dispalure (**218**) or its enantiomer **219**[70] (Scheme 32).

When **210** reacts with a variety of Grignard reagents in THF at −80°C, the corresponding ketones **220** are obtained. Because of their tendency to rapidly racemize, these ketones are reduced with L-Selectride in THF at room temperature to afford almost exclusively the *syn*-isomer **221** in good yields.[71]

This highly stereoselective preparation of enantiomerically pure masked 1,2-diols (e.g., **224**) was successfully applied to the stereo- and enantiospecific synthesis of (+)-exobrevicomin (**228**), the principal aggregation pheromone of the female western pine beetle *Dendroctonus brevicomis.* D-Glutamic acid (**119**) is converted to ketone **223** in 58% overall yield. Reduction of **223** with K-Selectride affords diastereomerically pure alcohol **224.** This is then converted to the protected ketone **227,** which quantitatively cyclizes to **228** by treatment with diluted sulfuric acid[71] (Scheme 33).

The enantiospecific synthesis of (*R*)-(+)-δ-*n*-hexadecanol lactone (**234**), the pheromone isolated from the mandibules of the hornet *Vespa orientalis,* is achieved in eight steps from L-glutamic acid (**73**) by way of (*S*)-(+)-4-hydroxymethyl-4-butanolide (**229**)[72] (Scheme 34).

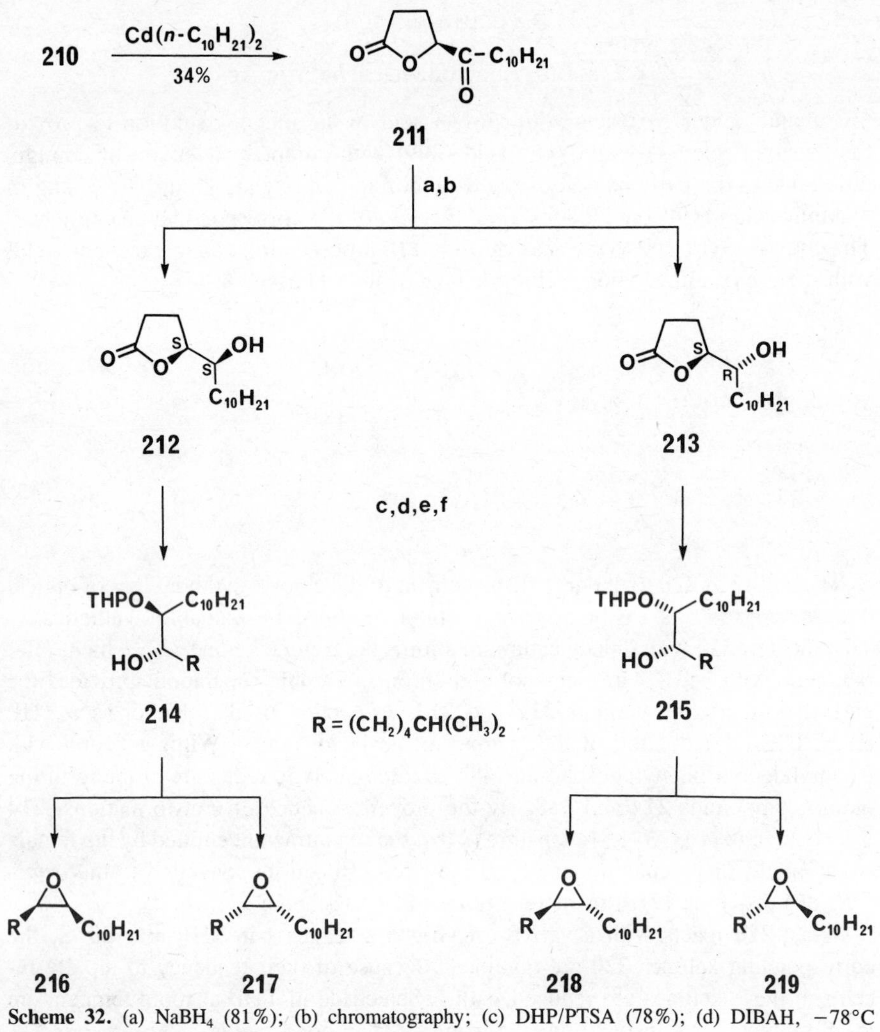

210 —$Cd(n\text{-}C_{10}H_{21})_2$, 34%→ 211 ($C\text{-}C_{10}H_{21}$, O)

211 —a,b→ 212 (S, S, OH, $C_{10}H_{21}$) + 213 (S, R, OH, $C_{10}H_{21}$)

c,d,e,f

212 → 214 (THPO, $C_{10}H_{21}$, HO, R); 213 → 215 (THPO, $C_{10}H_{21}$, HO, R)

$R = (CH_2)_4CH(CH_3)_2$

214 → 216 (O, R, $C_{10}H_{21}$) + 217 (O, R, $C_{10}H_{21}$); 215 → 218 (O, R, $C_{10}H_{21}$) + 219 (O, R, $C_{10}H_{21}$)

Scheme 32. (a) $NaBH_4$ (81%); (b) chromatography; (c) DHP/PTSA (78%); (d) DIBAH, −78°C (94%); (e) isobutylene, Ph_3P (77%); (f) H_2, PtO_2 (98%).

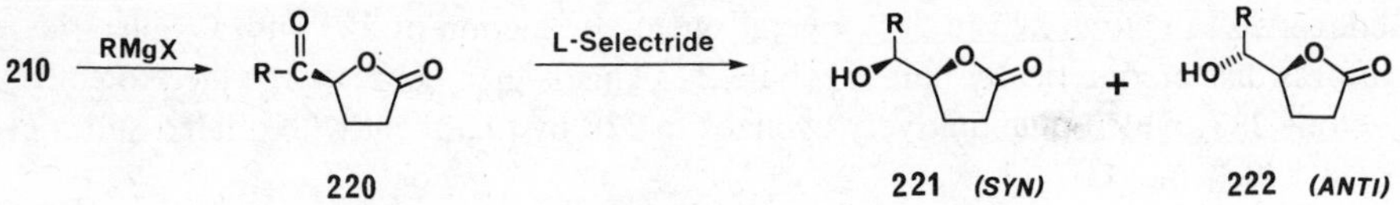

210 —RMgX→ 220 (R–C, O) —L-Selectride→ 221 (*SYN*) (R, HO, O) + 222 (*ANTI*) (R, HO, O)

R	221:222	Yield 221 (%)
C_2H_5	99:1	85
$i\text{-}C_3H_7$	95:5	71
$n\text{-}C_4H_9$	97:3	85
$n\text{-}C_5H_{11}$	96:4	82
$n\text{-}C_{10}H_{23}$	92:8	61

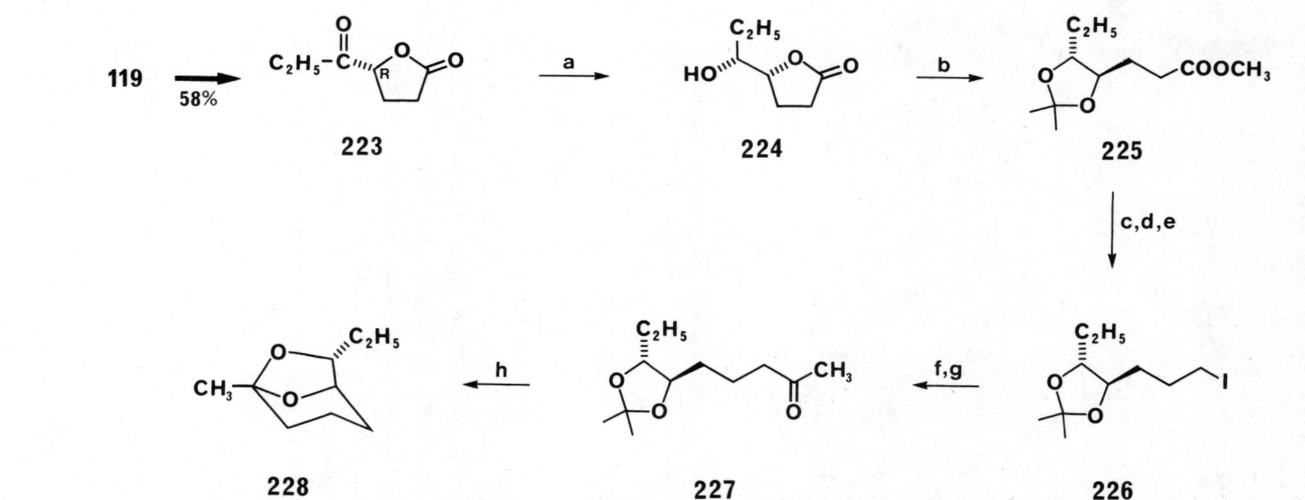

Scheme 33. (a) K-Selectride (92%); (b) DMP, Amberlyst-15, CH_3OH (92%); (c) $LiAlH_4$; (d) TsCl/pyridine; (e) NaI, acetone; (f) $Et_2NCHMeCN$, LDA; (g) $SiO_2 \cdot H_2O$ (72%); (h) 0.5 *M* H_2SO_4.

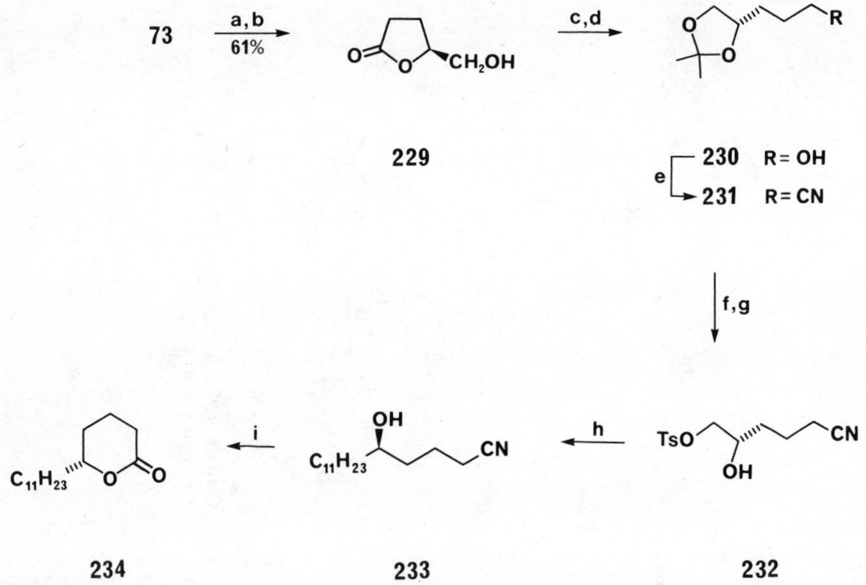

73 →(a,b; 61%) 229 →(c,d) 230 R = OH; 231 R = CN (e) →(f,g) 232 →(h) 233 →(i) 234

Scheme 34. (a) HNO_2; (b) $BH_3 \cdot Me_2S$; (c) DMP, Amberlyst-15, CH_3OH (82%); (d) $LiAlH_4$ (95%); (e) TsCl/pyridine; NaCN/HMPA; (f) H_2SO_4 (90%); (g) TsCl/pyridine (76%); (h) $(C_{10}H_{21})_2CuMgBr$; (i) NaOH/EtOH.

The sex pheromone of the female Japanese beetle *Popillia japonica* Newman has been identified as (*R*)-(*Z*)-5-(1-decenyl)dihydro-2(3*H*)-furanone (**236**). Whereas the pure **236** is a powerful attractant of males, the racemic mixture and the (*S*)-(*Z*)-enantiomer are strong inhibitors to male response.[73,74] Its synthesis from **119** proceeds through the acid chloride (*R*)-**210,** which is converted to the aldehyde **229** by the Rosenmund reduction. The crude aldehyde **235** is treated with nonylidene triphenylphosphorane at −60°C in THF–HMPA to afford (*R*)-**236** as a 9:1 mixture of (*Z*)/(*E*) isomers. These are readily separated by HPLC to provide pure **236.**[74]

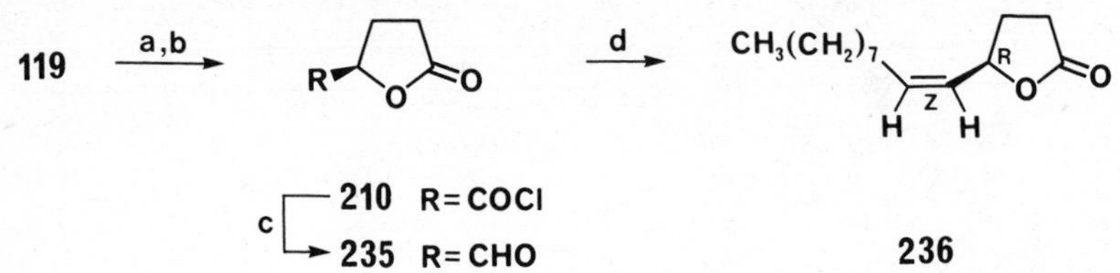

119 →(a,b) 210 R = COCl; 235 R = CHO (c) →(d) 236

(a) HNO_2 (75%); (b) $SOCl_2$ (87%); (c) H_2, $Pd/BaSO_4$, tetramethylthiourea; (d) $Ph_3P{=}CH(CH_2)_7CH_3$, then HPLC.

Optically active epoxyterpenes are interesting because of their biological activity as well as their intermediacy for the biosynthesis of cyclic terpenes and steroids. A synthesis of these molecules has been developed in which L-glutamic acid (**73**) provides the appropriate chirality.[75] Ring opening of ketolactone **237**, prepared from **73** in 32% overall yield, with piperidine leads to a partially racemized amide that can be rendered optically pure by recrystallization. The pure amide is converted to the (*S*)-acetonide **238**, which undergoes a Wittig reaction to afford a mixture of α,β-unsaturated esters [(*Z*):(*E*) = 26:74]. Reduction with $LiAlH_4$ and chromatographic separation yields the (*E*)-alcohol **239**.[75] It is a key intermediate for the synthesis of such epoxyterpenes as (*R*)-(+)-epoxygeraniol (**240**),[75] (*R*)-(+)-10,11-epoxyfarnesol (**241**),[76] and (*R*)-(+)-squalene-2,3-oxide (**242**)[76] (Scheme 35).

(*S*)-γ-Hydroxymethyl-γ-butyrolactone (**229**), easily prepared by the reduction of **209** with either $NaBH_4$[69,77,78] or more conveniently with borane–dimethylsul-

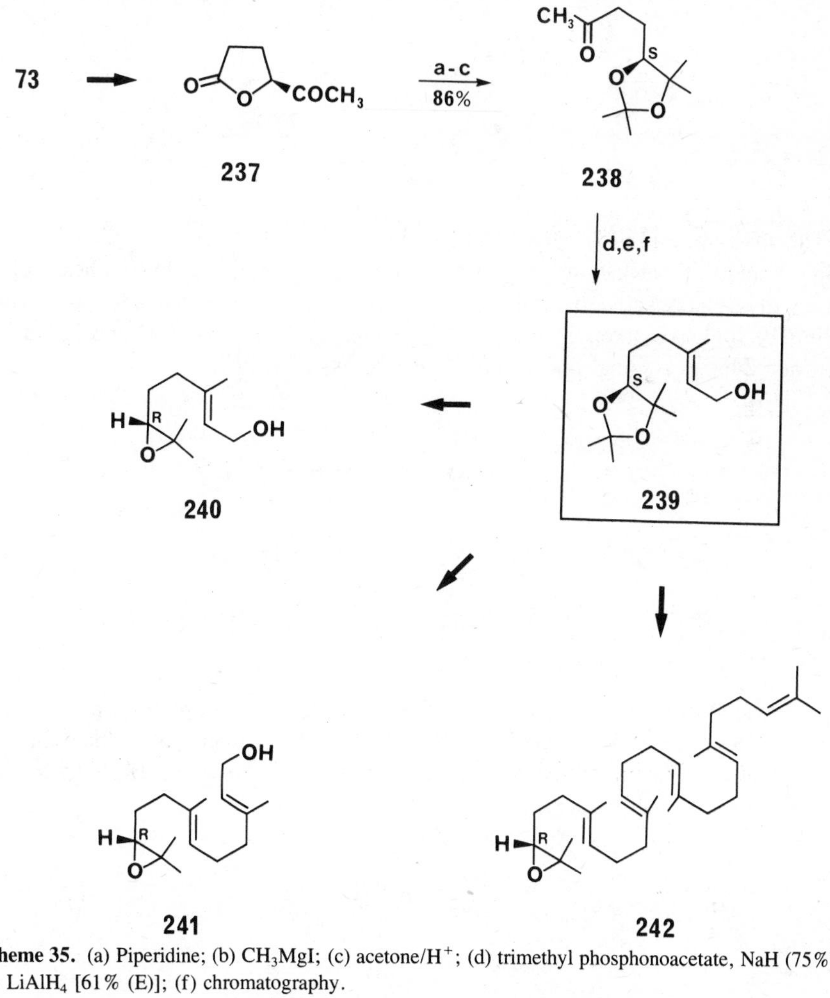

Scheme 35. (a) Piperidine; (b) CH_3MgI; (c) acetone/H^+; (d) trimethyl phosphonoacetate, NaH (75%); (e) $LiAlH_4$ [61% (E)]; (f) chromatography.

229 $\xrightarrow[Py]{TsCl}$ **243**

fide,[72,78] can be tosylated to provide a crystalline derivative **243,** which may be recrystallized to ensure the highest optical purity.

Whereas the same sequence employing (*R*)-glutamic acid **119** provides the corresponding (*R*)-**245,** a more practical and inexpensive alternate route is available. Quantitative conversion of **243** to epoxide **244** followed by lactone formation proceeds with complete inversion of configuration to provide **245** in 82% yield from **229.**[79]

243 $\xrightarrow[CH_3OH]{CH_3ONa}$ **244** $\xrightarrow[-10°C,\ 86\%]{CF_3COOH}$ **245**

The ability of **243** to undergo nucleophilic substitution reactions without loss of chirality and without lactone ring opening can be exploited for the syntheses of a variety of optically active natural products. Displacement of the tosylate group with lithium dialkyl or dialkenylcuprates[78,80] provides 4-substituted-γ-lactones **246a–c,** of which **246c** is a social pheromone of the black-tailed deer.[81]

243 $\xrightarrow{R_2CuLi}$ **246**

(a) R=CH_3 (66%); (b) R=n-C_4H_9 (41%); (c) R=$CH_3(CH_2)_4CH\overset{Z}{=}CH$ (12%)

The availability of **246a** in high optical purity allows for its use in the preparation of the spirocyclic compound 2-ethyl-1,6-dioxospiro[4.4]nonane (**248**), the principal component of the aggregation pheromone of the beetle *Pityognes chalcographus,* a pest of the Norway spruce.[82]

246a $\xrightarrow{LiC\equiv CCH_2OTHP}$ **247** $\rightarrow$ **248**

Tosylate **243** is smoothly converted to the iodide **249** when treated with lithium iodide in acetone. Reductive dehalogenation with Raney nickel provides (*R*)-(+)-γ-methylbutyrolactone (**250**), which, after reduction to a lactol with DIBAH, undergoes a Wittig reaction with isopropylidene triphenylphosphorane to afford (*R*)-(+)-sulcatol (**251**), a population aggregation pheromone produced by males of the desirable ambrosia beetle *Gnathotrichus sulcatus.*[77] Its optical antipode can be similarly prepared from **119.**

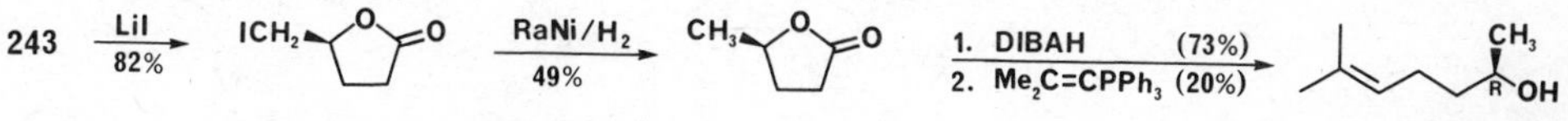

243 → (LiI, 82%) → 249 → ($RaNi/H_2$, 49%) → 250 → (1. DIBAH (73%); 2. $Me_2C{=}CPPh_3$ (20%)) → 251

Treatment of **249** with Na_2CO_3 in methanol affords the epoxide **244,** which condenses with lithium diisopropylidenecuprate to give lactone **252.** This is converted to the butenolide **253,** which undergoes a stereospecific 1,4-addition with lithium dimethyl cuprate to furnish the (3*S*, 4*R*)-isomer of eldanolide (**254**), the wing gland pheromone, and aphrodisiac secretion of the abdominal hairpencils of the African sugar caneborer *Eldana saccharina*[83,84] (Scheme 36).

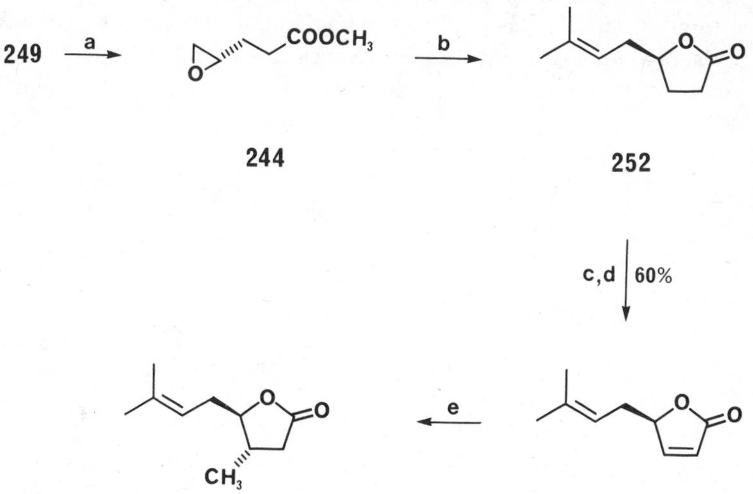

249 → (a) → 244 → (b) → 252 → (c,d 60%) → 253 → (e) → 254

Scheme 36. (a) Na_2CO_3/CH_3OH (95%); (b) $[CH_3C(=CH_2)]_2CuLi$ (64%); (c) LDA, PhSeBr; (d) H_2O_2; (e) Me_2CuLi (66%).

When tosylate **255** (prepared from (*R*)-**245**) is reduced with $LiAlH_4$ in THF, (4*S*)-(+)-pentane-1,4-diol (**256**) is produced in 76% yield. Differential protection of the primary and secondary hydroxyl groups provides **257,** which is transformed to iodo derivative (**258**). This chiral synthon, when coupled with the chiral acetylene **259** (derived from D-mannitol), provides the key step in the convergent stereoselective synthesis of (+)-brefeldin A (**260**), an antibiotic fungal metabolite of biological interest[85,86] (Scheme 37).

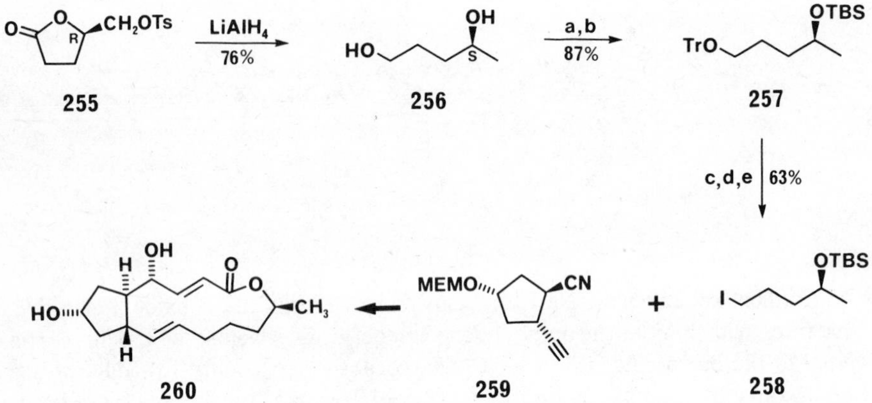

Scheme 37. (a) TrCl/pyridine; (b) TBS–Cl; (c) Na/NH_3; (d) TsCl/pyridine; (e) NaI, acetone.

A milder reduction of **243** with lithium borohydride provides tosylate **261** in 67% yield. This is selectively protected and converted to the epoxide **262,** which, when allowed to react in the presence of CuI–catalyst with the Grignard reagent derived from **263** in THF at −30°C for 1 hr and at 0°C for 4 hr, affords **264.** Further elaboration of **264** provides **265,** which is a synthon for the construction of the macrocyclic moieties of the cytochalasins A, B, and F and desoxaphomin[87] (Scheme 38).

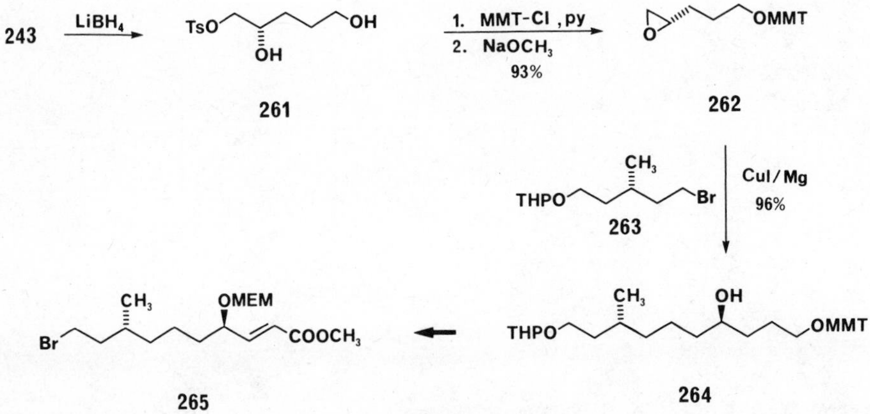

Scheme 38

Displacement of the tosylate of **243** with sodium azide in DMF gives 5-azido-4-pentanolide (**266**) in 87% yield. Catalytic reduction of the azido function with concomitant lactam formation furnishes (*S*)-5-hydroxy-2-piperidinone (**267**), which undergoes successful reduction with borane in THF to give (*S*)-(−)-3-piperidinol (**268**) in good yield.[88]

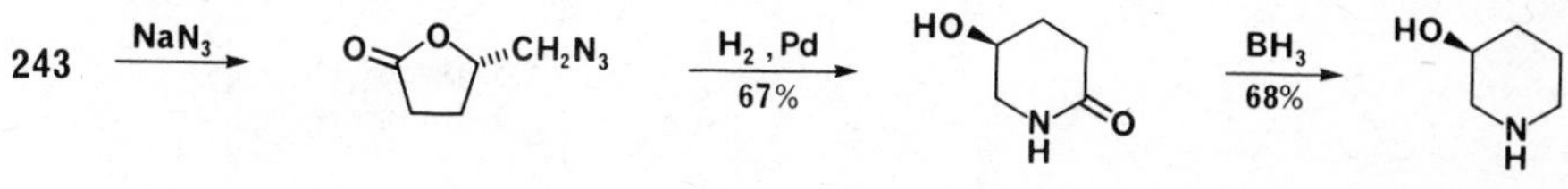

243 —NaN_3→ 266 —H_2, Pd, 67%→ 267 —BH_3, 68%→ 268

When **243** is reacted with the lithium enolate of *t*-butyl propionate in HMPA/THF (1:1), the (*E*)-alcohol **270** is formed. The inverted stereochemistry at C-6 occurs by way of the epoxide **269,** which is formed by an intramolecular displacement of tosylate. Hydrogenation of **270** provides alcohol **271** in 55% overall yield. This key alcohol is converted to the (Z)-olefin **272,** which undergoes a highly stereoselective addition reaction with NBS in DMSO followed by a reduction with n-Bu_3SnH to furnish a separable mixture of 8-*epi*-*t*-butyl nonactate (**273**) and *t*-butyl nonactate (**274**) in a ratio of 4:1[89] (Scheme 39).

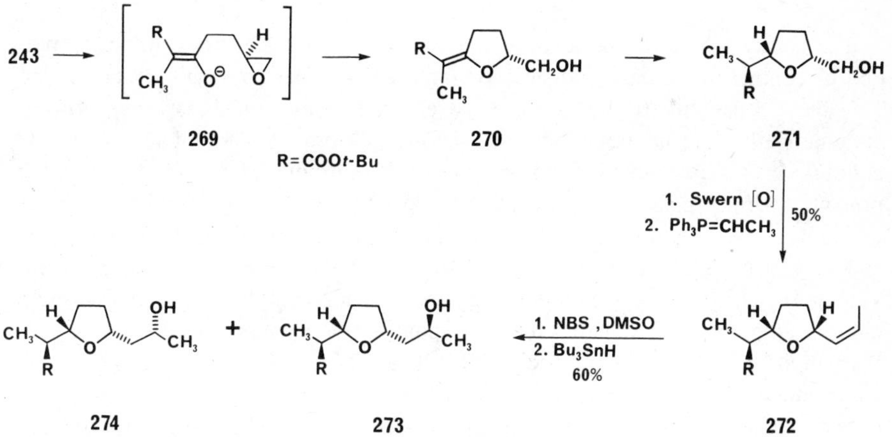

243 → [269] → 270 → 271; R = COO*t*-Bu

271 —1. Swern [O]; 2. Ph_3P=$CHCH_3$; 50%→ 272 —1. NBS, DMSO; 2. Bu_3SnH; 60%→ 273 + 274

Scheme 39

If the primary hydroxy group of **229** is converted to the silyl ether **275** and treated with the lithium anion of *t*-butyl propionate, no intramolecular displacement can occur, and the stereochemistry at C-6 is maintained. Hydrogenation then affords the enantiomeric alcohol **276.** This then provides the key intermediate for the synthesis of the enantiomeric nonactates **278** and **279**[89] (Scheme 40). Both (+)-**274** and (−)-**279** have 90% ee.

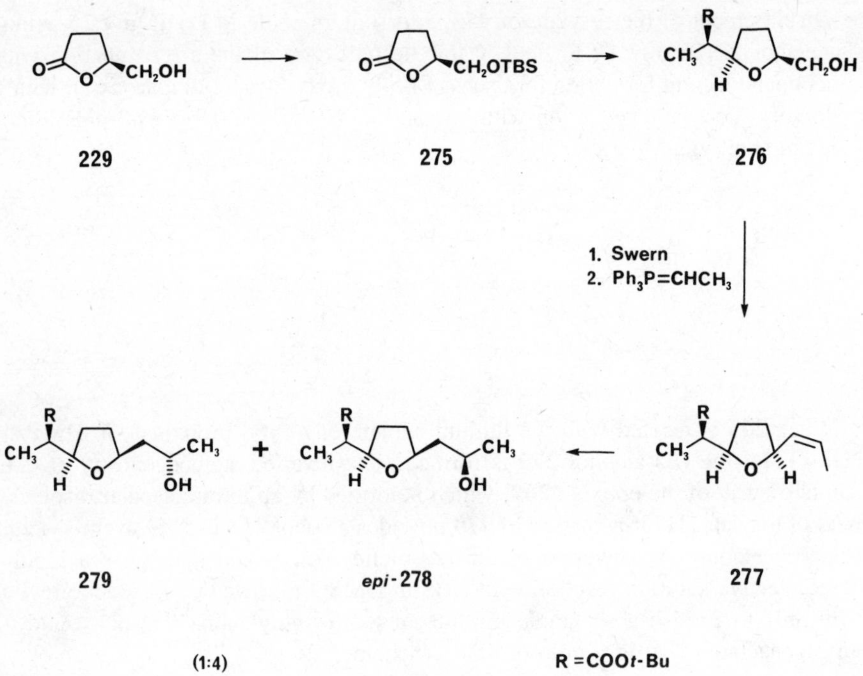

229 → 275 → 276

1. Swern
2. $Ph_3P{=}CHCH_3$

279 + *epi*-278 ← 277

(1:4)

R = COO*t*-Bu

Scheme 40

The ability to control the relative and absolute stereochemistry in the construction of contiguous tertiary and quarternary carbon centers by a highly controlled 1,3- and 1,4-asymmetric induction is extremely valuable for the synthesis of enantiomerically pure natural products. Lactone **229** provides a template where the chirality of C-5 maintains a fixed orientation. By introduction of steric bulk at the primary hydroxy position, an unsymmetrical environment is created that directs incoming groups from the least hindered face (diastereofacial selectivity).

The group preferred for introduction of steric bulk is the trityl group (triphenylmethyl), which is easily introduced by the reaction of **229** with triphenylmethyl chloride in pyridine.[90] Whereas the corresponding (*R*)-enantiomer **282** can be similarly prepared from D-glutamic acid (**119**), it can also be prepared from **280** through a convenient inversion route[91] (Scheme 41).

Treatment of **280** with 1 equivalent of LDA at −78°C generates the lithium enolate, which undergoes a highly stereoselective alkylation with allyl bromide to afford the chiral lactone **283** in good yield. In the total synthesis of the Corynanthe-type indole alkaloid (−)-antirhine (**288**), **283** is converted to the versatile chiral synthon formylmethyl(vinyl)tetrahydropyranone (**287**) in respectable overall yield[92] (Scheme 42).

Lactone **283** has also been used to enantioselectively prepare the β-lactam **291,** which has been successfully converted to the carbapenam ring systems **292** and **293**[93] (Scheme 43).

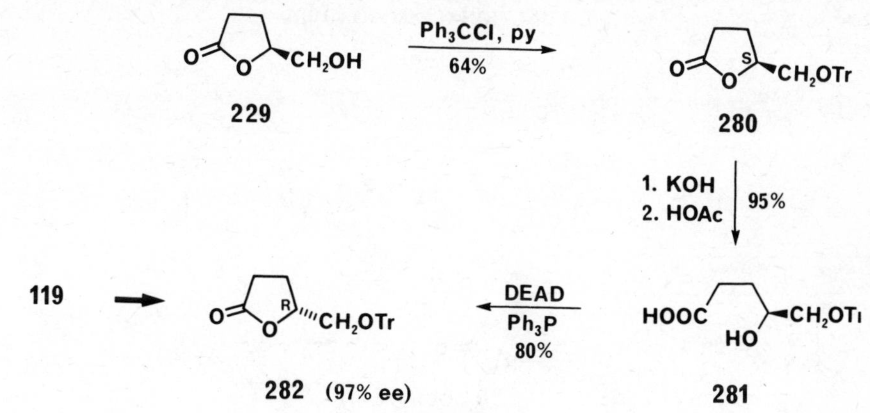

Scheme 41

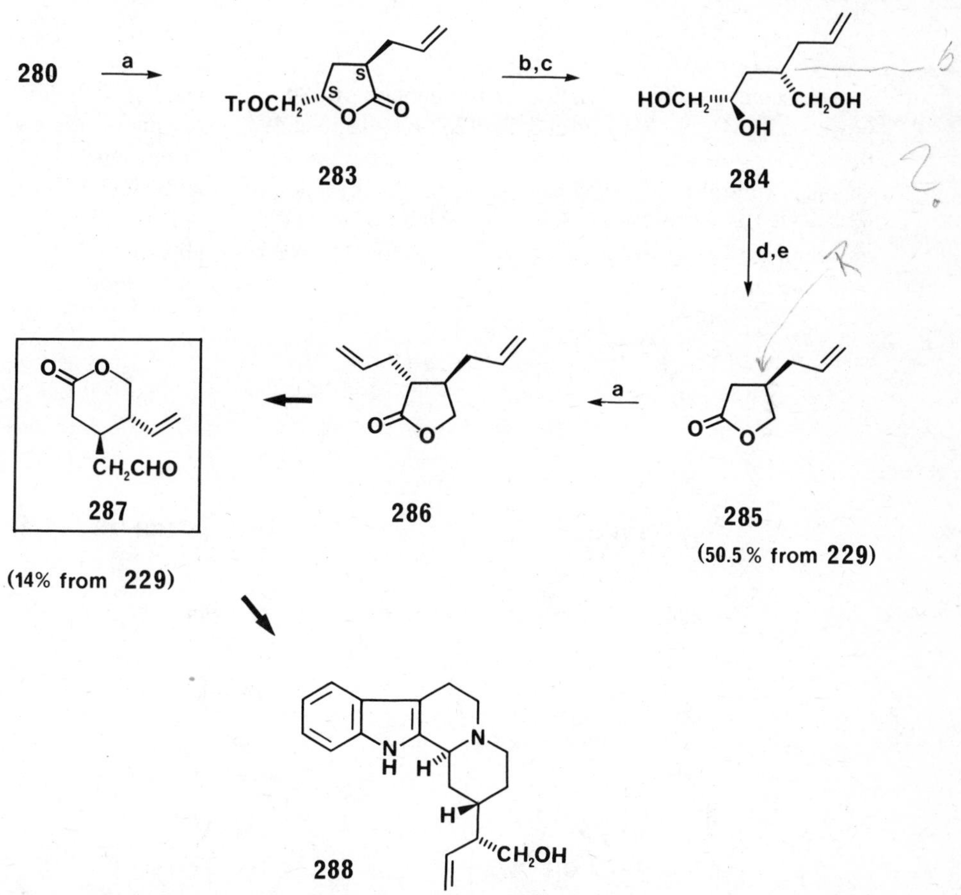

Scheme 42. (a) $CH_2{=}CHCH_2Br$, LDA, $-78°C$; (b) $LiAlH_4$; (c) CH_3OH/H^+; (d) $NaIO_4$; (e) CrO_3/H_2SO_4.

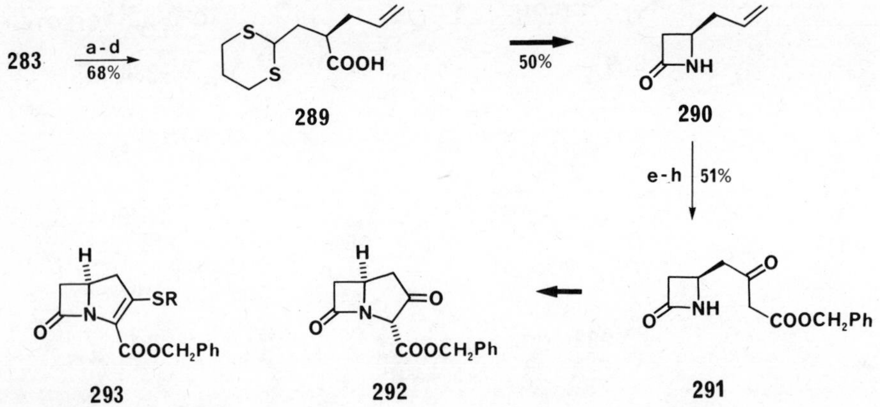

Scheme 43. (a) HCl/CH_3OH; (b) KOH/CH_3OH; (c) $NaIO_4$; (d) propane-1,3-dithiol, $BF_3 \cdot Et_2O$; (e) TBS–Cl, TEA; (f) OsO_4, $NaIO_4$; (g) $(EtO)_2P(O)CH_2COOCH_2Ph$, NaH; (h) *t*-BuOOH, $Na_2Pd(Cl)_4$, HOAc.

The stereoselective alkylation of the enolate of **280** with 2-ethylallyl bromide proceeds in 64% yield to furnish the (2*R*)-alkyllactone **294,** a key intermediate in the synthesis of the indole alkaloid (−)-velbanamine (**296**).[94] This same stereochemical control is exploited to bring about the inversion of chirality at C-3 of **294** to furnish **295** (ratio **294**:**295** = 1:9), which is utilized for the preparation of the enantiomeric (+)-velbanamine (**297**)[95] (Scheme 44). Such chirality control by kinetic protonation occurs most efficiently by using saturated aqueous sodium sulfate as the proton donor.

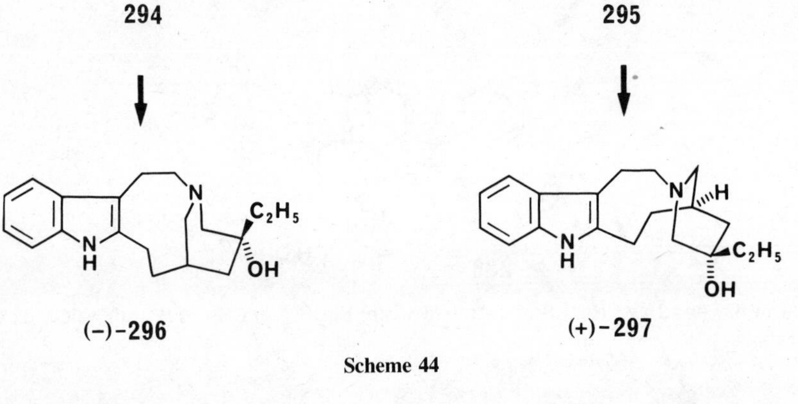

Scheme 44

The enolate of **283** can undergo a second stereoselective alkylation with ethyl bromide to furnish **298** having a chiral quarternary center. This feature is present in the indole alkaloids (+)-quebrachamine (**299**),[96] (−)-quebrachamine (**300**),[97] and (+)-eburnamine (**301**),[98] which are enantioselectively synthesized starting with **298.**

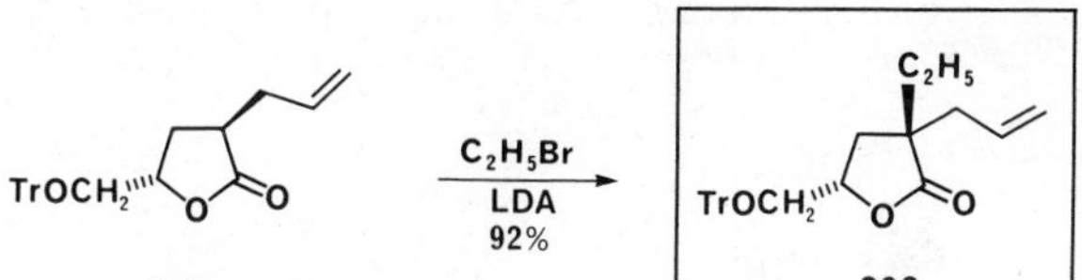

C_2H_5Br / LDA / 92%

TrOCH2 **283** → TrOCH2, C_2H_5 **298**

C_2H_5 **299** C_2H_5 **300** HO, H, N, C_2H_5 **301**

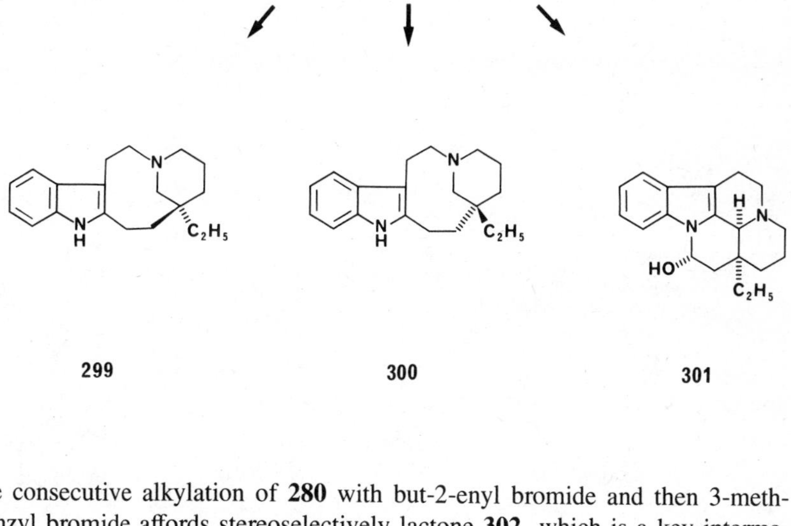

The consecutive alkylation of **280** with but-2-enyl bromide and then 3-methoxybenzyl bromide affords stereoselectively lactone **302,** which is a key intermediate in the synthesis of the gibbane **303.**[99]

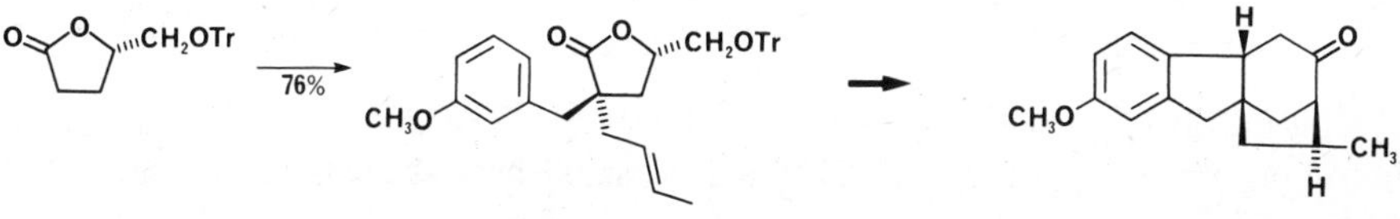

CH_2OTr **280** —76%→ CH_3O, CH_2OTr **302** → CH_3O, H, CH_3, H **303**

If **280** is initially converted to the corresponding alkylidene or arylidene **304** and then catalytically hydrogenated, reduction of the olefin occurs stereoselectively from the less hindered face to afford **305.** This is beautifully illustrated in the synthesis of the antileukemic lignan (−)-podorhizon (**306**).[101] By using the "self-immolative" technique[100] in which the original chiral center at C-4 is destroyed after contributing in the creation of the new asymmetric center at C-2, one can introduce the second chiral center at C-3. On the other hand, a stereoselective alkylation of **280** provides a route to (+)-podorhizon (**309**)[100] (Scheme 45). Recrystallization of the intermediates or final product ensures high optical purity.[102]

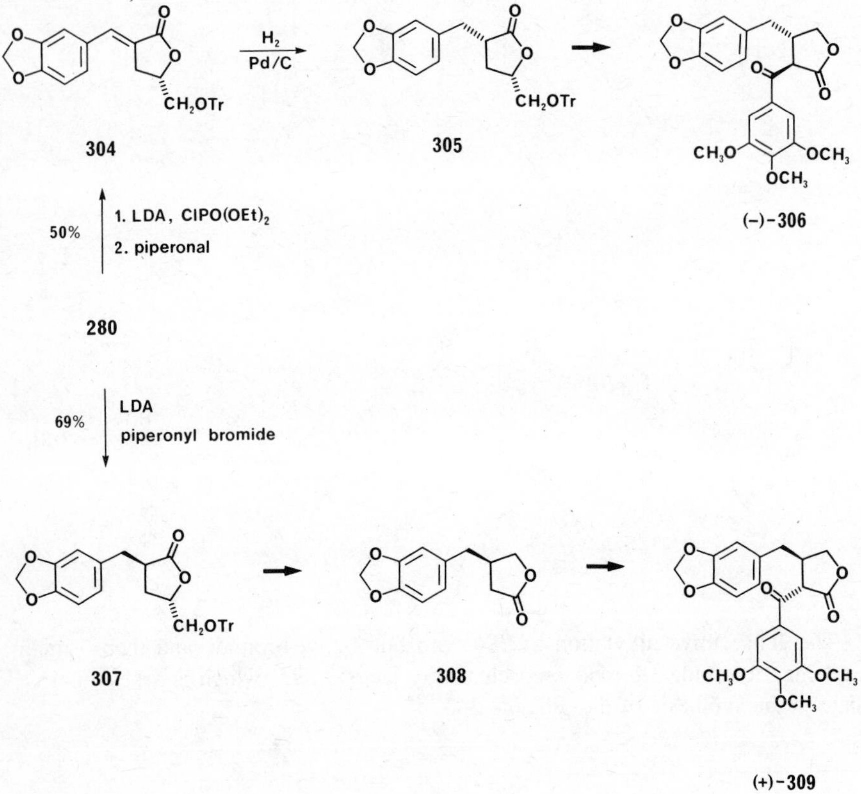

CH_2OTr **304** → H_2, Pd/C → CH_2OTr **305** → CH_3O OCH_3 OCH_3 (−)-**306**

280 → 1. LDA, $ClPO(OEt)_2$; 2. piperonal (50%) → **304**

280 → LDA, piperonyl bromide (69%) → **307**

CH_2OTr **307** → **308** → CH_3O OCH_3 OCH_3 (+)-**309**

Scheme 45

This methodology has also been successfully used to synthesize the antileukemic lignan precursor (−)-steganone (**312**) and to establish its absolute configuration[103] (Scheme 46).

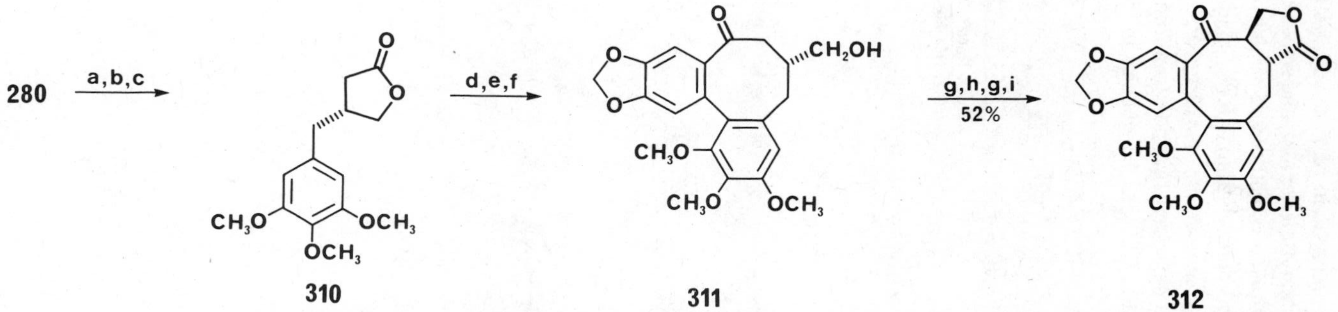

280 $\xrightarrow{a,b,c}$ 310 $\xrightarrow{d,e,f}$ 311 $\xrightarrow[52\%]{g,h,g,i}$ 312

Scheme 46. (a) LDA, $ArCH_2Br$ (78%); (b) $LiAlH_4$, then H_2, Pd/C (91%); (c) $NaIO_4$, then CrO_3 (91%); (d) 2-bromopiperonal, I_2, CF_3COOAg (100%); (e) Cu, 230°C; (f) LiHMDS (67%); (g) CrO_3; (h) CH_2O/KOH; (i) xylene, reflux.

The chiral butenolide **313**, prepared in 65% overall yield from **280** by way of a thermal *syn* elimination of phenylselenoxide, undergoes a highly specific asymmetric 1,4-addition with the lithiated trimethoxybenzaldehyde dithioacetal (**314**) to afford (after desulfurization and detritylation) a 65% yield of optically pure crystalline **315.** No base-induced racemization of **313** occurs. This then undergoes a stereoselective alkylation with piperonyl bromide to afford **316,** which is used to prepare either (−)-isostegane (**317**)[104] or (+)-steganacin (**318**)[105] (Scheme 47).

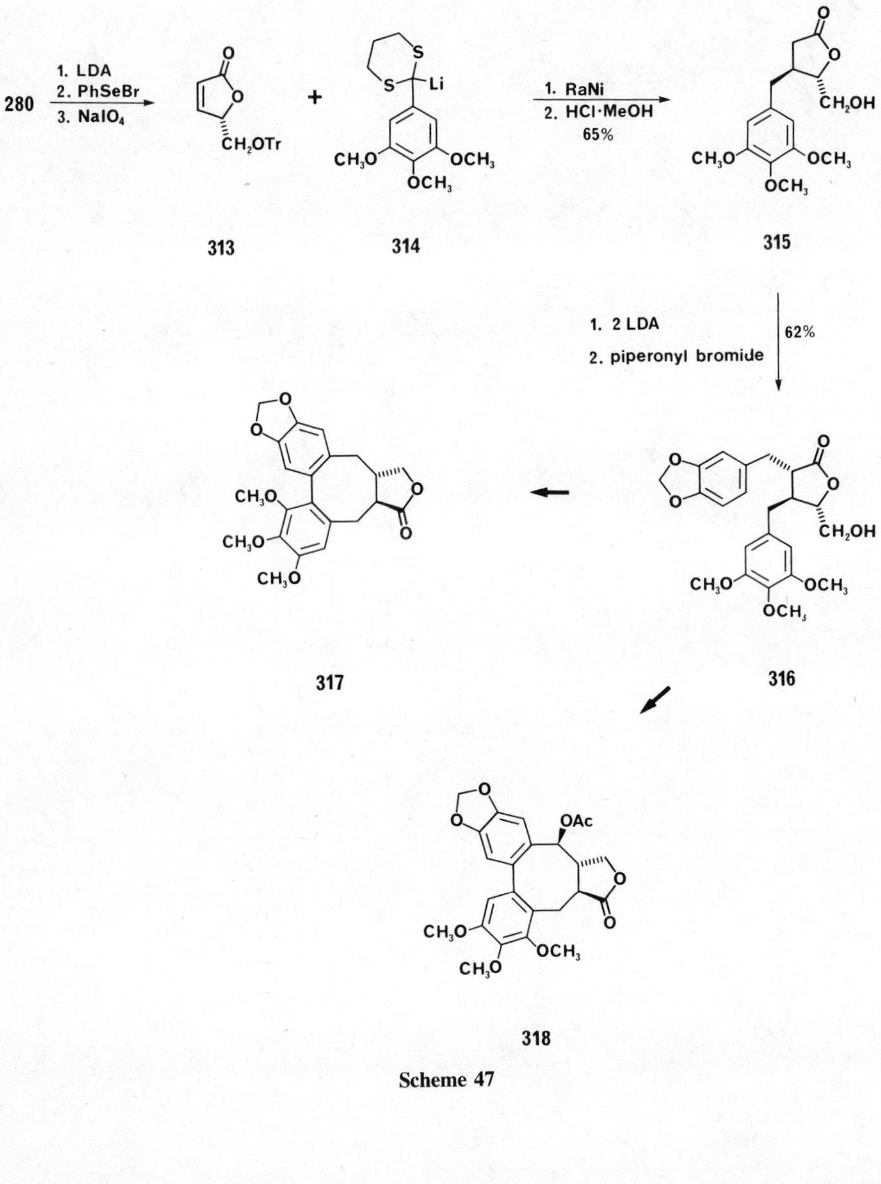

Scheme 47

Moreover, this highly controlled consecutive 1,4- and 1,3-asymmetric induction process can be used to establish the correct relative and absolute stereochemistry of the contiguous tertiary and quarternary carbon centers in the 4-methoxy-6,6-dialkylcyclohexenone portion of the cytotoxic neolignan (−)-megaphone (**321**).[106] In the enantiospecific total synthesis of **321**, one starts with the *exo*-alkylidene lactone **319** (Scheme 48).

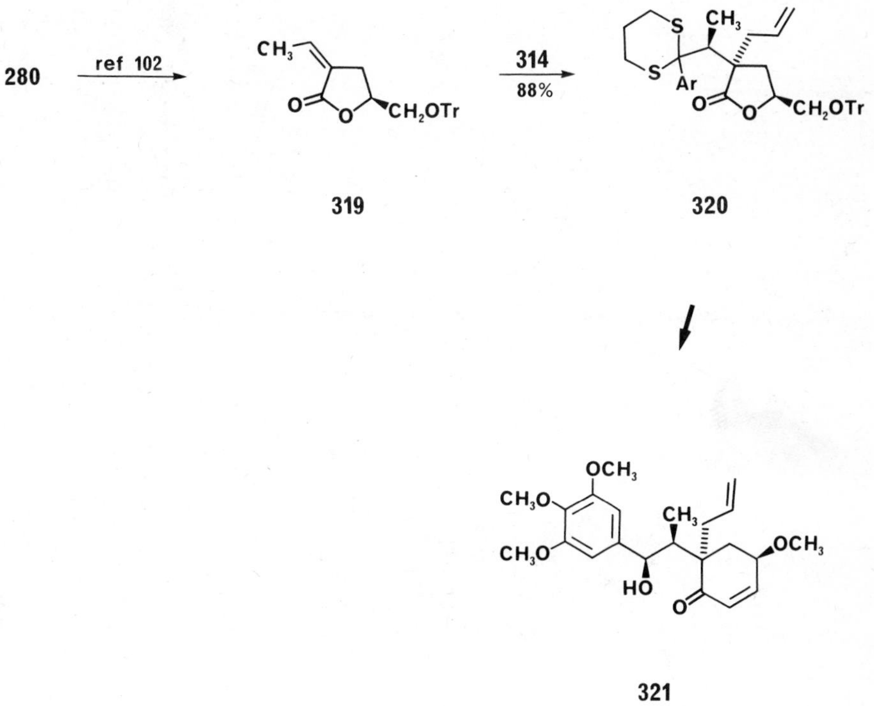

280 → (ref 102) 319 → (314, 88%) 320 → 321

Scheme 48

Hanessian et al.[107–109] have exploited chiral butyrolactone **229** as a replicating template (chiron) to construct acyclic seven-carbon chains that have a predictable substitution pattern. In this method, the steric effect of the anchored hydroxymethyl substituent in **229** controls the stereoselectivity in both the alkylation and conjugate addition reactions. Destruction of the original chiral center (a self-immolative process) and replication of a new chiral butyrolactone provide the template required for a second stereoselective transformation. This process leads to acyclic stereoselection having predictable 1,5-*C*-methyl substituents (**326** and **328**)[107] (Scheme 49), 1,3-*C*-methyl substituents (**333** and **335**),[108] (Scheme 50) and 1,3-*C*-hydroxy substituents (**338** and **340**)[109] (Scheme 51).

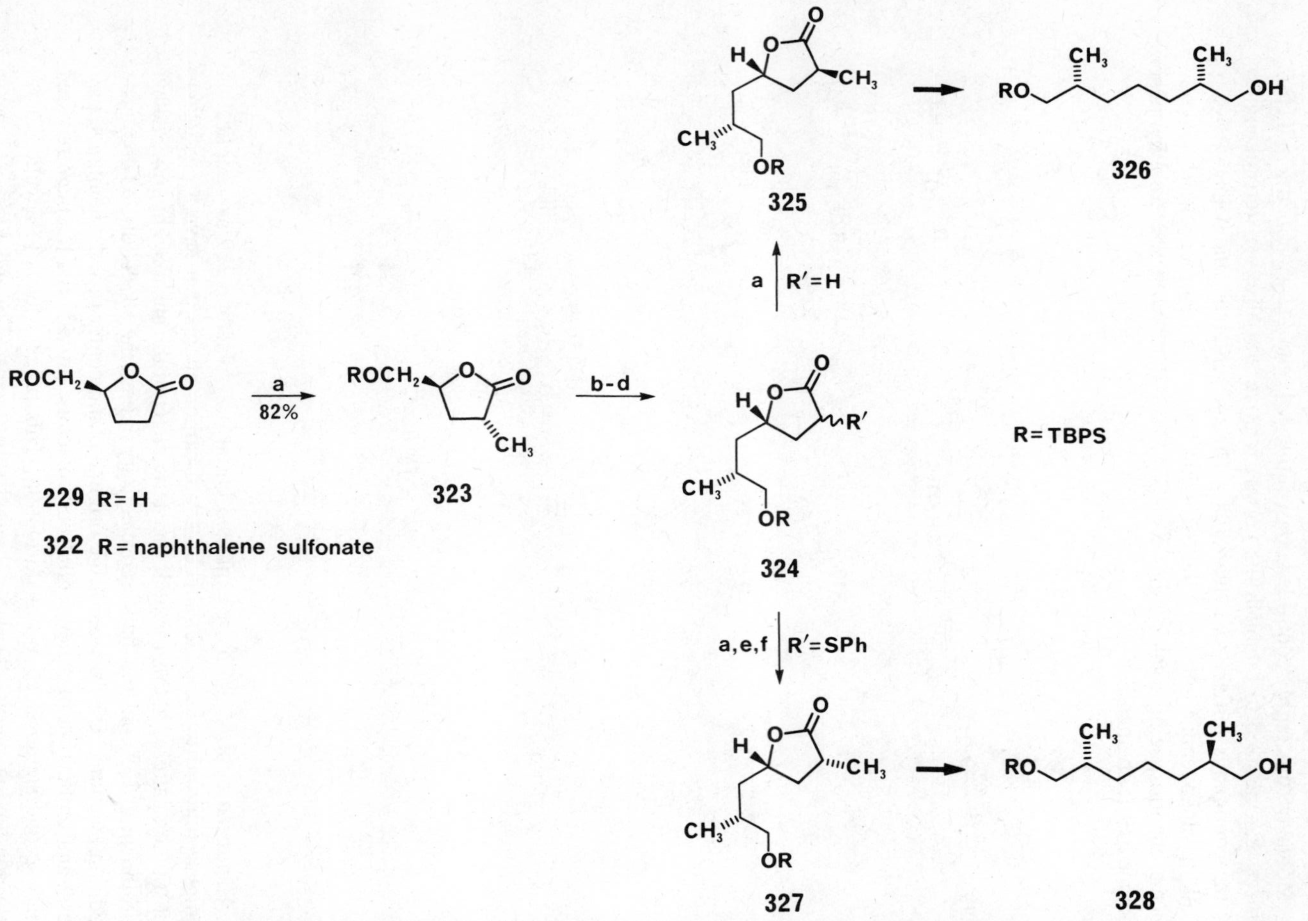

229 R = H

322 R = naphthalene sulfonate

a, 82%

323

b-d

324

a | R′ = H

325

326

a, e, f | R′ = SPh

327

328

R = TBPS

Scheme 49. (a) LiHMDS, −78°C, CH_3I (82%); (b) $BH_3 \cdot Me_2S$, then TBPSCl (85%); (c) $NaOCH_3/CH_3OH$ (100%); (d) LiHMDS, $PhSCH_2COOH$, then ethyl (dimethylaminoethyl)carbodiimide, DMAP (75%).

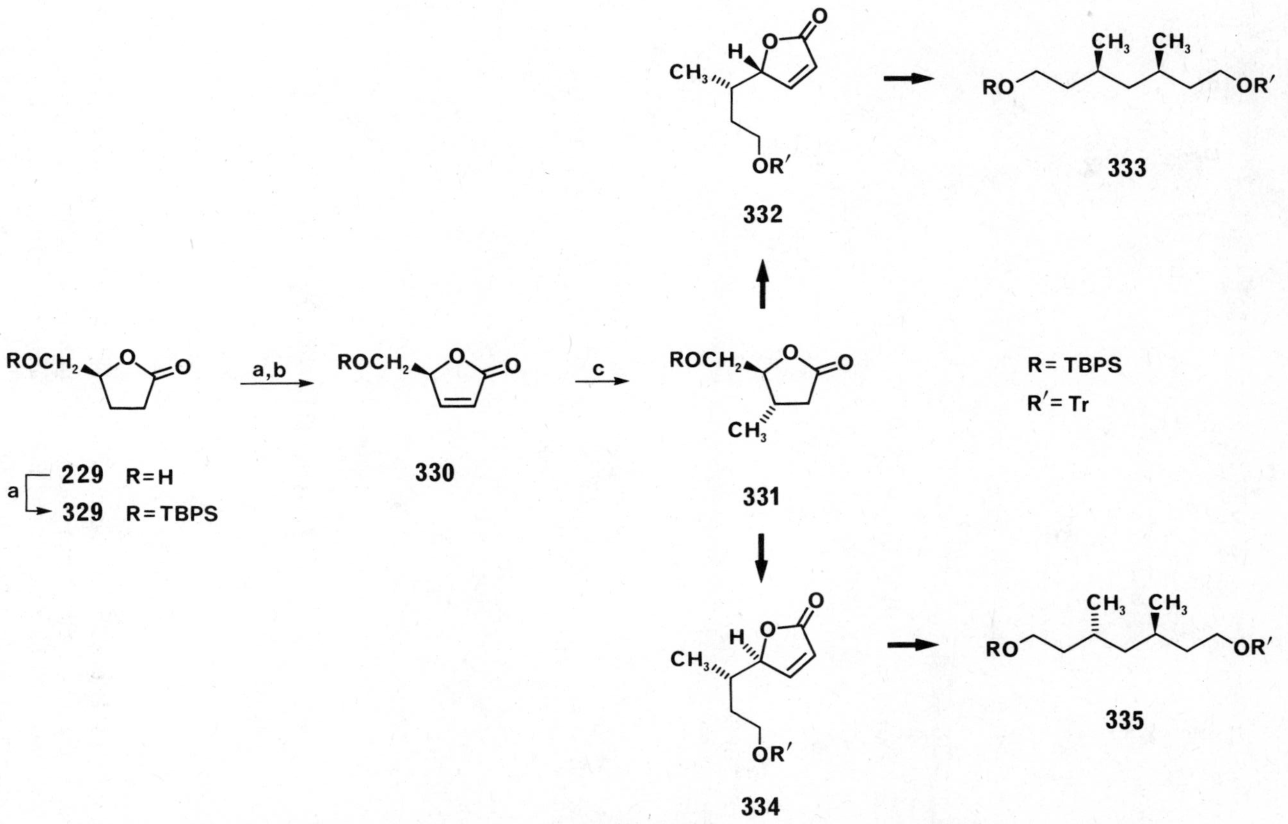

R = TBPS
R′ = Tr

229 R=H
329 R=TBPS

Scheme 50. (a) TBPSCl, DMAP (85%); (b) LiHMDS, PhSeCl, then H_2O_2, 0°C (50%); (c) Me_2CuLi, −20°C (87%).

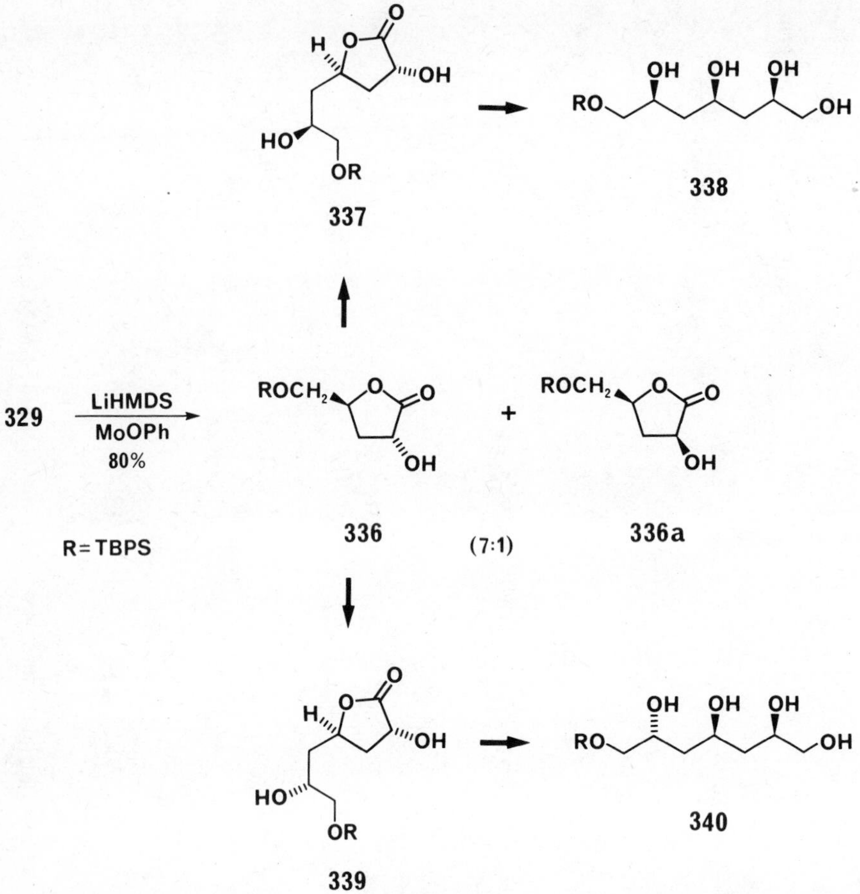

337 → 338

329 —LiHMDS, MoOPh, 80%→ 336 + 336a (7:1)

R = TBPS

336 → 337

336 → 339 → 340

Scheme 51

7.3 ORNITHINE

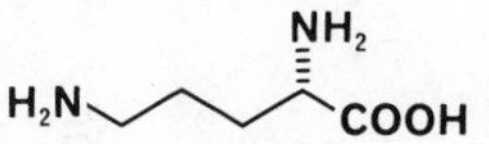

(2*S*)-2,5-Diaminovaleric acid (**341**)

L-Ornithine (**341**) is a nonproteogenic amino acid, which in mammals is an intermediate in the urea cycle and an intermediate in the biosynthesis of arginine.[110]

Derivatives of **341** readily form chiral cyclic lactams. Hydrogenolysis and esterification of N^{α}-Ts-N^{δ}-Cbz–L-ornithine (**342**)[111] provides **343,** which cyclizes to 3-*p*-toluenesulfonyl-L-piperid-2-one (**345**) in 82% yield when treated with ammonia in chloroform. Alternatively, **342** can be esterified and converted to the amide **334,** which quantitatively cyclizes to **345** under reductive conditions[112] (Scheme 52).

Ornithine is also used to prepare 18-membered macrocycles, which have attracted interest from physiologists and biologists because of their ion-binding abil-

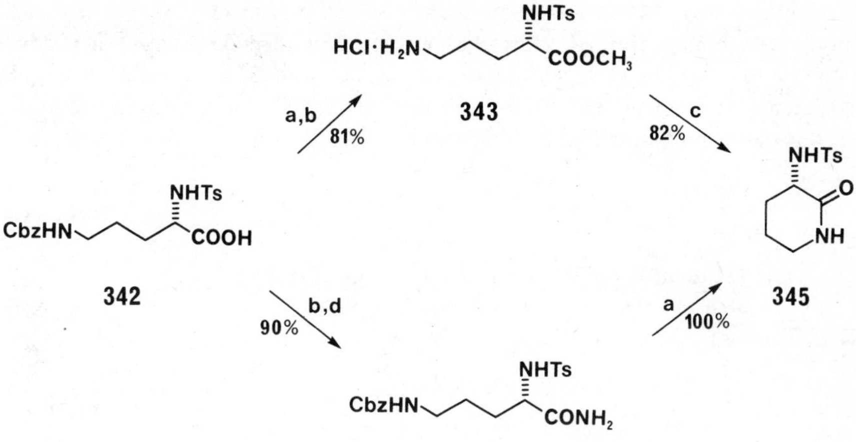

Scheme 52. (a) H_2, Pd; (b) CH_3OH, PTSA, benzene; (c) NH_3, $CHCl_3$; (d) NH_3, EtOH.

ity. Whereas these cyclic peptides can be prepared by an appropriate linear sequence followed by cyclization, this is not feasible for cyclic hexapeptides because of a preferred diketopiperazine formation. However, if the dipeptide bonds are forced to remain in the *trans* configuration, this can be avoided.[113] Selective deprotection of N^{α}-Boc–N^{δ}-Cbz–L-ornithine (**346**)[114] and *in situ* reductive alkylation of the resultant primary amine with glyoxylic acid afford the diacid **347,** which is quantitatively cyclized to the lactam (*S*)-3-amino-2-oxo-1-piperidineacetic acid (**348**). Cyclotrimerization of **348** using the diphenylphosphoryl azide (DPPA) procedure gives the cyrstalline hexapeptide **349** in 25% yield[113,115] (Scheme 53).

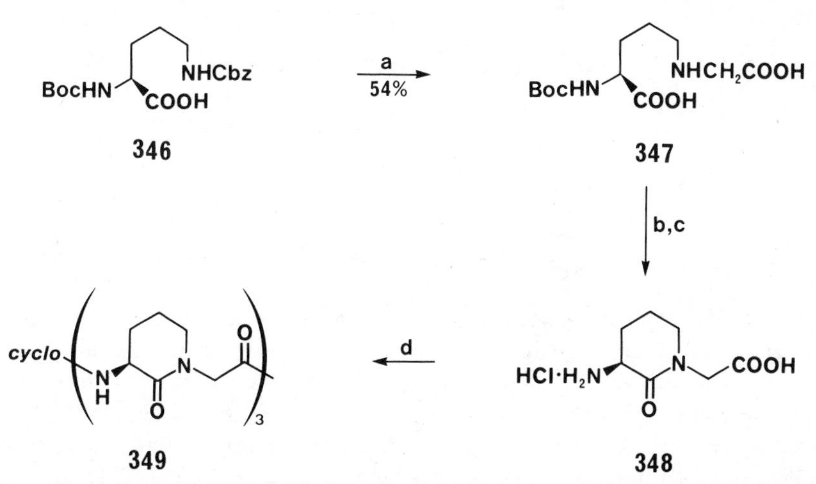

Scheme 53. (a) H_2, Pd/C, OHCCOOH, CH_3OH; (b) DMF, heat; (c) HCl, EtOAc; (d) DPPA, TEA, DMF, $-20°C$ (25%).

Conversion of **341** to the distosylate, followed by reduction of the carboxylic group with borane in THF, furnishes alcohol **350** in good yield. When **350** is allowed to react under high dilution conditions with **351,** one obtains the 18-membered macrocycle **352,** which can be deprotected with hydrogen bromide in acetic acid to provide chiral **353**[116] (Scheme 54).

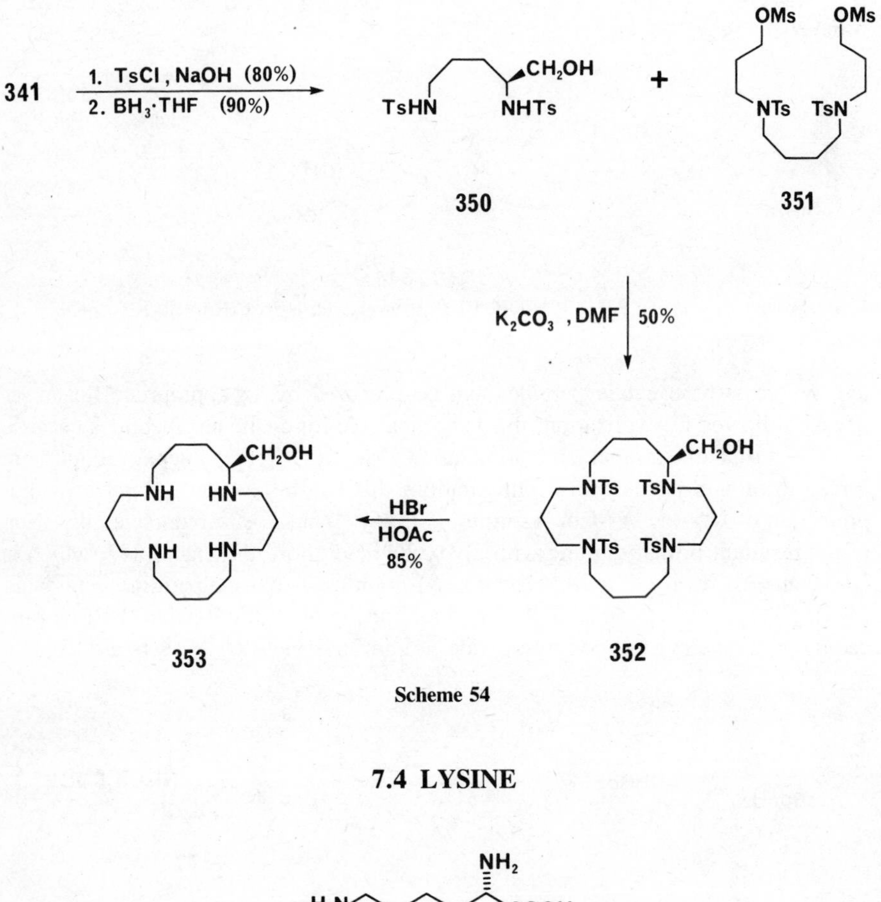

Scheme 54

7.4 LYSINE

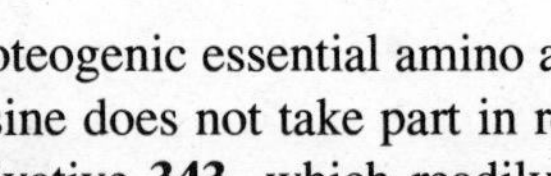

(2*S*)-2,6-Diaminocaproic acid (**354**)

L-Lysine (**354**) is a basic proteogenic essential amino acid needed for bone growth, particularly in children. Lysine does not take part in reversible transaminations.[117]

Unlike the ornithine derivative **343,** which readily cyclizes to the lactam **345,** the corresponding lysine derivative **355**[111] cyclizes in poor yield to the seven-membered lactone **357.** Esterification to **356,** followed by a reductive cyclization to **357,** proceeds in only 7% yield[112] (Scheme 55).

L-Piperidine-2-carboxylic acid (**361**) (L-pipecolic acid) is a widespread nonpro-

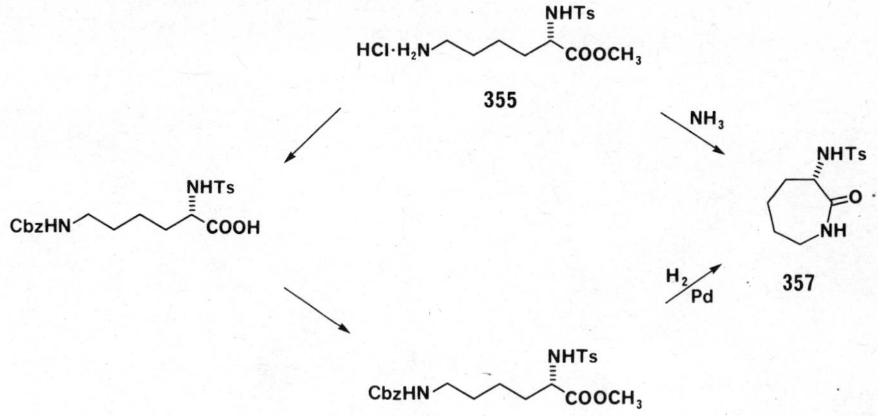

NHTs
HCl·H$_2$N COOCH$_3$
355
NH$_3$
NHTs
O
NH
357
H$_2$ Pd
NHTs
CbzHN COOH
NHTs
CbzHN COOCH$_3$
356

Scheme 55

teinogenic amino acid found in plants. It can be prepared from L-lysine (**354**) in two different ways: (1) bromination of the terminal amino group of **358,** followed by cyclization of the ester **359** with NaH in DMF, gives the *N*-tosylate ester **360;** and (2) detosylation and ester hydrolysis affords **361** in good yield.[118] Alternatively, **361** is prepared in 39% yield in one step by the reaction of L-lysine (**354**) with disodium nitrosyl–pentacyanoferrate(II)[119] (Scheme 56).

The direct anodic oxidation and methoxylation of the lysine derivative **362** afford the α'-methoxylated cyclic carbamate **363.** Heating **363** with a catalytic amount of ammonium chloride yields the α',β'-unsaturated carbamate **364,** which can be converted to **361** without the occurrence of any racemization.[120]

Moreover, **363** undergoes a stereoselective nucleophilic substitution with trimethylsilyl cyanide in the presence of titanium tetrachloride at $-50°C$ to afford

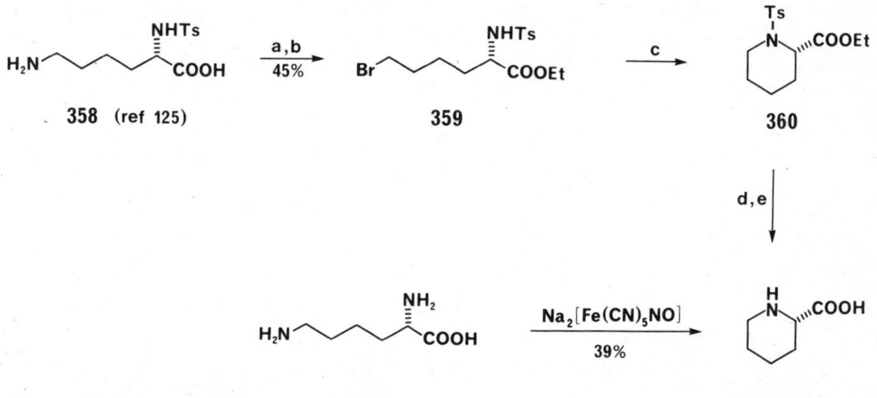

NHTs
H$_2$N COOH
358 (ref 125)
a,b 45%
NHTs
Br COOEt
359
c
Ts
N COOEt
360
d,e
NH$_2$
H$_2$N COOH
354
Na$_2$[Fe(CN)$_5$NO] 39%
H
N COOH
361

Scheme 56. (a) KBr, HBr, $NaNO_2$; (b) HCl, EtOH; (c) NaH, DMA (100%); (d) NaOH, CH_3OH (82%); (e) 0.5 *N* H_2SO_4 (76%).

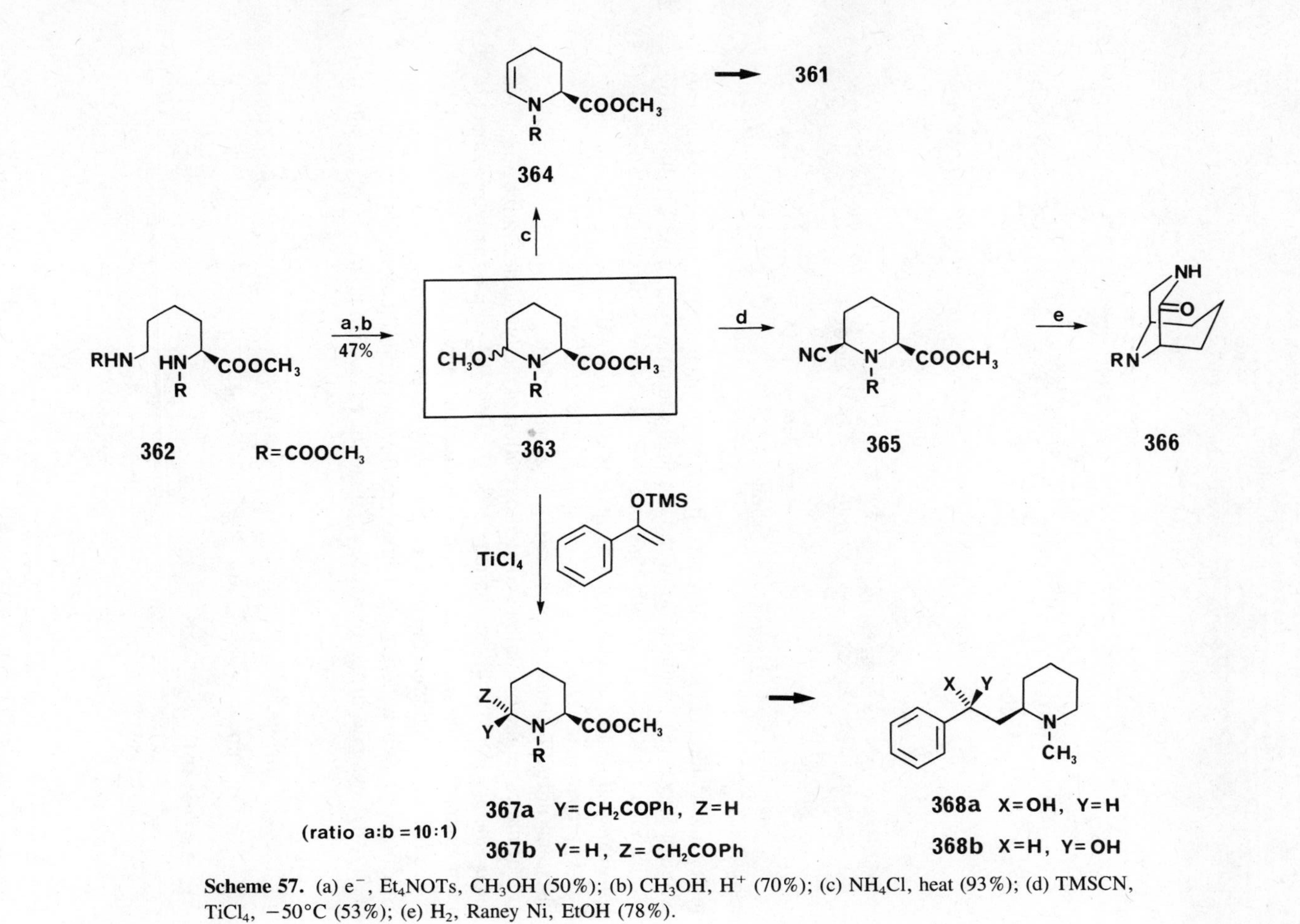

Scheme 57. (a) e^-, Et_4NOTs, CH_3OH (50%); (b) CH_3OH, H^+ (70%); (c) NH_4Cl, heat (93%); (d) TMSCN, $TiCl_4$, $-50°C$ (53%); (e) H_2, Raney Ni, EtOH (78%).

the single diastereomer **365,** which has the *cis* stereochemistry. Catalytic reduction of **365,** followed by a spontaneous cyclization, furnishes the diazabicyclo[3.3.1]nonan-2-one **366.** Intermediate **363** has also been used to prepare (+)-sedamine (**368a**) and (+)-allosedamine (**368b**)[121] (Scheme 57).

The enantiomeric D-pipiecolic acid (**370**) can be prepared from **354** by a Walden inversion. Thus, treatment of the hydrochloride of **354** with nitrosyl bromide provides the bromide **369,** which, under basic conditions, cyclizes to **370** in 34% yield.[122]

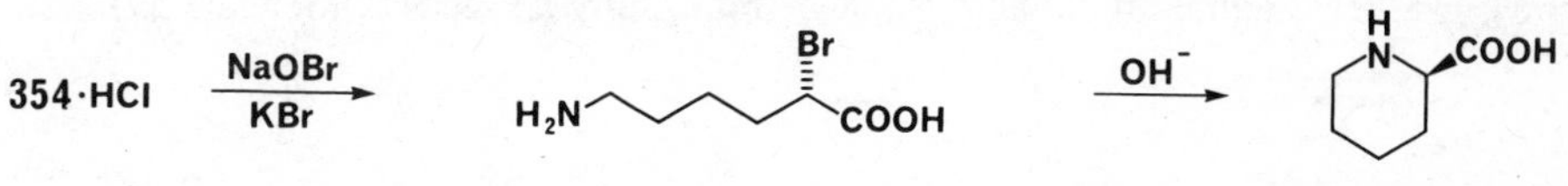

354·HCl —NaOBr, KBr→ **369** —OH^-→ **370**

γ-Lactam-constrained dipeptides serve as analogs of glycyl dipeptides restricted to turn conformations. The synthesis of the tryptophyllysine lactams **374a** and **374b** from lysine is illustrated in Scheme 58. The reaction of commercially available lysine ester **371** with isopropylidene cyclopropane-1,1-dicarboxylate (**372**) provides

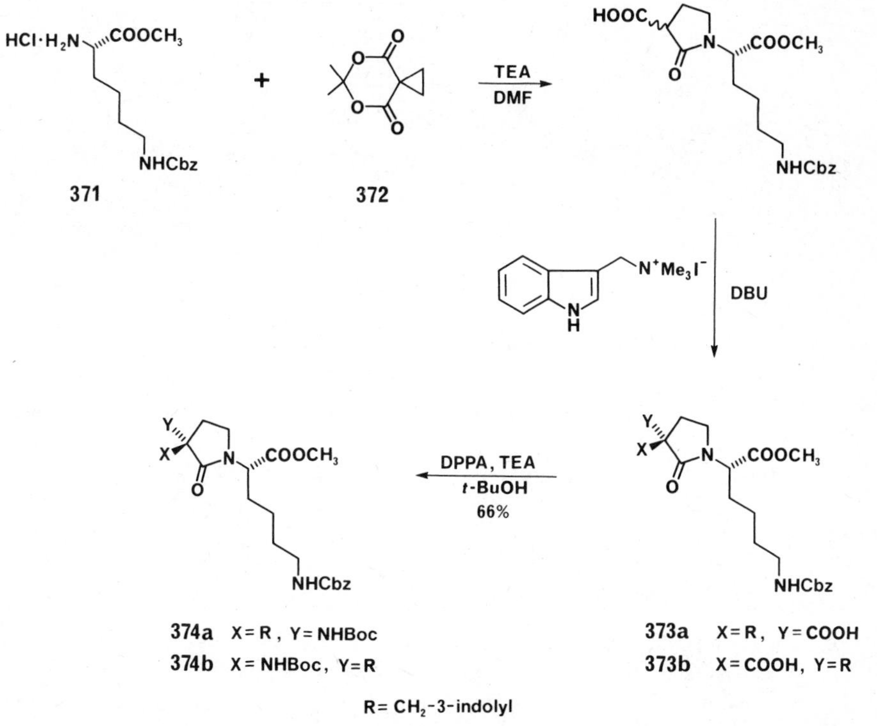

371 + **372** —TEA, DMF→ —DBU, indolylmethyl-$N^+Me_3I^-$→ **373a/373b** —DPPA, TEA, *t*-BuOH, 66%→ **374a/374b**

374a X = R, Y = NHBoc
374b X = NHBoc, Y = R

373a X = R, Y = COOH
373b X = COOH, Y = R

R = CH_2-3-indolyl

Scheme 58

a diastereomeric mixture of acids that is reacted crude with gramine methiodide in DMF to furnish a 1 : 1 mixture of **373a** and **373b.** Readily separated, each diastereomer is converted stereospecifically to **374a** and **374b** by a modified Curtius rearrangement with DPPA in *t*-butanol[123] (Scheme 58).

The photochlorination of lysine (**354**) in strong acid followed by the reaction of the resultant γ-chloro-L-lysine with silver acetate provides *threo*-γ-hydroxy-L-lysine (**375**), which is cyclized with nitrosyl chloride to a 80 : 20 mixture of *cis*-4-hydroxy-L-pipecolic acid (**376**) and *trans*-4-hydroxy-L-pipecolic acid (**377**). Moreover, **375** can be readily converted to *threo*-γ-hydroxyhomo-L-arginine (**378** ⇄ **379**), a new natural amino acid isolated from *Lathyrus* seeds[124] (Scheme 59).

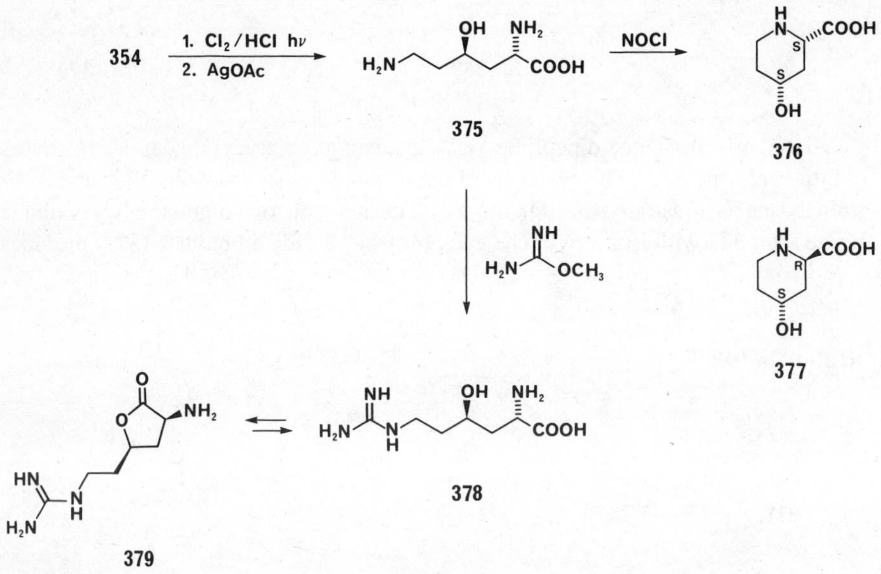

354 —(1. Cl_2/HCl hν; 2. AgOAc)→ 375 —(NOCl)→ 376 + 377

375 —(H_2NC(=NH)OCH_3)→ 378 ⇄ 379

Scheme 59

REFERENCES

1. T. Scott and M. Brewers, *Concise Encyclopedia of Biochemistry*, Walter de Gruyter, New York, 1983, p. 41.
2. L. I. Krimen, *Org. Syn.*, **50,** 1 (1970).
3. D. Seebach and D. Wasmuth, *Angew. Chem., Int. Ed.*, **20,** 971 (1981).
4. M. Lapidus and M. Sweeney, *J. Med. Chem.*, **16,** 163 (1973).
5. J. D. McDermed, G. M. McKenzie, and A. P. Phillips, *J. Med. Chem.*, **18,** 362 (1975).
6. J. E. Nordlander, M. J. Payne, F. G. Njoroge, V. M. Vishwanath, G. R. Han, G. D. Laikos, and M. A. Balk, *J. Org. Chem.*, **50,** 3619 (1985).
7. G. J. McGarvey, R. N. Hiner, Y. Matsubara, and T. Oh, *Tetrahedron Lett.*, 2733 (1983).

8. M. Larcheveque and Y. Petit, *Tetrahedron Lett.*, 3705 (1984).
9. H. Schwarz, F. M. Bumpus, and I. H. Page, *J. Am. Chem. Soc.*, **79,** 5697 (1957).
10. H. Gregory, J. S. Morley, J. M. Smith, and M. J. Smithers, *J. Chem. Soc.* (London), C, 715 (1968).
11. S. Fushiya, S. Nakatsuyama, Y. Sato, and S. Nozoe, *Heterocycles*, **15,** 819 (1981).
12. J. Faust, K. Schreiber, and H. Ripperger, *Z. Chem.*, **24,** 330 (1984).
13. T. N. Salzmann, R. W. Ratcliffe, and B. G. Christensen, *Tetrahedron Lett.*, 1193 (1980).
14. L. Zervas, M. Winitz, and J. P. Greenstein, *J. Org. Chem.*, **22,** 1515 (1957).
15. T. N. Salzmann, R. W. Ratcliffe, B. G. Christensen, and F. A. Bouffard, *J. Am. Chem. Soc.*, **102,** 6161 (1980).
16. P. J. Reider and E. J. J. Grabowski, *Tetrahedron Lett.*, 2293 (1982).
17. R. Labia and C. Morin, *Chem. Lett.*, 1007 (1984).
18. N. Ikota, H. Shibata, and K. Koga, *Heterocycles*, **14,** 1077 (1980).
19. N. Ikota, H. Shibata, and K. Koga, *Chem. Pharm. Bull.*, **33,** 3299 (1985).
20. E. Sandrin and R. A. Boissonnas, *Helv. Chim. Acta*, **46,** 1637 (1963).
21. E. Fischer, *Chem. Ber.*, **34,** 433 (1901).
22. T. Scott and M. Brewer, *Concise Encyclopedia of Biochemistry*, Walter de Gruyter, New York, 1983, p. 41.
23. M. Wilchek, S. Ariely, and A. Patchornik, *J. Org. Chem.*, **33,** 1258 (1968).
24. P. M. Hardy and D. J. Samworth, *J. Chem. Soc., Perkin Trans. I*, 1954 (1977).
25. I. P. Karrer and A. Schlosser, *Helv. Chim. Acta*, **6,** 411 (1923).
26. F. Schneider, *Liebigs Ann. Chem.*, **529,** 1 (1937).
27. J. Rudinger, K. Poduska, and M. Zaoral, *Collect. Czech. Chem. Commun.*, **25,** 2022 (1960).
28. M. Otsuka, A. Kittaka, T. Iimori, H. Yamashita, S. Kobayashi, and M. Ohno, *Chem. Pharm. Bull.*, **33,** 509 (1985).
29. Y. Umezawa, H. Morishima, S. Saito, T. Takita, H. Umezawa, S. Kobayashi, M. Otsuka, M. Narita, and M. Ohno, *J. Am. Chem. Soc.*, **102,** 6630 (1980).
30. M. Otsura, M. Narita, M. Yoshida, S. Kobayashi, M. Ohno, Y. Umezawa, H. Morishima, S. Saito, T. Takita, and H. Umezawa, *Chem. Pharm. Bull.*, **33,** 520 (1985).
31. B. D. Christie and H. Rapoport, *J. Org. Chem.*, **50,** 1239 (1985).
32. T. Scott and M. Brewers, *Concise Encyclopedia of Biochemistry*, Walter de Gruyter, New York (1983), p. 180.
33. J. D. Aebi and D. Seebach, *Helv. Chim. Acta*, **68,** 1507 (1985).
34. R. B. Silverman and M. W. Holladay, *J. Am. Chem. Soc.*, **103,** 7357 (1981).
35. B. H. Lee, G. J. Gerfen, and M. J. Miller, *J. Org. Chem.*, **49,** 2418 (1984).
36. R. K. Olsen, K. Ramasamy, and T. Emergy, *J. Org. Chem.*, **49,** 3527 (1984).
37. B. H. Lee and M. J. Miller, *Tetrahedron Lett.*, 927 (1984).
38. G. Benz, *Liebigs Ann. Chem.*, 1424 (1984).
39. W. Oppolzer and K. Thirring, *J. Am. Chem. Soc.*, **104,** 4978 (1982).
40. S. Hanessian and S. P. Sahoo, *Tetrahedron Lett.*, 1425 (1984).
41. M. Itoh, *Chem. Pharm. Bull.*, **17,** 1679 (1969).

42. D. H. R. Barton, D. Crich, Y. Herve, P. Potier, and J. Thierry, *Tetrahedron*, **41,** 4347 (1985).
43. J. Smith and L. Pease, *CRC Crit. Rev. Biochem.*, **8,** 315 (1980).
44. U. Nagi and K. Sato, *Tetrahedron Lett.*, 647 (1985).
45. P. M. Hardy, *Synthesis*, 290 (1978).
46. H. Gibian and E. Klieger, *Annalen*, **640,** 145 (1961).
47. J. Rudinger, *Coll. Czech. Chem. Comm.*, **19,** 365 (1954).
48. C. R. Harington and R. C. G. Moggridge, *J. Chem. Soc.* (London), 706 (1940).
49. T. Mukaiyama, M. Asami, J. Hanna, and S. Kobayashi, *Chem. Lett.*, 783 (1977).
50. S. Iriuchijima, *Synthesis*, 684 (1978).
51. U. Schmidt and R. Schölm, *Synthesis*, 752 (1978).
52. A. Pfaltz, N. Bühler, R. Neier, K. Hirai, and A. Eschenmoser, *Helv. Chim. Acta*, **60,** 2653 (1977).
53. R. B. Silverman and M. A. Levy, *J. Org. Chem.*, **45,** 815 (1980).
54. R. Fujimoto and Y. Kishi, *Tetrahedron Lett.*, 4197 (1981).
55. M. Roth, P. Dubs, E. Gotschi, and A. Eschenmoser, *Helv. Chim. Acta*, **54,** 710 (1971).
56. S. Danishefsky, E. Berman, L. A. Clizbe, and M. Hirama, *J. Am. Chem. Soc.*, **101,** 4385 (1979).
57. S. Danishfesky, J. Morris, and L. A. Clizbe, *J. Am. Chem. Soc.*, **103,** 1602 (1981).
58. Y. Ohfune and M. Tomita, *J. Am. Chem. Soc.*, **104,** 3511 (1982).
59. J. S. Petersen, G. Fels, and H. Rapoport, *J. Am. Chem. Soc.*, **106,** 4539 (1984).
60. P. Quitt, J. Hellerbach, and K. Vogler, *Helv. Chim. Acta*, **46,** 327 (1963).
61. K. Shiosaki and H. Rapoport, *J. Org. Chem.*, **50,** 1229 (1985).
62. N. Ikota and A. Hanaki, *Heterocycles*, **22,** 2227 (1984).
63. E. J. Corey and D. Enders, *Chem. Ber.*, **111,** 1337 (1978).
64. D. Enders, H., Eichenauer, and R. Pieter, *Chem. Ber.*, **112,** 3703 (1979).
65. W. Bartmann and B. M. Trost (Eds.), *Selectivity—A Goal for Synthetic Efficiency*, Verlag Chemie, Weinheim, 1984, pp. 65–86.
66. J. D. Morrison, *Asymmetric Synthesis*, Vol. 3, Academic, Orlando, 1984, pp. 275–339.
67. T. F. Buckley, III and H. Rapoport, *J. Org. Chem.*, **48,** 4222 (1983).
68. O. Cervinka and L. Hub, *Coll. Czech. Chem. Commun.*, **33,** 2927 (1968).
69. C. Eguchi and A. Kakuta, *Bell. Chem. Soc. Jpn.*, **47,** 1704 (1974).
70. S. Iwaki, S. Marumo, T. Saito, M. Yamada, and K. Katagiri, *J. Am. Chem. Soc.*, **96,** 7842 (1974).
71. M. Larcheveque and J. Lalande, *J. Chem. Soc., Chem. Commun.*, 83 (1985).
72. M. Larcheveque and J. Lalande, *Tetrahedron*, **40,** 1061 (1984).
73. J. H. Tumlinson, M. G. Klein, R. E. Doolittle, T. L. Ladd, and A. T. Proveaux, *Science*, **197,** 789 (1977).
74. R. E. Doolittle, J. H. Tumlinson, A. T. Proveaux, and R. R. Heath, *J. Chem. Ecol.*, **6,** 473 (1980).
75. S. Yamada, N. Oh-hashi, and K. Achiwa, *Tetrahedron Lett.*, 2557 (1976).

76. S. Yamada, N. Oh-hashi, and K. Achiwa, *Tetrahedron Lett.*, 2561 (1976).
77. K. Mori, *Tetrahedron*, **31,** 3011 (1975).
78. U. Ravid, R. M. Silverstein, and L. R. Smith, *Tetrahedron*, **34,** 1449 (1978).
79. P.-T. Ho and N. Davies, *Synthesis*, 462 (1983).
80. U. Ravid and R. M. Silverstein, *Tetrahedron Lett.*, 423 (1977).
81. R. G. Brownlee, R. M. Silverstein, D. Müller-Schwarze, and A. G. Singer, *Nature*, **221,** 284 (1969).
82. L. R. Smith, H. J. Williams, and R. M. Silverstein, *Tetrahedron Lett.*, 3231 (1978).
83. J. P. Vigneron, R. Meric, M. Larcheveque, A. Debal, G. Kunesch, P. Zagatti, and M. Gallois, *Tetrahedron Lett.*, 5051 (1982).
84. J. P. Vigneron, R. Meric, M. Larcheveque, A. Debal, J. Y. Lallemand, G. Kunesch, P. Zagatti, and M. Gallois, *Tetrahedron*, **40,** 3521 (1984).
85. T. Kitahara, K. Mori, and M. Matsui, *Tetrahedron Lett.*, 3021 (1979).
86. T. Kitahara and K. Mori, *Tetrahedron*, **40,** 2935 (1984).
87. J. Ackermann, N. Waespe-Sarcevic, and C. Tamm, *Helv. Chim. Acta*, **67,** 254 (1984).
88. R. K. Olsen, K. L. Bhat, R. B. Wardle, W. J. Hennen, and G. D. Kini, *J. Org. Chem.*, **50,** 896 (1985).
89. S. Batmangherlich and A. H. Davidson, *J. Chem. Soc., Chem. Commun.*, 1399 (1985).
90. M. Taniguchi, K. Koga, and S. Yamada, *Tetrahedron*, **30,** 3547 (1947).
91. S. Takano, M. Yonaga, and K. Ogasawara, *Synthesis*, 265 (1981).
92. S. Takano, N. Tamura, and K. Ogasawara, *J. Chem. Soc., Chem. Commun.*, 1155 (1981).
93. S. Takano, C. Kasahara, and K., Ogasawara, *Chem. Lett.*, 631 (1982).
94. S. Takano, M. Yonaga, K. Chiba, and K. Ogasawara, *Tetrahedron Lett.*, 3697 (1980).
95. S. Takano, W. Uchida, S. Hatakeyama, and K. Ogasawara, *Chem. Lett.*, 733 (1982).
96. S. Takano, K. Chiba, M. Yonaga, and K. Ogasawara, *J. Chem. Soc., Chem. Commun.*, 616 (1980).
97. S. Takano, M. Yonaga, and K. Ogasawara, *J. Chem. Soc., Chem. Commun.*, 1153 (1981).
98. S. Takano, M. Yonaga, M. Morimoto, and K. Ogasawara, *J. Chem. Soc., Perkin Trans. I*, 305 (1985).
99. S. Takano, C. Kasahara, and K. Ogasawara, *J. Chem. Soc., Chem. Commun.*, 637 (1981).
100. K. Tomioka and K. Koga, *Tetrahedron Lett.*, 3315 (1979).
101. K. Tomioka, H. Mizuguchi, and K. Koga, *Tetrahedron Lett.*, 4687 (1978).
102. K. Tomioka, H. Mizuguchi, and K. Koga, *Chem. Pharm. Bull.*, **30,** 4304 (1982).
103. J. P. Robin, O. Gringore, and E. Brown, *Tetrahedron Lett.*, 2709 (1980).
104. K. Tomioka, T. Ishiguro, and K. Koga, *J. Chem. Soc., Chem. Commun.*, 652 (1979).
105. K. Tomioka, T. Ishiguro, and K. Koga, *Tetrahedron Lett.*, 2973 (1980).
106. K. Tomioka, H. Kawasaki, Y. Iitaka, and K. Koga, *Tetrahedron Lett.*, 903 (1985).
107. S. Hanessian, P. J. Murray, and S. P. Sahoo, *Tetrahedron Lett.*, 5623 (1985).
108. S. Hanessian, P. J. Murray, and S. P. Sahoo, *Tetrahedron Lett.*, 5627 (1985).
109. S. Hanessian, S. P. Sahoo, and P. J. Murray, *Tetrahedron Lett.*, 5631 (1985).

110. T. Scott and M. Brewer, *Concise Encyclopedia of Biochemistry*, Walter de Gruyter, New York, 1983, p. 316.
111. B. C. Barrass and D. T. Elmore, *J. Chem. Soc.*, 3134 (1957).
112. B. C. Barrass and D. T. Elmore, *J. Chem. Soc.*, 4830 (1957).
113. R. M. Freidinger, D. A. Schwenk, and D. F. Veber, *Peptide Structure and Biologic Function, Proceedings of the 6th American Peptide Symposium*, 1979, p. 703.
114. T. Kato and N. Izumiya, *Bull. Chem. Soc. Jpn.*, **39,** 2242 (1966).
115. D. F. Veber and R. M. Freidinger, U.S. Patent 4,192,875 (1980).
116. B. Lavery, T. R. Wagler, and G. J. Burrows, *189th ACS National Meeting*, Miami, FL, 1985.
117. T. Scott and M. Brewer, *Concise Encyclopedia of Biochemistry*, Walter de Gruyter, New York, 1983, p. 258.
118. T. Fujii and M. Miyoshi, *Bull Chem. Soc. Jpn.*, **48,** 1341 (1975).
119. L. Kisfaludy and F. Korenczki, *Synthesis*, 163 (1982).
120. T. Shono, Y. Matsumura, and K. Inoue, *J. Chem. Soc., Chem. Commun.*, 1169 (1983).
121. K. Irie, K. Aoe, T. Tanaka, and S. Saito, *J. Chem. Soc., Chem. Commun.*, 633 (1985).
122. U. Schiedt and H. G. Hoss, *Z. Physiol. Chem.*, **308,** 179 (1957).
123. R. M. Freidinger, *J. Org. Chem.*, **50,** 3631 (1985).
124. Y. Fujita, J. Kollonitsch, and B. Witkop, *J. Am. Chem. Soc.*, **87,** 2030 (1965).
125. B. C. Barrass and D. T. Elmore, *J. Chem. Soc.*, 4830 (1957).

CHAPTER EIGHT

THE PROLINE FAMILY

8.1 PROLINE AND ITS ESTERS

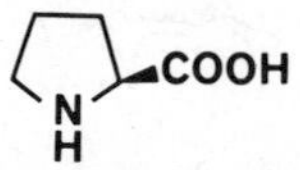

COOH
N
H

L-Proline; (*S*)-pyrrolidine-2-carboxylic acid

L-Proline is a proteogenic amino acid that is very soluble in water and has a rotation of $[\alpha]_D^{25} = -86.2$. Because it is an imino acid, it forms a yellow color, and not the characteristic purple color, with ninhydrin.[1] It is prepared by fermentation or by isolation from protein hydrolysates and is relatively inexpensive. Whereas D-proline is significantly more costly, its synthetic availability from D-glutamic acid[2] provides an affordable alternative. Because it is a chiral pyrrolidine possessing two important functional groups, proline offers the necessary substrate for many synthetic applications leading to more complicated chiral products.

(−)-Odorine (**4**) is a bisamide in which both nitrogen atoms are attached to the same chiral center at C-2′. The stereochemistry of (+)-odorine, its naturally occurring enantiomer isolated from the leaves of *Aglaia odoata* Lour,[3] was established by the synthesis of **4** from L-proline.[4] The key step in the synthetic sequence is the thermal Wolff rearrangement of the azide **2,** which occurs with strict retention of configuration to provide the isocyanate **3.** Addition of 2-butylmagnesium bromide at −78°C affords a separable mixture of **4** and its diastereomer (−)-epiodorine (**5**) (Scheme 1).

Pumiliotoxin B (**13**), first isolated from the Panamanian poison frog *Dendrobates* *pumilio*,[5] is a member of the dendrobatid alkaloids that have in common the unusual (Z)-6-alkylidineindolizidine (1-azobicyclo[4.3.0]nonane) ring system. Pumiliotoxin B is a powerful myotonic and cardiotonic agent.[6] It was originally believed to be a steroidal alkaloid, and its structure was determined from the X-ray structure elucidation of Pumiliotoxin 251D (**10**), which is the major component of the basic skin extracts of the Ecuadorian poison frog, *Dendrobates tricolor*.[7] The total synthesis of enantiomerically pure **10** is accomplished by a

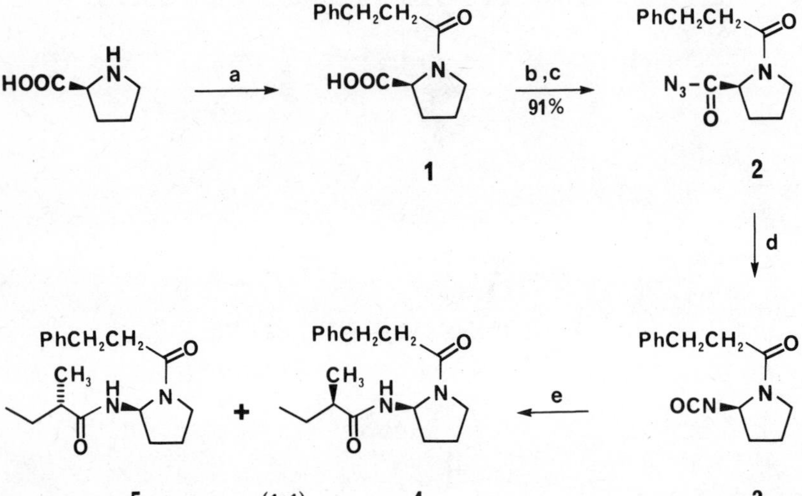

Scheme 1. (a) $PhCH_2COCl$, NaOH (85%); (b) ClCOOEt, Et_3N; (c) NaN_3; (d) reflux THF; (e) $CH_3CH_2CH(CH_3)MgBr$ (72%).

remarkably elegant chiral convergent synthesis in nine-steps from methyl *N*-(Cbz)–L-prolinate (**6**), easily prepared from L-proline.[8] Note that the reaction of **6** with methylmagnesium iodide is preferred since methyllithium deprotonates at C-2′ and also reacts with the carbamate. Subsequent dehydration and a nonstereoselective epoxidation provide an easily separatable 1:1 mixture of diastereomers **7a** and **7b.** The reaction of **7a** with the vinylalanate of **8a** provides the bicyclic carbamate **9** in 38% yield. Utilization of an iminium ion vinylsilane cyclization results in stereospecific assembly of the required (Z)-6-alkylidene indolizidine system. Thus enantiomerically pure **10** is prepared from **6** in an overall yield of 3.6%.[9] Likewise, Pumiliotoxin B (**13**) is prepared from **7a** by a reaction with the vinylalanate of **8b** prepared from 3-(benzyloxy)propionitrile. Elaboration to the aldehyde **11** and a subsequent Wittig reaction with **12,** prepared from ethyl L-lactate, provide the ketone, which undergoes a *threo*-selective reduction with excess lithium aluminum hydride to afford **13** as a 15:1 ratio of *threo* to *erythro*. Thus, from **6,** this sequence provides **13** in an overall yield of 1.8%[10] (Scheme 2).

L-Proline can also be successfully applied to the asymmetric synthesis of threonine (**17**). The condensation of the hydrochloride of *N*-benzylprolyl chloride (**14**) with *o*-aminoacetophenone furnishes **15** in moderate yields. The resulting Schiff base derived

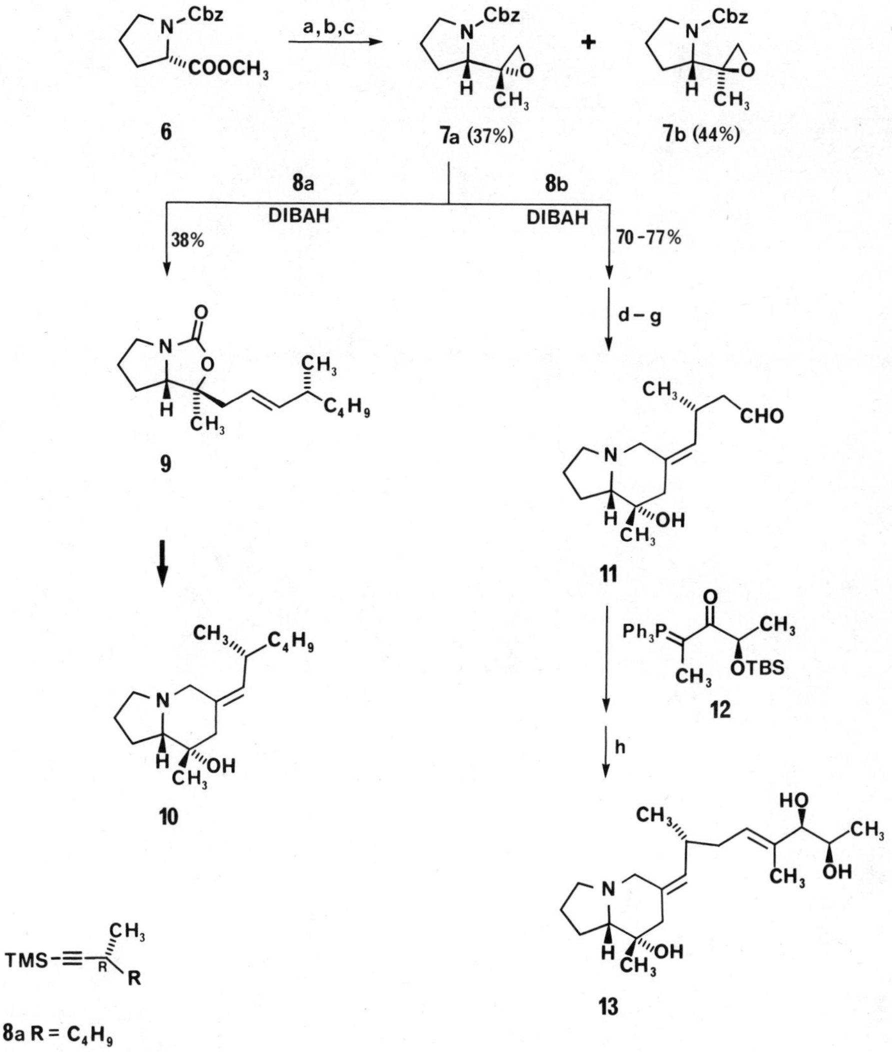

6 a,b,c 7a (37%) + 7b (44%)

8a DIBAH 38% 9 → 10

8b DIBAH 70-77% d-g 11; 12; h 13

8a R = C_4H_9

8b R = $(CH_2)_2OCH_2Ph$

Scheme 2. (a) MeMgI (86%); (b) $SOCl_2$, pyridine (54% from **6**); (c) MCPBA; (d) (i) KOH, EtOH, (ii) 37% HCHO, MeOH; (e) camphorsulfinic acid (52%); (f) NH_3, Li (61%); (g) oxalyl chloride, DMSO (88%); (h) $LiAlH_4$ (74%)

from the reaction of **15** with glycine gives the copper complex **16** when reacted with copper(II) sulfate. The reaction of this complex with acetaldehyde affords a mixture from which threonine is isolated in 80–100% yields with an optical purity of 97–100%. The chiral reagent (*N*-benzylproline) is recovered without loss of optical purity[11a,b] (Scheme 3).

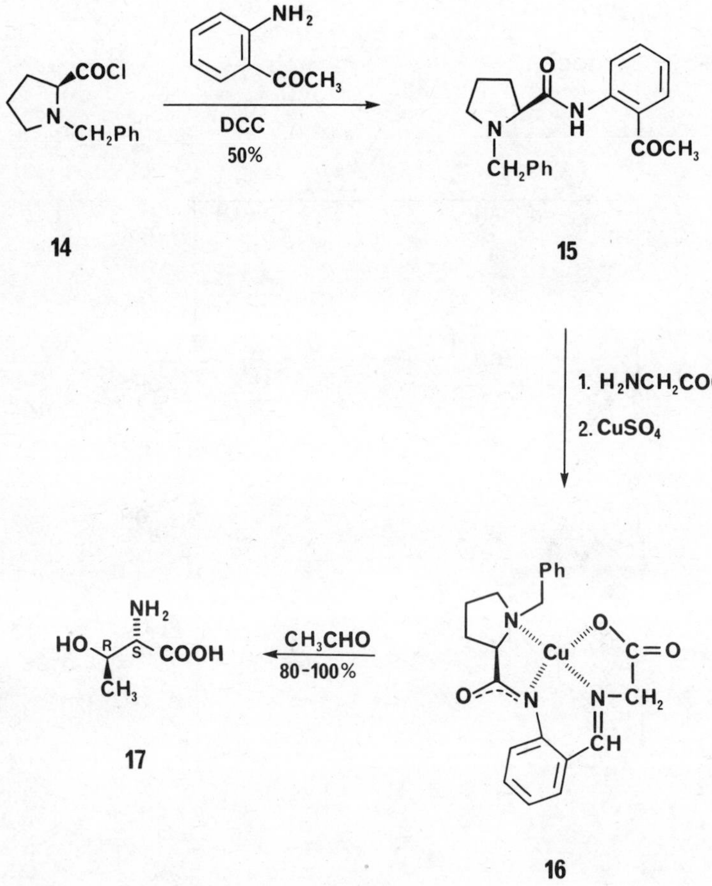

Scheme 3

The asymmetric reduction of prochiral ketones by use of the proline modified borohydride reagent **19** provides chiral alcohols with only modest optical purity (15–50%). However, the asymmetric reduction of ketone **18** followed by deprotection affords the cardiotonic agent (−)-4-hyroxy-α-(3,4-dimethoxyphenethylaminoethyl)benzyl alcohol hydrochloride (**20**) in excellent yield (62% optical purity). The use of D-proline results in the alcohol with inverted configuration.[12]

1. (19: prolinate COONa, N–BH$_3$) 2. H_2, Pd/C 80%

18 → **20**

Recently, chiral polymeric reagents prepared from polymeric (*S*)-prolinols and borane have been successfully used for the enantioselective reduction of prochiral ketones to afford chiral alcohols with up to 80% optical purity.[13]

Another approach based on the redox reaction catalyzed by NADH-dependent dehydrogenases[15] is the nonenzymatic asymmetric reduction of prochiral ketones with chiral NADH model compounds bearing L-prolinamide (**21a**) or L-prolinol (**21b**). The enantiomeric excess depends on the specific blockage of one of the diastereomeric faces of the dihydropyridine nucleus. With magnesium perchlorate as the source of metal ion, the reduction of ethyl benzoylformate (**22**) provides (−)-ethyl mandelate (**23**) in good chemical yield with 83.2% optical purity. The use of other metal ions such as cobalt or zinc results in a significant reduction in optical yields.[14,15]

Feedback effects of the reaction product on the product stereochemistry also contribute to higher optical yields.[16]

Alternatively, the asymmetric reduction of the α-ketoamides **24,** prepared from the reaction of L-proline methyl ester with an α-ketoacid in the presence of DCC, and with $NaBH_4$, produces optically active α-hydroxyacids, but with only moderate optical purity. However, a striking solvent effect is observed when a 99 : 1 mixture of THF and MeOH is used. In this case,

Ph-CO-CO-OEt **22**, $Mg(ClO_4)_2$, 84%

23

21a R = $CONH_2$

21b R = CH_2OH

(*S*)-(+)-mandelic acid (**25**) is obtained with a 64% enantiomeric excess.[17] Protected homoallylic alcohols **26** with diastereomeric excesses of up to 89% are obtained by the addition of allyltrimethylsilane to **24**[18] (Scheme 4).

A novel and convenient entry to various optically active alkaloids is available by the asymmetric reduction of cyclic imines **27** with chiral sodium triacyloxy borohydride **28,** easily obtained from the reaction of sodium borohydride with three equivalents of *N*-benzyloxycarbonyl–L-proline in THF. In this

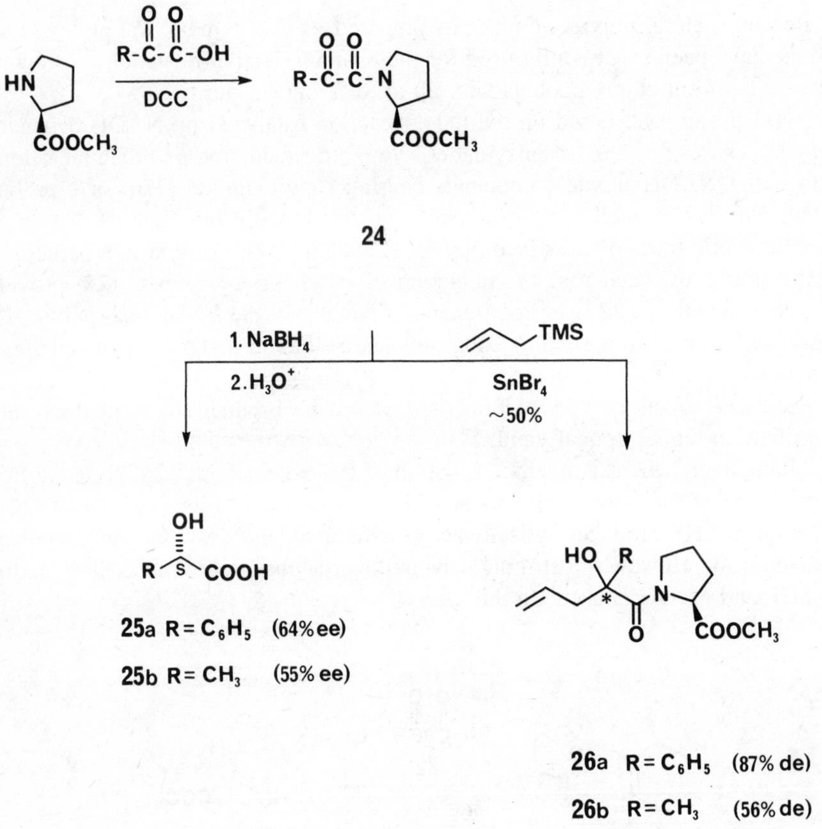

24

1. $NaBH_4$ 2. H_3O^+

TMS, $SnBr_4$, ~50%

25a R = C_6H_5 (64% ee)

25b R = CH_3 (55% ee)

26a R = C_6H_5 (87% de)

26b R = CH_3 (56% de)

Scheme 4

manner, (*S*)-(+)-norlandanosine hydrochloride (**29a**), salsolidine (**29b**), norcryptostyline I (**29c**), and norcryptostyline II (**29d**) are produced in excellent chemical and optical yields.[19]

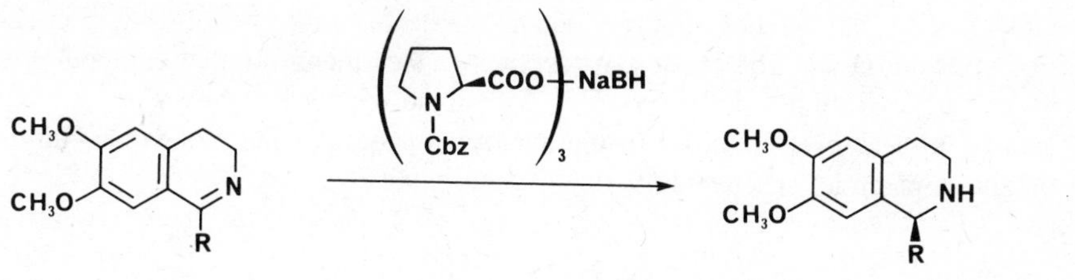

29	R	Chemical Yield (%)	Optical Yield (%)
a	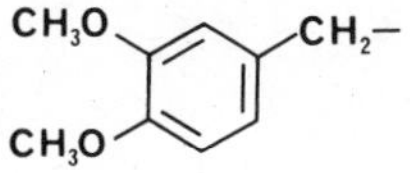 CH_3O, CH_3O, CH_2—	68	60
b	CH_3	85	70
c	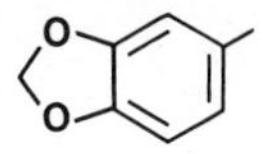 O, O	90	86
d	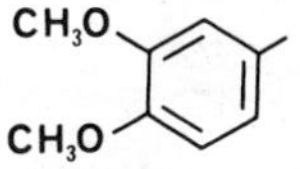 CH_3O, CH_3O	87	73

Analogously, the reduction of 1-methyl-3,4-dihydro-β-carboline (**30**) with **28** in THF proceeds smoothly to furnish tetrahydroharman (**31**) in 85% chemical yield with 79% enantiomeric excess.[19]

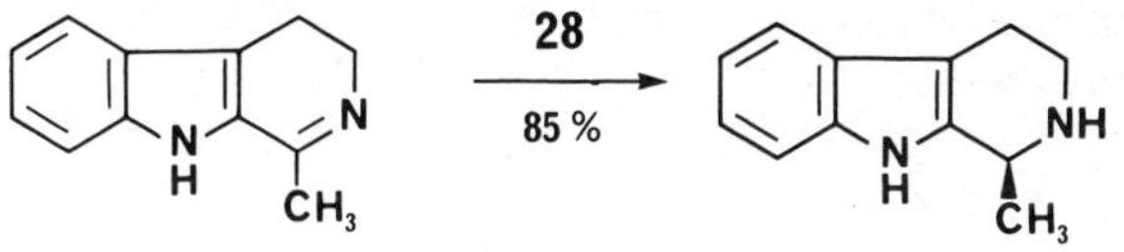

The reductive amination of ethyl 2-thienylacetoacetate (**32**) with methyl L-prolinate and sodium cyanoborohydride provides a separable mixture of diastereomeric amino diesters **33** and **34** in 92% yield. Dieckmann cyclization and hydrolysis of each afford (−)-(1*R*,8*S*,13*S*)-9-aza-1-hydroxy-5-thiatetracyclo[6.5.1.0.^{2,6}0^{9,13}]tetradeca-2(6)-3-diene (**35**) and the (+)-(1*S*,8*R*,13*R*)-enantiomer **36** in 63 and 46% yields, respectively. Treatment of either compound with excess *n*-butyllithium and trimethylcholorosilane results in a novel fragmentation leading to 4-(2-pyrrolidino)-2-(trimethylsilyl)benzothiophene **37**, which, unfortunately, is racemic[20] (Scheme 5).

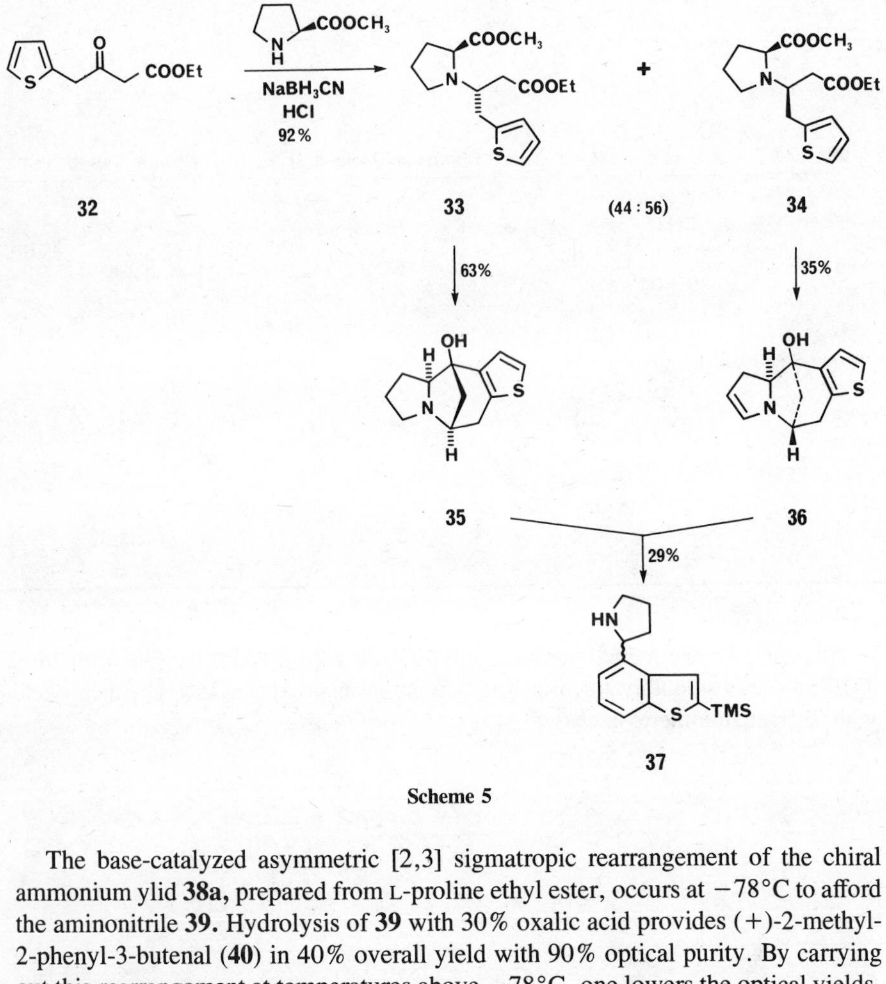

COOCH$_3$ N H NaBH$_3$CN HCl 92%

COOEt S O **32** → COOCH$_3$ N COOEt S **33** + COOCH$_3$ N COOEt S **34** (44 : 56)

33 63% → **35**; **34** 35% → **36**

35 + **36** 29% → **37** (HN, S, TMS)

Scheme 5

The base-catalyzed asymmetric [2,3] sigmatropic rearrangement of the chiral ammonium ylid **38a,** prepared from L-proline ethyl ester, occurs at −78°C to afford the aminonitrile **39.** Hydrolysis of **39** with 30% oxalic acid provides (+)-2-methyl-2-phenyl-3-butenal (**40**) in 40% overall yield with 90% optical purity. By carrying out this rearrangement at temperatures above −78°C, one lowers the optical yields. Hydrogenation of **40** over Raney nickel furnishes (−)-2-methyl-2-phenyl-1-butanol (**41**). A similar rearrangement of ylid **38b,** prepared from (*R*)-(−)-2-methylpyrrolidine, affords **40** in 30% yield having only 16% ee[21] (Scheme 6).

L-Proline methyl ester serves as a recoverable chiral auxiliary in the efficient asymmetric synthesis of L-α-amino acids and L-*N*-methyl-α-amino acids. Coupling with an α-ketoacid in the presence of DCC, followed by cyclization with ammonia or methylamine, affords stereoselectivity the (*S*)-hydroxydiketopiperazine **42.** The use of prototropic solvents rather than DME yields a diastereomeric mixture. Facile dehydration with trifluoroacetic acid or thionyl chloride–pyridine, followed by hydrogenation with Adam's

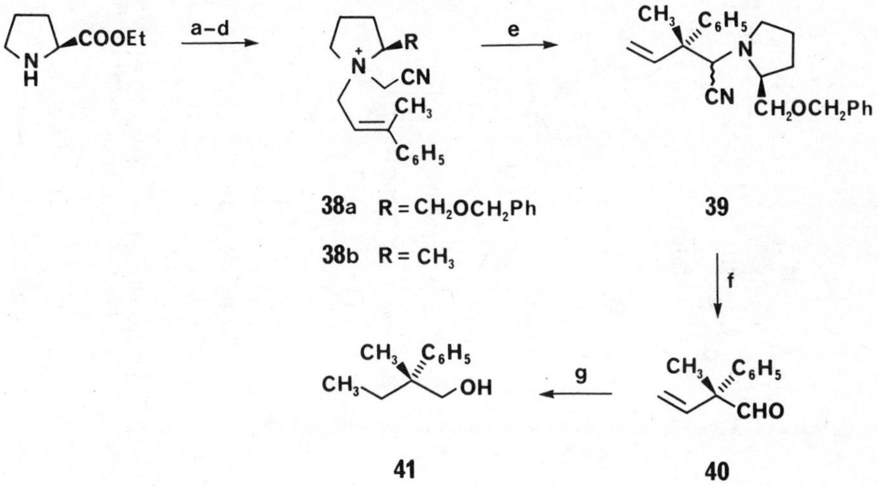

COOEt a–d R e CH_3 C_6H_5 N CN CN CH_3 C_6H_5 CH_2OCH_2Ph

38a R = CH_2OCH_2Ph **39**

38b R = CH_3

f

CH_3 C_6H_5 CH_3 OH g CH_3 C_6H_5 CHO

41 **40**

Scheme 6. (a) (*E*)-β-Methylcinnamic acid; (b) $LiAlH_4$; (c) NaH, $PhCH_2Br$ (overall 67%); (d) Cl-CH_2CN; (e) *t*-BuOK, THF–DMSO; (f) 30% oxalic acid; (g) H_2, Raney Ni.

catalyst, affords the (*S*,*S*)-cyclopeptide **43** in excellent yields with the asymmetric induction exceeding 90%. Lower inductions are observed when L-phenylalanine or L-alanine is used as the chiral auxiliary. Acidic hydrolysis converts **43** to the appropriate L-amino acids **44.**[22] Aromatic amino acids are also prepared in this manner. Thus the reaction of benzylideneoxazoline **45** with L-proline affords a dipeptide that is cyclized to diketopiperazine **46.** Hydrogenation occurs with 90% optical induction. Acidic hydrolysis provides the *N*-methyl derivatives of L-phenylalanine (**47a**) or L-DOPA (**47b**)[23] (Scheme 7).

Optical induction is also observed during the biomimetic formation of cysteine from the model *N*-benzyloxycarbonyl–dehydroalanyl–L-proline methyl amide (**48**).[24,25] The addition of methanethiol gives almost exclusively (optical yield $\leq$ 85%) *N*-benzyloxycarbonyl–*S*-methyl–D-cysteinyl–L-proline methylamide (**49**).[26,27]

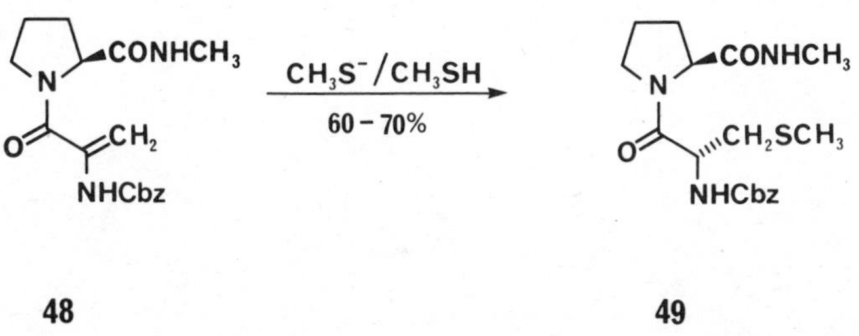

$CONHCH_3$ CH_3S^-/CH_3SH 60 – 70% $CONHCH_3$

N O CH_2 NHCbz N O CH_2SCH_3 NHCbz

48 **49**

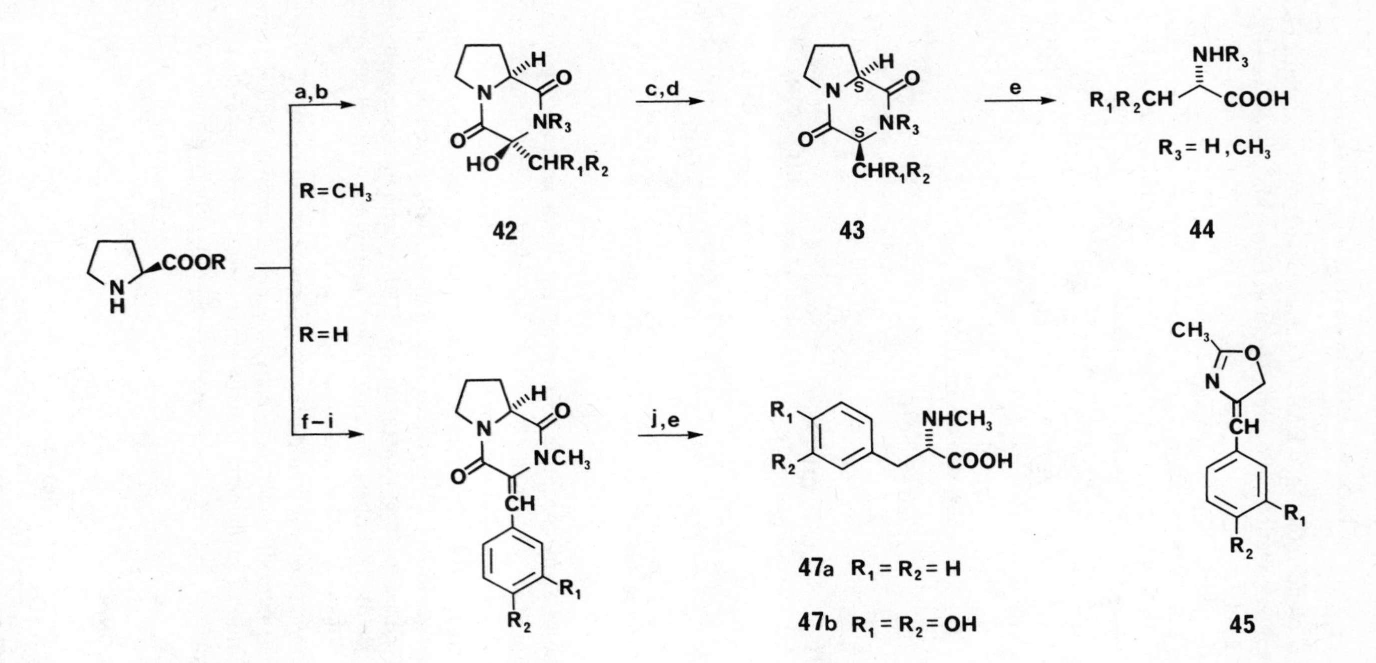

Scheme 7. (a) α-Ketoacid, DCC (75–80%); (b) NH_2R_3; (c) CF_3COOH; (d) H_2, Adam's catalyst (100%); (e) H_3O^+; (f) **45**, NaOH; (g) DCC (75%); (h) $C_6H_5NH_2$ (92%); (i) NaH, MeI (67%); (j) PdO_2, AcOH (62%).

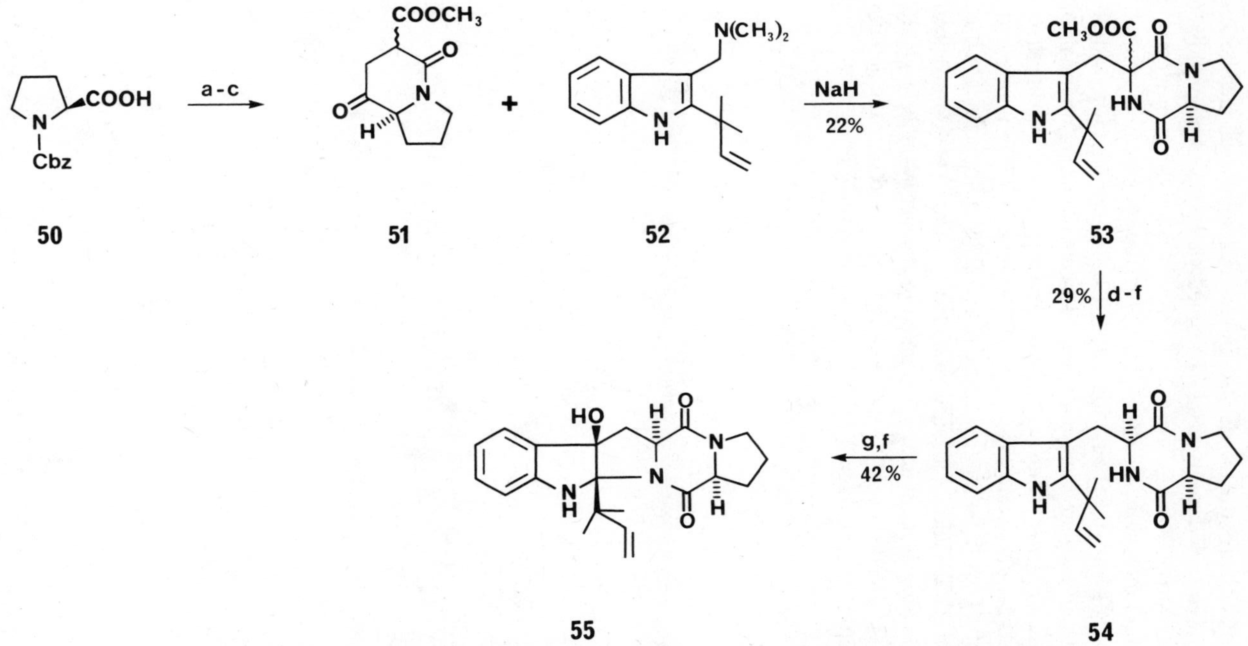

Scheme 8. (a) $SOCL_2$, C_6H_6; (b) $H_2NCH(COOMe)_2$, Na_2CO_3 (69% from **50**); (c) (i) H_2, 20% Pd/C, (ii) 2-hydroxypyridine (93%); (d) NaOH, MeOH; (e) 60–65°C, dioxane; (f) chromatography; (g) Rose Bengal, *hν*.

The diketopiperazine structure is also present in some fungal metabolites. Deoxybrevianamide [(*E*)-**54**] L-prolyl-2-(1′,1′-dimethylallyl)-L-tryptophyldiketopiperazine is a metabolite of *Aspergillus ustus*[28,29] and a degradation product of brevianamide [(*E*)-**55**], a metabolite of *Penicillium brevicompactum*.[30] The Schotten–Baumann reaction of the acid chloride of *N*-Cbz-L-proline (**50**) with dimethylaminomalonate gives a urethane that, after debenzylation and thermal cyclization (catalyzed by 2-hydroxypyridine),[32] affords (−)-methyl 1,4-dioxoperhydropyrrolo[1,2-a]pyrazine-3-carboxylate (**51**) in excellent yield.[31,32] The condensation of the sodium salt of **51** with 3-dimethylaminomethyl-2-(1′,1′-dimethylallyl)indole (**52**) provides **53** in 22% yield as a diastereomeric mixture. These are demethoxycarbonylated and chromatographically separated to furnish **54** in 29% yield.[31-34] The dye-sensitized photooxidation[31,32] of **54** to **55** is preferred over the aerial oxidation with iodosobenzene[34] (Scheme 8).

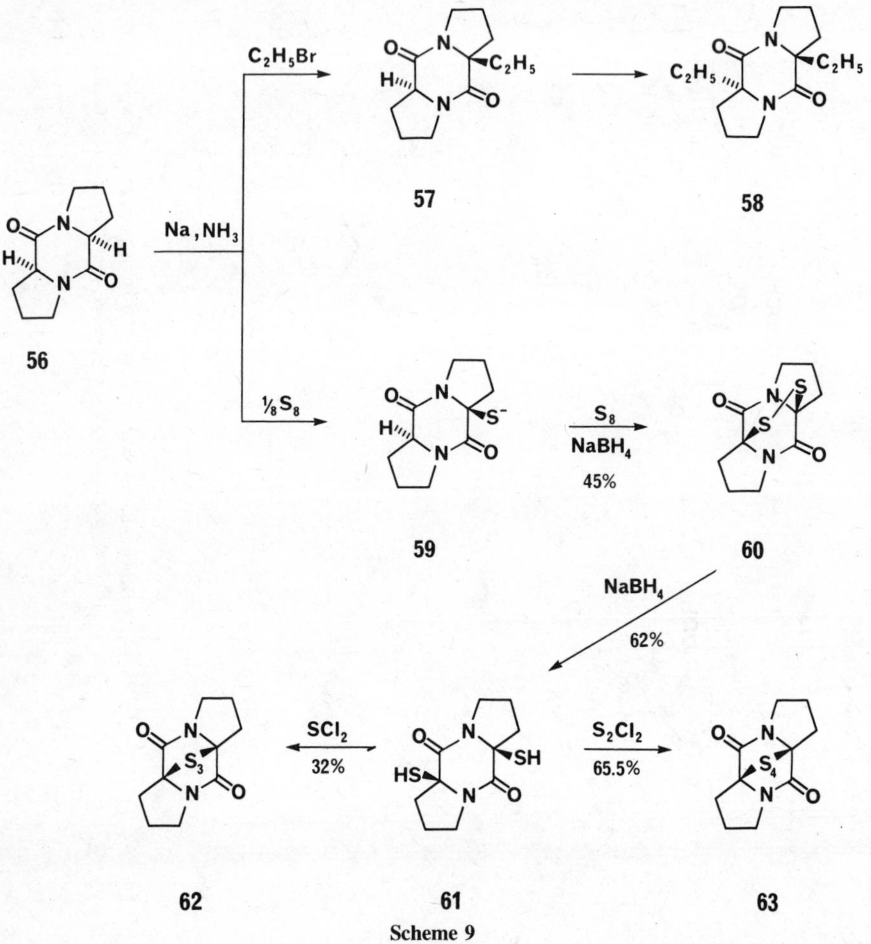

Scheme 9

The dioxopiperazine L-prolyl–L-proline anhydride (**56**) readily prepared from L-proline ethyl ester in 85% yield[35] undergoes metalation to form a monoanion, which, when alkylated, results in the formation of the optically active 3-substituted derivative **57.** Further metalation and alkylation lead to the optically inactive *trans*-(*meso*)-3,6-disubstituted product **58.**[35] Electrophilic introduction of sulfur at C-3 proceeds analogously, affording **59.** However, the more nucleophilic mercapto ion reacts with sulfur faster than the C-6 carbanion to form the *cis*-anhydride **60** in 44.9% yield after reduction.[36] This is the simplest model corresponding in conformation and configuration to the naturally occurring antibiotics gliotoxin, sporidesmin, aranotin, and chaetocin IV.[36] The reduction of **60** with sodium borohydride affords the *cis*-3,6-dimercapto-L-prolyl–L-proline anhydride (**61**), which reacts with

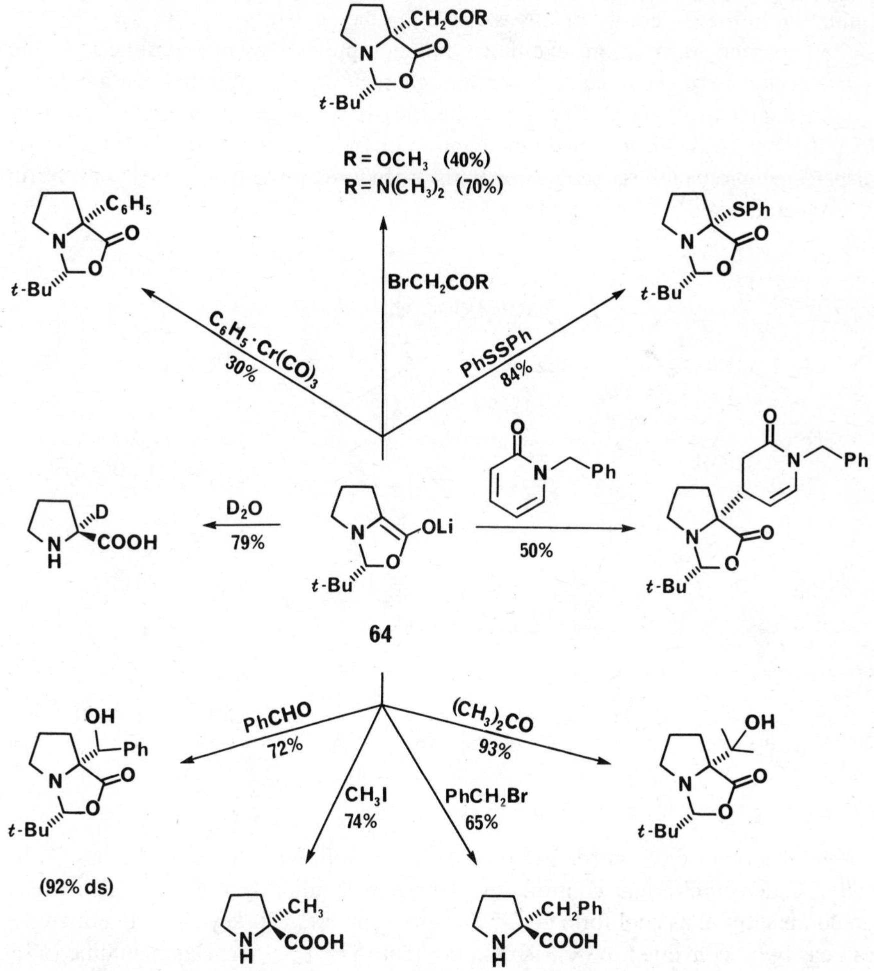

Scheme 10. (Reprinted from *J. Org. Chem.* **48** (26), 1983 (outside back cover) with permission Aldrich Chemical Co, Inc., Milwaukee, WI.)

either sulfur dichloride (SCl_2) or sulfur monochloride (S_2Cl_2) to give epitrithio- and epitetrathio-L-prolyl–L-proline anhydrides **62** and **63,** respectively[37] (Scheme 9).

Another bicyclic system, the chiral mixed acetal **64** derived from the condensation of L-proline with trimethylacetaldehyde (pivalaldehyde) under acidic catalysis, provides a synthetic method for the α-alkylation of proline without racemization. This overall enantioselective alkylation without the use of a chiral auxiliary is termed "self-reproduction of chirality."[38] When **64** is treated with LDA at $-78°C$, a nonracemic enolate results that can undergo reactions with a variety of electrophiles *cis* to the *t*-butyl group (*re*-facial). Although chemical yields are moderate to good, the diastereoselectivity is 90–100% in most cases. Subsequent removal of the acetal moiety provides the optically active α-substituted L-proline derivatives possessing the same sense of chirality as the starting prolines. At present, difficulty with acetal cleavage of some of the alkylated products presents a limitation to the generality of this method[39] (Scheme 10).

Perhaps one of the more exciting synthetic applications of L-proline is to the chiral synthesis of (8α)-necine bases that characterize the pyrrolizidine alkaloids. Commercially available *N*-Cbz-L-proline (**50**) is converted with equal facility to **65** either by an Arndt–Eistert homologation or by a reduction–tosylation–cyanide displacement–ethanolysis sequence. Both procedures provide optically pure **65** in 70% overall yield[40] (Scheme 11a).

Scheme 11a. (a) $COCl_2$, DMF (100%); (b) CH_2N_2 (83%); (c) Ag_2O, EtOH (92%); (d) H_2, Pd/C (70%); (e) BH_3–Me_2S (99%); (f) TsCl, pyridine (96%); (g) NaCN, DMSO (91%); (h) HCl, EtOH (70%).

The reaction of **65** with ethyl bromoacetate, followed by a Dieckmann cyclization under equilibrium control, gives the pyrrolizidine keto ester, which exists predominantly in its enol form **66** (75% yield from **65**). This key chiral intermediate can easily be converted to (−)-isoretronecanol (**67**), (−)-trachelanthamidine (**68**), or (−)-supinidine (**69**), all of which have better than 80% optical purity.[40] Moreover, catalytic reduction of **66** and subsequent chromatographic separation of the resultant mixture affords in 36% yield (1*S*, 2*S*, 8*S*)-1-ethoxycarbonylpyrrolizidine-2-ol (**70**), which, after reduction with lithium aluminum hydride, yields optically pure petasinecine (**71**)[41] (Scheme 11b).

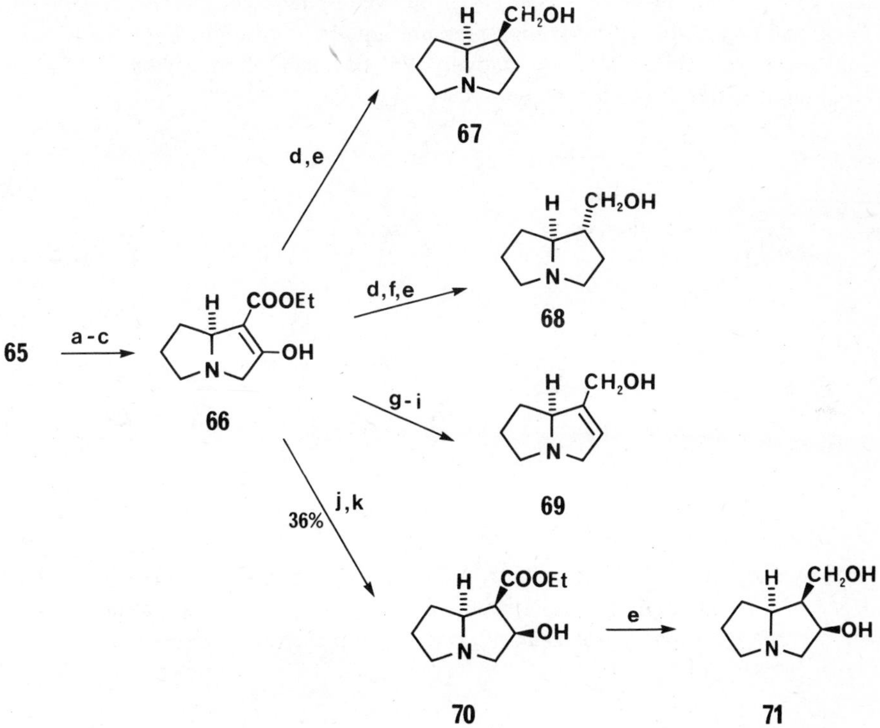

Scheme 11b. (a) $BrCH_2COOEt$ (92%); (b) NaOAc (82%); (c) AcOH, H_2O (99%); (d) Rh/Al_2O_3, AcOH (60–80%); (e) $LiAlH_4$; (f) NaOEt (70%); (g) $NaBH_3CN$ (57%); (h) MsCl, Et_3N (73%); (i) DIBAH (64%); (j) H_2, Pt, AcOH; (k) chromatography.

Proline is also able to act as a carrier of chiral information to other centers of a complex molecule. Peptide cyclization of **72** affords a single chiral cyclopeptide diastereomer, (5*S*,9*R*)-9-isopropyl-5,6-trimethylene-8-deamino-1,2-dihydro-*p*-phencyclopeptine (**73**), in 18% yield.[42]

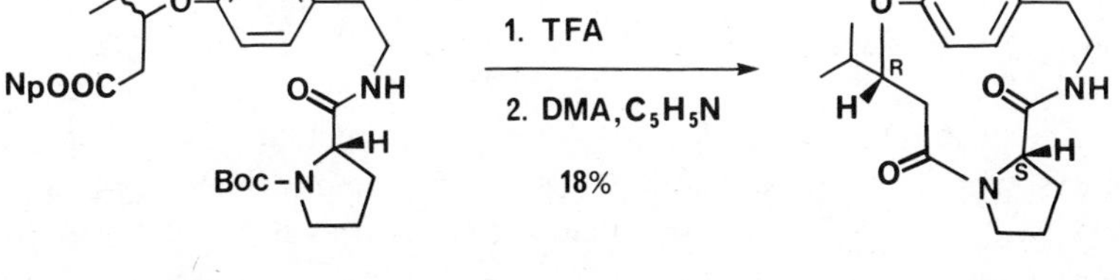

Angiotensin-converting enzyme (ACE) inhibition produces a fall in blood pressure and thus offers a control to hypertension. An orally active potent ACE inhibitor, (4*S*,9a*S*)-hexahydro-4-methyl-1*H*,5*H*-pyrrolo-[2,1-c] [1,4]thiazepine-1,5-dione (**75**) is prepared from Boc-L-proline by condensation with 3-mercapto-2-methylpropanoic acid *t*-butyl ester (**74**), followed by deprotection and cyclization. Fractional recrystallizations or high-pressure liquid chromatographic (HPLC) separation provides chiral **75.**[43] Alternatively, **75** is available from captopril[44] (**76**) by an intramolecular thiolactonization.

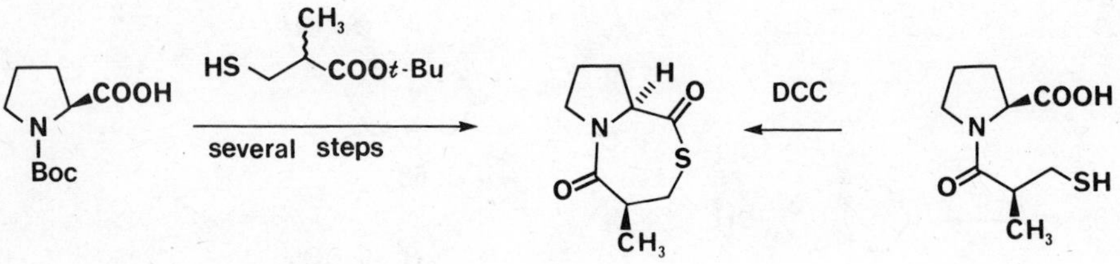

74 **75** **76**

A second powerful ACE inhibitor, 6,7-dihydro-8*H*-pyrrolo[2,1-c] [1,4]thiazine-1(8a*H*), 4(3*H*)-dione (**78**), is prepared from 1-(mercaptoacetyl)-L-proline (**77**) by the reaction with diphenylphosphorylazine in the presence of triethylamine.[45]

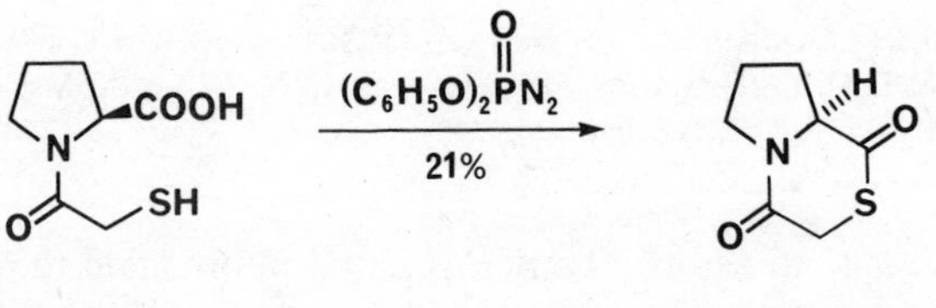

77 **78**

Continuing interest in this highly important area of antihypertensive drug development has provided a large variety of acyclic analogs, many of them proline derivatives.[46,47]

A trace alkaloid of *Cynanchum vincetoxicum* contains the hexahydrobenzo[f] pyrrolo[1,2-b]isoquinoline structure, which can be obtained from L-proline methyl ester. A mild Friedel–Crafts-type intramolecular acylation of the *N*-naphthalenyl

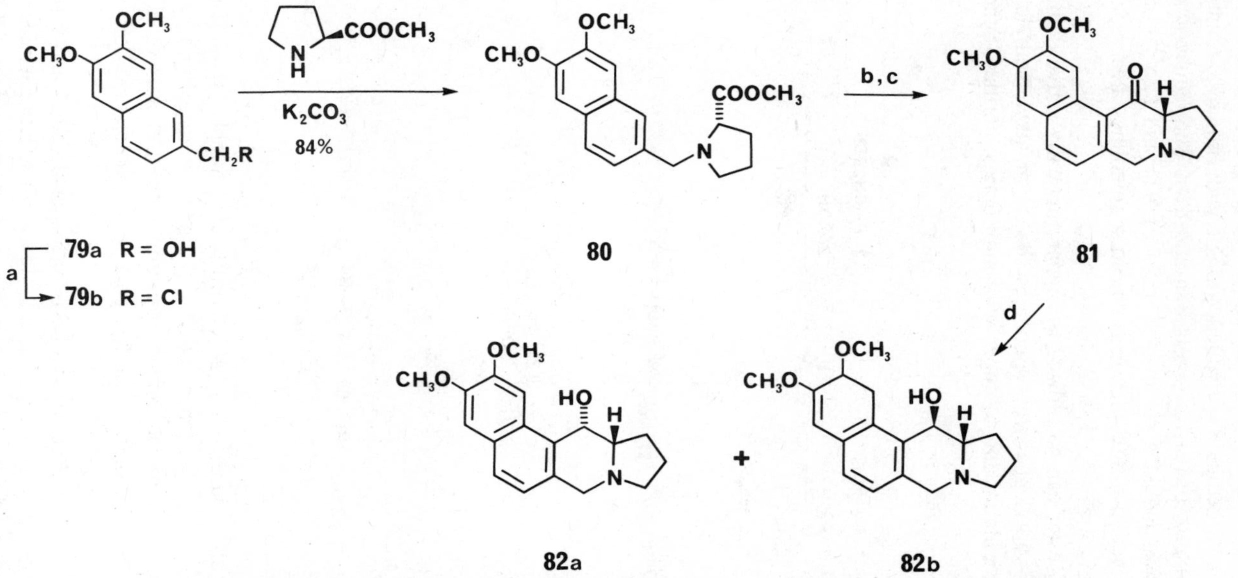

79a R = OH

79b R = Cl

80

81

82a + 82b

204 **Scheme 12.** (a) $SOCl_2$, $CHCl_3$ (65%); (b) NaOH, MeOH (94%); (c) PPA (58%); (d) $LiAlH_4$.

205

ester **80** affords chiral ketone **81.** Reduction of **81** with lithium aluminum hydride provides a 4:1 mixture of diastereomers **82a** and **82b.** The reduction of **81** with sodium borohydride provides only a 2:1 ratio of **82a** and **82b**[48] (Scheme 12). Various optically active hexahydrobenzopyrroloisoquinolines are prepared in this manner.[49]

L-Proline ethyl ester can be utilized to prepare optically active dialkyl phenylphosphates (**83**),[50] alkyl phenylphosphonates (**84**),[51] or alkyl methylphenylphosphinates (**85**).[52] Yields are fair to good, with high optical purities (Scheme 13).

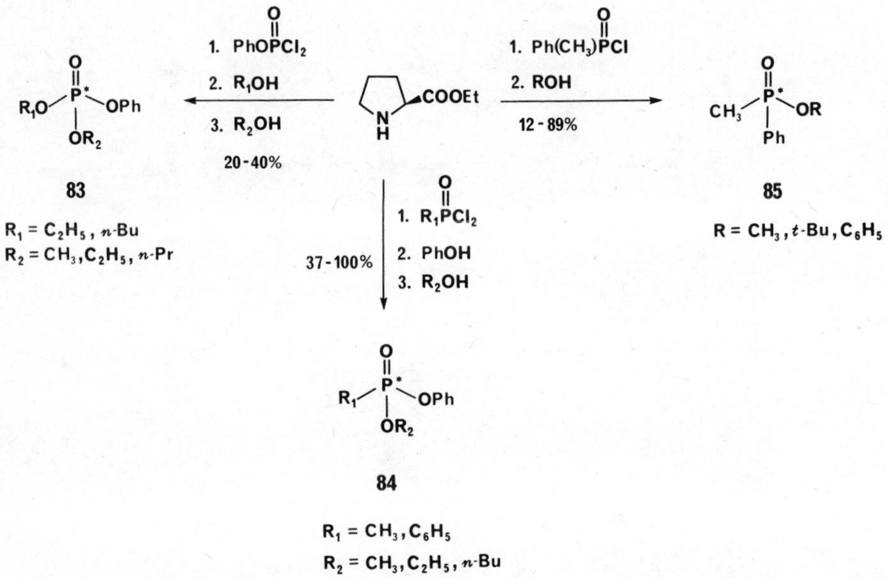

1. $PhOP(O)Cl_2$ 2. R_1OH 3. R_2OH 20-40%

1. $Ph(CH_3)P(O)Cl$ 2. ROH 12-89%

1. $R_1P(O)Cl_2$ 2. PhOH 3. R_2OH 37-100%

83 $R_1 = C_2H_5$, *n*-Bu; $R_2 = CH_3, C_2H_5$, *n*-Pr

85 $R = CH_3$, *t*-Bu, C_6H_5

84 $R_1 = CH_3, C_6H_5$; $R_2 = CH_3, C_2H_5$, *n*-Bu

Scheme 13

Diastereomerically pure ethyl *N*-[chloro(phenyl)thiophosphonyl]-L-prolinate (**86**), prepared from the reaction of ethyl L-prolinate with phenylthiophosphonic dichloride followed by chromatographic separation, provides a useful substrate for the preparation of various optically active phenylphosphonic acid derivatives such as **87, 88,** or **89**[53] (Scheme 14).

A very interesting and effective synthesis of 5-substituted L- and D-tryptophans from proline is accomplished by a new route involving electrochemical oxidation. The anodic oxidation of the *N*-methoxycarbonyl derivative of methyl L-prolinate (**90**) provides the hydroxyproline **91,** which as an aldehyde equivalent undergoes the Fischer indole synthesis with substituted phenylhydrazines to afford N_b-protected L-tryptophans. Removal of the protection groups with trimethyliodosilane yields the optically active 5-substituted L-tryptophans **92.** The D-isomer of **92** is obtained by using D-prolinate.[54]

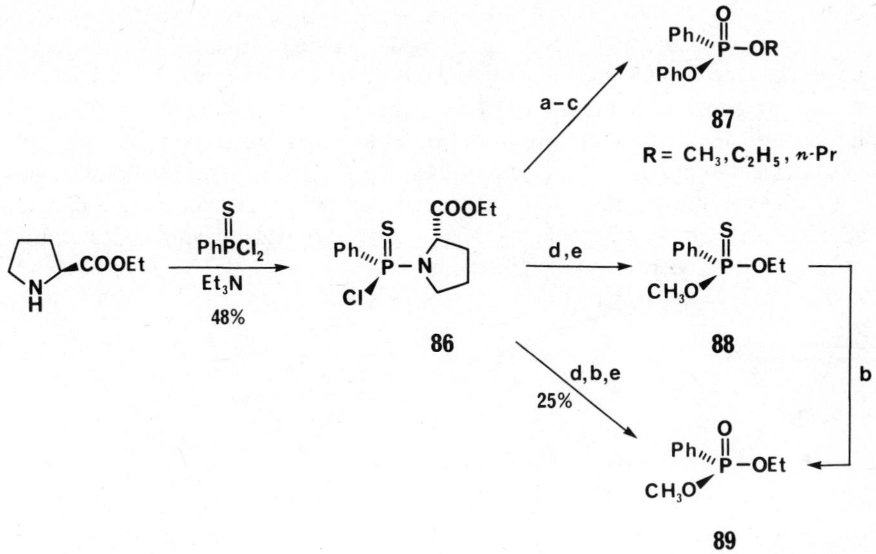

Scheme 14. (a) PhONa, THF (90%); (b) MCPBA; (c) ROH, H^+; (d) MeOH, Et_3N (86%); (e) EtOH, H_2SO_4 (40%).

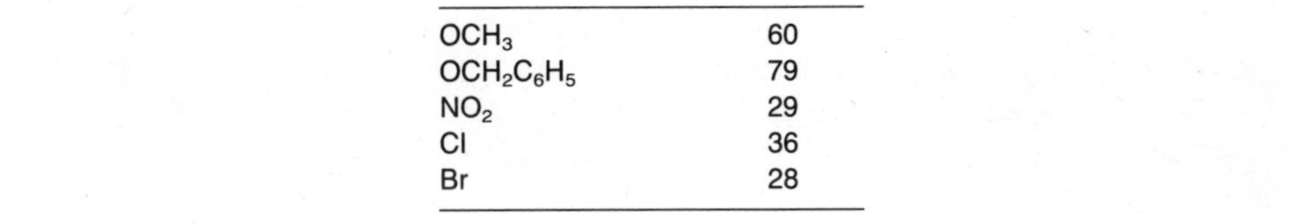

X	Yield **92** (%)
OCH_3	60
$OCH_2C_6H_5$	79
NO_2	29
Cl	36
Br	28

8.1.1 Asymmetric Halolactonization

Optically active α,α-disubstituted α-hydroxy acids and their corresponding ketones are extremely useful substrates for the synthesis of chiral natural products such as camptothecin[55] and several anthracycline antibiotics. Exploitation of a new asymmetric synthesis utilizing the halolactonization reaction as a key step provides, with high degrees of stereoselectivity and regiospecificity, the intermediates for these chiral α-hydroxy derivatives.[56,57] Thus the halolactonization of (*S*)-*N*-(α,β-unsaturated)carbonyl prolines **93** with NBS proceeds to give a mixture of diastereomeric halolactones **94a** and **94b,** whose ratio depends on the R groups. In the case where $R_1 = R_2 = CH_3$, **93** undergoes the halolactonization in 84% yield to afford a 94.5:5.5 mixture of **94a:94b.** However, when $R_1 = C_6H_5$ and $R_2 = CH_3$, it is not the free acid but the potassium salt of **93** that undergoes the halolactonization in 91% yield to provide a 99:1 ratio of **94a:94b.**[57] Reductive removal of the halide with tri-*n*-butyltin hydride, followed by acidic hydrolysis, affords the optically active (*R*)-α-hydroxy acids **95a** and **95b** in excellent yields, having optical purities of 89 and 98%, respectively[56,57] (Scheme 15).

95a $R_1=R_2=CH_3$ (97%)

95b $R_1=Ph$, $R_2=CH_3$ (91%)

Scheme 15

This high stereo- and regioselectivity suggests transition state **96,** in which the two alkyl groups R_1 and R_2 are located at the α and β positions in a *cis* relationship. Free rotation of the bond between the asymmetric carbon and nitrogen is prohibited by the cyclic structure.[58] Interestingly, the β,β-disubstituted case (R_2 = H, $R_1 = R_3 = CH_3$) leads exclusively to a mixture of seven-membered ring halolactones, which, unfortunately, cannot be used to prepare chiral β-hydroxy-β,β-disubstituted acids.[59]

96

The anthracycline antibiotics daunorubicin (**102**) and adriamycin (**103**) both possess a chiral α-hydroxy ketone structure at C-9. Because of their promising antineoplastic activity, various racemic syntheses have been developed. However, the efficient asymmetric halolactonization described above provides these in chiral form. The successful synthesis of the model compound (*R*)-2-hydroxy-1,2,3,4-tetrahydro-2-naphthoic acid (**100a**) and its conversion to the (*R*)-ketone **101,** having 92% optical purity, established the feasibility of this approach.[60,61] Thus the halolactonization of the potassium salt of **97b** with NBS in DMF provides a mixture of halolactones **98b** and **99b**.

A recrystallization affords pure **98b.** Subsequent debromination and hydrolysis afford (*R*)-**100b** in excellent yield and having an optical purity of 92%. Conversion of **100b** by previously developed synthetic routes completes the preparation of chiral **102** and **103**[61,62] (Scheme 16).

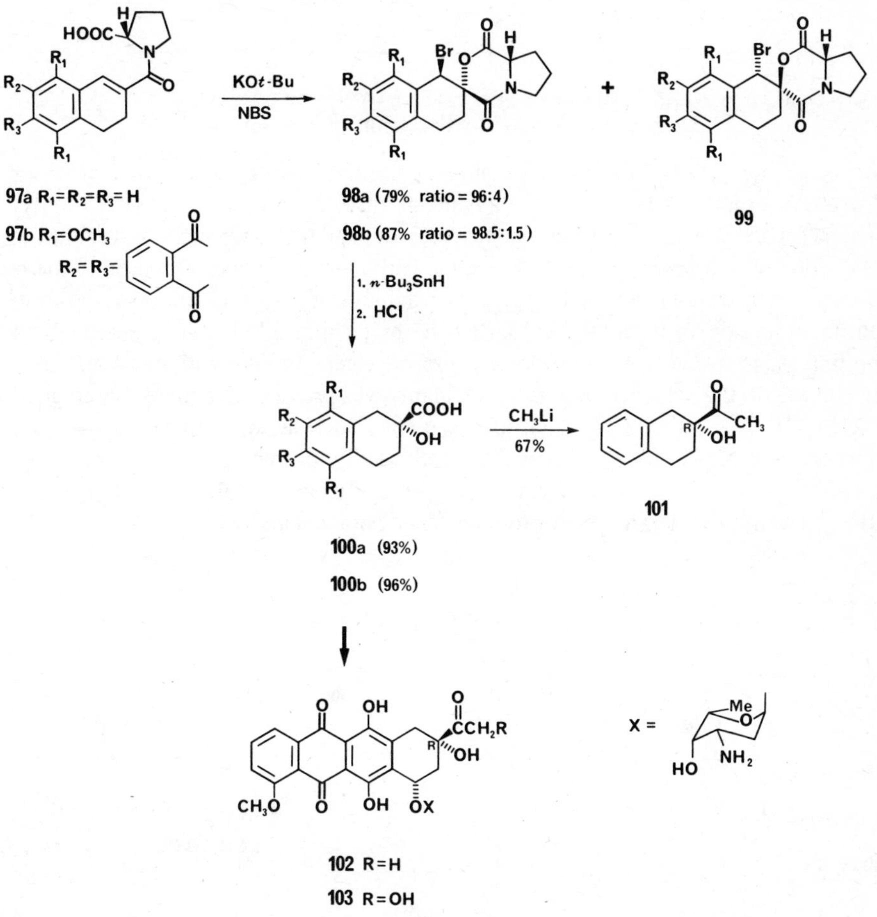

Scheme 16

Bromolactone **94a** can be utilized to prepare optically active (2*R*, 3*S*)-epoxyaldehydes (**104**) having 84–98% ee. Treatment of **94a** with sodium methoxide provides the epoxy esters that undergo reductive cleavage of the proline moiety with sodium bis(2-methoxyethoxy)aluminum hydride (Vitride) to afford **104.**[63,64]

1. $NaOCH_3$
2. Vitride

94a **104**

R_1	R_2	Yield **104** (%)	Percent ee
C_6H_5	CH_3	72	98
n-C_6H_{13}	CH_3	68	98
— $(CH_2)_4$ —		58	98

8.1.2 Asymmetric Alkylation of Enamines

Enamines are carbonyl derivatives that can be utilized as "enolate equivalents" or "masked enolates." They offer the advantage of high regio- and stereoselectivity. The alkylation of the imino anion provides one of the best methods for the α-alkylation of ketones.[65–67] Enantioselective carbon–cabon bond formation in the α position of prochiral carbonyl compounds is achievable with enamines derived from chiral amines.[68] Yamada et al. reported the first chiral alkylation of enamines of proline esters.[69] The condensation of L-proline esters **105a–c** with cyclohexanone in benzene in the presence of 4-Å molecular sieves provides enamines **106** in good yields.[69–71] Alkylation with electrophilic olefins such as acrylonitrile and methyl acrylate or strongly electrophilic halides such as allyl bromide, ethyl bromoacetate, and benzyl bromide succeeds in producing optically active 2-alkyl cyclohexanones **107.**[70] Methyl iodide and ethyl bromide show poor reactivity.

70 – 90% 30 – 50%

105a R = CH_3
105b R = C_2H_5
105c R = *t*-Bu

106

107
R = $CH_2CH_2COOCH_3$, CH_2Ph, CH_2CH_2CN

The highest optical yields (≈59%) are obtained with L-proline *tert*-butyl ester **105c.** There is also a solvent dependence: nonpolar aprotic solvents such as dioxane and benzene produce poor chemical yields but good optical yields, whereas polar protic solvents produce reverse results.[69,70] Elevated temperatures should be avoided.[70] Interestingly, chiral bromination of **106** results in the formation of 2-bromocyclohexanone **107** (R = Br), but only in 30–37% optical yield and 48% chemical yield.[72]

The enamine **109** derived from 2-phenylpropanal and the pyrrolidine derivative of L-proline (**108**) reacts with methyl vinyl ketone to provide, after acidic hydrolysis, (*R*)-(+)-4-methyl-4-phenyl-2-cyclohexenone (**110**) in 40–50% yield with an optical purity of 36.5%. Other proline derivatives provide significantly lower optical yields.[73] Extensive variations result in a maximum optical yield of 56%.[74–76]

1. MVK
2. AcOH
48%

108 **109** **110**

The successful chiral synthesis of (+)-mesembrine (**113**) illustrates the application of this method to a challenging natural product. The key intermediate, *N*-formyl-2-(3′4′-dimethoxyphenyl)-4-methylaminobutyraldehyde (**111**), reacts with **108** to form an enamine that is alkylated with methyl vinyl ketone and then hydrolyzed with aqueous acetic acid to provide **112** in 38% overall yield. Transformation of **112** to **113** is easily accomplished with ethanolic hydrochloric acid. Repeated recrystallization of this partially optically active product affords chirally pure **113.**[77,78]

1. **108**
2. MVK
3. AcOH
38%

HCl
C_2H_5OH
70%

111 **112** **113**

The acid-catalyzed biogenetic-type cyclization of optically active citralenamine **115,** prepared from the reaction of **108** with citral (**114**), results in the formation of (*R*)-α-cyclocitral (**116**) with an optical yield of 33% and a chemical yield of 8%. The treatment of **116** with 1-lithio-1-propene, followed by an oxidation with CrO_3 in pyridine, affords the famous perfume (*R*)-(+)-*trans*-α-damascone (**117**) in 27.5% optical yield.[79] Clearly, these results are not synthetically attractive. Nevertheless, they do provide useful information for the development of better chiral reagents.

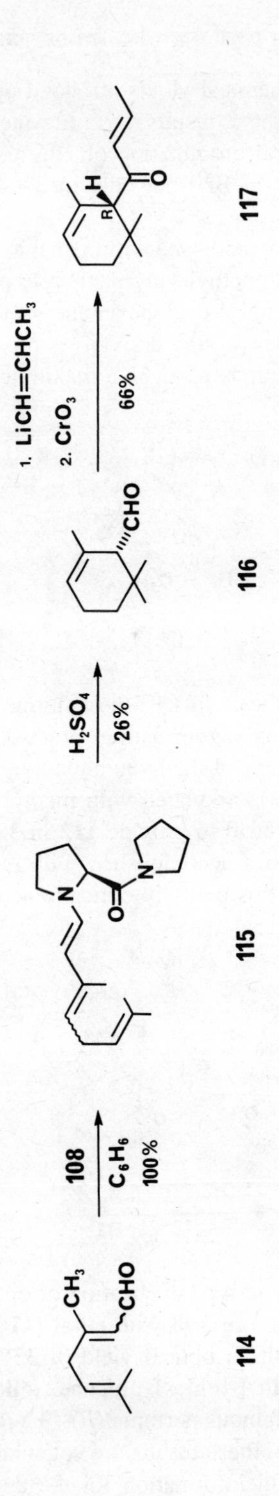
CH3
CHO
114
108
C6H6
100%
N
O
N
115
H2SO4
26%
CHO
116
1. LiCH=CHCH3
2. CrO3
66%
H
R
O
117

8.1.3 Asymmetric Aldol Reactions

The existence of a prochiral center in a molecule provides an opportunity for an optically active asymmetric reagent to differentiate between the enantiotopic groups, with the result that enantioselective syntheses are possible. Such reasoning has led to the total synthesis of optically active steroids by the exploitation of an asymmetrically biased intramolecular aldol cyclization.[80–82] In the presence of only 3 mol % of L-proline in DMF, the asymmetric cyclization of 2-methyl-2-(3-oxobutyl)-1,3-cyclopentanedione (**118**) results in a near quantitative yield and 93% ee of (+)-(3a*S*,7a*S*)-3a,4,7,7a-tetrahydro-3a-hydroxy-7a-methyl-1,5(6*H*)-indandione (**119**), which on dehydration provides almost quantitatively (with 95% ee) (+)-(7a*S*)-7,7a-dihydro-7a-methyl-1,5(6*H*)-indandione (**120**).[81] The direct heating of **118** with 40 mol % of L-proline and traces of perchloric acid in acetonitrile yields **120** directly in 86.6%, but with an enantiomeric excess of only 84%.[80]

COOH
N
H
DMF
100%
PTSA
C_6H_6
90 - 94%
Me O
OH
S

118 **119** **120**

Accessible on a 50–250-g scale[83,84], **120** is selectively reduced with sodium borohydride followed by an acid-catalyzed alkylation of the resulting alcohol with isobutylene to afford **121** in essentially quantitative yield. The very important chiral intermediate **121** is used for the total synthesis of estradiol (**122**)[85,86] and (+)-19-nortestosterone (**123**).[87] Direct carbonation of **121** with magnesium methyl carbonate, followed by a decarboxylative Mannich reaction, affords **124,** which is essential for the chiral synthesis of (+)-estr-4-ene-3, 17-dione (**125**),[83] (+)-estrone-3-methyl ether (**126**),[88] and (+)-equilenine-3-methyl ether (**127**)[88] (Scheme 17).

In his pyridine route to the total synthesis of optically active estrone and the biologically active 19-norsteroids, Danishefsky achieved disappointing results when attempting an asymmetric cyclization with L-proline. However, induction of asymmetry does occur with high enantioselectivity when phenylalanine is used instead[89,90] (see Chapter 2).

The success observed for the chiral synthesis of **120** suggests a similar asymmetric synthesis of (*S*)-8a-methyl-3,4,8,8a-tetrahydro-1,6(2*H*,7*H*)-naphthalenedione (**129**) (the Wieland–Miescher ketone), which is a versatile building block in the synthesis of steroids and terpenes. Consequently, the cyclization of 2-methyl-2-(3-oxobutyl)-1,3-cyclohexanedione (**128**) in the presence of 5 mol % L-proline and 1 *N* perchloric acid in acetonitrile affords **129** in 80% chemical yield with an optical purity of 68%.[80,91] Scale up to 1 mol provides a 56.8% chemical yield of

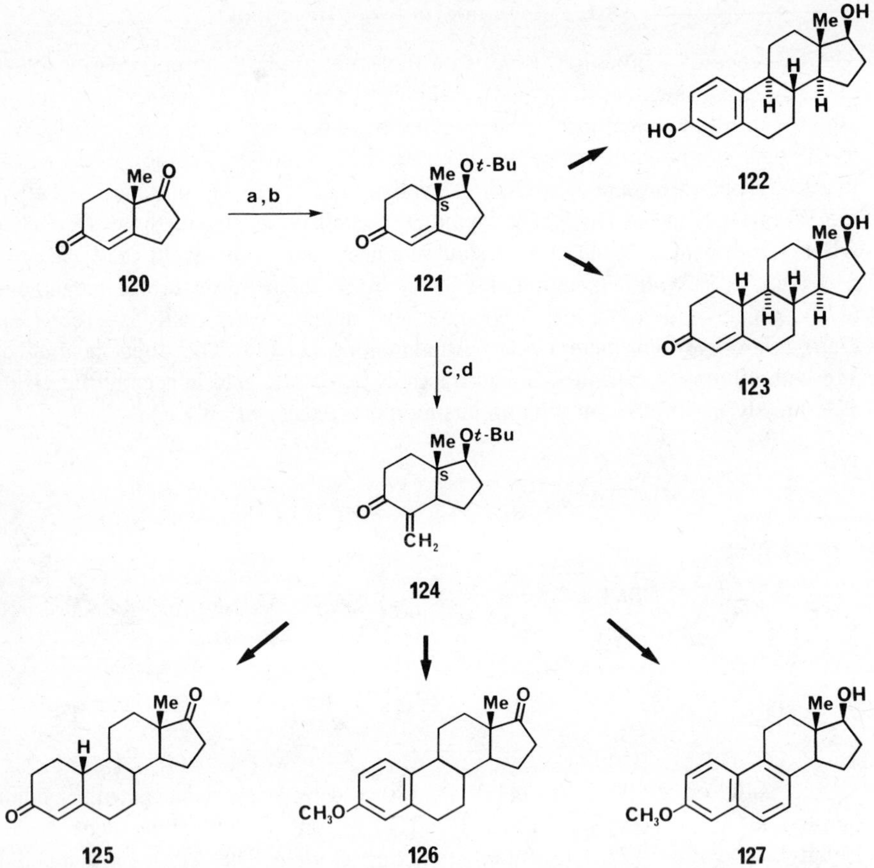

Scheme 17. (a) $NaBH_4$; (b) isobutylene, H_3PO_4, BF_3–etherate (100%); (c) MMC, DMF (64%); (d) HCHO, DMSO, piperidine (95%).

129, with almost 100% enantiomeric purity following several recrystallizations.[92,93]

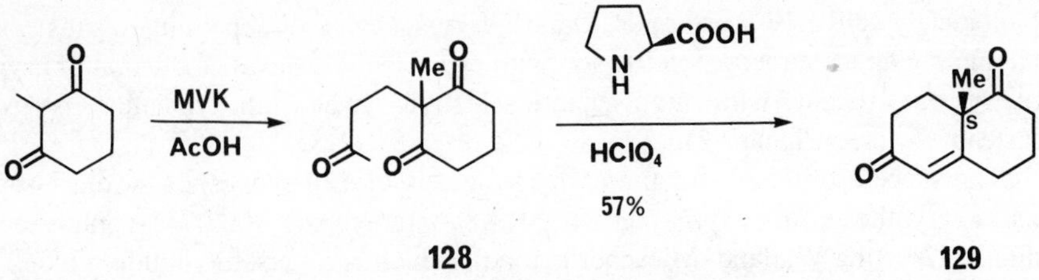

The synthesis of both (1*S*, 8a*S*)-(+)-1,8a-dihydro-1-methoxy-8a-methylazulene (**130**) and (1*S*, 8a*S*)-(+)-1,8a-dihydro-1-methoxy-6,8a-dimethylazulene (**131**) from **129** illustrates the application of readily available chiral precursor **129** to the determination of absolute stereochemistry of (+)-1,8a-dihydro-3,8-dimethylazulene, a labile biosynthetic intermediate for 1,4-dimethylazulene.[94]

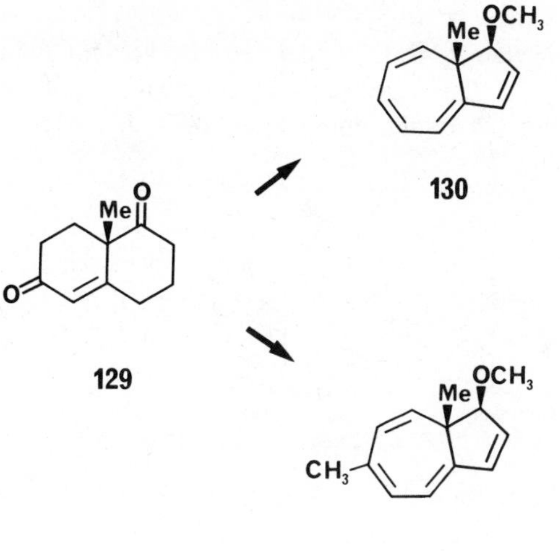

Me OCH₃

130

Me O

O

129

Me OCH₃

CH_3

131

Several useful generalizations appear from these various asymmetric aldol cyclizations:[82]

1. L-Amino acids induce the *S* configuration, whereas D-Amino acids induce the *R* configuration.
2. Amino acids are more effective than their amides or esters in producing high optical yields.
3. For substrates unsubstituted in the α position, L-proline is preferred. However, for larger groups ($R \neq H$), phenylalanine in conjunction with a mineral acid is preferred.
4. Polar aprotic solvents such as DMF and acetonitrile at room temperature allow for the isolation of the intermediate alcohols in high chemical and optical yields. Addition of mineral acids in trace amounts leads directly to enone with somewhat reduced optical yields.
5. Temperature does not seem to be a critical factor.
6. With amino esters or amides, the most favorable induction observed is with the use of toluene as a solvent.

The efficient chiral intramolecular aldolizations of these triketones suggests a synthesis of more functionalized chiral products possessing units suitable for the construction of certain tetracyclic triterpenes, such as gibberellins and kaurenes. When the symmetric triketone **132,** derived from indan-1,3-dione by successive alkylations with 2-propynylbromide and methyl vinyl ketone, is treated with 0.1 equivalent of L-proline in DMF at room temperature for 2 days, followed by dehydration of the unstable ketol with *p*-toluenesulfonic acid in benzene, enantiomerically pure enone **133** is obtained in 66% yield after a single recrystallization from methanol.

The enone **133** is converted with mercuric oxide to a separable mixture of diastereomeric gibbanes, of which **134a** is isolated in 38% yield. Acetylation to the corresponding acetate **134b** and reduction with sodium hydrotelluride provides *cis*-**135a** in good yield. On the other hand, sequential reduction with sodium borohydride and sodium in liquid ammonia followed by a Jones oxidation affords *trans*-**135b**[95] (Scheme 18).

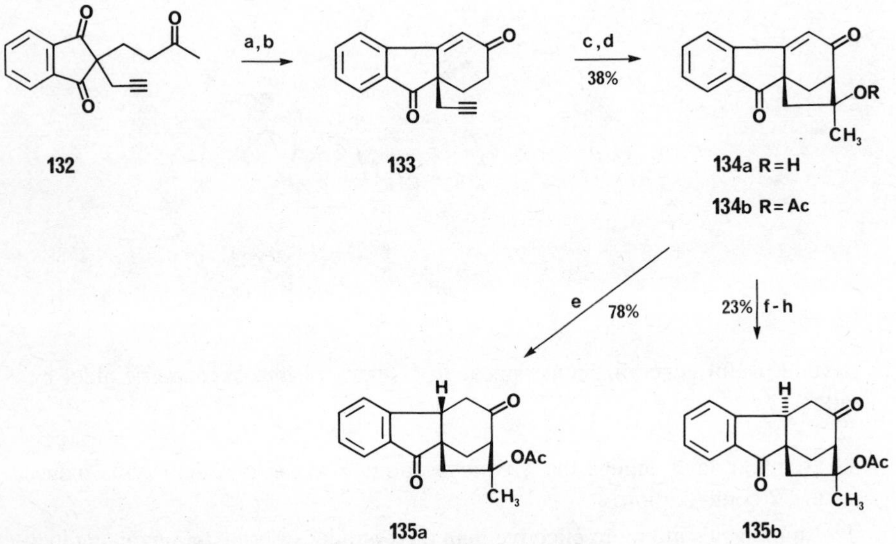

a,b
c,d 38%
OR
CH_3
132
133
134a R=H
134b R=Ac
e 78%
23% f-h
H
OAc
CH_3
135a
135b

Scheme 18. (a) L-Proline, DMF (89%); (b) PTSA (92%); (c) HgO, H_2SO_4, chrmoatography; (d) Ac_2O, BF_3–etherate; (e) NaTeH; (f) $NaBH_4$; (g) Na, NH_3; (h) Jones oxidation.

Erythronolide A (**142**), the aglycone of the macrolide antibiotic erythromycin, is a marvelously complex molecule possessing a 14-membered ring lactone with 10 asymmetric centers. Recognition that the substitution and stereochemistry of the chiral centers at C-4, C-5, and C-6 are idential with the chiral centers at C-10, C-11, and C-12 suggests the *cis*-fused chiral dithiadecaline **138** as a common intermediate. Interestingly, the asymmetric aldolization of **137** in the presence of L-proline results in a virtually racemic mixture of aldols, whereas with D-proline the aldols **138a** and **138b** are provided in good yield with high optical purity (>80%). Elaboration of **138b** to **139** provides the common intermediate for the synthesis of aldehyde **140** and ketone **141.** These are then taken through several chemical transformations to the final product **142**[96] (Scheme 19).

Whereas chiral α-amino acids such as L-proline and L-phenylalanine catalyze asymmetric aldolization to provide products having the same absolute configurations as natural steroids at C-13, an inverse asymmetric course is observed with the use of optically active β-amino acids.[97,98] (*S*)-Homoproline (**143**) is

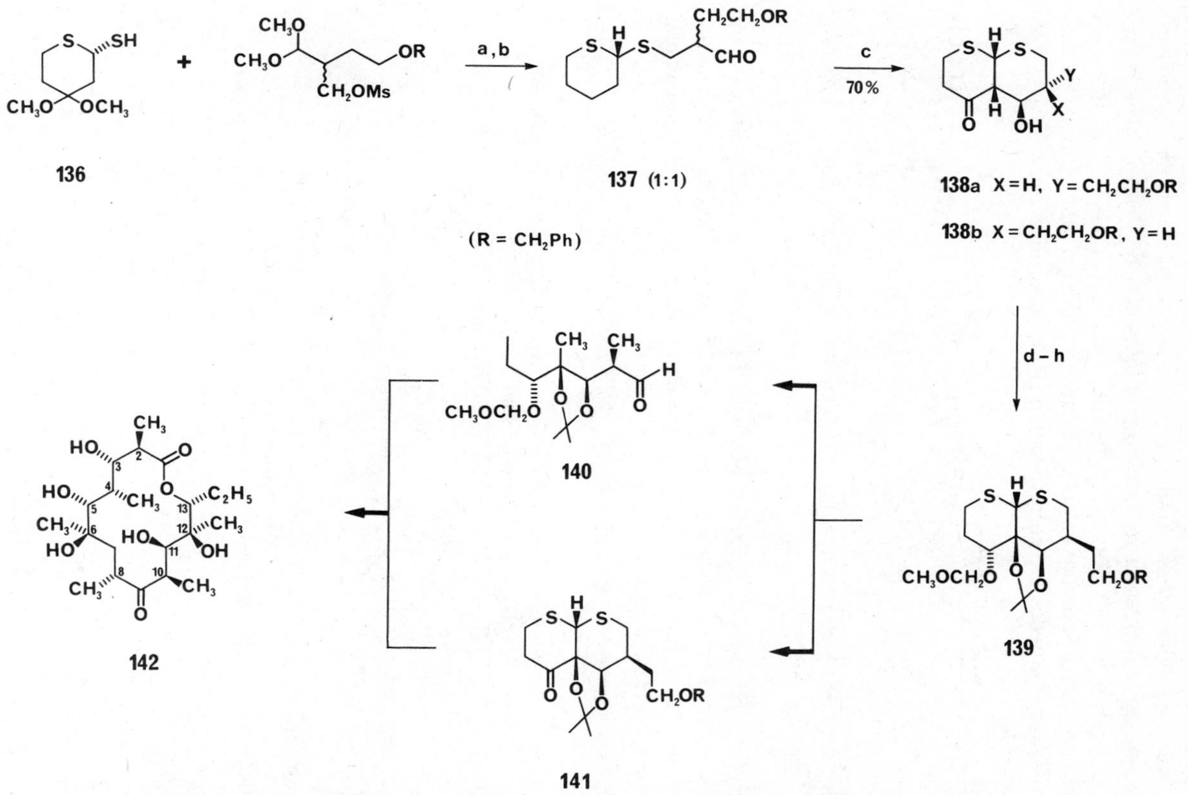

136

137 (1:1)

(R = CH_2Ph)

138a X = H, Y = CH_2CH_2OR

138b X = CH_2CH_2OR, Y = H

139

140

141

142

Scheme 19. (a) NaH, DMSO; (b) Ac_2O, H_2O; (c) (i) D-proline, PhMe, MeOH, (ii) chromatography; (d) MsCl, pyridine; (e) alumina; (f) $NaBH_4$, MeOH; (g) $MeOCH_2I$, KH (10–12% overall from **136**); (h) (i) OsO_4, (ii) $Me_2C(OMe)_2$, PTSA (74%).

prepared from L-proline by an Arndt–Eistert sequence in which the Wolff rearrangement proceeds with strict retention of configuration.[99–104] It can also be prepared in 57% yield from (*S*)-2-oxo-5-pyrrolidine–acetic acid (**144**)[102] by treatment with triethyloxonium fluoroborate followed by a reduction with sodium borohydride.[98] D-Proline provides the corresponding (*R*)-homoproline.

50 —(a–e)→ **143** ←(f,g)— **144**

(a) $(COCl)_2$, DMF; (b) CH_2N_2 (65%); (c) PhCOOAg, Et_3N (98%); (d) K_2CO_3 (85%); (e) H_2, Pd/C (90%); (f) $(EtO)_3BF_4$; (g) $NaBH_4$.

The asymmetric aldol of triketone **118** with **143** under the conditions used for the L-proline cyclizations affords the enedione **120** having the (*R*) configuration. The reaction is solvent dependent, and although chemical yields are comparable, optical yields decrease significantly, as is shown in Table 8.1.[97]

TABLE 8.1. Asymmetric Aldol Cyclizations Mediated by Chiral Proline and Homoproline[a]

118 —($HClO_4$, catalyst)→

Chiral Catalyst	Solvent	Chemical Yield (%)	Optical Yield (%)	Configuration
L-proline (COOH)	DMF	97	94	S
	CH_3CN	90	82.2	S
(S)-homoproline (CH_2COOH)	DMF	99	57.6	R
	CH_3CN	90	11.2	R
D-proline (COOH)	DMF	87	91.8	R
(R)-homoproline (CH_2COOH)	DMF	99	57.9	S

[a]Data obtained from Reference 97.

8.2 PROLINOL AND ITS DERIVATIVES

The reduction of L-proline with borane–dimethylsulfide–BF_3[103,104] or lithium aluminum hydride[105,106] affords good yields of L-prolinol (**145**) with little or no racemization.[104] Subsequent formylation and *O*-alkylation yields (*S*)-(−)-2-methoxymethyl-1-formylpyrrolidine (**146**), which readily deformylates with base to afford (*S*)-(+)-2-methoxymethylpyrrolidine (**147**).[105]

[H]; 1. HCOOEt (85%); 2. NaH, CH_3I (80%); KOH 65%

145 146 147

(*R*)-(−)-2-Methylpyrrolidine (**148**), easily prepared from **145** in moderate chemical yield with an enantiomeric excess of 95%, reacts with propionaldehyde to provide an enamine (not isolated) that undergoes a [2 + 2] cycloaddition with sulfene (generated *in situ* from methanesulfonyl chloride and triethylamine) to afford, after a controlled Hofmann elimination, (*R*)-(−)-4-methylthiete-1, 1-dioxide (**149**). The asymmetric induction observed for this sulfene–enamine cycloaddition is only 25%.[107]

145 —a, b→ 148 —c–f→ 149

(a) $SOCl_2$, HCl (80%); (b) $LiAlH_4$ (52%); (c) CH_3CH_2CHO, K_2CO_3; (d) MsCl, Et_3N (56%); (e) MeI (95%); (f) AgO, $CaSO_4$ (20%).

A similar asymmetric induction is observed in the Diels–Alder [4 + 2] cycloaddition of sulfines **150** prepared from **147.** The reaction of **150** with 2,3-dimethyl-1,3-butadiene furnishes the dihydrothiopyran *S*-oxides **151a** and **151b,** having about 40% optical purity at best.[108]

When **145** reacts with phenylphosphonic or phenylthiophosphonic dichloride, a mixture of diastereomeric bicyclic oxazaphospholes **152** and **153**

147 —a–c→ 150 —(2,3-dimethyl-1,3-butadiene), −78°C, 57%→ 151a + 151b

(a) RCH_2SO_2Cl, Et_3N (49–83%); (b) *n*-BuLi, TMSCl (83–89%); (c) *n*-BuLi, SO_2.

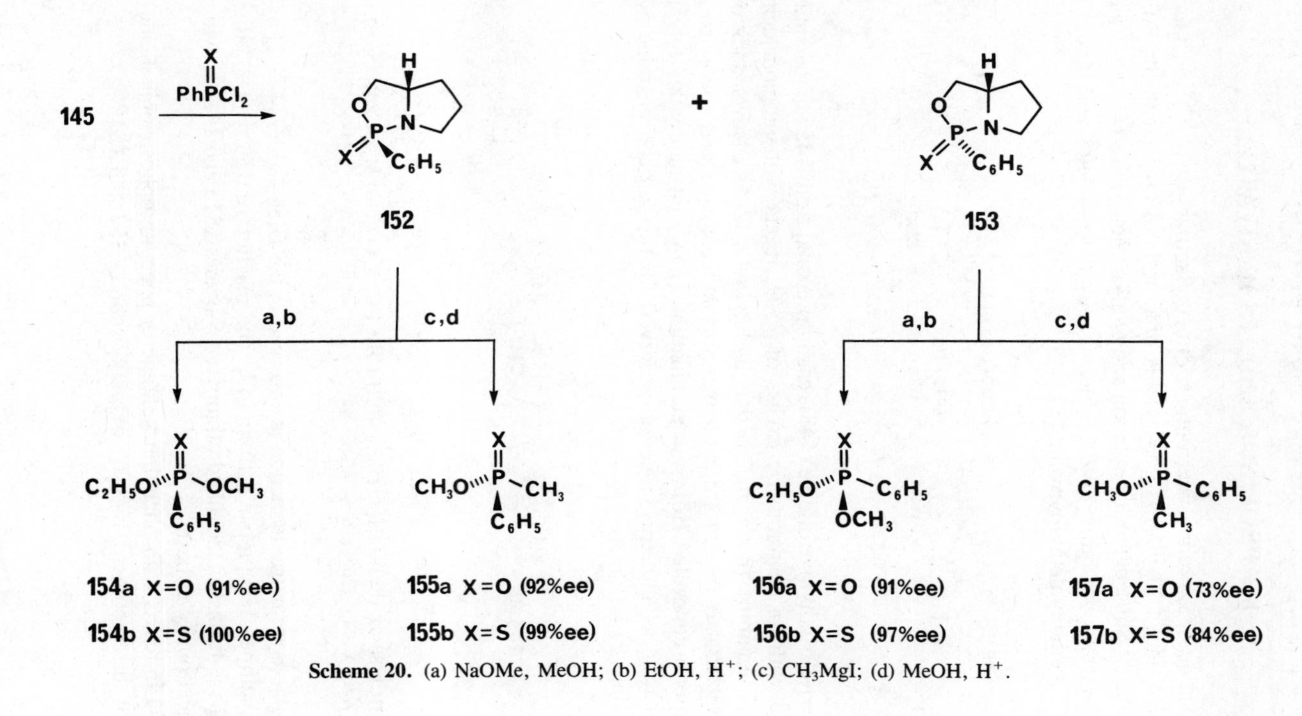

Scheme 20. (a) NaOMe, MeOH; (b) EtOH, H^+; (c) CH_3MgI; (d) MeOH, H^+.

is obtained in 80–90% yield. Easily separated by flash chromatography, these pure diastereomers undergo base-catalyzed methanolysis exclusively with P–O bond cleavage and complete inversion of configuration to form the corresponding phosphonates **154** or thiophosphonates **156.**[109]

The bicyclic oxazaphospholes also react well with Grignard reagents such as methylmagnesium iodide to provide chiral phosphinic or thiophosphinic esters **155** and **157,** respectively, with excellent optical purity.[110] This, too, occurs with inversion of configuration (Scheme 20).

Asymmetric induction in the Michael-type addition of organocuprates to prochiral α,β-unsaturated carbonyl compounds would be a valuable addition to methods available for the formation of carbon–carbon bonds. The ability of (*S*)-1-methyl-2-hydroxymethylpyrrolidine (**158**) to induce such a highly enantioselective addition is demonstrated by the synthesis of 1,3-diphenyl-1-butanone **159.** This is obtained in 88% chemical yield, having an optical purity of 61%. Among the various attempts, this represents the best result.[111]

CH_2OH; N; CH_3 — 1. CH_3MgBr, CuBr; 2. PhCH=CHCOPh; 88% → CH_3, O, S, C_6H_5, C_6H_5

158 **159**

Diastereoselective conjugate addition of Grignard reagents to the α,β-unsaturated amides **160a–d** derived from L-prolinol (**145**) and its derivatives affords 3-substituted carboxylic acids **161** having good enantiomeric excesses (50–89%) when carried out in the presence of such tertiary amines as 1,8-diazabicyclo[5.4.0]undec-7-ene (DBU), 1,5-diazabicyclo[4.3.0]non5-ene (DBN), *N,N,N',N'*-tetramethylethylenediamine (TMEDA), or (−)-sparteine. In the absence of these tertiary amines, enantiomeric excesses are low to moderate. There is also a drastic decrease in the asymmetric induction when the amide derived from **147** is used. This is attributed to a loss of chelation of the tertiary amine to magnesium of the magnesium alkoxide.[112]

A useful synthetic application of **145** and **147** is to the asymmetric alkylation of amide anions to generate chiral α-substituted carboxylic acids. Treatment of **145** or **147** with an acid chloride in the presence of triethylamine provides the corresponding amides **162** and **163.** Alternatively, **163** is easily derived from **162** by treatment with *n*-butyllithium followed by *O*-alkylation with methyl iodide. Deprotonation with either *t*-butyllithium[113] at room temperature or LDA[113,114] at room temperature generates stereoselectively (>97%) chiral amide enolates pos-

1. R_4MgX
3° Amine
2. HCl

160 → 161

a. $R_1 = Ph$, $R_2 = CH_3$, $R_3 = H$

b. $R_1 = Ph$, $R_2 = H$, $R_3 = H$

c. $R_1 = Ph$, $R_2 = H$, $R_3 = CH_3$

d. $R_1 = CH_3$, $R_2 = CH_3$, $R_3 = H$

Amide	R_4	3°-Amine (solvent)	Yield **161** (%)	Percent ee
160a	n-C_4H_9	— (THF)	33	16
160a	n-C_4H_9	— (toluene)	52	37
160a	n-C_4H_9	DBU (THF)	39	89
160a	n-C_4H_9	DBU (toluene)	81	88
160a	C_2H_5	DBU (toluene)	66	82
160b	n-C_4H_9	DBU (toluene)	29	84
160c	n-C_4H_9	DBU (toluene)	23	4
160d	C_2H_5	DBU (toluene)	27	69
160d	n-C_6H_{13}	DBU (toluene)	23	68
160d	n-C_6H_{13}	TMEDA (toluene)	22	66

sessing (*Z*)-enolate geometry.[114] Strikingly, the alkylation of the anion of **163** leads to products having the opposite configuration from that obtained from the alkylation of the dianion of **162.**[113,114] Hydrolysis of the amides by a two-phase system (concentrated HCl–hexane–reflux) produces the chiral acids **166** or **167** without affecting the configuration. The enantiomeric excesses are moderate, ranging from 12 to 85% (Scheme 21).

An improvement of this procedure relies on increasing the steric bulk adjacent to the asymmetric center on proline. Conversion of proline methyl ester (**105a**) to the amide with an appropriate acid chloride, followed by treatment with methylmagnesium iodide, furnishes the tertiary alcohol **168.** Deprotonation with LDA and alkylation at −100°C affords a diastereomeric mixture (ratios between 88:12 and 95:5) that is easily separated by column chromatography to yield pure **169.** Subsequent hydrolysis of **169** provides the chiral carboxylic acids **170** in 55–82% chemical yields having very high ee percentages.[115]

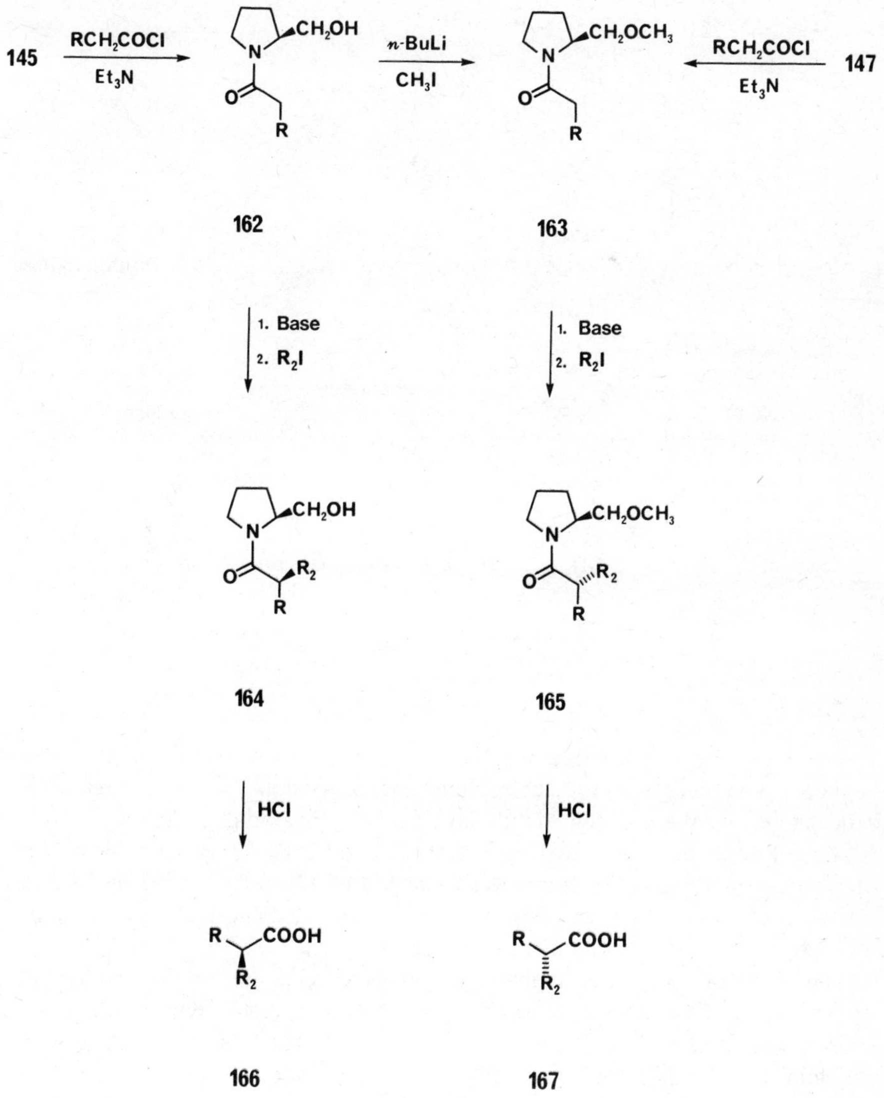

145 $\xrightarrow[Et_3N]{RCH_2COCl}$ 162 $\xrightarrow[CH_3I]{n\text{-BuLi}}$ 163 $\xleftarrow[Et_3N]{RCH_2COCl}$ 147

162 $\xrightarrow{1.\ Base;\ 2.\ R_2I}$ 164 $\xrightarrow{HCl}$ 166

163 $\xrightarrow{1.\ Base;\ 2.\ R_2I}$ 165 $\xrightarrow{HCl}$ 167

$R = CH_3, C_2H_5, n\text{-}C_8H_{17}$

$R_2 = CH_3, C_2H_5, n\text{-Bu}, i\text{-Bu}, PhCH_2$

Scheme 21

105 → (a,b) → 168 → (c) → 169 → (d) → 170

(a) R_1CH_2COCl, Et_3N (82–95%); (b) CH_3MgI (82–86%); (c) LDA, R_2x (72–96%); (d) HCl, dioxane (74–86%).

R_1	R_2	Percent ee (**170**)	Configuration
CH_3	C_2H_5	96	(*R*)
CH_3	*n*-C_4H_9	87	(*R*)
n-C_4H_9	CH_3	>99	(*S*)
CH_3	*n*-C_8H_{17}	90	(*R*)
n-C_8H_{17}	CH_3	78	(*S*)
CH_3	$C_6H_5CH_2$	87	(*R*)
$C_6H_5CH_2$	CH_3	>99	(*S*)

Enantiomerically pure (*R*)- and (*S*)-α-hydroxyketones such as **173** and **174** can be prepared by an asymmetric nucleophilic carbamoylation.[116] The reaction of **146** with ketones by the addition of lithium tetramethylpiperidide yields a separable mixture of diastereomeric α-hydroxyamides **171** and **172.** Subsequent treatment of each with methyllithium results in the formation of **173** (or **174**) and the *vicinal*-diol (**175** or **176**), which can be separated chromatographically. The crystalline diols after two recrystallizations provides ee values of >90% (Scheme 22).

Equally as important as chiral α-alkyl alkanoic acids are the β-substituted aldehydes **179.** Allylamines **177,** easily prepared in 80% yield from the reaction of cinnamyl bromide with **147,** undergo deprotonation with *t*-butyllithium to afford the chiral homoenolate equivalent **178.** This undergoes alkylation with high diastereoselectivity, furnishing, after acidic hydrolysis, the chiral β-substituted aldehydes **179** with enantiomeric excesses of up to 66%. It should be noted that the enantiomeric excesses vary with the gegenion, solvent, the temperature and that the tightest ion pair is expected to provide the highest asymmetric induction. The enantiomeric excess is also dependent on the size of the alkylating reagent.[117]

When **147** is incorporated as the amino group of α-aminonitriles **180,** high diastereoselectivity resulting from kinetic control is observed in the reaction with Grignard reagents. The ratio of the two amines produced may be reversed simply by an appropriate choice of the order of introduction of the R_1 and R_2 groups.[118]

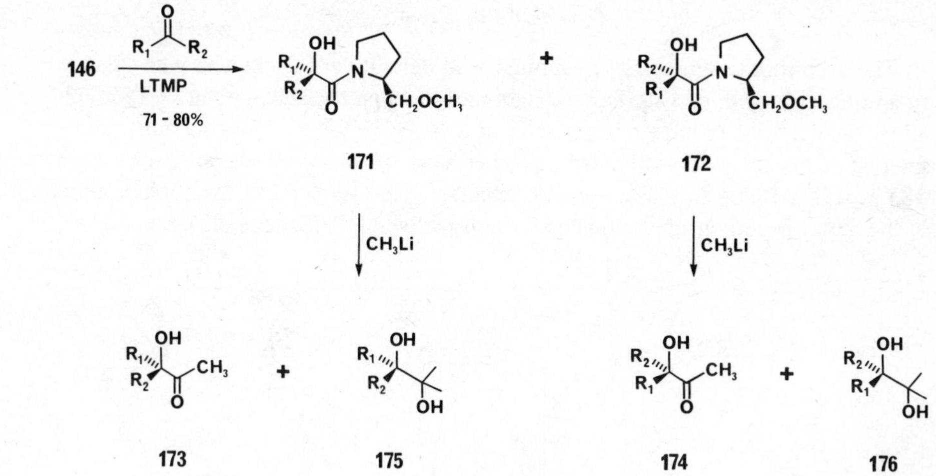

146 $\xrightarrow[\text{71 - 80\%}]{R_1COR_2,\ \text{LTMP}}$ 171 + 172

171: R_1, R_2, OH, O, N, CH_2OCH_3

172: R_2, R_1, OH, O, N, CH_2OCH_3

171 $\xrightarrow{CH_3Li}$ 173 + 175

172 $\xrightarrow{CH_3Li}$ 174 + 176

173: R_1, R_2, OH, O, CH_3

175: R_1, R_2, OH, OH

174: R_2, R_1, OH, O, CH_3

176: R_2, R_1, OH, OH

$R_1 = C_6H_5$, *t*-Bu

$R_2 = CH_3$, *i*-Pr

Scheme 22

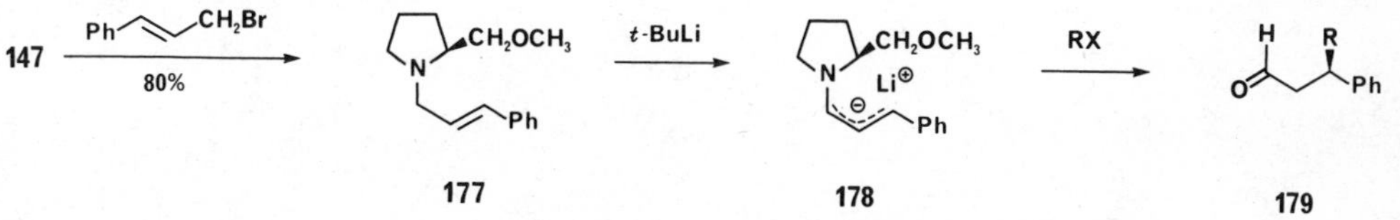

---

147 $\xrightarrow[\text{80\%}]{Ph\text{-}CH{=}CH\text{-}CH_2Br}$ 177 $\xrightarrow{t\text{-BuLi}}$ 178 $\xrightarrow{RX}$ 179

177: N, CH_2OCH_3, Ph

178: N, CH_2OCH_3, $Li^{\oplus}$, $\ominus$, Ph

179: H, O, R, Ph

R	Yield **179** (%)	Percent ee	Configuration
CH_3	50	65	(*R*)
C_2H_5	34	65	(*R*)
i-C_3H_7	73	66	(*S*)
n-C_4H_9	89	48	(*R*)
$CH_2{=}CHCH_2$	49	51	(*R*)

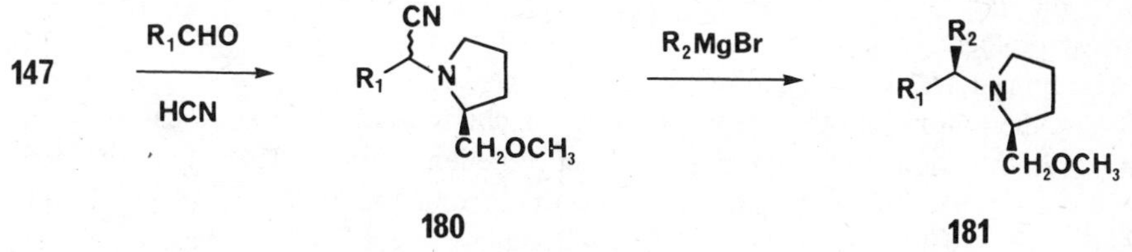

147 $\xrightarrow[\text{HCN}]{R_1CHO}$ 180 $\xrightarrow{R_2MgBr}$ 181

180: CN, R_1, N, CH_2OCH_3

181: R_2, R_1, N, CH_2OCH_3

R_1	R_2	Yield (%)	(*R, S*)	(*S, S*)
CH_3	C_6H_5	88	10	90
C_6H_5	CH_3	89	90	10

The asymmetric synthesis of α-amino acid derivatives is achieved with the use of amide derivatives of **147.** Deprotonation of the hippuric acid derivative **182** with LDA followed by an alkylation results in a 60–96% yield of **184** having a diastereomeric excess of 21–46%. On the other hand, alkylation of the silyl derivative **183** provides **184** with diastereomeric excesses of 51–60%. The (*S*) configuration at the chiral α-carbon predominates in both pathways[119] (Scheme 23).

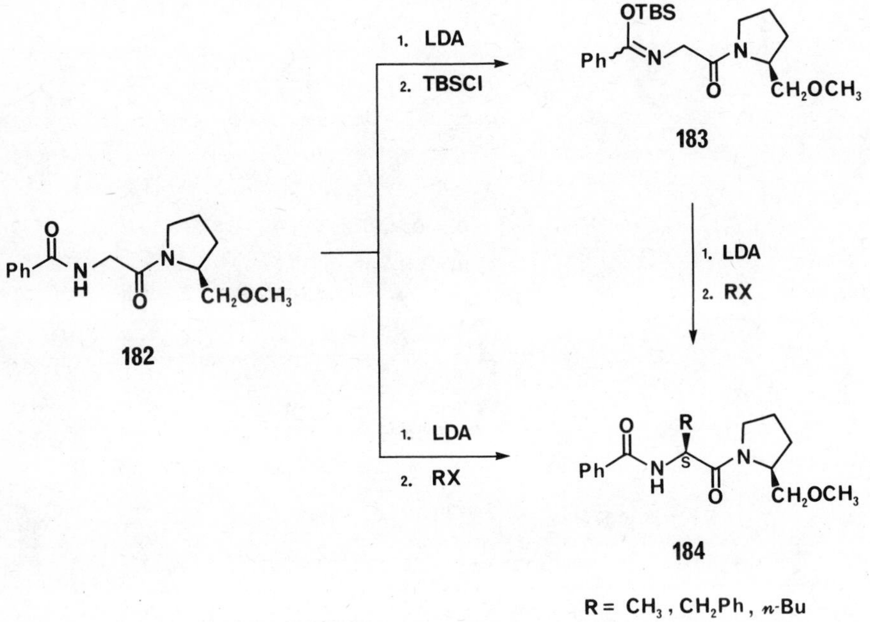

Ph-C(=O)-NH-CH2-C(=O)-N(pyrrolidine)-CH2OCH3 → 1. LDA 2. TBSCl → **183** (OTBS, Ph, N, CH2OCH3)

182 → 1. LDA 2. RX → **184**; **183** → 1. LDA 2. RX → **184**

R = CH_3, CH_2Ph, *n*-Bu

Scheme 23

(+)-Cannabispirenone A (**187**), recently investigated for its possible role as an estrogen-potentiating substance and/or analgesic, prossesses a chiral quarternary spiro carbon as a key feature. The Michael reaction of enamine **186** (prepared from **147** and aldehyde **185**) with methyl vinyl ketone, followed by acidic hydrolysis and deprotection, provides **187,** with an optical purity of 50%. The absolute configuration of **187** is assigned as (*R*)[120] (Scheme 24).

Another natural product possessing such a chiral quarternary spiro carbon is the novel nitrogenous monoterpenoid insect repellent produced by the millipede *Polyzonium rosalbum.*[121] Known as (+)-polyzonimine (**192**), it is (+)-6,6-dimethyl-2-azaspiro[4.4]non-1-ene. Its total synthesis utilizes the asymmetric [2,3] sigmatropic rearrangement of the ammonium ylide **190,** which is prepared from the reaction of **188** with L-benzyloxyprolinol **189,** followed by quarternization with cyanomethyl benzenesulfonate in acetonitrile. The chiral aldehyde **191** is then converted through a series of standard transformations to **192** in 60% optical purity[122] (Scheme 25).

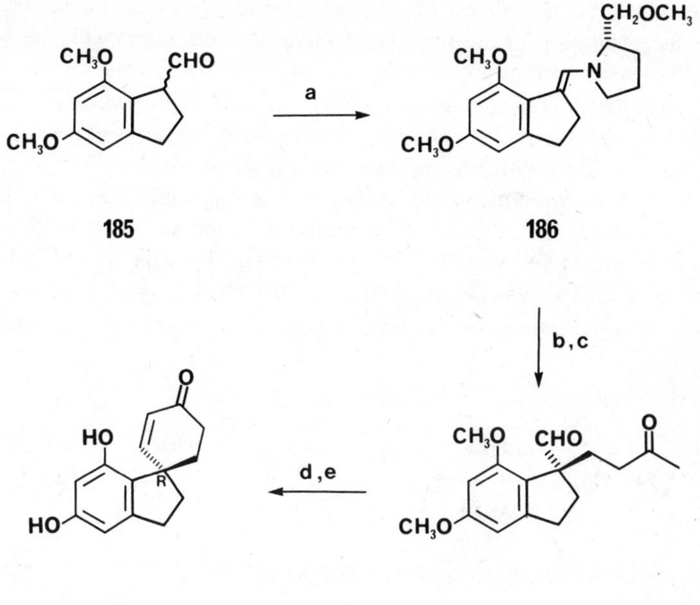

Scheme 24. (a) **147,** toluene; (b) MVK; (c) NaOAc, EtOH; (d) reflux (e) *t*-BuSLi, HMPA.

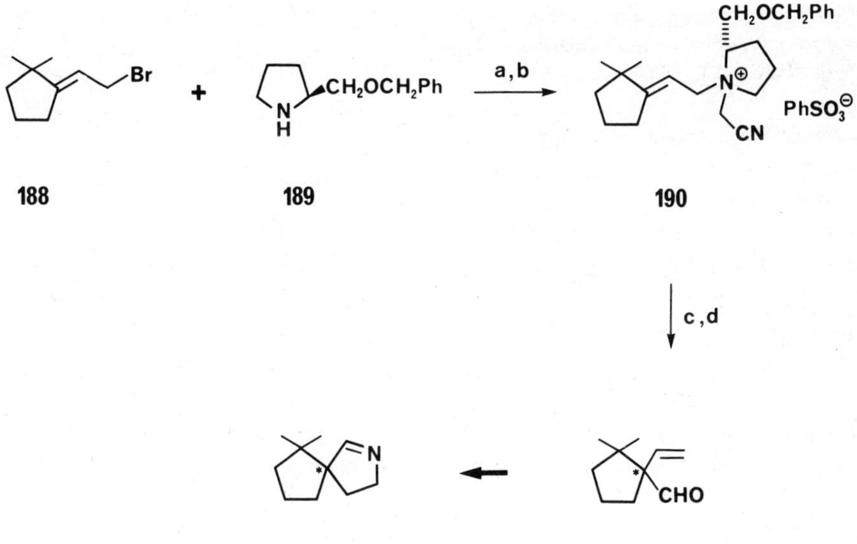

Scheme 25. (a) K_2CO_3, DMSO; (b) $PhSO_3CH_2CN$, CH_3CN; (c) *t*-BuOK, THF, DMSO, −78°C; (d) $CuSO_4$, EtOh, H_2O (61% overall yield).

A novel method developed for the preparation of optically active aldehydes involves the resolution of the racemic aldehyde by reacting it with the secondary amine **145** to generate diastereomeric oxazolidine derivatives that can be separated. In this way, (*R*)-(+)-α-cyclocitral (**193a**) and (*S*)-(+)-6-hydroxy-α-cyclocitral (**193b**) are prepared with 19 and 63% optical purity, respectively. Base-catalyzed condensation of **193b** with acetone gives a 23% yield of (*S*)-(+)-6-hydroxy-α-ionone (**194**), which is oxidized with *t*-butyl chromate (prepared by the successive addition of chromium trioxide and acetic anhydride to *t*-butanol) to afford (*S*)-(+)-dehydrovomifoliol (**195**). The plant growth regulator (*S*)-(+)-abscisic acid (**196**) can be prepared from **195.** Optical yields are approximately 40%, and chemical yields for this sequence are generally poor[123] (Scheme 26).

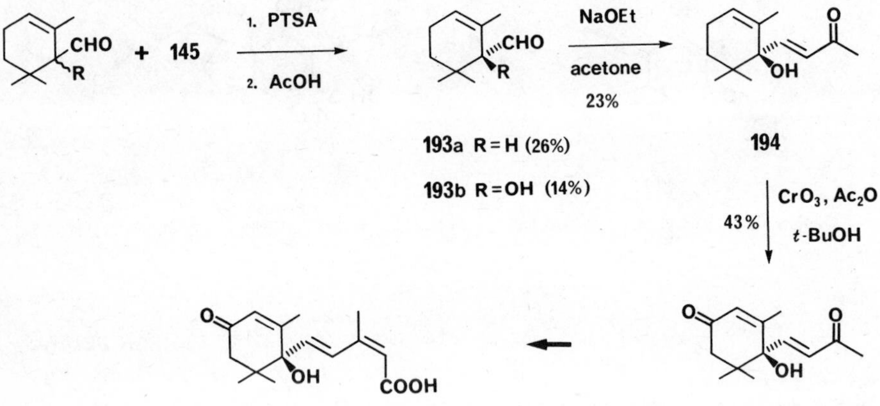

CHO R + 145 — 1. PTSA, 2. AcOH → CHO R — NaOEt, acetone, 23% → OH O

193a R = H (26%)

193b R = OH (14%)

194

CrO_3, Ac_2O, *t*-BuOH, 43%

196 OH COOH ← 195

Scheme 26

8.3 CHIRAL AUXILIARIES DERIVED FROM PROLINE

The ability to form carbon–carbon bonds α to the carbonyl group of aldehydes and ketones is one of the most important synthetic operations in organic chemistry. Reactive enolate equivalents such as metalated imines, oximes, and hydrazones (azaenolates) successfully allow regiochemical control and prevent unwanted di- or polyalkylation as well as aldol-type self-condensations.[68] Moreover, with the advent of chiral auxiliaries that often offer excellent enantioselectivity, asymmetric carbon–carbon bond formation is now becoming the preferred method for the synthesis of chiral natural products.

Enders and co-workers[124–126] have developed the chiral auxiliaries (*S*)- and (*R*)-1-amino-2-methoxymethylpyrrolidine (**197** and **199**, respectively), known more familiarly as SAMP and RAMP, respectively. Whereas SAMP is easily prepared from (*S*)-prolinol (**145**) by two separate routes in about 50% overall yield (Scheme 27),[127, 128] RAMP, although similarly available from (*R*)-prolinol, is more practically and economically prepared in six steps from D-glutamic acid (**198**) in an overall yield of 35% (Scheme 28).[129, 130]

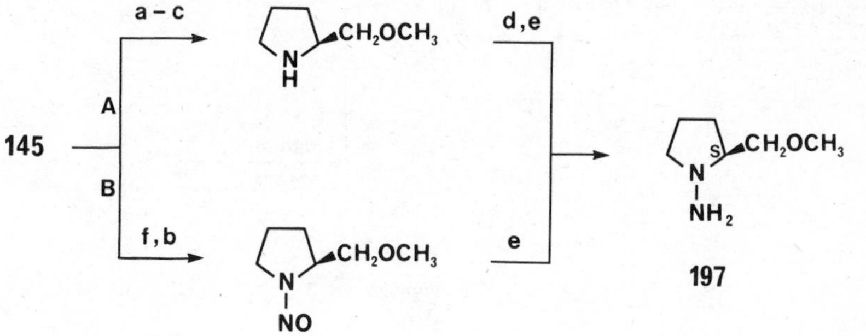

Scheme 27. (a) $HCOOCH_3$ (100%); (b) NaH, CH_3I (100%); (c) 10% KOH, 130°C (88%); (d) HONO (86%); (e) $LiAlH_4$ (85%); (f) EtONO (92%).

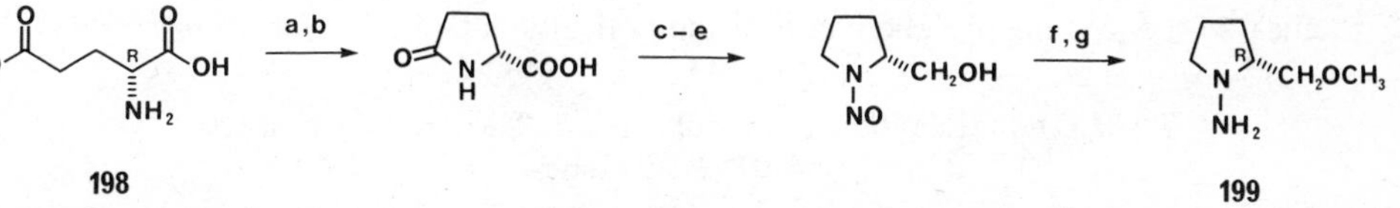

Scheme 28. (a) H_2O, reflux; (b) ion-exchange column; (c) CH_2N_2; (d) $LiAlH_4$; (e) RONO; (f) NaH, Ch_3I; (g) $LiAlH_4$.

Both are colorless oils that can be stored safely for months in a refrigerator. Chiral SAMP– and RAMP–hydrazones are prepared by mixing a pure aldehyde (at 0°C) or ketone (at 25°C) with either SAMP or RAMP. Sterically hindered ketones, aromatic or α,β-unsaturated ketones require azeotropic removal of water. The hydrazones are isolated by either distillation or chromatography as stable colorless to pale yellow oils.[124, 127, 131]

Optically active α-substituted aldehydes **201** are available from primary aldehydes *via* alkylation of SAMP–hydrazones **200.**[124, 128, 131] In order to obtain the highest levels of asymmetric induction (>90% ee), ether is preferred as the solvent over THF.[124] Moreover, reductive amination of these hydrazones **200** provides an enantioselective route to β-chiral primary amines **202** in 94–99% ee[132] (Scheme 29).

By simply changing the Cahn–Ingold–Prelog (CIP) priorities of R_1 and R_2, it is possible to prepare both enantiomers of a single aldehyde using only SAMP. Known as "opposite enantioselectivity through synthon control,"[128] this procedure provides a useful trick in asymmetric syntheses. Tables 8.2 and 8.3 illustrate this nicely.

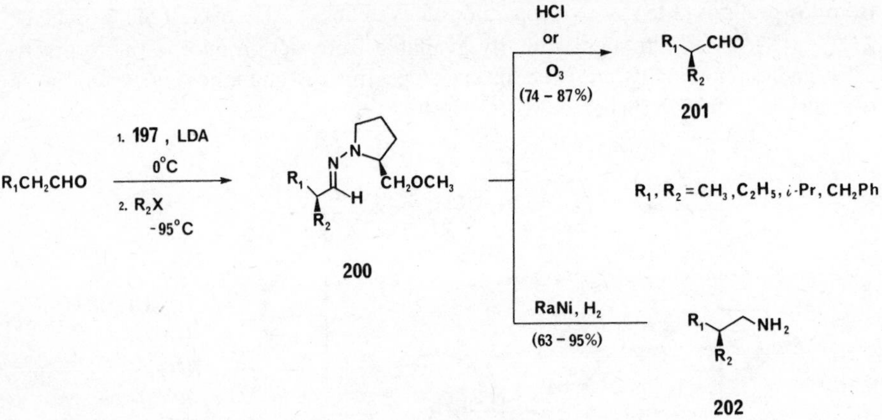

HCl or O_3 (74 – 87%)

R_1CH_2CHO → 1. **197**, LDA, 0°C; 2. R_2X, −95°C → **200** → **201** ($R_1CH(R_2)CHO$)

$R_1, R_2 = CH_3, C_2H_5, i\text{-Pr}, CH_2Ph$

RaNi, H_2 (63 – 95%) → **202** ($R_1CH(R_2)CH_2NH_2$)

Scheme 29

Racemization-free transformations of the chiral α-substituted aldehydes listed in Table 8.2 provide a variety of important chiral substrates[124] (Scheme 30).

The total synthesis of ionophore antibiotic X-14547A (**208**) features as its key step the efficient coupling of chiral tetrahydropyran **207** (prepared from diethyl D-tartrate) with chiral tetrahydroindan **206.** The enantioselective synthesis of **206** relies on the asymmetric alkylation of the optically active SAMP hydrazone **203.**

TABLE 8.2. Enantiomeric α-Substituted Aldehydes Derived from SAMP–Hydrazones[a]

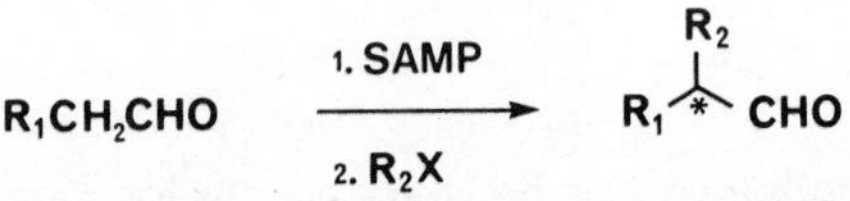

R_1CH_2CHO → 1. SAMP; 2. R_2X → $R_1C^*H(R_2)CHO$

R_1	R_2	Yield (%)	Percent ee	Configuration
CH_3	C_2H_5	71	77	(*S*)
C_2H_5	CH_3	65	62	(*R*)
CH_3	$C_6H_5CH_2$	61	82–>95	(*S*)
$C_6H_5CH_2$	CH_3	63	>90	(*R*)
CH_3	n-C_6H_{13}	52	>95	(*S*)
n-C_6H_{13}	CH_3	61	95	(*R*)
C_2H_5	t-$C_4H_9(CH_3)_2Si(CH_2)_3$	71	82–95	(*S*)
i-C_3H_7	CH_3	60	57	(*R*)
$CH_2{=}CH(CH_2)_7$	CH_3	51	95	(*R*)
C_6H_5	CH_3	80	31	(*R*)

[a]Data obtained from references 124, 128, and 131.

TABLE 8.3. Enantiomeric Primary Amines Derived from SAMP–Hydrazones[a]

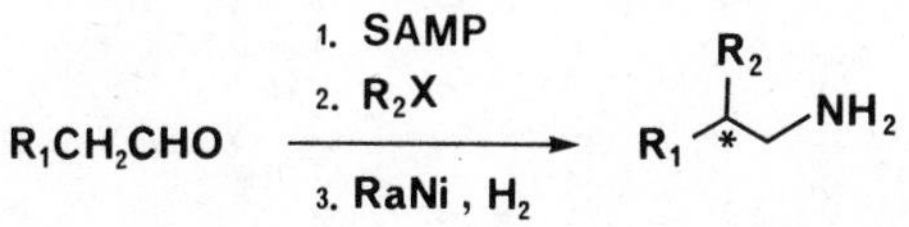

R_1CH_2CHO → 1. SAMP; 2. R_2X; 3. RaNi, H_2 → R_1–*CH(R_2)–CH_2NH_2

R_1	R_2	Yield (%)	Percent ee	Configuration
CH_3	n-C_6H_{13}	63	>95	(*S*)
n-C_6H_{13}	CH_3	56	>90	(*R*)
CH_3	$C_6H_5CH_2$	52	>90	(*S*)
$C_6H_5CH_2$	CH_3	41	>95	(*R*)
C_5H_5	CH_3	61	95	(*R*)

[a]Data obtained from reference 132.

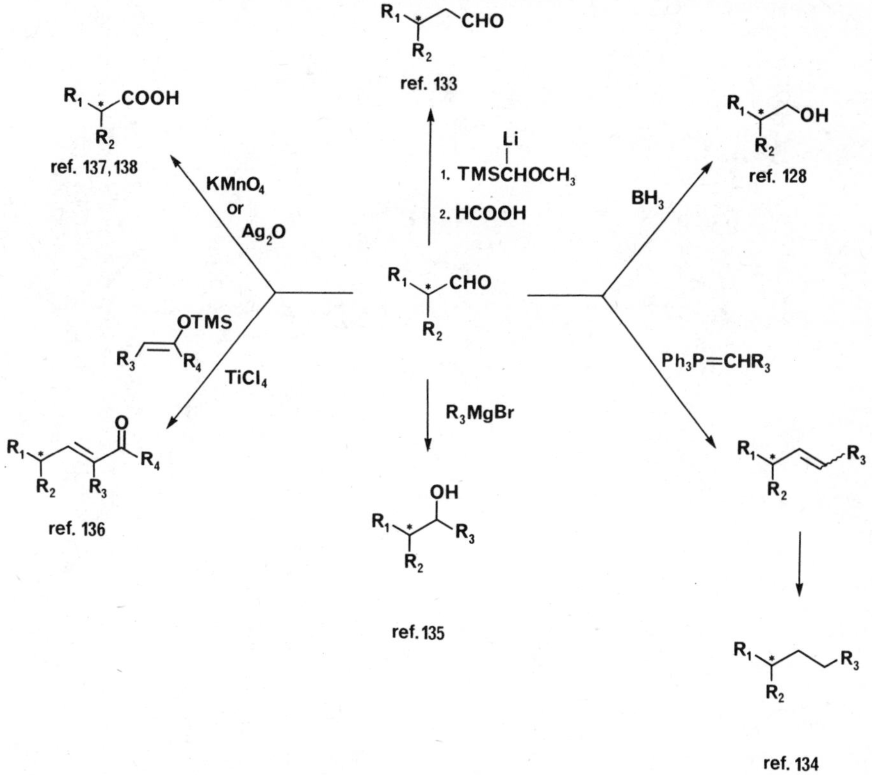

R₁CH(R₂)CHO

1. TMSCH(Li)OCH₃; 2. HCOOH → R₁CH(R₂)CH₂CHO, ref. 133

KMnO₄ or Ag₂O → R₁CH(R₂)COOH, ref. 137, 138

BH₃ → R₁CH(R₂)CH₂OH, ref. 128

OTMS enol ether (R₃, R₄), TiCl₄ → ref. 136

R₃MgBr → R₁CH(R₂)CH(OH)R₃, ref. 135

Ph₃P=CHR₃ → R₁CH(R₂)CH=CHR₃ → R₁CH(R₂)CH₂CH₂R₃, ref. 134

Scheme 30. (Reprinted with permission from reference 124. Copyright 1984 Academic Press.)

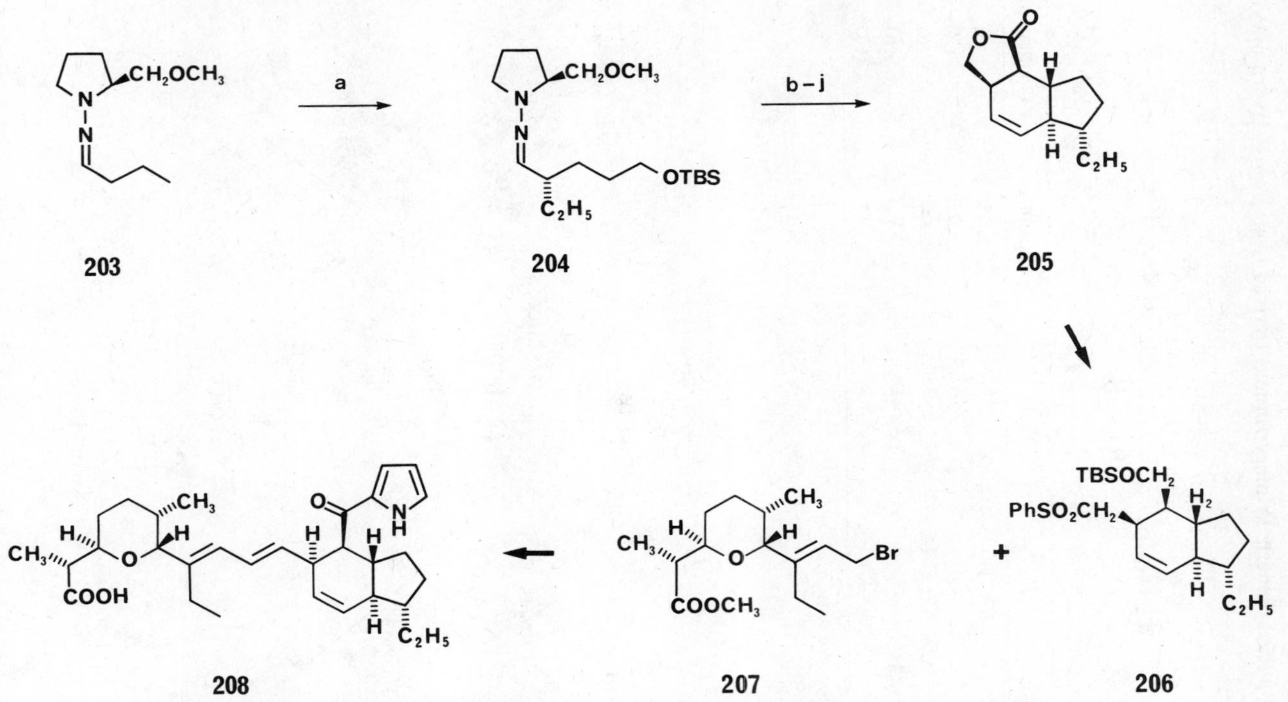

203 → (a) → 204 → (b–j) → 205 → 206 + 207 → 208

Scheme 31. (a) LDA, $I(CH_2)_3OTBS$, Et_2O (85%); (b) O_3, $-78°C$, CH_2Cl_2 (100%); (c) $(MeO)_2P(O)CH_2CH{=}CHCOOMe$, LDA (95%); (d) DIBAH, CH_2Cl_2 (99%); (e) t-$BuPh_2SiCl$, imidazole, DMF (95%); (f) AcOH, THF, H_2O (74%); (g) CrO_3, pyridine, CH_2Cl_2 (95%); (h) $Ph_3P{=}CHCOOMe$ (90%); (i) 130°C, toluene (70%); (j) n-Bu_4NF (100%).

This proceeds in 85% chemical yield with a 95% diastereomeric purity. An intramolecular Diels–Alder reaction establishes the remaining stereochemical centers of the tetrahydroindan[139–141] (Scheme 31).

Acyclic ketones, easily converted to their SAMP hydrazones, undergo deprotonation (LDA at 0°C) and alkylation with excellent selectivity. Good chemical yields with overall enantiomeric excesses of 94–95% often result when diethyl ether is used as the solvent.[124, 127, 128, 142] In the case of an unsymmetrical ketone, a small amount of the regioisomer may result.[142]

The synthetic utility of this method is illustrated by the synthesis of various natural products derived from the SAMP–hydrazone of diethylketone **209.** Almost

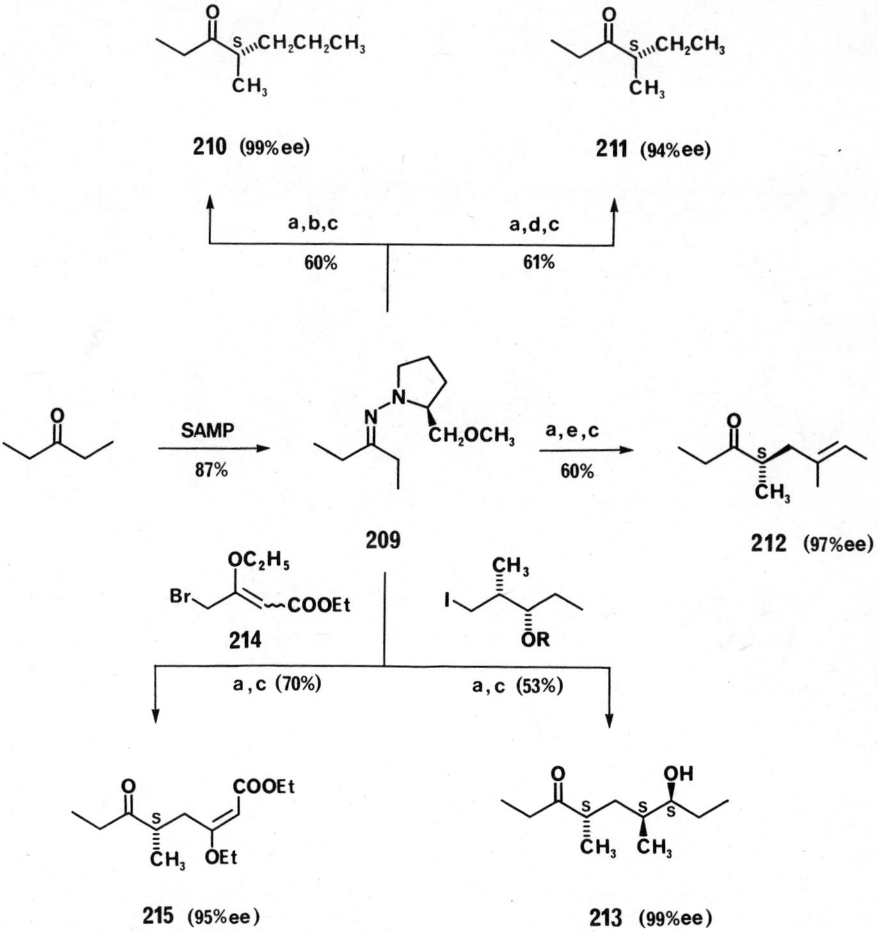

Scheme 32. (a) LDA, Et_2O, 0°C; (b) n-C_3H_7I, −110°C (85%); (c) CH_3I, 3 N HCl, pentane (82%); (d) C_2H_5I, −110°C; (e) (E)-l-bromo-2-methyl-2-butene, −110°C (85%).

TABLE 8.4. α-Chiral Cycloalkanones and Cycloalkenones of *R* Configuration[a]

Cyclic Ketone	Yield (%)	ee (%)
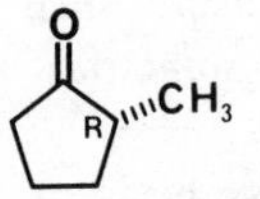 O, R, CH_3	66	86
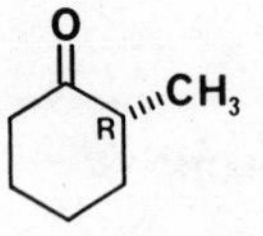 O, R, CH_3	71	98
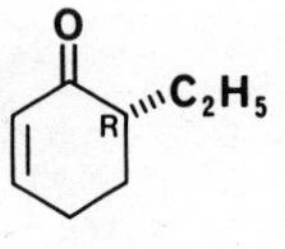 O, R, C_2H_5	57	61–75
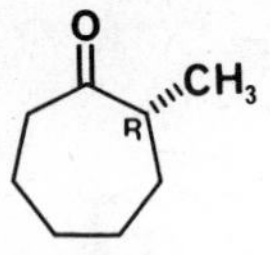 O, R, CH_3	59	94
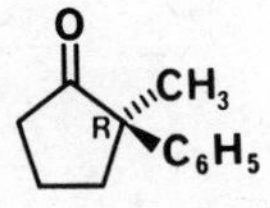 O, R, CH_3, C_6H_5	45	77
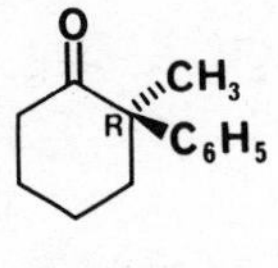 O, R, CH_3, C_6H_5	43	93

[a] Data obtained in part from reference 124.

complete asymmetric induction in observed in the synthesis of (+)-(*S*)-4-methyl-3-heptanone (**210**), the active principal alarm pheromone of the leaf-cutting ant *Atta texana.*[142] A second ant alarm pheromone of the genus *Manica*, (+)-(*S*)-4-methyl-3-hexanone (**211**), is prepared with a >95% enantiomeric excess.[143]

The acyclic enone (*E*)-4,6-dimethyl-6-octene-3-one (**212**) is the major component of the defense secretion of ''daddy longlegs'' *Leobunum vittatum* and *L. calcar* (Opilliones) and is also prepared with virtually complete asymmetric induction.[144]

Serricornin[(4*S*, 6*S*, 7*S*)-4,6-dimethyl-7-hydroxy-3-nonane] (**213**), the sex pheromone of the cigarette beetle *Lasioderma senicorne* F., is prepared by a chiral convergent synthesis with >97% optical purity.[145] Most importantly, bi- and polyfunctional electrophiles such as **214** undergo successful alkylation with subsequent oxidative cleavage of the hydrazone to provide **215** with excellent enantioselectivity and having the enol ether moiety intact[124] (Scheme 32).

Extension of the scope of the reaction to include SAMP–hydrazones of cyclic ketones allows asymmetric electrophilic substitution α to the carbonyl group to provide α-substituted cyclic ketones with 73–99% ee
and good overall chemical yield[124, 127, 128] (see Table 8.4). Note that it is possible to prepare chiral quarternary carbon centers by this method.[124]

The regioselective alkylation of the SAMP–hydrazone of cyclohexenone **216** is used to prepare (+)-eremophilenolide (**218**), a naturally occurring eremophilane-type sesquiterpene. The overall yield is 7%, with an optical purity of 89%.[146]

1. LDA
2. Br
3. CH_3I
26%

216 **217** (89% ee) **218**

When this method is applied to the asymmetric aldol reaction, one obtains β-chiral ketols in good to excellent overall chemical yields (51–82%), but with only moderate enantiomeric excesses (31–62%).[147] The enantioselective synthesis of the major pungent principle of ginger (*S*)-(+)-[6]-gingerol (**221**) requires, as a key step, the aldol reaction of the chiral RAMP–hydrazone **219** with *n*-hexanal. Unfortunately, **221** is obtained with only a 39% enantiomeric excess. Using the SAMP–hydrazone **220**, one obtains the (*R*)-(−) antipode **222** also in modest optical yield (36% ee)[148] (Scheme 33).

Virtually complete 1,6-asymmetric induction is observed in the asymmetric Michael addition of lithiated methylketone–SAMP–hydrazones **223** to α,β-unsaturated esters. This provides a simple and efficient method for the formation of β-substituted δ-ketoesters **224** in 45–62% overall yield and 96–100% ee.[149]

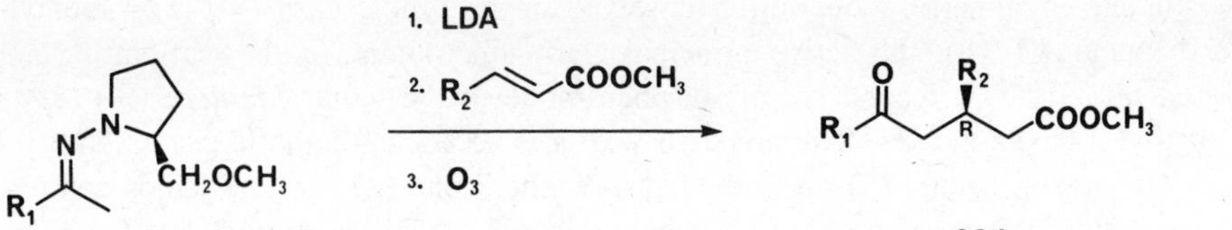

223

224

$R_1 = CH_3, C_2H_5, i\text{-}Pr, n\text{-}C_5H_{11}$

$R_2 = CH_3, C_6H_5$

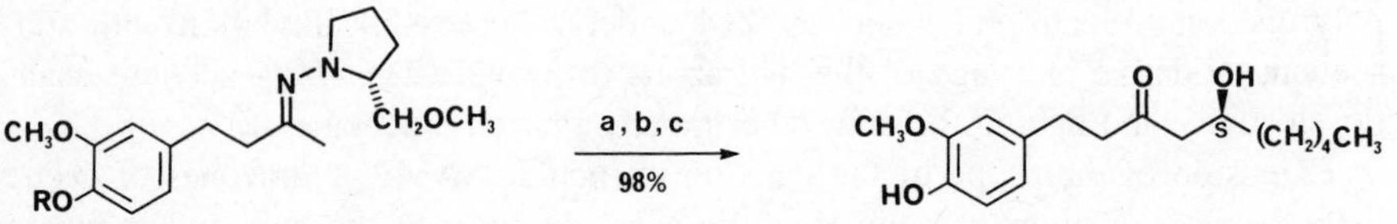

219

(R = CH_2Ph)

221 (39% ee)

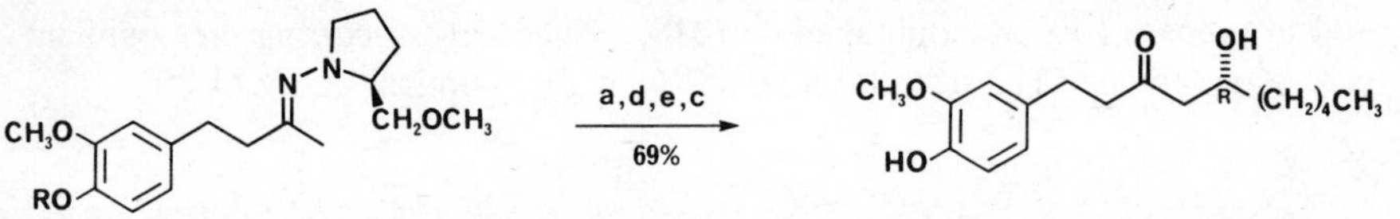

222 (36% ee)

220

Scheme 33. (a) *n*-Buli, $-78°C$, *n*-C_5H_{11} CHO (99%); (b) $NaIO_4$, MeOH, pH7, 25°C (98%); (c) Pd/C, MeOH (98%); (d) TMSCl, $-78°C$ (98%); (e) 30% H_2O_2, MeOH, pH7 (86%).

Mukaiyama and co-workers[149,150] have developed a variety of highly stereoselective reactions based on the concept of "synthetic control." This concept is characterized by the utilization of chelate complexes of common metals for the intra- or intermolecular interactions of reacting species, leading to highly stereospecific or entropically advantageous reactions.[149–151]

This chelation is observed when lithium aluminum hydride reacts with the chiral diamines (*S*)-2-(*N*-substituted aminomethyl)pyrrolidines **225** to produce chiral hydride reagents that possess rigidity due to their *cis*-fused five-membered ring. This translates into high enantiofacial selectivity in the reduction of prochiral ketones to optically active secondary alcohols.[152] A wide variety of these 2-(*N*-substituted aminomethyl)pyrrolidines are easily prepared from L-proline in four steps.[153] However, the best optical purities are observed with (*S*)-2-(anilinomethyl)[152,153] and (*S*)-2-(2,6-xylidenemethyl)pyrrolidine[154] (see first two entries). These asymmetric reductions are usually carried out at $-100°C$ in ether.[153,154]

This observed rigidity resulting from the coordination of **225** with organometallic reagents should enable the highly stereoselective asymmetric addition of anions to aldehydes. Indeed, the enantioselective addition of organolithium com-

225 226

(a) CbzCl, $NaHCO_3$; (b) ClCOOEt, RNH_2; (c) H_2, Pd/C; (d) $LiAlH_4$.

Chiral Hydride Reagent

R	Yield **226** (%)	Optical yield (%)
2,6-diMeC_6H_3	87	95
C_6H_5	84	84
1-Naphthyl	67	77
i-C_3H_7	65	47
4-Pyridyl	64	7
2-Pyridyl	88	2
2-MeOC_6H_4	18	0

pounds at −123°C, using (2*S*,2′*S*)-2-hydroxymethyl-1-[(1-methylpyrrolidine-2-yl)methyl]pyrrolidine (**227**)[156] as the chiral ligand provides chiral secondary alcohols, but with only moderated optical purities.[155] However, very high optical purities (>95%) are attained when the addition is carried out a −123°C in a 1:1 mixture of dimethoxyethane and ether.[157]

Either (*S*) or (*R*) configurations result, depending on the size of the alkyllithium. If the reaction is performed with dialkylmagnesium reagents, alcohols possessing the (*R*) configuration are obtained.[158]

227

RM	R_1	Yield (%)	Optical Yield (%)	Configuration
MeLi	C_6H_5	82	21	(*R*)
EtLi	C_6H_5	32	39	(*R*)
n-BuLi	C_6H_5	60	72	(*S*)
n-BuLi	*i*-C_3H_7	47	56	(*S*)
PhLi	*n*-C_4H_9	46	11	(*R*)
Me_2Mg	C_6H_5	56	34	(*R*)
Et_2Mg	C_6H_5	74	92	(*R*)
$(n\text{-}C_4H_9)_2Mg$	C_6H_5	94	88	(*R*)
$(n\text{-}C_4H_9)_2Mg$	*i*-C_3H_7	70	22	(*R*)

By reacting the lithium salts of methyl phenylsulfide, 2-methylthiothazoline, acetonitrile, or *N*-nitrosodimethylamine with lithiated **227,** one may obtain optically active oxiranes **228,**[159] thiirane **229,**[159] and amino alcohols **230** and **231**[159] with low to moderate enantiomeric excesses. In the case of **228,** the best optical yield (72%) could be obtained by using a variation of **227** wherein the *N*-methyl group is replaced by a neopentyl group[159] (Scheme 34).

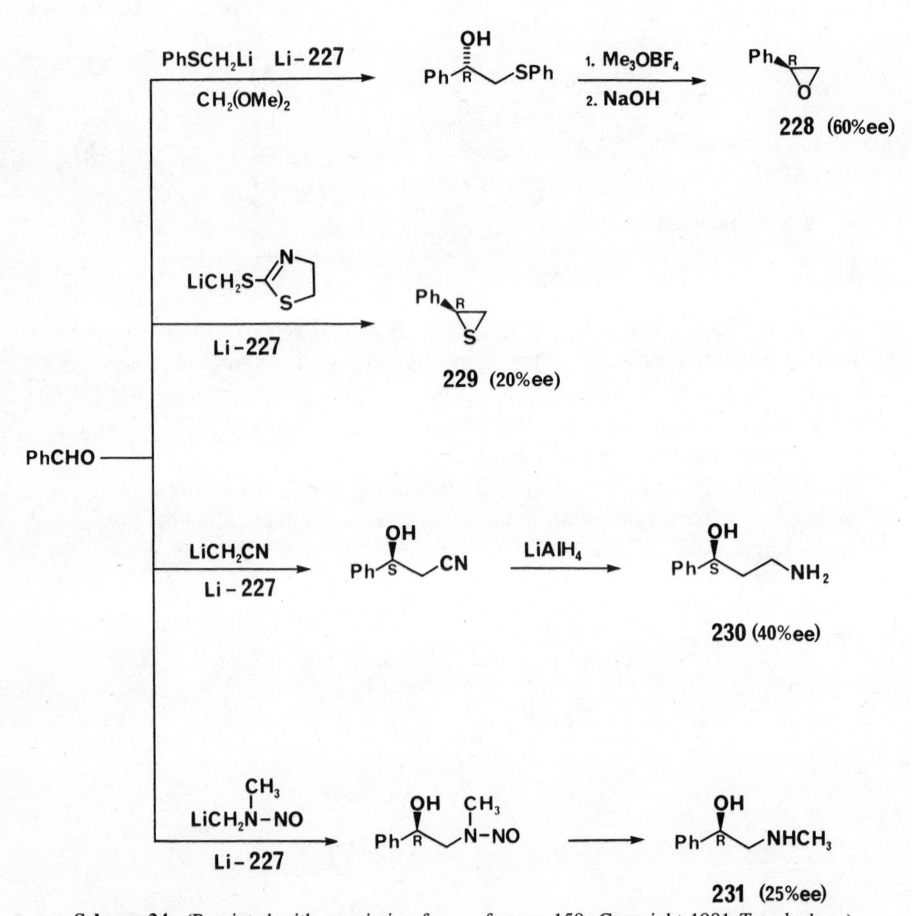

Scheme 34. (Reprinted with permission from reference 150. Copyright 1981 *Tetrahedron.*)

The enantioselective addition of lithium trimethylsilylacetylides to aldehydes in the presence of **227** allows preparation of chiral alkynyl alcohols **232** in good chemical yields, with 40–92% ee.[160] Successful conversion of **232** to 5-substituted-2(5H)-furanones **233** provides useful chiral intermediates for the total synthesis of such antifungal agents as avenaciolide (**234**), isoavenaciolide (**235**), or protolichesterinic acid (**236**)[161] (Scheme 35).

In the presence of **227,** the magnesium salt of allyltoylsulfone **237** undergoes an asymmetric addition to acetone to afford the corresponding (*S*)-β-hydroxysulfone **238** in 80% ee. This represents the first example of ligand-controlled chiral induction to the α-carbon of allyl sulfone by an electrophile. The corresponding (*R*)-**238** is available in 69% ee with the use of the modified chiral ligand **239** also prepared from L-proline.[162]

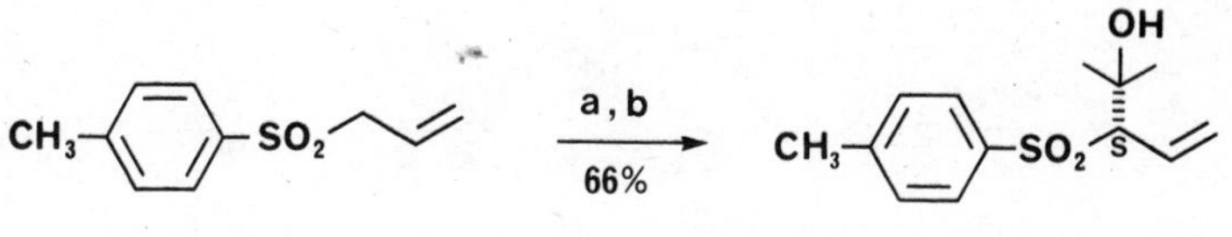

CH_3–C6H4–SO_2 (237) → a, b, 66% → CH_3–C6H4–SO_2 (S), OH (238)

237

238

CH_2N

N

OH

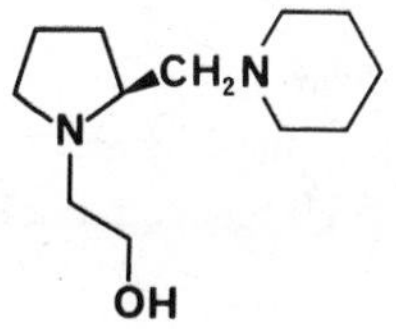

239

(a) EtMgBr, $-100°C$, **227**; (b) acetone.

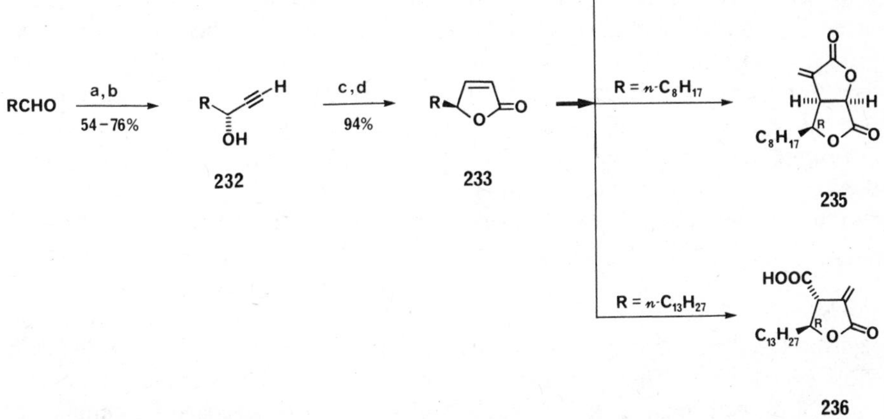

RCHO → a,b (54–76%) → R, H, OH (232) → c,d (94%) → R, O, O (233)

R= *n*-C_8H_{17} → O, O, H, H, R, C_8H_{17}, O, O (234)

R = *n*-C_8H_{17} → O, O, H, H, R, C_8H_{17}, O, O (235)

R = *n*-$C_{13}H_{27}$ → HOOC, R, $C_{13}H_{27}$, O, O (236)

232

233

234

235

236

Scheme 35. (a) TMSC≡CH, *n*-BuLi, **227**; (b) NaOH; (c) *n*-BuLi, CO_2; (d) H_2, $Pd/BaSO_4$.

The rigid *cis*-fused bicyclic ring observed in these chelation reactions can also be generated when **225** (R = C_6H_5) is first cyclized with phenylglyoxal to the aminal **240.** Subsequent treatment of **240** with Grignard reagents followed by hydrolysis affords α-alkyl-α-hydroxy aldehydes **241** in good chemical yields with optical purities greater than 94%.[163]

225 R = C_6H_5 → ($PhCOCHO$) → 240 → (1. R_1MgX; 2. HCl; 67–82%) → 241

$R_1 = CH_3, C_2H_5$, *i*-Pr, CH=CH$_2$

The aminal **243,** prepared from the reaction of **225** (R = C_6H_5) with fumaraldehydic acid methyl ester (**242**), undergoes a stereoselective Michael addition of a Grignard reagent to afford 3-alkylsuccinaldehydic acid methyl esters **244** with 85–93% ee.[164]

225 R = C_6H_5 + 242 (OHC–CH=CH–COOCH$_3$, THF) → 243 → (1. R_1MgBr, CuI; 2. HCl; 38–83%) → 244

$R_1 = C_2H_5$, *i*-Pr, *n*-Bu, *n*-C_5H_{11}, $PhCH_2$

A more general and versatile method for the preparation of chiral α-hydroxyaldehydes involves the use of aminal **245,** which undergoes Grignard additions with the same enantiofacial selectivity. The configurations obtained can be controlled by the order of the introduction of the two desired substituents.[165] This is beautifully illustrated by the synthesis of (*S*)-frontalin (**246**) (overall yield of 40%), the active pheromone of several species of beetles belonging to the genus *Dendroctinus.* A reversal of the addition of the Grignard reagents provides (*R*)-frontalin (**247**) in a 47% overall chemical yield[166] (Scheme 36).

The marine antibiotic (−)-malyngolide (**249a**), isolated from the blue-green alga *Lyngbya majuscula* Gomont, is prepared in high optical purity by employing the chiral diol **248.** Aminal **245** is treated sequentially at −100°C with 4-pentenylmagnesium bromide and nonylmagnesium bromide to afford a keto aminal that is hydrolyzed and immediately reduced with sodium borohydride to furnish **248** in 52% overall yield. Through a series of transformations, **248** is converted to a 2:1 ratio of diastereomers **249a** and **249b,** which are easily separated by chromatography. The optical purity of **249a** is 95%.[167]

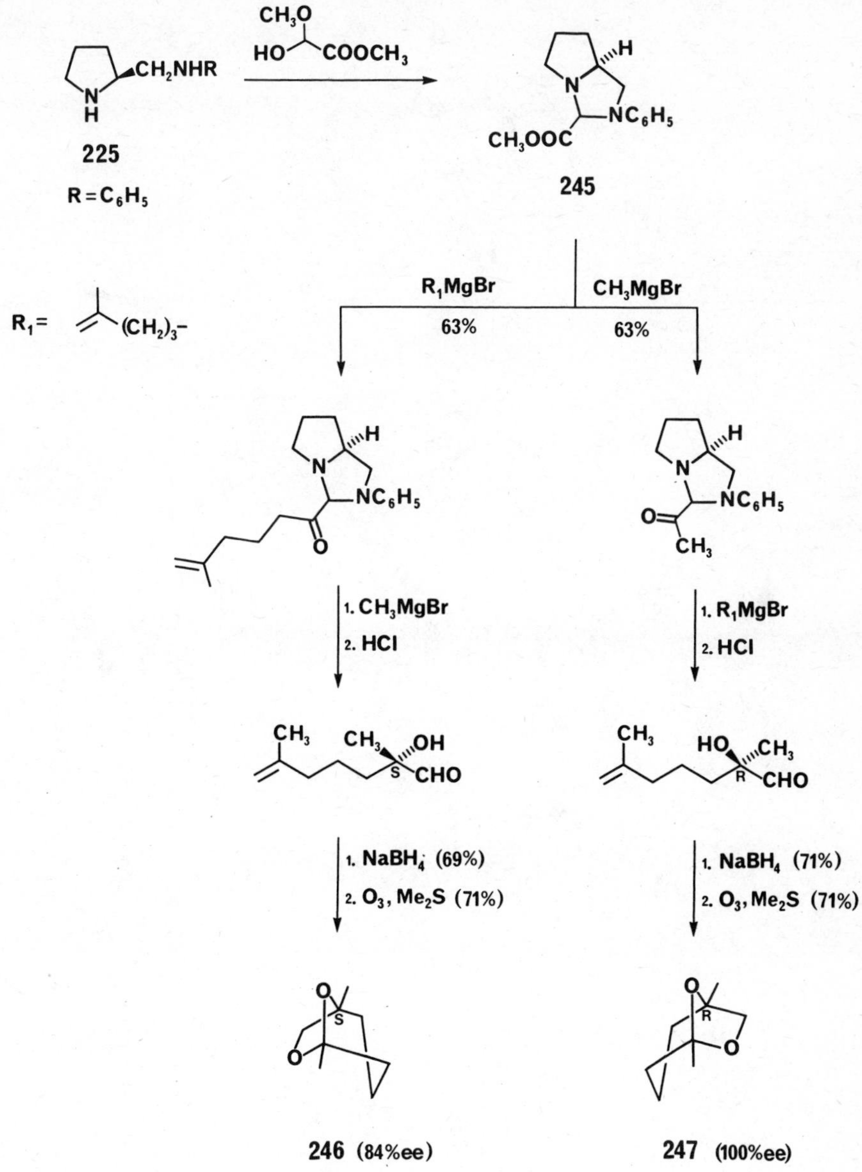

Scheme 36

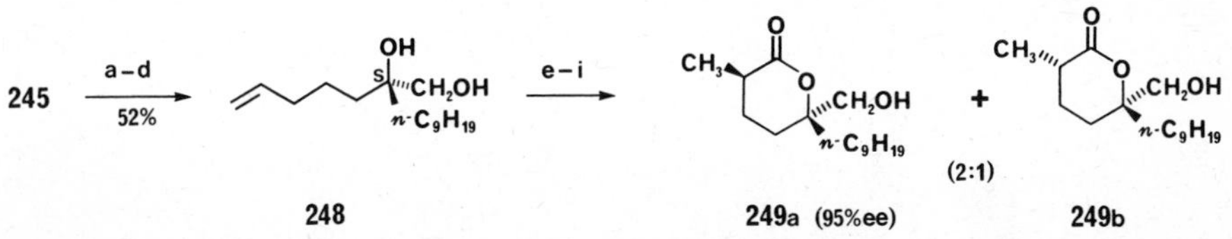

(a) 4-Pentenyl MgBr, $MgCl_2$; (b) n-$C_9H_{19}MgBr$; (c) 2% HCl; (d) $NaBH_4$; (e) TBSCl (98%); (f) O_3, $(CH_3)_2S$ (69%); (g) PDC, DMF (100%); (h) LDA, HMPA, CH_3I (74%); (i) n-Bu_4NF (58%).

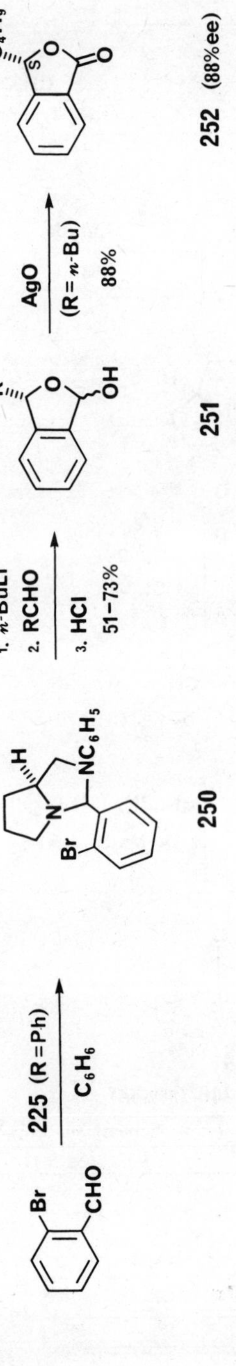

Br, CHO

225 (R = Ph)

C_6H_6

Br, N, H, NC_6H_5

250

1. n-BuLi

2. RCHO

3. HCl

51–73%

R, O, OH

251

AgO

(R = n-Bu)

88%

C_4H_9, S, O, O

252 (88%ee)

When aminal **250**—prepared by the reaction of **225** (R = C_6H_5) with *o*-bromobenzaldehyde—is treated with *n*-butyllithium, a chiral aryllithium reagent is obtained. This reacts enantioselectively with aliphatic aldehydes to provide (on hydrolysis) various 3-alkyl-1-hydroxy-2-oxaindanes (**251**). Silver oxide oxidation of **251** (R = *n*-C_4H_9) affords (*S*)-3-butylphthalide (**252**), which is an essential oil of celery.[168]

Chiral β-formyl-β-hydroxy esters **254** can be prepared by the reaction of metal enolates with keto aminals **253.** Whereas zinc enolates (Reformatsky reaction) react with **253a** to furnish (*S*)-**254,** lithium enolates react with **253a** to give (*R*)-**254.** Introduction of an *ortho*-methoxyl group to enhance the lithium coordination to the nitrogen (i.e., **253b**) results in the formation of (*S*)-**254.** An extra advantage of this approach over that of the Reformatsky reaction is the realization of significantly better optical and chemical yields of **254.**[169]

1. $CH_2{=}C(OM)OC_2H_5$
2. HCl
(50–85%)
M = Li or Na

254 (84–92%ee)

$R_1 = CH_3, C_2H_5$, *i*-Pr, Ph

253a $R_2 = C_6H_5$

253b R_2 = o-$CH_3OC_6H_4$

A highly enantioselective cross-aldol reaction between aromatic ketones and various aldehydes is achieved via divalent tin enolates (generated *in situ* by treating the ketone with stannous triflate) using the chiral diamine **255** as a ligand. Although a variety of chiral ligands may be employed, use of **255** provides the highest optical purities.[170]

$C_6H_5COR_1$ + **255**

1. $Sn(OTf)_2$
2. R_2CHO

255

R_1	R_2	Ratio erythro : threo	Yield (%)	Optical purity (%)
CH_3	C_6H_5	6:1	74	80
CH_3	*p*-MeC_6H_4	8:1	72	80
CH_3	*p*-ClC_6H_4	6:1	72	85
CH_3	*p*-$MeOC_6H_4$	8:1	78	80
C_2H_5	C_6H_5	5:1	72	75
CH_3	*t*-C_4H_9	10:0	57	90
CH_3	cyclo-C_6H_{11}	4:1	67	80

On the other hand, the cross-aldol reaction between aliphatic ketones and various aldehydes proceeds with 50–80% ee when the chiral ligand **256** is used.[171] The *erythro* geometry predominates in all cases.[170,171]

R–CO–CH₂–R₁ + **256** → 1. Sn(OTf)₂, 2. R₂CHO → erythro

256

R	R_1	R_2	Ratio *erythro*:*threo*	Yield (%)	Optical purity (%)
C_2H_5	CH_3	C_6H_5	35:25	60	55
C_2H_5	CH_3	p-$MeOC_6H_4$	89:11	69	50
t-C_4H_9	CH_3	C_6H_5	95:5	29	80
t-C_4H_9	CH_3	p-ClC_6H_4	95:5	32	80
t-C_4H_9	CH_3	p-$MeOC_6H_4$	95:5	24	77

When the asymmetric aldol reaction is extended to the reaction of 3-acetylthiazolidine-2-thione (**257**) with various aldehydes in the presence of chiral ligand **255**, chiral β-hydroxy derivatives **258** having the (*S*) absolute configuration are obtained.[171,172] Interestingly, this same stannous triflate-mediated aldol reaction with 3-(2-benzyloxyacetyl)thiazolidine-2-thione (**259**) in the presence of **255** affords predominantly *threo*-**260** (ratio >8:2). High asymmetric induction (~90% ee) is achieved with every aldehyde.[173]

1. Sn(OTf)₂ 2. **255**; (R=H) R_1CHO 65–85%; (R=OCH₂Ph) R_1CHO 68–93%

257 R=H

259 R=OCH₂Ph

258 (65–90%ee)

260 (87–94% ee)

2-Alkylmalic acids are incorporated as the carboxylic component in a majority of pyrrolizidine alkaloids possessing biological activity. In the presence of the chiral diamine (2*S*)-1-methyl-2-[(*N*-1-naphthylamino)methyl] pyrrolidine (**261**), the tin(II) enolate of **257** reacts with various α-ketoesters to afford the corresponding 2-substituted malates **262** with extremely high asymmetric induction.[174]

257 + 261 → 1. $Sn(OTf)_2$; 2. $RCOCOOCH_3$; 65–80% → 262 (85–95% ee)

$R = CH_3$, *i*-Pr, *i*-Bu, C_6H_5

Utilizing the coordination of a chiral ligand to give an unsymmetrical environment to a symmetrical molecule, Mukaiyama and co-workers[175] have been able to prepare optically active glycerol derivatives **267a** by the enantioselective acylation of the two primary hydroxyl groups of the prochiral glycerol **263.** When **263** is treated with either 1,1′-dimethylstannocene (**264**) or (methylcyclopentadienyl)tin(II) chloride (**265**), a tin(II) alkoxide forms. This compound interacts with chiral diamine (*S*)-1-methyl-2-(morpholinomethyl)pyrrolidine **266** to generate a complex in which the environments around the primary hydroxyls are different. Consequently, the addition of benzoyl chloride leads to a selective acylation. Moreover, with the use of **265** and excess benzoyl chloride, the optical purity of **267a** is increased to 84% ee as a result of the occurrence of an additional kinetic resolution step.

α-Substituted α-amino acids **270a** or propargylamines **270b** are easily prepared by chirally directed alkylation of the amidine function of **269.**[176,177] The chiral reagent needed for the synthesis of **269** is (*S*)-(−)-dimethoxymethyl-2-methoxymethylpyrrolidine (SDMP) (**268**),[176] which, in turn, is prepared from **146.** Optical purities are poor to moderate.[177]

8.4 4-HYDROXYPROLINE

trans-4-Hydroxy-L-proline ((4*R*)-hyroxy-2-(*S*)-pyrrolidineacetic acid) (**271**) is an important component of animal supportive tissue,[1,178] The hydrogen bonding ability of the 4-hydroxy group with neighboring carbonyl groups is believed to stabilize the collagen structure.[179] Possessing two chiral centers bearing versatile functionality as well as the basic imino moiety, **271** is an attractive chiral synthon for a variety of synthetic challenges.

The conversion of the 4-hydroxyl group into a good leaving group, such as the

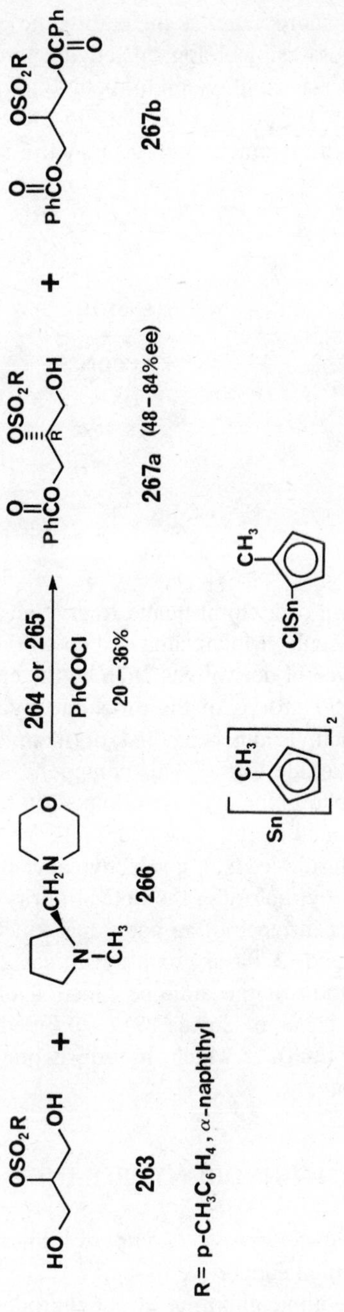

OSO_2R
HO OH
263

+

CH_2N O
N
CH_3
266

264 or 265
PhCOCl
20 – 36%

O
PhCO OSO_2R OH
R
267a (48 – 84%ee)

+

O
PhCO OSO_2R OCPh
O
267b

R = p-$CH_3C_6H_4$, α-naphthyl

Sn[CH_3]$_2$
264

ClSn CH_3
265

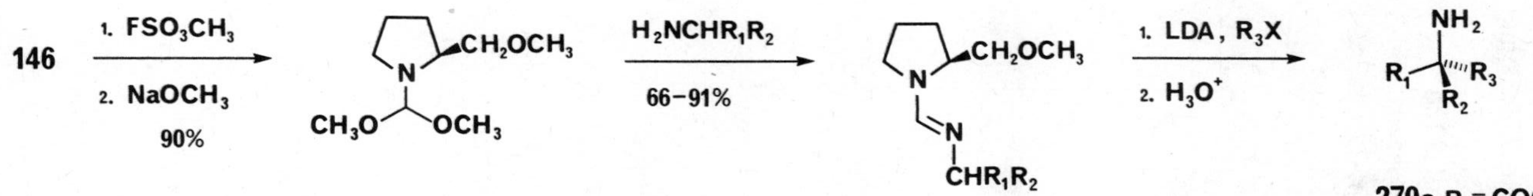

146 → (1. FSO_3CH_3; 2. $NaOCH_3$; 90%) → 268 → ($H_2NCHR_1R_2$; 66–91%) → 269 → (1. LDA, R_3X; 2. H_3O^+) → 270

CH_2OCH_3 N CH_3O OCH_3

268

CH_2OCH_3 N N CHR_1R_2

269

NH_2 R_1 R_3 R_2

270a $R_1 = COOH$ (15–51%ee)

270b $R_1 = C{\equiv}CH$ (9–84%ee)

$R_2 = H, CH_3$

$R_3 = CH_2C_6H_5$, alkyl

tosylate in **272a,**[179] provides an opportunity for the direct displacement by a variety of nucleophiles. Heating of **272a** with potassium fluoride in diethylene glycol, followed by saponification and deprotection, affords **273**[180] (83% *cis*). Likewise, the *cis*-4-iodo derivative, **274,** can be prepared by treating **272a** with sodium iodide in acetone. Mixtures of *cis* and *trans* isomers are usually obtained; however, if the reaction conditions are carefully controlled, **274** having up to 90% *cis* purity can be obtained.[178,182]

Displacement of the hydroxyl group of **272b**[181] with either phosphorus pentachloride or phosphorus pentabromide proceeds with complete inversion to provide, after saponification and deprotection, **275** and **276,** respectively.[181]

HO, COOH, N, H
271
R_1O, $COOCH_3$, N, R_2
272a R_1=Ts , R_2= Cbz
272b R_1=H , R_2=Ts
1. KF
2. NaOH
62%
F, COOH, N, H
273
NaI
acetone
I, $COOCH_3$, N, Cbz
274
PX_5
X, COOH, N, H
275 X=Cl (77%)
276 X=Br (53%)

Displacement of the tosylate of **272a** with either methyl mercaptide[179] or benzylmercaptide[183] (in this case using **272d,** in which R_1 = Ts and R_2 = Ac) affords, after saponification and deprotection, **277a** or **277b,** respectively, without any observed racemization. Metal–ammonia reduction of **277b** provides the free sulfhydryl, which lactonizes in the presence of DCC to the very unstable lactone **278.**[183]

TsO, $COOCH_3$, N, R
272c R=H
272d R=Ac
$CH_3S^{\ominus}$
or
$PhCH_2S^{\ominus}$
R_1S, COOH, N, H
277a $R_1=CH_3$
277b $R_1=CH_2Ph$
1. Na, NH_3
(44%)
2. DCC
S, =O, N, H
278

The azide ion, a potent nucleophile, displaces the tosyl group of **272e** (R_1 = Ts, R_2 = Ts) with inversion to provide stable, crystalline *N*-tosyl-*cis*-4-azido methyl ester **279a** in 88% yield. Catalytic reduction of the azide to an amino group, followed by saponification and reductive cleavage of the *N*-tosyl residue, provides *cis*-4-amino-L-proline **279b**, which is crystalline but hygroscopic and discolors in air. Cyclization of **279b** with the water-soluble *N*-cyclohexyl-*N'*-(β-morpholinyl-4-ethyl)carbodimide methyl *p*-toluenesulfonate produces the crystalline lactam **280.**[181]

The Jones oxidation of Cbz-protected **271**[179] provides 4-keto-L-proline (**281a**), which is stereospecifically reduced with $NaBH_4$ to afford 4-*allo*-hydroxy-L-proline (**282**) in good yield.[179] Lactonization of **282** with tosyl chloride in anhydrous pyridine furnishes **283** in 87% yield. Note that conversion of **282** to the tosylate **284** provides the corresponding *cis*-4-hydroxy derivative needed to prepare by direct displacements the corresponding *trans*-4-substituted proline derivatives.[181–183] Reduction of **281a** with tritium-labeled $NaBH_4$ furnishes [4-^{3}H]*allo*-hydroxy-L-proline (**285a**). On the other hand, a similar reduction of 4-keto-L-proline **281b** provides a 3:1 mixture of separable diastereomeric **285b** and **286** (Scheme 37).[184] Note that all transformations described for the 4-hydroxy-L-proline derivatives can be analogously performed on the 4-hydroxy-D-proline derivatives.

The tritosyl-L-prolinol derivative **287,** which is readily prepared from **271** by a five-step sequence involving *O*-tosylation, *N*-tosylation, esterification, lithium borohydride reduction, and *O*-tosylation,[185] is an extremely useful intermediate for synthesis of chiral bicyclic molecules. Reactions with nucleophiles such as amines,[185,186] thioacids,[190] and carbanions[185–187,191] produce a variety of interesting bicyclics **288–295** in good yields (Scheme 38).

N-Benzoyl-4-tosylhydroxy-L-proline methyl ester (**296**)[188] serves as the useful intermediate for the synthesis of chiral *N*-benzoyl-2-oxa-5-azabicyclo[2.2.1]heptane (**297**)[188] and *N*-benzoyl-2-oxa-5-azabicyclo[2.2.1]heptane-3-one (**298**).[188,189] Selective reduction of the methyl ester with $LiBH_4$, followed by a base-catalyzed displacement of the tosyloxy group by the newly formed primary alcohol, affords **297.** Saponification of **296** under controlled conditions provides **298.** Both possess the (2*S*) configuration. Recently, a one-step synthesis of **298** from *N*-benzoyl-4-hydroxy-L-proline (**299**) using diethylazodicarboxylate (DEAD) and triphenylphosphine (Mitsunobu reaction[193]) has been reported[192] (Scheme 39).

Optically active pyrrolizidine bases with 1-substituents are widely distributed in a number of plant families. A general method for the preparation of the pyrrolizidine nucleus involves a regiospecific 1,3-dipolar cycloaddition of an alkyne to the mesoionic oxazolone derived from L-proline.[194,195] By use of 4-hyroxy-L-proline instead, the stereochemistry of the hydroxyl group controls the formation of the two chiral centers at C-1 and C-8 in the newly formed pyrrolizidines.[196] The regiospecific 1,3-dipolar cycloaddition of ethyl propiolate with (−)-(2*S*,4*R*)-*N*-formyl-4-formylproline (**300**)[196] affords ethyl (+)-(2*R*)-2-formyl-2,3-dihydro-1*H*-pyrrolizine-7-carboxylate **301** in 80% yield. Catalytic hydrogenation of the deformylated ester occurs stereospecifically from the less hindered β face (*cis* addition). Subsequent displacement of the hydroxy group with chlorine, followed by hydro-

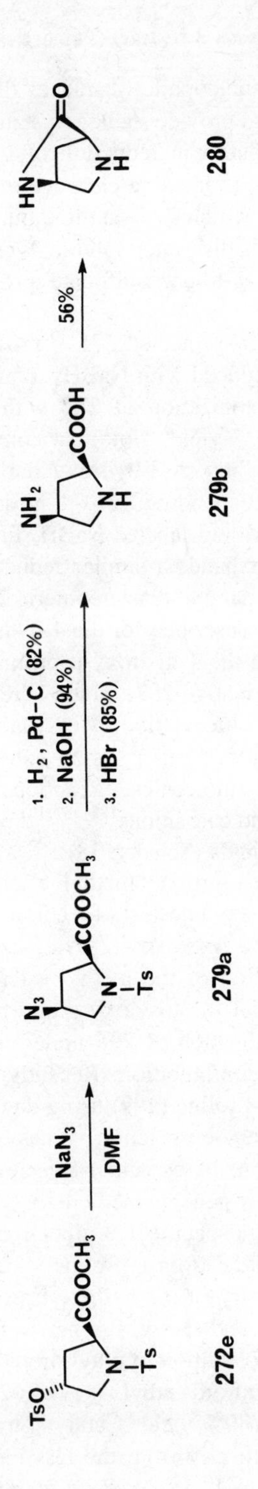
TsO
COOCH$_3$
N
Ts
272e
NaN$_3$
DMF
N$_3$
COOCH$_3$
N
Ts
279a
1. H$_2$, Pd–C (82%)
2. NaOH (94%)
3. HBr (85%)
NH$_2$
COOH
N
H
279b
56%
HN
O
N
H
280

CrO_3, H_2SO_4, 68%

281a R = Cbz

281b R = H

$NaBH_4$, 77%

282

TsCl, Py, 87%

283

1. CH_2N_2 2. TsCl Py, 87%

TsO … $COOCH_3$, Cbz

284

$NaBH^3_4$

285a R = Cbz

285b R = H

\+

286

Scheme 37

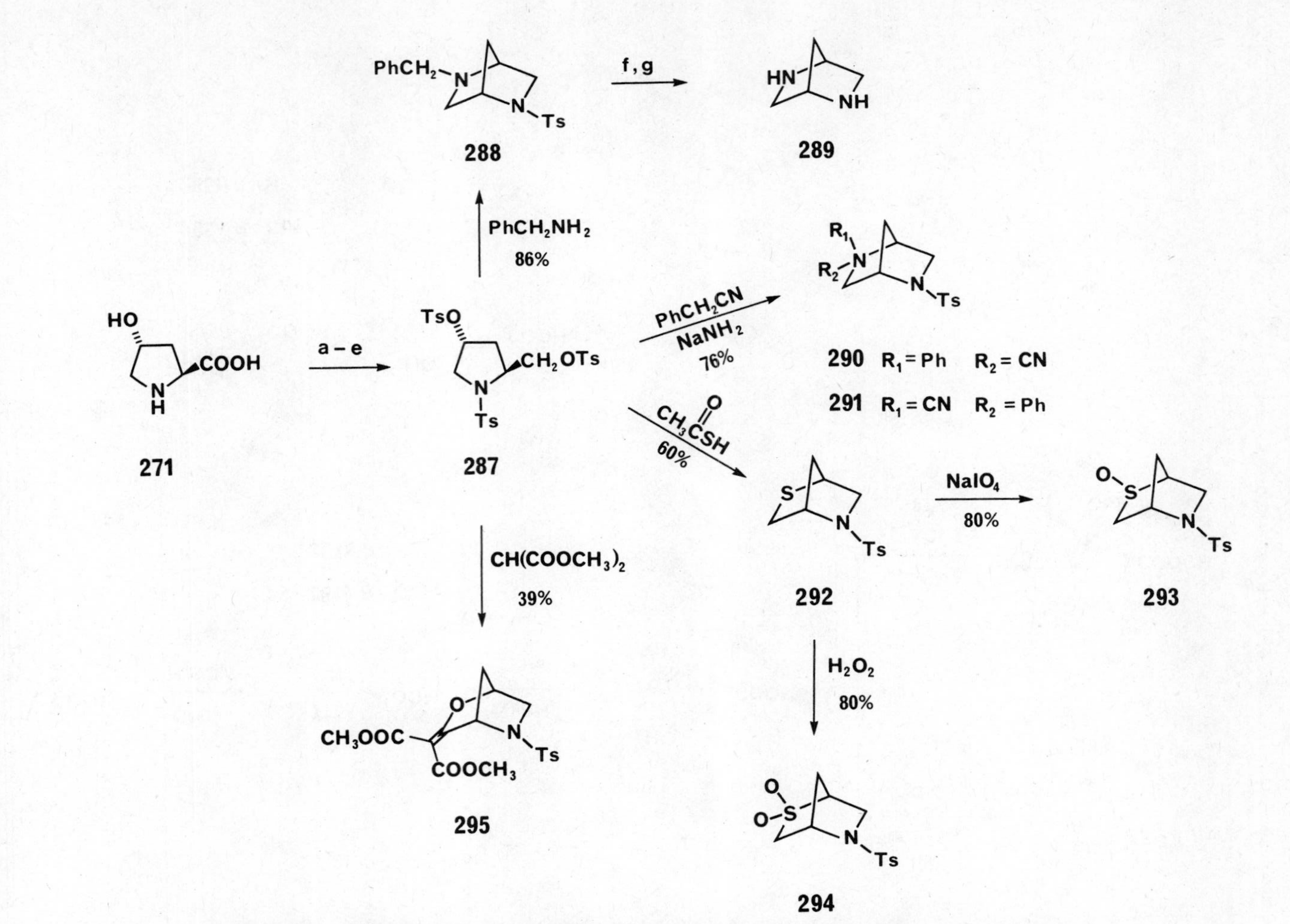

Scheme 38. (a) NaOH, pTSCl (82%); (b) NaOH, pTsCl (47%); (c) CH_2N_2 (92%); (d) $LiBH_4$ (86%); (e) pTsCl, pyridine (87.6%); (f) P(red), AcOH (76%); (g) AgCl, $NaHSO_3$, then 10% Pd/C, H_2 (85%).

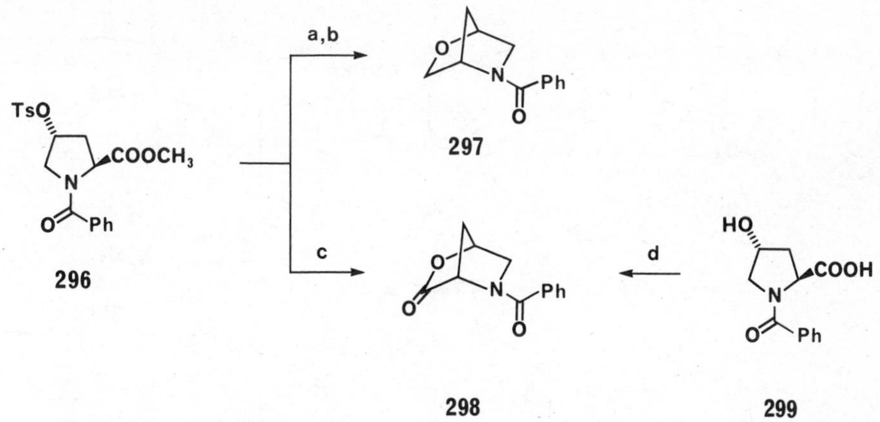

Scheme 39. (a) $LiBH_4$ (93%); (b) $NaOCH_3$ (90%); (c) NaOH (70%); (d) DEAD, Ph_3P (93%).

genolysis of the chlorine, results in a good yield of **302.** Conversion **302** to the 8β-pyrrolizidine bases (+)-isoretronecanol (**303**), (+)-laburnine (**304**), and (+)-supinidine (**305**) confirms the stereochemical assignments.[196,197]

In order to prepare the corresponding 8α-pyrrolizidine bases, it is necessary to effect an epimerization of **301.** This is accomplished through a three-step sequence in which **301** is deformylated, converted to the tosylate, and nucleophilically displaced by the formate anion. Subsequent deformylation, catalytic hydrogenation, and removal of the 6β-hydroxy group provides **306.** Following similar synthetic condition, one can use **306** to prepare (−)-isoretronecanol (**307**), (−)-trachelanthamidine (**308**), and (−)-supinidine (**309**). The optical purities of all six synthetic bases exceeeds 80%[196] (Scheme 40).

Chiral bisphosphine ligands prepared from **271** have been developed for use in asymmetric homogeneous catalysis.[198] One of the first, (2*S*,4*S*)-*N*-Boc-4-diphenylphosphino-2-diphenylphosphinomethylpyrrolidine (BPPM) (**310**), is easily prepared in good overall yield from **271.**[199] Hydrogenation of pyruvate esters catalyzed by neutral rhodium(I) complexes of **310** in dry THF or benzene provides optically active lactates **311** in nearly quantitative yields with 65–76% ee.[200,201] In this manner, (*R*)-(−)-pantolactone (**312**), an important intermediate in the synthesis of pantheine and coenzyme A, is prepared with 86.7% ee when the reduction is done in benzene[202,203] (Scheme 41).

(*S*)-Salsolidine (**315**) can be prepared in 34% optical yield when the BPPM–rhodium complex described above is used. In this case, optical yields depend markedly on the hydrogenation conditions used.[204] If the same catalytic reduction is performed with PPPM (**313**), prepared from **310** by removal of the Boc group and acylation with pivaloyl chloride, the enantiomeric (*R*)-salsolidine (**316**) is obtained in 45% optical yield.[204]

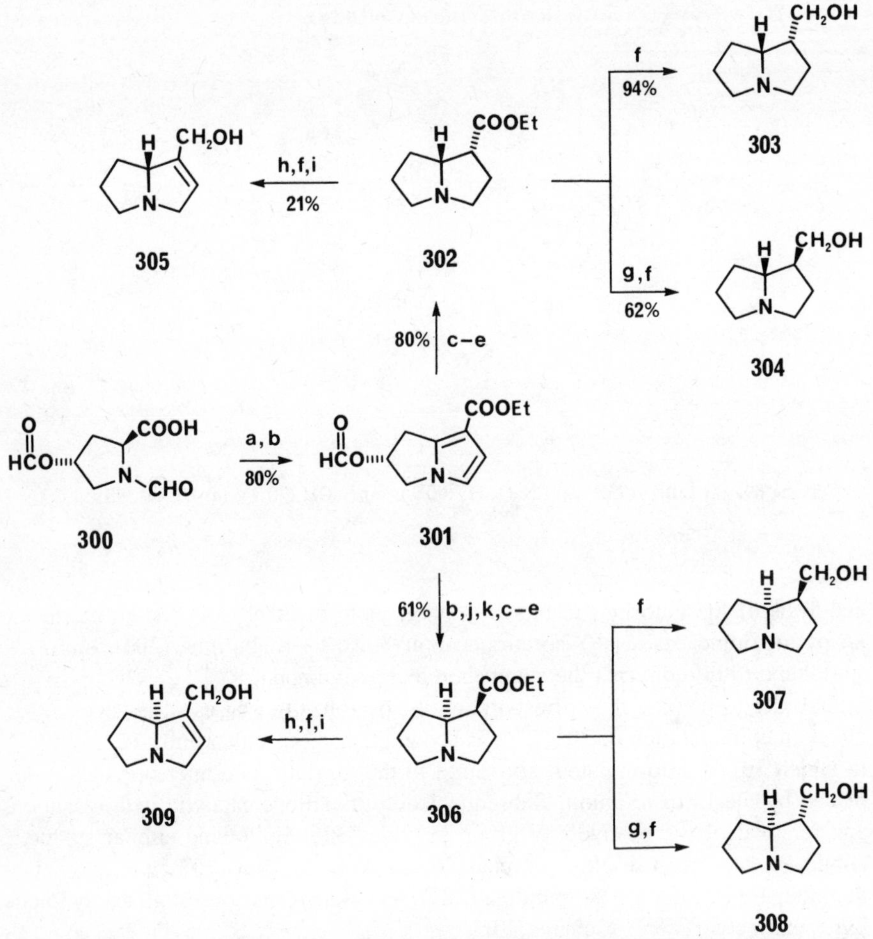

Scheme 40. (a) Ethyl propiolate, Ac_2O, 140°C (80%); (b) NH_4OH, EtOH (100%); (c) H_2, 10% Pd/C, AcOH (80%); (d) $SOCl_2$ (90%); (e) H_2, Raney Ni (90%); (f) $LiAlH_4$ (94%); (g) NaOEt (68%); (h) LDA, PhSeCl (62%); (i) H_2O_2 (55%); (j) pTsCl, pyridine (91%); (k) Et_4NCHO (84%).

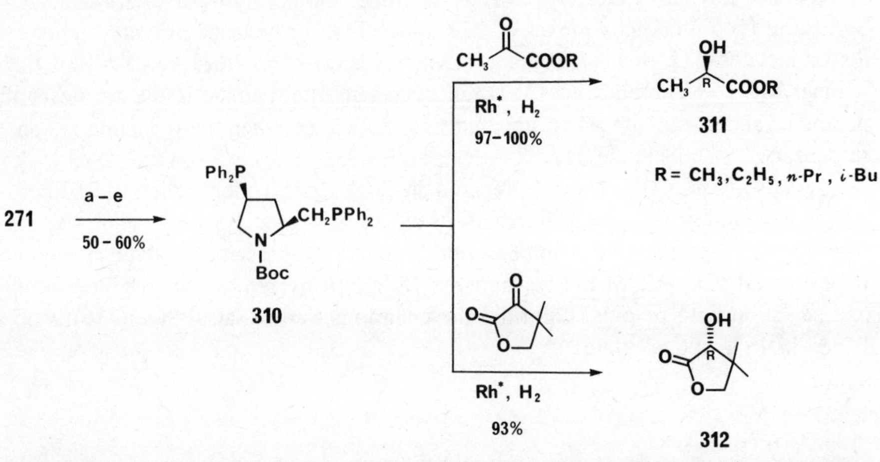

R = CH_3, C_2H_5, *n*-Pr, *i*-Bu

Scheme 41. (a) HCl, EtOH; (b) $(t\text{-BuO})_2CO$, $CHCl_3$; (c) $LiAlH_4$; (d) pTsCl, pyridine; (e) $NaPPh_2$.

(a) CF_3COOH; (b) $(CH_3)_3COCl$.

(*S*)-3,4-Dehydroproline (**321**), when incorporated into biologically active peptides, seldom results in a loss of activity and in some cases results in increased activity with reduced toxicity of the product.[205–207] The chiral synthesis of *N*-Boc-L-3,4-dehydroproline (**320a**) (X = Boc) from **317** (X = Boc) requires, as the key step, the Tchugaeff pyrolysis of the dithiocarbonate **318**. Although **320a** can be produced in a moderate yield (64%), under such harsh conditions both the 4,5-dehydro ester (3–5%) and **317** (5–9%) are produced as well.[206]

A milder thermal *syn* elimination of the labile selenoxides, prepared by the hydrogen peroxide oxidation of either the corresponding Boc–**319a**[207] or Cbz–**319b**[208] selenides, allows the reaction to be performed at room temperature with exclusive formation of **320a** or **320b** in 87% yield. The exchange of esters in the Boc series occurs during the selenide formation. Treatment of **320b** with trimethylsilyliodide in refluxing acetonitrile followed by cationic exchange chromatography furnishes **321** (99.8% ee) in 80% yield[208] (Scheme 42).

The Geissman–Waiss lactone [(+)-2-oxa-6-azabicyclo[3.3.0]octan-3-one] (**325**)[209,210] is an important chiral intermediate for the synthesis of some necine bases. It can be prepared from commercially available *N*-Cbz–4-hydroxy-L-proline (**322**). Homologation by means of a Wolff rearrangement furnishes **323**, which, when converted to a xanthate ester, undergoes the Tchugaeff pyrolysis to provide, after chromatographic purification, the 3,4-dehydro derivative **324**. Saponification of the ester and electrophilic lactonization with either iodine or benzeneselenyl chloride, followed by a reductive step, results in the formation of **325**.[209] Such necines as (+)-retronecine (**326**), (−)-platynecine (**327**), and (+)-croalbinecine (**328**) can be prepared in high optical yield (96–99%) from **325**[211] (Scheme 43).

Crotanecine (**333**), a pyrrolizidine triol isolated from *Crotalaria* species,[212] may be viewed as a 6β-hydroxy derivative of **326**. Introduction of this β-hydroxy group into the 5 position of **322** is accomplished by initially converting **322** to the dehydronitrile **329** via the mild phenylselenoxide elimination route (overall yield of 66% from **322**). Excellent regioselectivity is observed when **329** is treated with *N*-

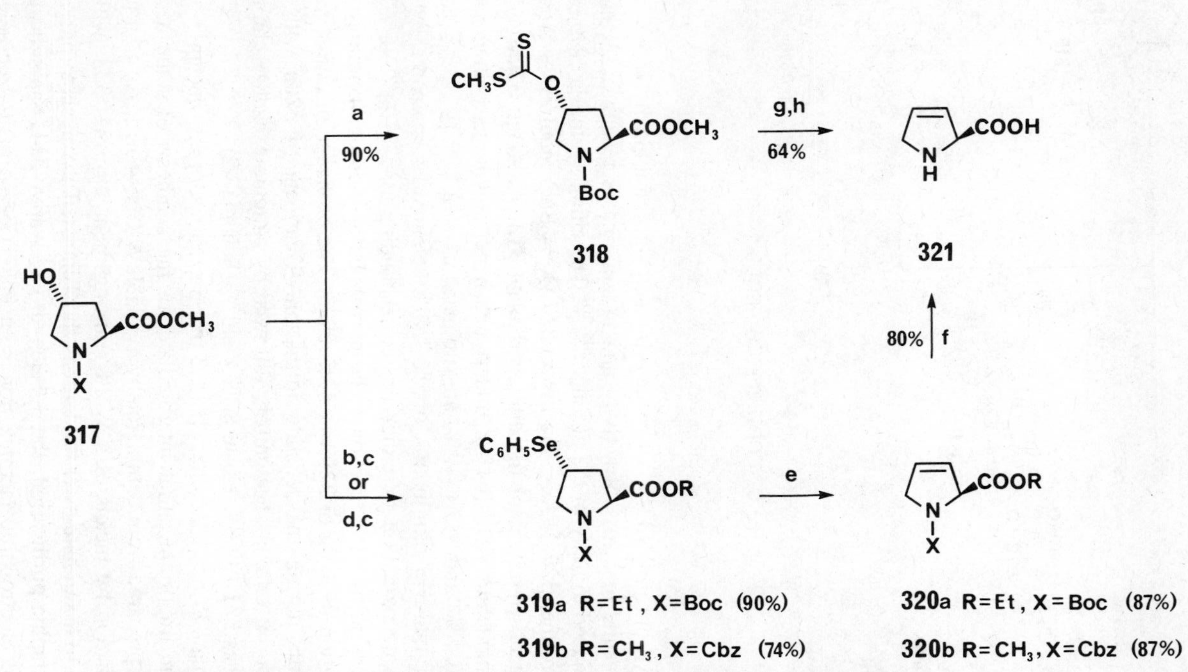

Scheme 42. (a) *n*-Bu_4NHSO_4, CS_2, NaOH; (b) DEAD, Ph_3P, CH_3I (90%); (c) PhSeNa, EtOH; (d) pTsCl, pyridine (91%); (e) H_2O_2; (f) TMSI, CH_3CN, then Dowex 50W-X8 resin; (g) 170–180°C, 12 Torr (64%); (h) 50% NaOH, H_2O/dioxane.

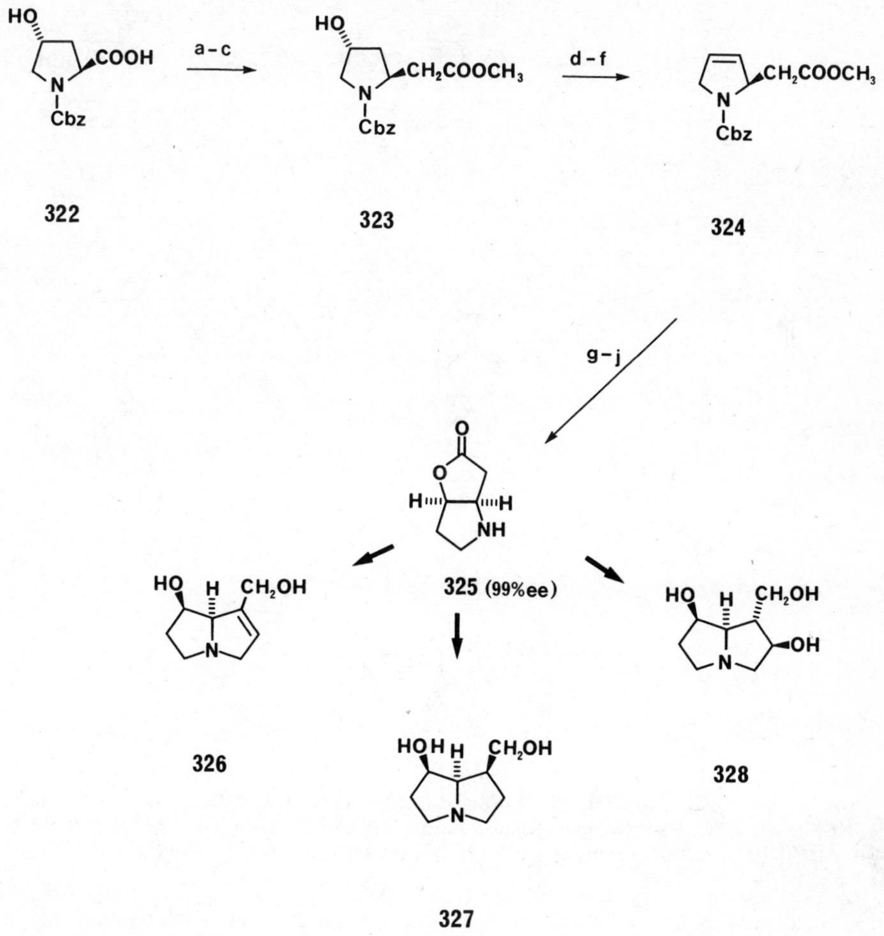

322 323 324

325 (99%ee)

326 327 328

Scheme 43. (a) TBSCl, imidazole, DMF; (b) *i*-BuOCOCl, Et_3N, CH_2N_2; (c) Ag_2O, MeOH; (d) *n*-Bu_4NHSO_4, CS_2, NaOH; (e) 170–180°C, 12 Torr (100% as 6:1 mixture); (f) chromatography; (g) Na_2CO_3, MeOH; (h) I_2, KI, $NaHCO_3$, or PhSeCl; (i) *n*-Bu_3SnH or Ph_3SnH; (j) 5% Pd/C.

iodosuccinimide in acetic acid to provide the iodoacetate **330.** Subsequent solvolysis in wet acetic acid containing silver acetate affords the *cis*-diol **331.** The crude mixture is treated with methanol containing hydrogen chloride to give a separable mixture of hydroxy lactones **332a** and **332b.** Elaboration of **332a** then provides **333** in an overall yield of 4% from **322.** The optical purity of the final product exceeds 80%[213] (Scheme 44).

The alkylation of the enolate of 2-(*tert*-butyl)-4-oxo-3-oxa-1-azabicylco[3.3.0]oct-7-ylacetate (**334**) with alkyl halides followed by cleavage of the products furnishes enantiomerically pure 2-substituted 4-hydroxyprolines **335a–c.**

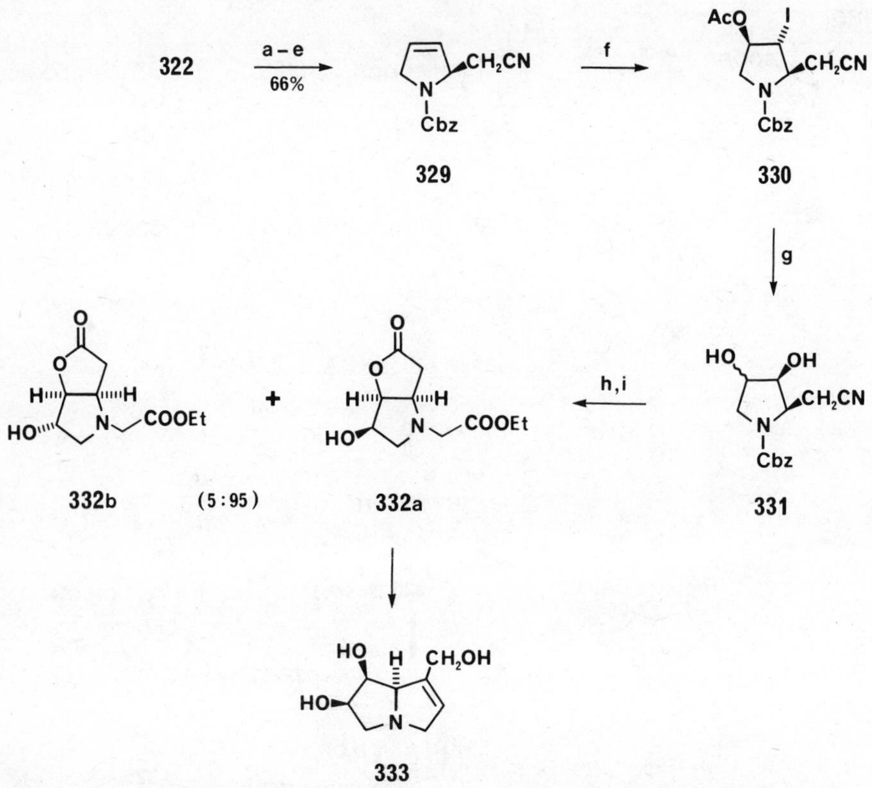

322 $\xrightarrow[66\%]{a-e}$ 329 $\xrightarrow{f}$ 330 $\xrightarrow{g}$ 331 $\xrightarrow{h,i}$ 332b + 332a (5 : 95); 332a → 333

Scheme 44. (a) BH_3–Me_2S, THF, 0°C; (b) pTsCl, pyridine; (c) NaCN, DMF, 95°C; (d) PhSeNa, THF–MeOH, 65°C; (e) H_2O_2, room temperature; (f) NIS, AcOH; (g) AgOAc, AcOH–H_2O, 95°C; (h) MeOH–HCl; (i) $BrCH_2COOEt$, Na_2CO_3, 75°C; pure **332a** 63% overall from 329.

With aldehydes and unsymmetrical ketones, a mixture of diastereomers at the prochiral center are formed (e.g., benzaldehyde = 4 : 1 ratio).[214]

The chiral auxiliary (2*S*,4*S*)-2-(anilinomethyl)-1-ethyl-4-hydroxypyrrolidine (**336**), prepared in five steps from **322,** catalyzes the highly enantioselective Michael addition of aromatic thiols to 2-cyclohexenone. The (*R*) absolute configura-

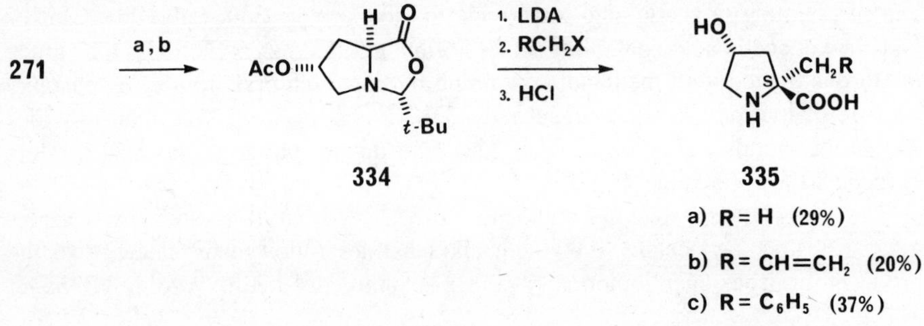

271 $\xrightarrow{a,b}$ 334 $\xrightarrow{\text{1. LDA; 2. } RCH_2X\text{; 3. HCl}}$ 335

a) R = H (29%)

b) R = $CH{=}CH_2$ (20%)

c) R = C_6H_5 (37%)

(a) AcOH, $HClO_4$, H_2O (64%); (b) *t*-BuCHO, TFA, CH_2Cl_2 (70%).

tion observed for the predominantly formed enantiomer **337** results from a preferred *re*-face attack by the thiolate.[215]

322 $\xrightarrow[43\%]{a-d}$ **336** + 4-R-C_6H_4SH + cyclohexenone $\xrightarrow{C_6H_5CH_3}$ **337**

(a) Ac_2O, pyridine (75%); (b) pivaloyl chloride, $C_6H_5NH_2$ (75%); (c) H_2, Pd/C (100%); (d) $LiAlH_4$ (80%).

R	Yield (%)	Optical yield (%)
H	83	77
CH_3	75	73
Cl	84	47
t-C_4H_9	74	88
CH_3O	75	83

A common feature of anthramycin (**342**),[216] Z–tomaymycin (**343**)[217,218] and chicamycin A (**341**)[219] is the pyrrolo[1,4]benzodiazepine nucleus. Because of their antitumor activity, these antibiotics have attracted considerable synthetic interest. The total synthesis of **342** invokes the elaboration of the polysubstituted benzene **338a,** which reacts smoothly with hydroxyproline methyl ester to afford **339** in 94% yield. Reduction of the nitro group with sodium dithionite and cyclization of the resulting amine affords lactam **340.** This important intermediate is then elaborated in several steps to **342.**[216] Utilization of this approach with **338b** allows one to prepare the corresponding intermediates for the total synthesis of **341** and **343** (Scheme 45).

Alternatively, the pyrrolo[1.4]benzodiazepine nucleus is accessible through a palladium-catalyzed carbonylation reaction. Insertion of carbon monoxide into the amine **345** in the presence of palladium acetate affords the optically active diazepinone **346** in 27% yield. This is then converted through established routes to **342**[220] (Scheme 46).

Recently, a novel synthesis of the key intermediate **348** has been developed in which *trans*-4-hydroxy-L-proline (**271**) condenses smoothly with 3-methoxy-4-methylisatoic anhydride (**347**). Anthramycin (**342**) is prepared in an 11% overall yield by this method, which offers promise in the synthesis of other members of this interesting class of natural products.[221]

R_3 R_2 NO_2 R_1 COCl + $COOCH_3$ HN OH

TEA, $-20°C$

R_3 R_2 NO_2 $COOCH_3$ R_1 N O OH

338

a) $R_1 = H$, $R_2 = CH_3$, $R_3 = OCH_2Ph$

b) $R_1 = OCH_3$, $R_2 = PNBO$, $R_3 = H$

339a (94%)

339b (84%)

1. $Na_2S_2O_4$; 2. HCl

or

1. H_2, Pd–C; 2. 110°C

H OCH_3 N– H HO N OH CH_3O O

341

H O R_3 N H R_2 N OH R_1 O

340a (86%)

340b (99%)

HO H OCH_3 N H CH_3 N $CONH_2$ O

342

H OCH_3 N H HO N H CH_3O O CH_3

343

Scheme 45

CH_3O CH_3 NH_2 Cl Br + COCl RN OR

344

($R = COCF_3$)

a–c

CH_3O R_1 O H CH_3 N HN OR_1 Cl Br

345

27% $Pd(OAc)_2$ CO

R_1 O CH_3O N H CH_3 N OR_1 Cl O

342

346

($R_1 = CH_2OCH_3$)

Scheme 46. (a) $NaHCO_3$ (79%); (b) NaH, $ClCH_2OCH_3$, TFA–HMPA (70%): (c) K_2CO_3, MeOH (92%).

271
65%
342

347 348

REFERENCES

1. T. Scott and M. Brewers, *Concise Encyclopedia of Biochemistry,* Walter de Gruyter, New York, 1983, p. 369.
2. D. Enders, H. Eichenauer, and R. Pieter, *Chem. Ber.*, **112,** 3703 (1979).
3. D. Shiengthong, A. Ungphakorn, D. E. Lewis, and R. A. Massy-Westropp, *Tetrahedron Lett.*, 2247 (1979).
4. P. J. Babidge, R. A. Massy-Westropp, S. G. Pyne, D. Shiengthong, A. Ungphakorn, and G. Veerachat, *Aust. J. Chem.*, **33,** 1841 (1980).
5. J. W. Daly and C. W. Myers, *Science* **156,** 970 (1967).
6. J. W. Daly, E. T. McNeal, L. E. Overman, and D. H. Ellison, *J. Med. Chem.* **28,** 482 (1985).
7. J. W. Daly, T. Tokuyama, T. Fujiwara, R. L. Hight, and I. L. Karle, *J. Am. Chem. Soc.*, **102,** 830 (1980).
8. A. Ito, R. Takahashi, and Y. Baba, *Chem. Pharm. Bull.*, **23,** 3081 (1975).
9. L. E. Overman and K. L. Bell, *J. Am. Chem. Soc.*, **103,** 1851 (1981).
10. L. E. Overman, K. L. Bell, and F. Ito, *J. Am. Chem. Soc.*, **106,** 4192 (1984).
11. (a) Y. N. Berlokon, I. E. Zel'tzer, M. G. Ryzhov, M. B. Saporovskaya, V. I. Bakhmutov, and V. M. Belikov, *J. Chem. Soc., Chem. Commun.*, 180 (1982); (b) Y. N. Belokon, A. G. Bulychev, S. V. Vitt, Y. T. Struchkov, A. S. Batsanov, T. V. Timofeeva, V. A. Tsyryapkin, M. G. Ryzhov, L. A. Lysova, V. I. Bakhmutov, and V. M. Belikov, *J. Am. Chem. Soc.*, **107,** 4252 (1985).
12. N. Umino, T. Iwakuma, and N. Itoh, *Chem. Pharm. Bull.*, **27,** 1479 (1979).
13. S. Itsuno, K. Ito, A. Hirao, and S. Nakahama, *J. Chem. Soc., Perkin Trans. I,* 2887 (1984).
14. N. Baba, J. Oda, and Y. Inouye, *J. Chem. Soc., Chem. Commun.*, 815 (1980).
15. M. Amano, N. Baba, J. Oda, and Y. Inouye, *Bioorg. Chem.*, **12,** 299 (1984).
16. N. Baba, M. Amano, J. Oda, and Y. Inouye, *J. Am. Chem. Soc.*, **106,** 1481 (1984).
17. K. Soai, K. Komiya, Y. Shigematsu, H. Hasegawa, and A. Ookawa, *J. Chem. Soc., Chem. Commun.*, 1282 (1982).
18. K. Soai and M. Ishizaki, *J. Chem. Soc., Chem. Commun.*, 1016 (1984).
19. K. Yamada, M. Takeda, and T. Iwakuma, *Tetrahedron Lett.*, 3869 (1981).
20. C. H. Heathcock, R. A. Jennings, and T. W. Vongeldern, *J. Org. Chem.*, **48,** 3428 (1983).
21. K. Hiroi and K. Nakazawa, *Chem. Lett.*, 1077 (1980).
22. B. W. Bycroft and G. R. Lee, *J. Chem. Soc., Chem. Commun.*, 988 (1975).
23. H. Poisel and U. Schmidt, *Chem. Ber.*, **106,** 3408 (1973).

24. U. Schmidt, A. Perco, and E. Öhler, *Chem. Ber.*, **107,** 2816 (1974).
25. M. Sokolovsky, T. Sadeh, and A. Patchornik, *J. Am. Chem. Soc.*, **86,** 1212 (1964).
26. U. Schmidt and E. Öhler, *Angew. Chem., Int. Ed.*, **15,** 42 (1976).
27. F. Öhler, E. Prantz, and U. Schmidt, *Chem. Ber.*, **111,** 1058 (1978).
28. P. S. Steyn, *Tetrahedron Lett.*, 3331 (1971).
29. P. S. Steyn, *Tetrahedron,* **29,** 107 (1973).
30. A. J. Birch and J. J. Wright, *Tetrahedron,* **26,** 2329 (1970).
31. T. Kametani, N. Kanaya, and M. Ihara, *J. Am. Chem. Soc.*, **102,** 3974 (1980).
32. T. Kametani, N. Kanaya, and M. Ihara, *J. Chem. Soc., Perkin Trans. I,* 959 (1981).
33. R. Ritchie and J. E. Saxton, *J. Chem. Soc., Chem. Commun.*, 611 (1975).
34. R. Ritchie and J. E. Saxton, *Tetrahedron,* **37,** 4295 (1981).
35. H. Poisel and U. Schmidt, *Chem. Ber.*, **105,** 625 (1972).
36. E. Öhler, H. Poisel, F. Tataruch, and U. Schmidt, *Chem. Ber.*, **105,** 635 (1972).
37. E. Öhler, F. Tataruch, and U. Schmidt, *Chem. Ber.*, **105,** 3658 (1972).
38. D. Seebach and R. Naef, *Helv. Chim. Acta,* **64,** 2704 (1981).
39. D. Seebach, M. Boes, R. Naef, and W. B. Schweizer, *J. Am. Chem. Soc.*, **105,** 5390 (1983).
40. H. Rüeger and M. Benn, *Heterocycles,* **19,** 1677 (1982).
41. H. Rüeger and M. Benn, *Heterocycles,* **20,** 235 (1983).
42. D. Goff, J. C. Lagarias, W. C. Shih, M. P. Klein, and H. Rapoport, *J. Org. Chem.*, **45,** 4813 (1980).
43. D. H. Kim, *J. Heterocycl. Chem.*, **17,** 1647 (1980).
44. D. W. Cushman, H. S. Cheung, E. F. Sabo, and M. A. Ondetti, *Biochemistry,* **16,** 5484 (1977).
45. R. L. White, Jr., *U.S. Patent* 4209617 (1980); *Chem. Abstr.*, **94,** 15753n, (1981).
46. J. T. Suh, J. W. Skiles, B. E. Williams, R. D. Youssefyeh, H. Jones, B. Loev, E. S. Neiss, A. Schwab, W. S. Mann, A. Khandwala, P. S. Wolf, and I. Weinryb, *J. Med. Chem.*, **28,** 57 (1985).
47. W. J. Greenlee, P. L. Allibone, D. S. Perlow, A. A. Patchett, E. H. Ulm, and T. C. Vassil, *J. Med. Chem.*, **28,** 434 (1985).
48. H. Budzikiewicz, L. Faber, E.-G. Herrmann, F. F. Perrollaz, U. P. Schlunegger, and W. Wiegrebe, *Liebigs Ann. Chem.*, 1212 (1979).
49. E. Gellert and N. Kumar, *Aust. J. Chem.*, **37,** 819 (1984).
50. T. Koizumi, Y. Kobagashi, H. Amitani, and E. Yoshii, *J. Org. Chem.*, **42,** 3459 (1977).
51. T. Koizumi, H. Amitani, and E. Yoshii, *Tetrahedron Lett.*, 3741 (1978).
52. T. Koizumi, H. Amitani, and E. Yoshii, *Synthesis,* 110 (1979).
53. T. Koizumi, H. Takagi, and E. Yoshii, *Chem. Lett.*, 1403 (1980).
54. K. Irie, A. Ishida, T. Nakamura, and T. Oh-ishi, *Chem. Pharm. Bull.*, **32,** 2126 (1984).
55. A. G. Schultz, *Chem. Rev.*, **73,** 385 (1973).
56. S. Terashima and S-s. Jew, *Tetrahedron Lett.*, 1005 (1977).
57. S-s. Jew, S. Terashima, and K. Koga, *Tetrahedron,* **35,** 2337 (1979).

58. S. Terashima, S-s. Jew, and K. Koga, *Chem. Lett.*, 1109 (1977).
59. S-s. Jew, S. Terashima, and K. Koga, *Tetrahedron,* **35,** 2345 (1979).
60. S. Terashima, S-s. Jew, and K. Koga, *Tetrahedron Lett.*, 4507 (1977).
61. S. Terashima, S-s. Jew, and K. Koga, *Tetrahedron Lett.*, 4937 (1978).
62. S-s. Jew, S. Terashima, and K. Koga, *Chem. Pharm. Bull.*, **27,** 2351 (1979).
63. S. Terashima, M. Hayashi, and K. Koga, *Tetrahedron Lett.*, 2733 (1980).
64. M. Hayashi, S. Terashima, and K. Koga, *Tetrahedron,* **37,** 2797 (1981).
65. G. Stork and S. Dowd, *J. Am. Chem. Soc.*, **85,** 2178 (1963).
66. G. Wittig and H. Reiff, *Angew. Chem., Int. Ed.*, **7,** 7 (1968).
67. G. Wittig and P. Suchanck, *Tetrahedron,* **22,** 347 (1966).
68. E. Buncel and T. Durst (Eds.), *Comprehensive Carbanion Chemistry,* Part B, Elsevier, New York, 1984, p. 65ff.
69. S. Yamada, K. Hiroi, and K. Achiwa, *Tetrahedron Lett.*, 4233 (1969).
70. K. Hiroi, K. Achiwa, and S. Yamada, *Chem. Pharm. Bull.*, **20,** 246 (1972).
71. K. Hiroi and S. Yamada, *Chem. Pharm. Bull.*, **21,** 47 (1973).
72. K. Hiroi and S. Yamada, *Chem. Pharm. Bull.*, **21,** 54 (1973).
73. S. Yamada and G. Otani, *Tetrahedron Lett.*, 4237 (1969).
74. G. Otani and S. Yamada, *Chem. Pharm. Bull.*, **21,** 2112 (1973).
75. G. Otani and S. Yamada, *Chem. Pharm. Bull.*, **21,** 2125 (1973).
76. T. Sone, K. Hiroi, and S. Yamada, *Chem. Pharm. Bull.*, **21,** 2331 (1973).
77. S. Yamada, *Tetrahedron Lett.*, 1133 (1971).
78. G. Otani and S. Yamada, *Chem. Pharm. Bull.*, **21,** 2130 (1973).
79. M. Shibasaki, S. Terashima, and S. Yamada, *Chem. Pharm. Bull.*, **23,** 279 (1975).
80. U. Enders, G. Sauer, and R. Wiechert, *Angew. Chem., Int. Ed.*, **10,** 496 (1971).
81. Z. G. Hajos and D. R. Parrish, *J. Org. Chem.*, **39,** 1615 (1974).
82. N. Cohen, *Acct. Chem. Res.*, **9,** 412 (1976).
83. R. A. Micheli, Z. G. Hajos, N. Cohen, D. R. Parrish, L. A. Portland, W. Sciamanna, M. A. Scott, and P. A. Wehrli, *J. Org. Chem.* **40,** 675 (1975).
84. Z. G. Hajos and D. R. Parrish, *Org. Syn.* **63,** 26 (1984).
85. U. Eder, H. Gibian, G. Haffer, G. Neef, G. Sauer, and R. Wiechert, *Chem. Ber.*, **109,** 2948 (1976).
86. T. Kametani, H. Matsumoto, H. Nemoto, and K. Fukumoto, *J. Am. Chem. Soc.*, **100,** 6218 (1978).
87. J. Tsuji, I. Shimizu, H. Suzuki, and Y. Naito, *J. Am. Chem. Soc.*, **101,** 5070 (1979).
88. N. Cohen, B. L. Banner, W. F. Eichel, D. R. Parrish, and G. Saucy, *J. Org. Chem.*, **40,** 681 (1975).
89. S. Danishefsky and P. Cain, *J. Am. Chem. Soc.*, **97,** 5282 (1975).
90. S. Danishefsky and P. Cain, *J. Am. Chem. Soc.*, **98,** 4975 (1976).
91. J. Ruppert, U. Eder, and R. Wiechert, *Chem. Ber.*, **106,** 3636 (1973).
92. J. Gutzwiller, P. Buchschacher, and A. Fürst, *Synthesis,* 167 (1977).
93. P. Buchschacher and A. Fürst, *Org. Syn.*, **63,** 37 (1984).
94. N. Harada, J. Kohori, H. Uda, K. Nakanishi, and R. Takeda, *J. Am. Chem. Soc.*, **107,** 423 (1985).

95. S. Takano, C. Kasahara, and K. Ogasawara, *J. Chem. Soc., Chem. Commun.*, 635 (1981).
96. R. B. Woodward et al., *J. Am. Chem. Soc.*, **103,** 3210 (1981).
97. P. Buchschacher, J.-M. Cassal, A. Fürst, and W. Meier, *Helv. Chim. Acta,* **60,** 2747 (1977).
98. T. Wakabayashi, K. Watanabe, and Y. Kato, *Syn. Commun.*, **7,** 239 (1977).
99. L. Balaspiri, B. Penke, Gy. Papp, Gy. Dombi, and K. Kovacs, *Helv. Chim. Acta,* **58,** 969 (1975).
100. J.-M. Cassal, A. Fürst, and W. Meier, *Helv. Chim. Acta,* **59,** 1917 (1976).
101. R. Busson and H. Vanderhaeghe, *J. Org. Chem.*, **43,** 4438 (1978).
102. T. Wakabayashi, Y. Katu, and K. Watanabe, *Chem. Lett.*, 1283 (1976).
103. C. F. Lane, *U.S. Patent* 3,935,280; *Chem. Abstr.* **84,** 135101p (1976).
104. G. S. Poindexter and A. I. Meyers, *Tetrahedron Lett.*, 3527 (1977).
105. D. Seebach, H. Kalinowski, B. Bastani, G. Crass, H. Daum, H. Dörr, and M. Schmidt, *Helv. Chim. Acta,* **60,** 301 (1977).
106. F. P. Doyle, M. D. Metha, G. S. Sach, and J. L. Pearson, *J. Chem. Soc.*, 4458 (1958).
107. L. A. Paquette, J. P. Freeman, and S. Maiorana, *Tetrahedron,* **27,** 2599 (1971).
108. L. A. G. M. Van den Broek, P. A. T. W. Porskamp, R. C. Haltiwanger, and B. Zwanenburg, *J. Org. Chem.*, **49,** 1691 (1984).
109. T. Koizumi, R. Yanada, H. Tagaki, H. Hirai, and E. Yoshii, *Tetrahedron Lett.*, 477 (1981).
110. T. Koizumi, R. Yanada, H. Tagaki, H. Hirai, and E. Yoshii, *Tetrahedron Lett.*, 571 (1981).
111. T. Imamoto and T. Mukaigama, *Chem. Lett.*, 45 (1980).
112. K. Soai, H. Machida, and A. Ookawa, *J. Chem. Soc., Chem. Commun.*, 469 (1985).
113. P. E. Sonnet and R. R. Heath, *J. Org. Chem.*, **45,** 3137 (1980).
114. D. A. Evans and J. M. Takacs, *Tetrahedron Lett.*, 4233 (1980).
115. L. Guogiang, M. Hjalmarsson, H.-E. Högberg, K. Jernstedt, and T. Norin, *Acta Chem. Scand.*, **B38,** 795 (1984).
116. D. Enders and H. Lotter, *Angew. Chem., Int. Ed.*, **20,** 795 (1981).
117. H. Ahlbrecht, G. Bonnet, D. Enders, and G. Zimmermann, *Tetrahedron Lett.*, 3175 (1980).
118. N. Maigrot, J.-P. Mazalegrat, and Z. Welvart, *J. Chem. Soc., Chem. Commun.*, 40 (1984).
119. K. G. Davenport, D. T. Mao, C. M. Richmond, D. E. Bergbreiter, and M. Newcomb, *J. Chem. Res. (S),* 148 (1984).
120. N. R. Natale, B. E. Marron, E. J. Evain, and C. D. Dodson, *Syn. Commun.*, **14,** 599 (1984).
121. J. Smolanoff, A. F. Kluge, J. Meinwald, A. McPhail, R. W. Miller, K. Hicks, and T. Eisner, *Science,* **188,** 734 (1975).
122. T. Sugahara, Y. Komatsu, and S. Takano, *J. Chem. Soc., Chem. Commun.*, 214 (1984).
123. M. Shibasaki, S. Terashima, and S. Yamada, *Chem. Pharm. Bull.*, **24,** 315 (1976).

124. J. D. Morrison (Ed.), *Asymmetric Synthesis,* Vol. 3, Academic, 1984, p. 275 ff.

125. W. Bartmann and B. M. Trost (Eds.), *Selectivity—A Goal for Synthetic Efficiency,* Verlag Chemie, Weinheim, 1984, p. 65 ff.

126. D. Enders, *Chem. Technol.*, 504 (1981).

127. D. Enders and H. Eichenauer, *Angew. Chem., Int. Ed.*, **15,** 549 (1976).

128. E. Enders and H. Eichenauer, *Chem. Ber.*, **112,** 2933 (1979).

129. D. Enders, H. Eichenauer, and R. Pieter, *Chem. Ber.*, **112,** 3703 (1979).

130. P. M. Hardy, *Synthesis,* 290 (1978).

131. D. Enders and H. Eichenauer, *Tetrahedron Lett.*, 191 (1977).

132. D. Enders and H. Schubert, *Angew. Chem., Int. Ed.*, **23,** 365 (1984).

133. P. Magnus and G. Roy, *Organometallics,* **1,** 553 (1980).

134. P. Salvadori, S. Bertozzi, and R. Lazzaroni, *Tetrahedron Lett.*, 195 (1977).

135. F. A. A. Elhafez and D. J. Cram, *J. Am. Chem. Soc.*, **74,** 5846 (1952).

136. K. Banno and T. Mukaiyama, *Chem. Lett.*, 279 (1976).

137. H. Matsushita, Y. Tsujino, M. Noguchi, M. Saburi, and S. Yoshikawa, *Bull. Chem. Soc. Jpn.*, **51,** 201, 862 (1978).

138. P. Pino, S. Pucci, I. Piacenti, and G. Dell'Amico, *J. Chem. Soc. C,* 1640 (1971).

139. K. C. Nicolaou, D. P. Papahatjis, D. A. Claremon, and R. E. Dolle, III, *J. Am. Chem. Soc.*, **103,** 6967 (1981).

140. K. C. Nicolaou, D. A. Clareman, D. P. Papahatjis, and R. L. Magolda, *J. Am. Chem. Soc.*, **103,** 6969 (1981).

141. K. C. Nicolaou, D. P. Papahatjis, D. A. Claremon, R. L. Magoldfa, and R. E. Dolle, *J. Org. Chem.*, **50,** 1440 (1985).

142. D. Enders, H. Eichenauer, U. Baus, H. Schubert, and K. A. M. Kremer, *Tetrahedron,* **40,** 1345 (1984).

143. D. Enders and H. Eichenauer, *Angew. Chem., Int. Ed.*, **18,** 397 (1979).

144. D. Enders and U. Baus, *Liebigs Ann. Chem.*, 1439 (1983).

145. K. Mori, H. Nomi, T. Chuman, M. Kohno, K. Kato, and M. Noguchi, *Tetrahedron,* **24,** 3705 (1982).

146. S. I. Pennanen, *Acta Chem. Scand.*, **B35,** 555 (1981).

147. H. Eichenauer, E. Friedrich, W. Lutz, and D. Enders, *Angew. Chem., Int. Ed.*, **17,** 206 (1978).

148. D. Enders, H. Eichenauer, and R. Pieter, *Chem. Ber.*, **112,** 3703 (1979).

149. T. Mukaiyama, "Synthetic Control Leading to Natural Products," in *Asymmetric Reactions and Processes in Chemistry,* E. Eliel and S. Otsuka (Eds.), American Chemical Society, Washington, DC, 1982, pp. 21–36.

150. T. Mukaiyama, *Tetrahedron,* **37,** 4111 (1981).

151. T. Mukaiyama, *Lectures in Heterocyclic Chem.* (supplement to *J. Heterocycl. Chem.*), **7,** 53 (1984).

152. T. Mukaiyama, M. Asami, J.-i. Hanna, and S. Kobayashi, *Chem. Lett.*, 783 (1977).

153. M. Asami, H. Ohno, S. Kobayashi, and T. Mukaiyama, *Bull. Chem. Soc. Jpn.*, **51,** 1869 (1978).

154. M. Asami and T. Mukaiyama, *Heterocycles,* **12,** 499 (1979).

155. T. Mukaiyama, K. Soai, and S. Kobayashi, *Chem. Lett.*, 219 (1978).
156. T. Mukaiyama, K. Soai, T. Sato, H. Shimizu, and K. Suzuki, *J. Am. Chem. Soc.*, **101,** 1455 (1979).
157. K. Soai and T. Mukaiyama, *Chem. Lett.*, 491 (1978).
158. T. Sato, K. Soai, K. Suzuki, and T. Mukaiyama, *Chem. Lett.*, 601 (1978).
159. K. Soai and T. Mukaiyama, *Bull. Chem. Soc. Jpn.*, **52,** 3371 (1979).
160. T. Mukaiyama, K. Suzuki, K. Soai, and T. Sato, *Chem. Lett.*, 447 (1979).
161. T. Mukaiyama and K. Suzuki, *Chem. Lett.*, 255 (1980).
162. T. Akiyama, M. Shimizu, and T. Mukaiyama, *Chem. Lett.*, 611 (1984).
163. T. Mukaiyama, Y. Sakito, and M. Asami, *Chem. Lett.*, 1253 (1978).
164. M. Asami and T. Mukaiyama, *Chem. Lett.*, 569 (1979).
165. T. Mukaiyama, Y. Sakito, and M. Asami, *Chem. Lett.*, 705 (1979).
166. Y. Sakito and T. Mukaiyama, *Chem. Lett.*, 1027 (1979).
167. Y. Sakito, S. Tanaka, M. Asami, and T. Mukaiyama, *Chem. Lett.*, 1223 (1980).
168. M. Asami and T. Mukaiyama, *Chem. Lett.*, 17 (1980).
169. Y. Sakito, and M. Asami, and T. Mukaiyama, *Chem. Lett.*, 455 (1980).
170. N. Iwasawa and T. Mukaiyama, *Chem. Lett.*, 1441 (1982).
171. T. Mukaiyama, N. Iwasawa, R. W. Stevens, and T. Haga, *Tetrahedron,* **40,** 1381 (1984).
172. N. Iwasawa and T. Mukaiyama, *Chem. Lett.*, 297 (1983).
173. T. Mukaiyama and N. Iwasawa, *Chem. Lett.*, 753 (1984).
174. R. W. Stevens and T. Mukaiyama, *Chem. Lett.*, 1799 (1983).
175. J. Ichikawa, M. Asami, and T. Mukaiyama, *Chem. Lett.*, 949 (1984).
176. M. Kolb and J. Barth, *Tetrahedron Lett.*, 2999 (1979).
177. M. Kolb and J. Barth, *Liebigs Ann. Chem.*, 1668 (1983).
178. A. B. Mauger and B. Witkop, *Chem. Rev.*, **66,** 47 (1966).
179. A. A. Patchett and B. Witkop, *J. Am. Chem. Soc.*, **79,** 185 (1957).
180. A. A. Gottlieb, Y. Fujita, S. Udenfriend, and B. Witkop, *Biochemistry,* **4,** 2507 (1965).
181. R. H. Andreatta, V. Nair, A. V. Robertson, and W. R. J. Simpson, *Aust. J. Chem.*, **20,** 1493 (1967).
182. Y. Fujita, A. A. Gottlieb, B. Peterkovsky, S. Udenfriend, and B. Witkop, *J. Am. Chem. Soc.*, **86,** 4709 (1964).
183. V. Eswarakrishnan and L. Field, *J. Org. Chem.*, **46,** 4182 (1981).
184. A. V. Robertson, E. Katz, and B. Witkop, *J. Org. Chem.*, **27,** 2676 (1962).
185. P. S. Portoghese and A. A. Mikhail, *J. Org. Chem.*, **31,** 1059 (1966).
186. A.-M. Sepulchre, J. Cleophax, J. Hildesheim, and S. D. Gerro, *Comptes Rend. Acad. Sci., Ser. C.*, **269,** 849 (1969).
187. P. S. Portoghese, A. A. Mikhail, and H. J. Kupferberg, *J. Med. Chem.*, **11,** 219 (1968).
188. P. S. Portoghese and J. G. Turcotte, *Tetrahedron,* **27,** 961 (1971).
189. P. S. Portoghese and J. G. Turcotte, *Nature,* **230,** 457 (1971).
190. P. S. Portoghese and V. G. Telang, *Tetrahedron,* **27,** 1823 (1971).

191. P. S. Portoghese and D. T. Sepp, *J. Heterocycl. Chem.*, **8,** 531 (1971).
192. M. M. Bowers-Nemia and M. Jollie, *Heterocycles,* **20,** 817 (1983).
193. O. Mitsunobu, *Synthesis,* 1 (1981).
194. M. T. Pizzorno and S. M. Albonico, *J. Org. Chem.*, **39,** 731 (1974).
195. M. T. Pizzorno and S. M. Albonico, *Chem. Ind.*, (London), 349 (1978).
196. D. J. Robins and S. Sakdarat, *J. Chem. Soc., Perkin Trans. I,* 909 (1981).
197. D. J. Robins and S. Sakdarat, *J. Chem. Soc., Chem. Commun.*, 1181 (1979).
198. V. Caplar, G. Comisso, and V. Sunjic, *Synthesis,* 85 (1981).
199. K. Achiwa, *J. Am. Chem. Soc.*, **98,** 8265 (1976).
200. K. Achiwa, *Tetrahedron Lett.*, 3735 (1977).
201. I. Ojima, T. Kogure, and K. Achiwa, *J. Chem. Soc., Chem. Commun.*, 428 (1977).
202. K. Achiwa, T. Kogure, and I. Ojima, *Tetrahedron Lett.*, 4431 (1977).
203. I. Ojima, T. Kogure, T. Terasaki, and K. Achiwa, *J. Org. Chem.*, **43,** 3444 (1978).
204. K. Achiwa, *Heterocycles,* **8,** 247 (1977).
205. C. R. Botos, C. W. Smith, Y. L. Chan, and R. Walter, *J. Med. Chem.*, **22,** 926 (1979).
206. J.-R. Dormoy, B. Castro, G. Chappuis, U. S. Fritschi, and P. Grogg, *Angew. Chem., Int. Ed.*, **19,** 742 (1980).
207. J.-R. Dormoy, *Synthesis,* 753 (1982).
208. H. Rüeger and M. H. Benn, *Can. J. Chem.*, **60,** 2918 (1982).
209. H. Rüeger and M. H. Benn, *Heterocycles,* **19,** 23 (1982).
210. T. A. Geissman and A. C. Waiss,Jr., *J. Org. Chem.*, **27,** 139 (1962).
211. H. Rüeger and M. Benn, *Heterocycles,* **20,** 1331 (1983).
212. C. K. Atal, K. K. Kapur, C. C. J. Culvenor, and L. W. Smith, *Tetrahedron Lett.*, 537 (1966).
213. Y. K. Yadav, H. Rüeger, and M. Benn, *Heterocycles,* **22,** 2735 (1984).
214. T. Weber and D. Seebach, *Helv. Chim. Acta,* **68,** 155 (1985).
215. T. Mukaiyama, A. Ikegawa, and K. Suzuki, *Chem. Lett.*, 165 (1981).
216. W. Leimgruber, A. D. Batcho, and R. C. Czajkowski, *J. Am. Chem. Soc.*, **90,** 5641 (1968).
217. Z. Tozuka, H. Yazawa, M. Murata, and T. Takaya, *J. Antiobiot.*, **36,** 1699 (1983).
218. T. Kaneko, H. Wong, and T. W. Doyle, *Tetrahedron Lett.*, 5165 (1983).
219. T. Kaneko, H. Wong, and T. W. Doyle, *J. Antibiot.*, **37,** 300 (1984).
220. M. Ishikura, M. Mori, M. Terashima, and Y. Ban, *J. Chem. Soc., Chem. Commun.*, 741 (1982).
221. J. N. Reed and V. Snieckus, *Tetrahedron Lett.*, 5505 (1984).

CHAPTER NINE

TRYPTOPHAN

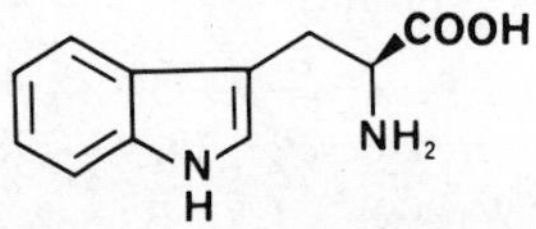

(*S*)-α-Amino-1*H*-indole-3-propanoic acid (**1**)

L-Tryptophan (**1**) is an aromatic, essential amino acid that is nutritionally very important, although it is present in relatively small amounts in proteins. Acid hydrolysis of proteins completely destroys tryptophan. Many physiologically important metabolites such as the actinomycins, indole alkaloids, phallotoxins, serotonin, and toad toxins are derived from tryptophan.[1]

Tryptoquivalines are potent tremorgenic toxins isolated from *Aspergillus clavatus* and *A. fumigatus* of mold-damaged rice. These novel metabolites possess a unique hexacylic skeleton (derived from tryptophan) that includes a spiro-γ-lactone.[2] L-Tryptophan (**1**) is readily converted in 50% overall yield to the quinazolinone derivative **3,** which undergoes oxidation with 2 equivalents of methanesulfonic anhydride–DMSO to give a good yield of spiro lactone **4** along with $<10\%$ of its epimer. Silylation of **4** with bis(trimethylsilyl)acetamide and condensation of the crude product with **5** in DMF furnishes an imide (not shown) that, when treated with triethylamine, gives the highly insoluble cyclol **6** in 81% overall yield from **4.** Deprotection to **7** and reduction with sodium cyanoborohydride furnishes a 4 : 1 mixture of **8** and **9.** After separation, **8** is reoxidized with DDQ and then treated with *m*-chloroperbenzoic acid to provide (−)-tryptoquivaline L (**10**). This can be made to undergo a contrathermodynamic epimerization to give (−)-tryptoquivaline G (**11**)[3] (Scheme 1).

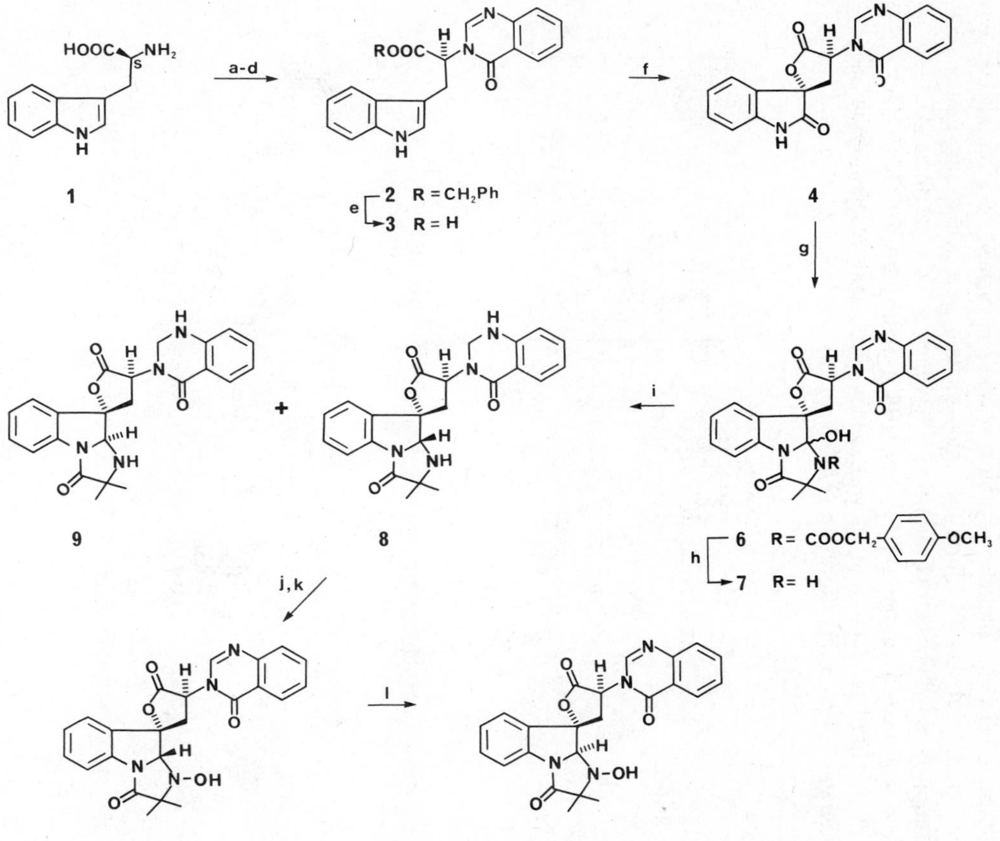

Scheme 1. (a) *o*-Nitrobenzoyl chloride; (b) $PhCH_2N_2$; (c) Fe/HCl; (d) PTSA/xylene; (e) H_2, Pd/C; (f) $(CH_3SO_2)_2O$–DMSO, CH_2Cl_2 (56–66%); (g) bis-TMS acetamide, TEA, then *N*-(*p*-methoxybenzylcarbonyl)methylalanine *p*-nitrophenyl ester (**5**); (h) CF_3COOH/EtOAc (76%); (i) $NaCNBH_3$; (j) DDQ (89%); (k) MCPBA (85%); (l) KH/THF/DMF.

An alternate synthesis of these mycotoxins takes advantage of a biomimetic oxidative double cyclization to form the imidazoindole spirolactone ring system. Condensation of **2** with ester **12** furnishes **13,** which is now set for a double-cyclization sequence. Treatment of **13** with NBS in refluxing CF_3COOH gives a mixture of cyclized products (not shown), which on reductive deprotection and chromatographic separation, affords **15** (21% from **13**) and **16** (14% from **13**). The

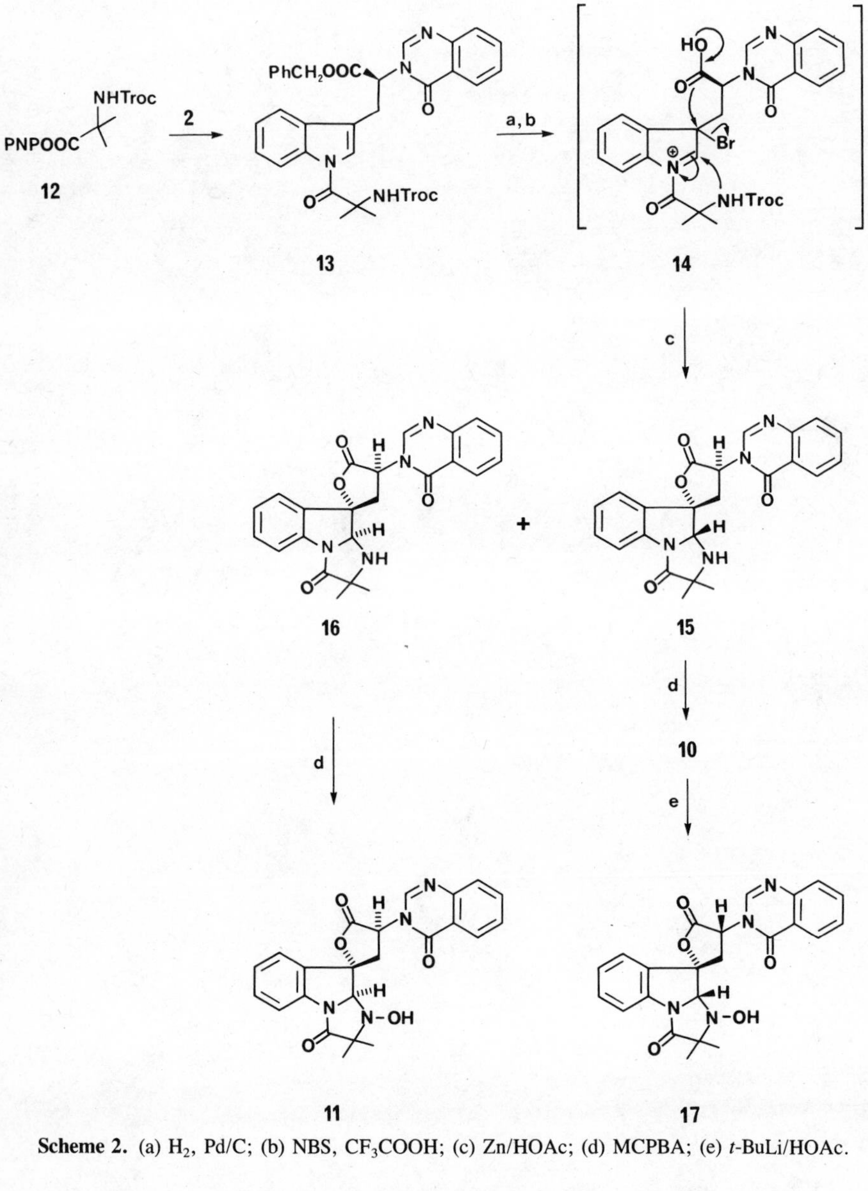

Scheme 2. (a) H_2, Pd/C; (b) NBS, CF_3COOH; (c) Zn/HOAc; (d) MCPBA; (e) *t*-BuLi/HOAc.

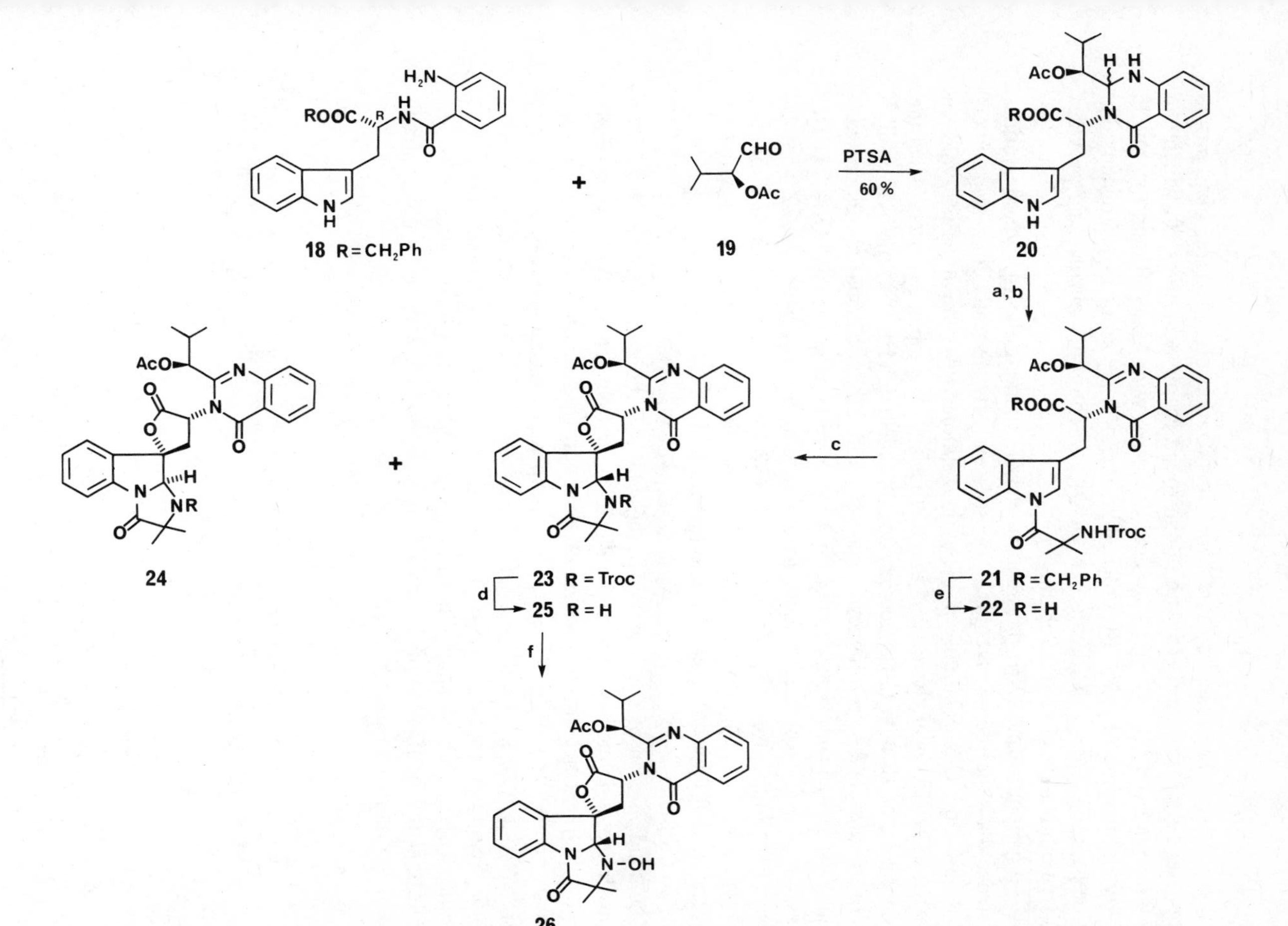

Scheme 3. (a) **12** (57%); (b) DDQ (91%); (c) NBS; (d) Zn/HOAc (66%); (e) H_2, Pd/C (100%); (f) MCPBA (93%).

products presumably arise from the double cyclization of the intermediate bromo compound **14.** Oxidation of **15** with *m*-chloroperbenzoic acid gives **10,** which can be epimerized at the original amino acid α-carbon with *t*-butyllithium to provide (+)-tryptoquivaline G (**17**). By proceeding analogously from D-tryptophan, one can prepare **17** directly. Oxidation of **16** with *m*-chloroperbenzoic acid provides (−)-tryptoquivaline G (**11**)[4] (Scheme 2).

(+)-Tryptoquivaline (**26**), the major metabolite among the 14 tryptoquivalines isolated and a tremorgenic mycotoxin, possesses highly sterically congested centers at the N-2 and C-3 positions of the quinazolinone ring. In order to avoid an epimerization step near the end of the synthesis, D-tryptophan is used. Thus the condensation of **18** with (*S*)-α-acetoxyisovaleraldehyde (**19**)[5,6] in methylene chloride in the presence of *p*-toluenesulfonic acid and 4-Å molecular sieves provides **20** in 60% yield. This is acylated with **12** and reoxidized with DDQ to **21.** Debenzylation of **21** followed by the oxidative double cyclization of **22** with *N*-bromosuccinimide gives **23** (16.3%) and **24** (32%). By using *N*-iodosuccinimide, one can increase the yield of **23** to 24.3%. Reductive deprotection of **23** with zinc in acetic acid followed by the oxidation of **25** with *m*-chloroperbenzoic acid provides **26**[7,8] (Scheme 3).

The condensation of a β-arylethylamine with a carbonyl compound to afford a tetrahydroisoquinoline is known as the Pictet–Spengler reaction.[9] By using optically active tryptophan as a chiral synthon, one can construct optically active 1,3-disubstituted β-carbolines. Generally, the Pictet–Spengler reaction is carried out in a protic solvent with acid catalysts. However, the ability to successfully employ aprotic conditions (refluxing benzene or toluene) allows for the preparation of carbolines derived from aldehydes containing acid-labile functionality. It is important that the boiling point of the aldehyde be higher than that of the aprotic solvent in order to ensure good yields of product.[10]

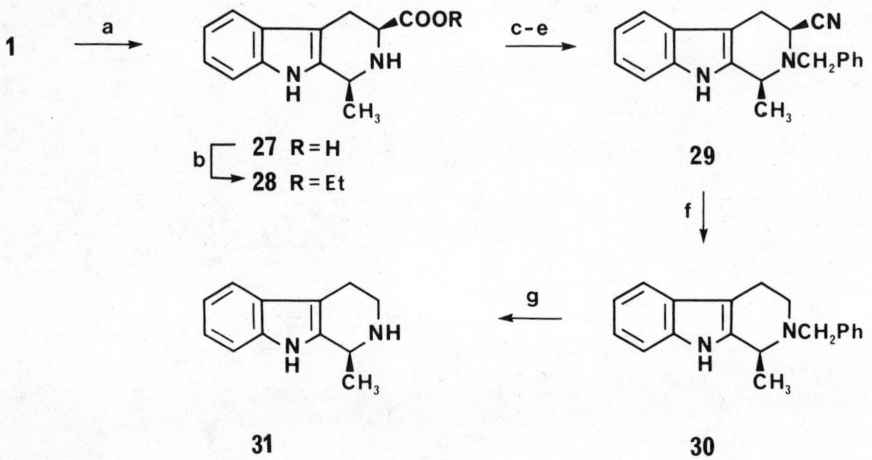

27 R = H

28 R = Et

29

31

30

Scheme 4. (a) CH_3CHO, H^+; (b) EtOH, H^+; (c) NH_3, CH_3OH; (d) $PhCH_2Br$; (e) $POCl_3$; (f) $NaBH_4$/EtOH/pyridine; (g) H_2, Pd/C.

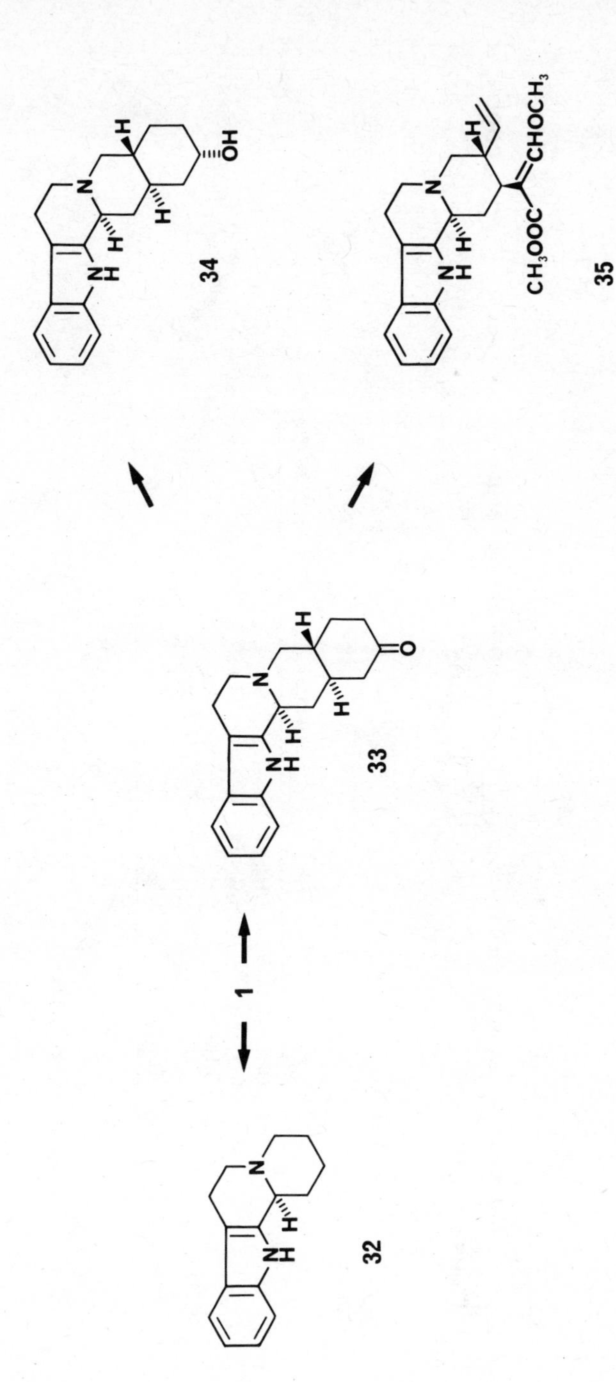
32
1
33
34
35
CH3OOC
CHOCH3

$$\mathbf{36} \xrightarrow{R_2X} \mathbf{37} \xrightarrow{NaBH_4} \mathbf{38}$$

R_1	R_2X	Yield **37** (%)	Yield **38** (%)
CH_3	CH_3I	90	90
CH_3	$C_6H_5CH_2Br$	74	91
C_6H_5	CH_3I	83	—
C_6H_{11}	CH_3I	80	—
$t\text{-}C_4H_9$	CH_3I	52	—

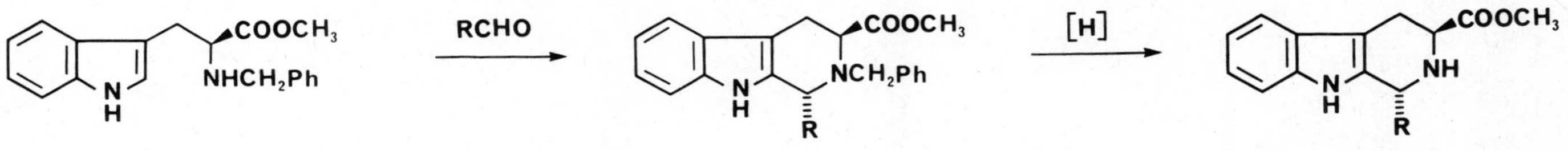

$COOCH_3$

$NHCH_2Ph$

39

RCHO

$COOCH_3$

NCH_2Ph

40

[H]

$COOCH_3$

NH

41

R	Yield **40** (%)	Yield **41** (%)
o-Hydroxyphenyl	97	75
Cyclohexyl	87	77
Ethyl	80	97
Formyldiethylacetal	75	99

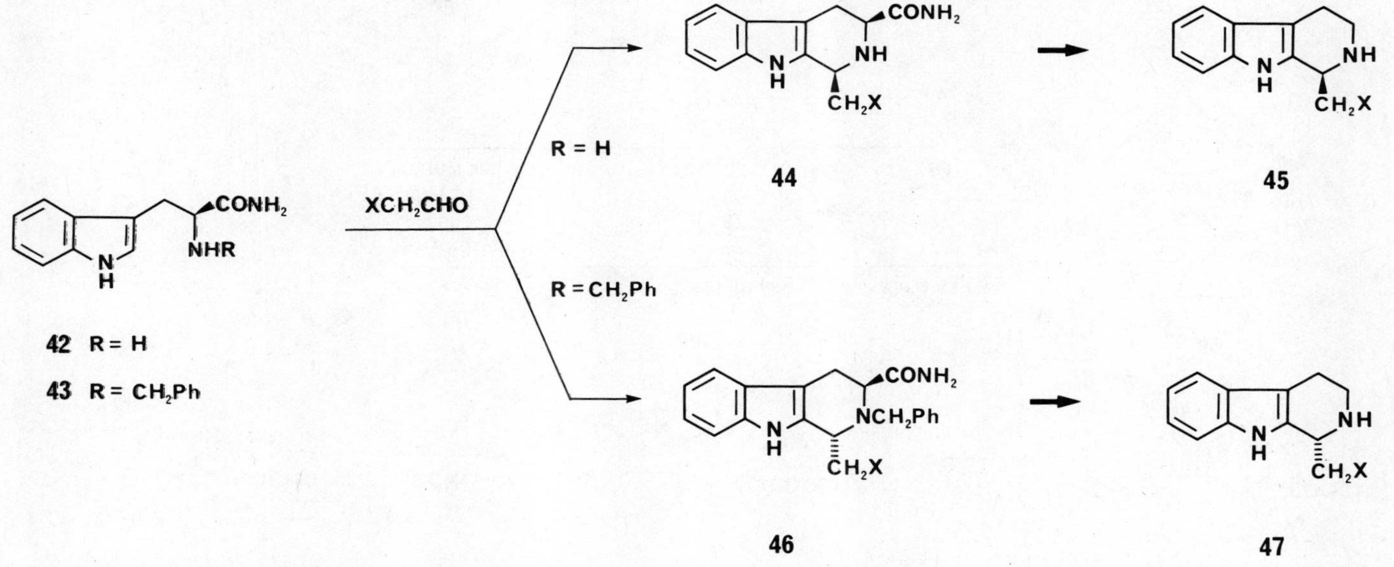
CONH₂
NHR
N
H
42 R = H
43 R = CH₂Ph
XCH₂CHO
R = H
R = CH₂Ph
44
45
46
47
NH
CH₂X
NCH₂Ph

$X = CH_3OOCC(SPh)_2CH_2-$

Scheme 5

The acid-catalyzed cyclization of L-tryptophan with acetaldehyde affords a 10:1 mixture of products, which, after purification by crystallization, provides the 1,3-*cis*-amino ester **28** as the major diastereomer.[11–13] This is easily converted to the amino nitrile **29,** which undergoes a reductive decyanization with either $NaBH_4$ in a mixture of ethanol and pyridine[11] or sodium in liquid ammonia[14] to provide **30.** Hydrogenolysis of the *N*-benzyl group gives the indole alkaloid (*S*)-(−)-tetrahydroharman (**31**)[11] (Scheme 4).

In a similar fashion, the indole alkaloids (*S*)-1,2,3,4,6,7,12,12a-octahydroindolo[2,3-a]quinolizine (**32**)[13] and (3*S*,15*S*,20*R*)-(−)-yohimbone (**33**)[15] can be prepared from L-tryptophan. The latter has been successfully converted to natural yohimbol (**34**)[16] and corynantheine (**35**).[17]

An annoying feature in these asymmetric Pictet–Spengler reactions is the production of a diastereomeric mixture of *cis*- and *trans*-1,3-disubstituted β-carbolines that require separation. To circumvent this situation, a method is available in which the cyclization of D- and L-N_b-methylthiocarbonyltryptophan methyl ester (**36**)[18] with an alkylating or acylating reagent in an aprotic solvent (acetone or methylene chloride) gives rise to the 3,4-dihydro-β-carbolines **37.** Reduction of **37** with sodium borohydride proceeds without racemization to provide stereospecifically the 1,3-*cis*-amino esters **38.**[19]

On the other hand, the stereospecific synthesis of *trans*-1,3-disubstituted-1,2,3,4-tetrahydro-β-carbolines **41** is accomplished in a two-step sequence that involves the Pictet–Spengler condensation of N_b-benzyltryptophan methyl ester (**39**) with aldehydes followed by catalytic hydrogenolysis of the N_b-benzyl moiety.[10,20,21] A stereoelectronic (antiperiplanar) attack of the 2,3-indole double bond on the intermediate benziminium ion combined with conformational effects has been suggested to account for this stereospecific cyclization.[21]

The exploitation of this stereoselectivity is beautifully illustrated in the synthesis of the two enantiomeric β-carbolines **45** and **47,** starting with tryptophanamide (**42**) and *N*-benzyltryptophanamide (**43**), respectively[22] (Scheme 5).

β-Carboline **45** has been successfully converted in 15 steps to (−)-ajmalicine (**48**), the only member of the heteroyohimbine family used therapeutically for the treatment of cardiovascular diseases.[23]

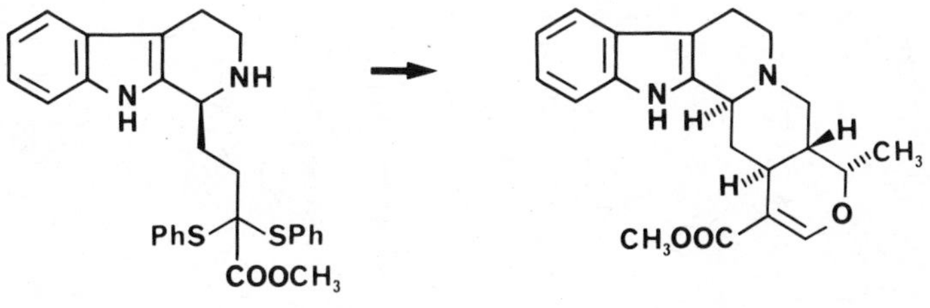

45 → **48**

The cyclization of tryptophan under Bischler–Napieralski conditions (trifluoroacetic acid, acetyl chloride) affords the 3,4-dihydro-β-carboline **49,** which is

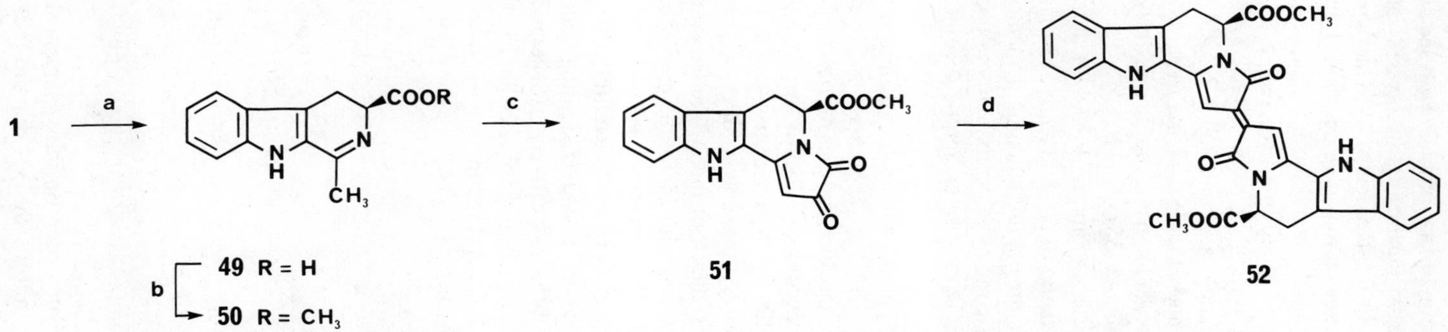

Scheme 6. (a) CF_3COOH, CH_3COCl; (b) $SOCl_2$, CH_3OH; (c) $(COCl)_2$, DME (82%); (d) $(CH_3O)_3P$, toluene, 110°C (63%).

converted to ester **50** with thionyl chloride–methanol. Condensation of **50** with oxalyl chloride in dimethoxyethane proceeds regioselectively to give the indolizino[8,7-b]indole derivative **51.** The deoxygenative dimerization of **51** with trimethyl phosphite proceeds smoothly and efficiently to provide the bis(indole)alkaloid trichotomine dimethyl ester (**52**), an unusual naturally occuring blue pigment isolated from the fruits of *Clerodendron trichotomum* and *Premna microphylla*[24,25] (Scheme 6).

Hydride reduction of **1** and subsequent tosylation gives a ditosylate **53,** which can be converted to the nitrile **54** with potassium cyanide. Deprotection, followed by formylation and cyclization with polyphosphoric ester (PPE), affords the unstable 3(*S*)-β-carboline **55.** This reacts with 3-methylenepentan-2-one (**56**) under acidic catalysis to provide a mixture of four isomeric indole[2,3-a]quinolizines (not shown), of which **57** (obtained in 21% yield from **55** by fractional crystallization) is isolated. This undergoes a base-catalyzed conversion to **58a** and its epimer **58b** (ratio = 2:1) having the sparagine skeleton. Base-catalyzed epimerization of **58b** occurs smoothly in 93% yield to augment the total yield of **58a.** Ring cleavage of **58a** and specific functionalization provide a route to 2-acylindol alkaloids such as dregamine (**59**) and epidregamine (**60**)[26] (Scheme 7).

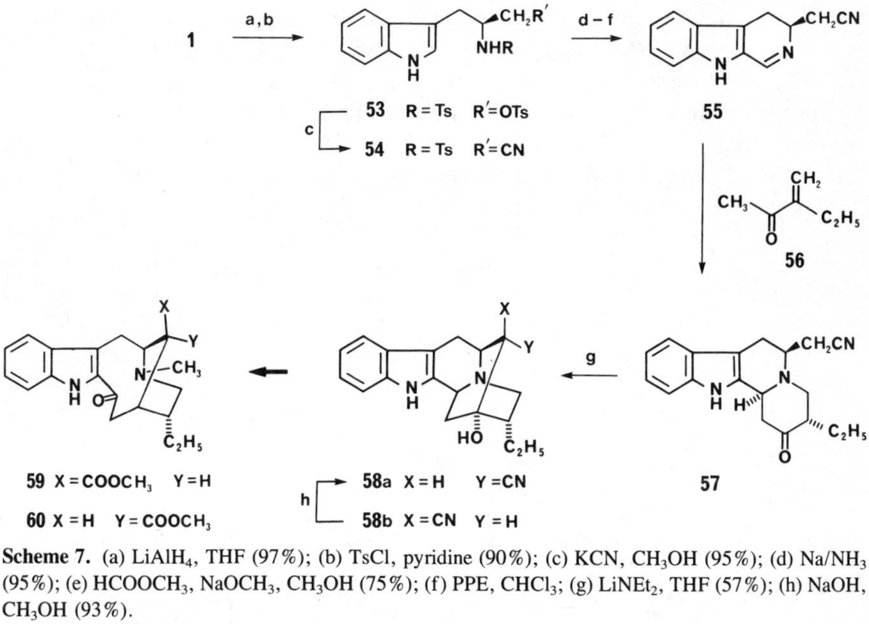

Scheme 7. (a) $LiAlH_4$, THF (97%); (b) TsCl, pyridine (90%); (c) KCN, CH_3OH (95%); (d) Na/NH_3 (95%); (e) $HCOOCH_3$, $NaOCH_3$, CH_3OH (75%); (f) PPE, $CHCl_3$; (g) $LiNEt_2$, THF (57%); (h) NaOH, CH_3OH (93%).

The cyclization of N_b-methoxycarbonyltryptophan methyl ester (**61**) in either 85% phosphoric acid[27] or trifluoroacetic acid[28] produces the crystalline *trans* cyclic tautomer **63a** in 85% yield. Brief treatment of **61** at lower temperature (−10°C) gives the less stable *cis* cyclic tautomer **62** (isolated as the N_a-acetyl derivative),[29] which can be converted to **63b** under equilibrium conditions.

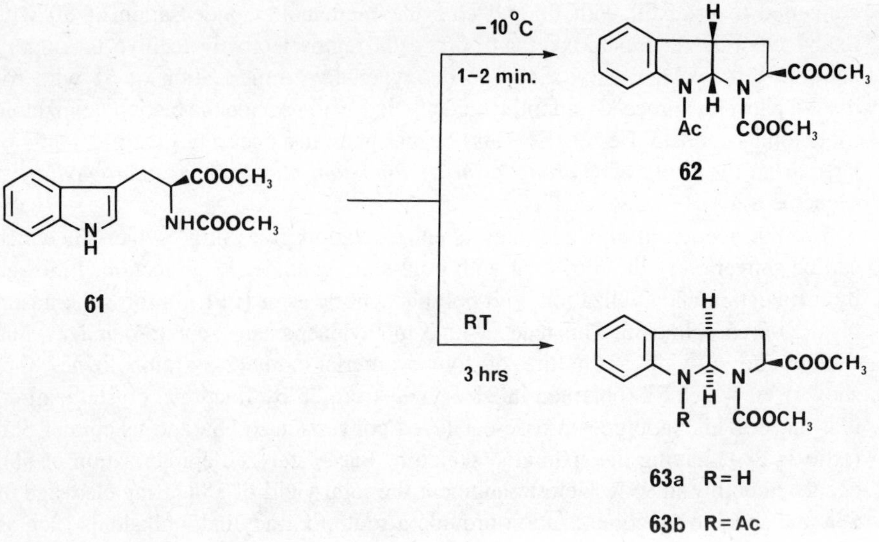

A similar cyclization of cyclo-L-tryptophan–L-proline (**64**)[30,36] in 85% phosphoric acid forms a single cyclic tautomer **66** in excellent yield. The other isomer **65**, obtained in 45–50% yield by dissolving **64** in trifluoroacetic acid at −10°C for 1–2 min, is the less stable kinetic product.[29] This result, opposite to that found in the tryptophan series (**61** → **63**), suggests that the folded structure of **65** is not preferred because of (1) steric hindrance between the proline ring and the aromatic (indole) residue and (2) strain introduced by the proline ring.[30]

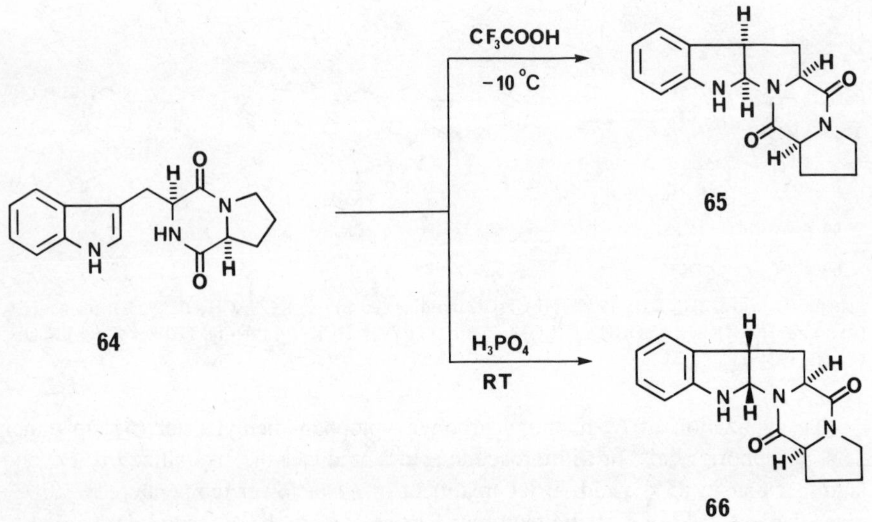

63 ⟶ 67 / 68 ⟶ (H_2SO_4) 69

X, H, N, COOCH$_3$, Ac, H, COOCH$_3$

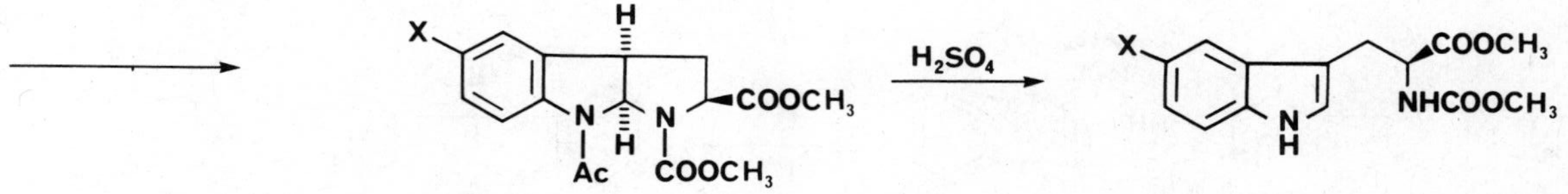

67 X = Cl (93%)

68 X = NO_2 (93%)

X, COOCH$_3$, NHCOOCH$_3$, N, H

69

66 ⟶ (1. $Pb(OAc)_4$; 2. Zn) 70

HO, H, N, H, O

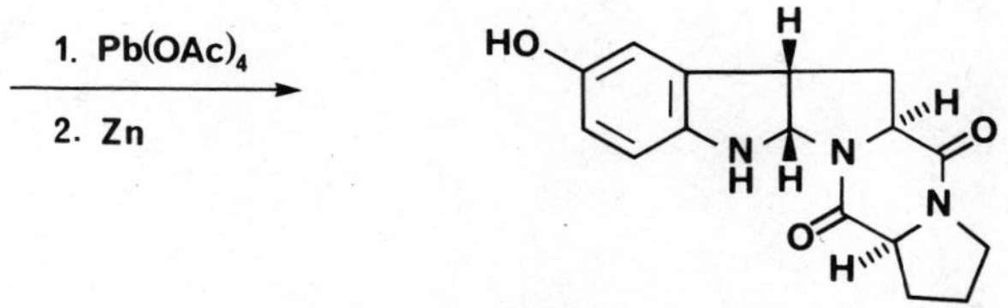

70

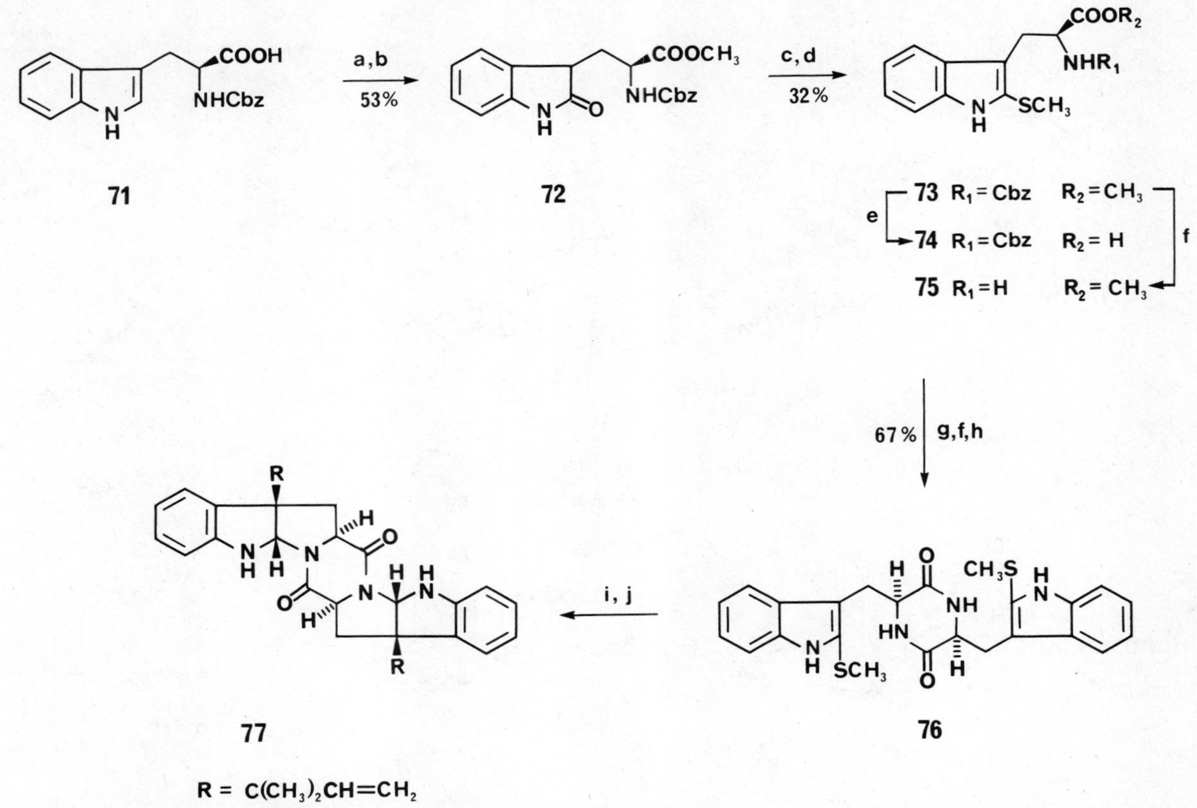

71 → (a,b; 53%) → 72 → (c, d; 32%) → 73–75

73 R_1 = Cbz $R_2 = CH_3$

74 R_1 = Cbz R_2 = H

75 R_1 = H $R_2 = CH_3$

73 → (e) → 74; 73 → (f) → 75

75 → (g,f,h; 67%) → 76 → (i, j) → 77

$R = C(CH_3)_2CH{=}CH_2$

Scheme 8. (a) HCl, DMSO; (b) HCl, CH_3OH; (c) P_2S_5/pyridine; (d) CH_3I, K_2CO_3; (e) NaOH, THF/CH_3OH (100%); (f) 30% HBr–HOAc (100%); (g) DCC, *N*-hydroxysuccinimide; (h) NH_3 gas, CH_3OH; (i) prenyl bromide, K_2CO_3 (18%); (j) $TiCl_4$, $LiAlH_4$ (15%).

Cyclic tautomers such as **63** and **66** can be regarded as protected forms of the corresponding indole. Electrophilic substitution at the 2 position is blocked, and these tautomers expected to react as indolines toward electrophiles, thus providing a simple method for preparing tryptophan derivatives carrying substituents on the benzene ring. Since the cyclic tautomers are easily converted to the open-chain form with dilute acid,[27] the preparation of substituted tryptophans **69** is allowed. Indeed, the reaction of **63b** with *N*-chlorosuccinimide[28] gives the 5-chloro derivative **67,** whereas the reaction with fuming nitric acid[31] provides the 5-nitro derivative **68.** Similarly, the oxidation of **66** with 2 equivalents of lead tetraacetate in trifluoroacetic acid, followed by a reduction with zinc, provides the 5-hydroxy derivative **70** in 73% yield.[29,31,32]

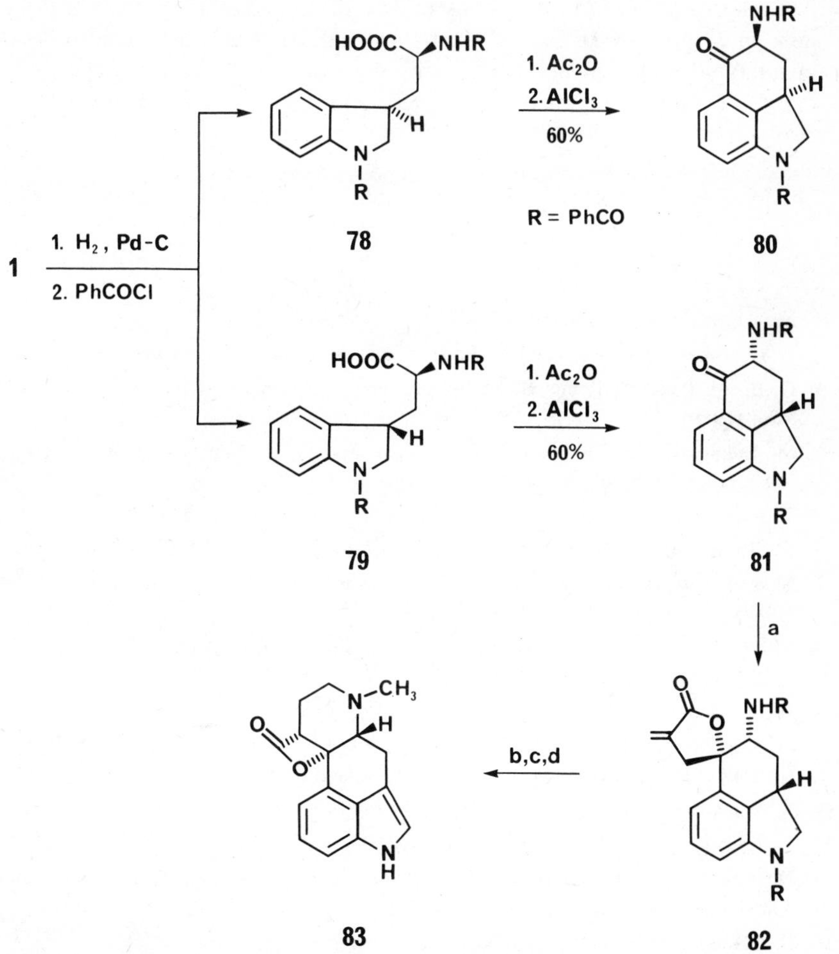

Scheme 9. (a) Ethyl α-(bromomethyl)acrylate, Zn, THF (83%); (b) NaH, CH_3I, DMF (80%); (c) HBr, CH_2Cl_2 (55%); (d) MnO_2, CH_2Cl_2 (75%).

Amauromine (**77**), a novel alkaloid possessing two reversed prenyl groups and manifesting vasodilating activity, can be viewed as a dimeric cyclic tautomer of **63a** in which the stereochemistry at the ring juncture is controlled by the diketopiperazine (as in **66**). *N*-Cbz–L-tryptophan (**71**) is converted to the 2-thiomethyl derivative **73** by way of the oxindole **72.** Coupling of the carboxylic acid **74** with the amine **75** (both quantitatively prepared from **73**), followed by diketopiperazine formation, provides **76.** Introduction of the two reversed prenyl groups is accomplished by a thio-Claisen rearrangement. Separation of the resulting mixture, followed by a reductive desulfurization, affords (**77**)[33] (Scheme 8).

Catalytic reduction of L-tryptophan with 10% Pd/C in 1 *N* HCl,[34] followed by separation and benzoylation of the resulting mixture, provides **78** (25% overall) and **79** (33% overall). Conversion of these compounds to their corresponding azalactones by brief heating with acetic anhydride, followed by a Friedel–Crafts reaction, affords **80** and **81.** As a consequence of a rapid epimerization of the azalactone of **79** prior to the Friedel–Crafts reaction, **81** is an enantiomer of **80.** The availability of either enantiomer permits the synthesis of optically active ergot alkaloids. This is exemplified by the preparation of (+)-10α-hydroxy-9,10-dihydroisolysergate γ-lactone (**83**) from **81**[35] (Scheme 9).

REFERENCES

1. T. Scott and M. Brewer, *Concise Encyclopedia of Biochemistry,* Walter de Gruyter, New York, 1985, p. 482.
2. T. Ohnuma, Y. Kimura, and Y. Ban, *Tetrahedron Lett.*, 4969 (1981).
3. G. Büchi, P. R. DeShong, S. Katsumura, and Y. Sugimura, *J. Am. Chem. Soc.*, **101,** 5084 (1979).
4. M. Nakagawa, M. Taniguchi, M. Sodeoka, M. Ito, K. Yamaguchi, and T. Hino, *J. Am. Chem. Soc.*, **105,** 3709 (1983).
5. T. Fujisawa, T. Mori, S. Tsuge, and T. Sato, *Tetrahedron Lett.*, 1543 (1983).
6. M. Taniguchi, K. Koga, and S. Yamada, *Chem. Pharm. Bull.*, **20,** 1438 (1972).
7. M. Nakagawa, M. Ito, Y. Hasegawa, S. Akashi, and T. Hino, *Tetrahedron Lett.*, 3865 (1984).
8. M. Nakagawa, M. Ito, Y. Hasegawa, S. Akashi, M. Taniguchi, and T. Hino, *Heterocycles,* **23,** 224 (1985).
9. W. M. Whaley and T. R. Govindachari, *Org. React.*, **6,** 151 (1975).
10. D. Soerens, J. Sandrin, F. Ungemach, P. Mokry, G. S. Wu, E. Yamanaka, L. Hutchins, M. DiPierro, and J. M. Cook, *J. Org. Chem.*, **44,** 535 (1979).
11. S. Yamada and H. Akimoto, *Tetrahedron Lett.*, 3105 (1969).
12. A. Brossi, A. Focella, and S. Teitel, *J. Med. Chem.*, **16,** 418 (1973).
13. H. Akinoto, K. Okamura, M. Yui, T. Shioiri, M. Kuramoto, Y. Kitugawa, and S. Yamada, *Chem. Pharm. Bull.*, **22,** 2614 (1974).
14. S. Yamada, K. Tomioka, and K. Koga, *Tetrahedron Lett.*, 61 (1976).
15. K. Okamura and S. Yamada, *Chem. Pharm. Bull.*, **26,** 2305 (1978).
16. B. Witkop, *Annalen,* **554,** 83 (1943).

17. R. L. Autrey and P. W. Schillard, *J. Chem. Soc., Chem. Commun.*, 841 (1961).
18. A. Ishida, T. Nakamura, K. Irie, and T. Oh-ishi, *Chem. Pharm. Bull.*, **30,** 4226 (1982).
19. T. Nakamura, A. Ishida, K. Irie, and T. Oh-ishi, *Chem. Pharm. Bull.*, **32,** 2859 (1984).
20. F. Ungemach, M. DiPierro, R. Weber, and J. M. Cook, *Tetrahedron Lett.*, 3225, (1979).
21. F. Ungemach, M. DiPierro, R. Weber, and J. M. Cook, *J. Org. Chem.*, **46,** 164 (1981).
22. G. Massiot and T. Mulamba, *J. Chem. Soc., Chem. Commun.*, 1147 (1983).
23. G. Massiot and T. Mulamba, *J. Chem. Soc., Chem. Commun.*, 715 (1984).
24. S. Iwadare, Y. Shizuri, K. Sasaki, and Y. Hirata, Japanese Patent 7,641,415; *Chem. Abstr.*, **85,** 25376v (1976).
25. G. Palmisano, B. Danieli, G. Lesma, and R. Riva, *J. Org. Chem.*, **59,** 3322 (1985).
26. J. P. Kutney, G. K. Eigendorf, H. Matsue, A. Murai, K. Tanaka, W. L. Sung, K. Wada, and B. R. Worth, *J. Am. Chem. Soc.*, **100,** 938 (1978).
27. M. Taniguchi and T. Hino, *Tetrahedron,* **37,** 1487 (1981).
28. T. Hino and M. Taniguchi, *J. Am. Chem. Soc.*, **100,** 5564 (1978).
29. T. Hino, M. Taniguchi, I. Yamamoto, K. Yamaguchi, and M. Nakagawa, *Tetrahedron Lett.*, 2565 (1981).
30. P. G. Sammes and A. C. Weedon, *J. Chem. Soc., Perkin Trans. I,* 3048 (1979).
31. T. Hino, M. Taniguchi, A. Gonsho, and M. Nakagawa, *Heterocycles,* **12,** 1027 (1979).
32. T. Hino, M. Taniguchi, and M. Nakagawa, *Heterocycles,* **15,** 187 (1981).
33. S. Takase, Y. Itoh, and I. Uchida, *Tetrahedron Lett.*, 847 (1985).
34. J. W. Daly, O. Mauger, V. K. Yonemitsu, K. Antonov, K. Takase, and B. Witcop, *Biochemistry,* **6,** 648 (1967).
35. J. Rebek, Jr., D. F. Tai, and Y.-K. Shue, *J. Am. Chem. Soc.*, **106,** 1813 (1984).
36. M. Taniguchi, I. Yamamoto, M. Nakagawa, and T. Hino, *Chem. Pharm. Bull.*, **33,** 4783 (1985).

AUTHOR INDEX

SUBJECT INDEX